Landscape Evolution in the United States

Landscape Evolution in the United States

An Introduction to the Geography, Geology, and Natural History

Joseph A. DiPietro
Department of Geology and Physics
University of Southern Indiana
Evansville

ELSEVIER

AMSTERDAM • BOSTON • HEIDELBERG • LONDON • NEW YORK • OXFORD • PARIS
SAN DIEGO • SAN FRANCISCO • SINGAPORE • SYDNEY • TOKYO

Elsevier
225 Wyman Street, Waltham, MA 02451, USA
525 B Street, Suite 1900, San Diego, CA 92101-4495, USA

First edition 2013

Notice
No responsibility is assumed by the publisher for any injury and/or damage to persons or property as a matter of products liability, negligence or otherwise, or from any use or operation of any methods, products, instructions or ideas contained in the material herein. Because of rapid advances in the medical sciences, in particular, independent verification of diagnoses and drug dosages should be made

Library of Congress Cataloging-in-Publication Data
DiPietro, Joseph A., 1957
    Landscape evolution in the United States: an introduction to the geography, geology, and natural history/Joseph DiPietro.
        p. cm.
    ISBN 978-0-12-397799-1
1. Landscapes–United States–History. 2. Landscape changes–United States–History.
3. Landscape assessment–United States–History. 4. United States–Geography. I. Title.
    QH76.D56 2013
    917.304'932–dc23
                            2012044288

British Library Cataloguing in Publication Data
A catalogue record for this book is available from the British Library

For information on all Elsevier publications
visit our web site at store.elsevier.com

ISBN: 978-0-12-397799-1

Contents

Part II
Structural Provinces

This book was written for those curious about the natural wonder and beauty of the United States, for those wishing to delve into how mountains form and evolve beyond the obligatory colliding of continents, and for those who seek insight on the reasoning and methods geologists use to interpret landscape evolution and geological history. It is a textbook appropriate for first-semester university students, a reference book for advanced geology students, and a book appropriate for a general audience or traveler who seeks a deeper understanding and appreciation of landscape.

As a general audience book, it introduces the physical characteristics of the United States, its structural framework, and its geological history. As a textbook, it shows how geography, geology, climate, and tectonics interact to shape the landscape of the United States. It could be used as a stand-alone text for an introductory course on the landscape and geology of the United States or, with proper supplementary material, it could serve as a text for a physical geology or national parks course. It would also be an excellent resource for a variety of other courses including historical geology, physical geography, geomorphology, structural geology, and tectonics. Additionally, the book would make excellent reference material for students and teachers at the K-12 level.

There are three parts to the book. Part I (Chapters 1–8) is concerned with understanding how rock type and rock structure combine with tectonic activity, climate, isostasy, and sea-level change to produce landscape and to predict how landscape will evolve. Concepts in Part I address why a landscape looks the way it does, how it formed, how it has evolved, how it will evolve, and how it may eventually reincarnate itself to form a new landscape. The concepts presented here can be applied to any landscape anywhere on Earth. The goal is for you to read the landscape wherever your travels take you.

Part II (Chapters 9–18) applies concepts introduced in Part I to the United States. The US is divided into nine structural provinces, each based on a similar rock type and rock structure. The discussion is on the distribution and origin of structural provinces with special emphasis on topography, rock type, rock structure, tectonic setting, climate, and recent uplift/erosion history. The final chapter in Part II is a discussion on the origin of the Grand Canyon.

Part III (Chapters 19–24) is a geological excursion into the criteria and tools geologists use to understand how compressional mountain systems form and evolve. Special emphasis is on the type of evidence geologists find, and the rationale they use, to interpret the rock record with reference to the US Appalachian and Cordilleran Mountain systems.

In a nutshell, Parts I and II deal specifically with present-day landscape, its evolution, and how the forces of nature affect it. The focus is on why the landscape looks the way it does. Part III deals with the rocks. The focus is on geological history, mountain building, and past landscapes. In this part of the book, the present-day landscape is of secondary importance.

The figures are designed to be both simple and informative. They are an integral part of the discussion and should be studied in detail. The figures are primarily of four types: Google Earth images, photographs, cross-section and map sketches, and landscape maps. The primary landscape map was hand-drawn by Erwin J. Raisz from field observations, aerial photographs, and satellite imagery. Raisz was a member of the Institute of Geographical Exploration at Harvard University for nearly 20 years beginning in 1931 and was one of the founding cartographers in the United States. The first edition of his seminal map, *Landforms of the United States*, was published in 1939. The sixth and last edition was completed in 1957. More than 50 years later it remains arguably the finest US landscape map ever produced. A high-quality copy of the map, as well as a variety of other maps, is available at www.raiszmaps.com. Most of the Raisz map figures are shown at a reduced scale where 1 inch on the map is equivalent to approximately 85 miles on the ground. I augment the maps with boundaries and contacts showing, for example, the distribution of physiographic provinces, rock types, structural provinces, climate zones, river systems, global wind patterns, glacial zones, and tectonic features. Ten uncolored maps are provided in the Appendix. These can be used for student exercises.

I refrain from using too much geological jargon in Parts I and II. My focus is on understanding the big picture and understanding basic concepts. For example, I purposely avoid using the geologic time scale. Instead, I use actual numbers, which are easier for a beginner in geology to place in proper order. My overarching goal is to present the novice reader with a deep understanding of the incredibly varied landscape of the US and to show how much fun it is to learn about geology. Part III is geological in the purest sense, and geological jargon and the geologic time scale are employed.

Whenever possible I use US customary units of measurement (inch, foot, mile) to cater to a primarily US audience. It is unfortunate that many readers in the US would not be able to convert meters to feet or kilometers to miles. If this book is to be used as a textbook, I strongly encourage the instructor to hammer away at students by instructing them to convert numbers to the metric system.

When discussing rates, I use 100 years as the common denominator. I do this because 100 years is approximately equivalent to a human lifetime, so the reader can quickly grasp the amount of change that occurs over the course of their existence. I apologize to the great states of Hawaii and Alaska as this book focuses on the landscape of the contiguous 48 states.

The book is an outgrowth of a nonmajor, introductory course I teach entitled Landscape and Geology of North America. I would like to thank the students who have taken this course for their input, interest, and attention. I especially would like to thank Amber King, Kristen Schmeisser, and James Wallace for reading and commenting on chapters and for help with the figures. I thank Tony Maria for commenting on several of the chapters, and Karen Sommer for help with the editing process. When writing this book I set out to be both logical and factual. Mistakes and inconsistences are my own and, for those, I appreciate your comments.

Listed below are some common units of measurement, conversions, and abbreviations for the reader's reference.

ABBREVIATIONS

millimeter (mm)	year (yr)
centimeter (cm)	million years (My)
meter (m)	million years ago (Ma)
kilometer (km)	billion years ago (Ga)
inch (in)	degrees Fahrenheit (°F)
feet (ft)	degrees Centigrade (°C)
mile (mi)	

CONVERSIONS

1 kilometer = 0.62 mile
1 mile = 1.61 kilometer
1 kilometer = 1000 meter = 3280 feet
1 mile = 5280 feet = 1610 meter
1 meter = 3.28 feet
1 foot = 0.305 meter
1 centimeter = 0.394 inches
1 inch = 2.54 centimeters
1 cm/yr = 10mm/yr = 10 km/My
1 in/yr = 15.78 miles/My
1 cm/yr = 39.4 in/100 years
1 in/yr = 254 cm/100 years
1 in/100 yr = 0.01 in/yr = 0.254 mm/yr
°C = (°F − 32) x 0.555
°F = (°C x 1.8) + 32

Part I

Keys to Understanding Landscape Evolution

The Tortoise and the Hare

From California to the coast of Maine, and from Florida to the coast of Washington, the contiguous United States has some of the most spectacular scenery on Earth including the Grand Canyon, the Rocky Mountains, and the majestic Appalachian Mountains. But the United States hasn't always looked like this. Thirty million years ago the San Andreas fault did not exist and the state of Nevada was only about half as wide as it is today. Yellowstone National Park has literally blown up three times during the past 2.2 million years, spreading volcanic ash as far east as Iowa and as far west as the Pacific Ocean. Periodically from about 2.5 million years ago to as recently as 18,000 years ago, a number of enormous ice sheets covered nearly all of Canada and a large part of the United States. During the height of these glacial advances, the shoreline of the eastern United States lay as much as 250 miles east of where it is today and the state of Florida was about twice its present day width. Areas of California, Oregon, and Washington have blown up repeatedly within the last few thousand years, including the volcanic explosion that created Crater Lake 7,700 years ago and the catastrophic 1980 explosion of Mount St. Helens. Fifteen thousand years ago giant lakes covered the desert regions of Nevada, California, and Utah, and tremendous floods poured through eastern Washington.

Clearly, many of the world's landforms are only a few thousand to several tens of millions of years old. This may seem ancient, but it amounts to only a small fraction of the 4.55-billion-year history of Earth. From this evidence alone we can surmise that landforms are ephemeral (lasting for a brief time) and are constantly in the process of change. So why is it that most people do not notice these changes? The answer is that most changes occur at a rate too slow for the average person to see. For example, the United States is blessed with many great rivers, all of which carry enormous amounts of sediment from the mountains to the sea. Do we notice that the mountain mass has been reduced?

The answer, of course, is no. The effect on the mountain is incremental and cumulative. If it takes 100,000 years to reshape a landscape, then a 100-year-old human would have witnessed only 0.1% of that change. Such a trivial amount likely would not directly impact our lives or our standard of living and therefore would not be noticed. There are, of course, catastrophic events that can shape a landform within a human lifetime. It took only a few minutes for more than 1000 feet of the Mount St. Helens volcano to blow away. Catastrophic changes are noticeable only because they occur rapidly and well within one person's lifetime. There is no doubt they contribute to the evolving landscape. But catastrophic events are periodic and, from a human perspective, generally do not repeat themselves for a long time.

In the legend of the tortoise and the hare, the tortoise was slow and steady; the hare was fast, but only for a short time. Spectators watching the race would have marveled at the rapid pace of the hare while perhaps not even noticing the tortoise as he passed by. But in the end the tortoise wins the race because of the cumulative effect of his slow and steady pace. Such are the processes of landscape evolution. Many processes effect only slow change on the landscape which can be steady or not so steady. But the hare is not out of the race completely. Periodic rapid changes occur, and these can completely change the look of a landscape within a human lifetime or within a few generations.

HOW SLOW IS SLOW?

We must now ask: How slow is slow? In Chapter 4 we will look at actual rates of change, but let us first put geological time in perspective. Let's say that you live to be 100 years old. In this case, one million years would seem like a long time. But what if you live to be 4.55 *billion* years old (the age of the earth)? From that perspective a million years seems trivial. If each year is counted as one second, then

4.55 billion seconds adds up to about 144 years. Using this time scale, the Earth would be 144 years old and an average human would be alive on Earth for less than two minutes; 100,000 years would pass in about 28 hours and one million years in about 11.6 days. Normally, nothing physically noticeable happens to a person in 28 hours or even 11.6 days. But, relatively speaking, the Earth could change enormously in 11 or 12 days. On this time scale, it might take the Earth less than a month to construct a large mountain range and only a few months to tear it down. So, in perspective, the Earth changes much faster than humans. The bottom line is that the Earth, and everything we see today, is constantly in the process of change. Some changes are rapid enough to notice. Others are not.

MAPS, CROSS-SECTIONS, AND SCALE

To discuss landscape, it is important to understand a few terms. Landforms are described by their *topography*, which is the shape and form of the Earth's surface as expressed in elevation above or below sea level. Simply stated, *topography* is the lay of the land. *Elevation* refers to the height above sea level, whereas *relief* refers to the difference in elevation between any two nearby points. For example, the greatest relief in the continental United States is in eastern California, along the eastern face of the Sierra Nevada, where Mt. Whitney, at an elevation of 14,495 feet, is only 75 miles from Death Valley, which is at an elevation of 282 feet below sea level. Relief between these two points is 14,777 feet.

$$14{,}495 \text{ feet} - (-282 \text{ feet}) = 14{,}777 \text{ feet}$$

It is also important that you understand the difference between a *map*, a *profile*, and a *cross-section*. All three are illustrated in Figure 1.1. Maps show the aerial extent of physiographic or geologic features as though you are looking at them from above, like the view from an airplane. To the uninitiated, maps are an underrated tool. But maps are like photographs: They hold an enormous amount of information that otherwise would be tedious and boring to convey in words. Maps give an instant visual perspective

of a landscape and also convey information regarding spatial relationships, size, location, topography, rock type, and rock structure. Maps have existed for hundreds of years. Before there was photography, one's vision and perspective of Earth was based largely on maps.

A *profile* is an outline of the shape of land as though looking from ground level. It is a "mug shot" that shows how topography changes along a straight line. It is like looking at the outline of a volcano. A cross-section is a profile that, in addition, shows the rock structure along a vertical slice through the interior of Earth. It is like slicing a volcano in half and looking inside.

All maps, profiles, and cross-sections have a scale that shows how distance on the map is related to distance on the ground. A fractional scale of 1:50,000 indicates that one unit on the map is equal to 50,000 units on the ground. A *unit* refers to any form of measurement such as an inch, a foot, or a centimeter. For example, at 1:50,000, one inch on a map is equal to 50,000 inches on the ground; one foot on the same map is equal to 50,000 feet on the ground. A *bar scale* shows the distance on a map relative to one or more miles (or kilometers) on the ground. Figure 1.2 compares the bar scale on a 1:24,000 map with the bar scale on a 1:250,000 map. One could easily see that the 1:24,000 map would show greater detail of a small area, whereas the 1:250,000 map would show less detail of a much larger area.

Map of a Volcano
It is as if you are looking down from an airplane.
You see the top of the volcano.

Profile of a Volcano
It is as if you are looking at the volcano from the side
You see the general outline (shape) of the volcano

Cross-Section of a Volcano
In addition to the profile, you see the type and distribution of rock layers below the surface within the volcano.

FIGURE 1.1   The relationship among map, profile, and cross-section.

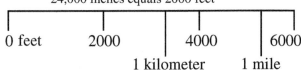

Bar scale for 1:24,000

1 inch on the map equals 24,000 inches on the ground
24,000 inches equals 2000 feet

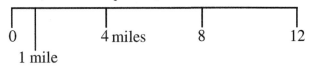

Bar scale for 1:250,000

1 inch on the map equals 250,000 inches on the ground
250,000 inches equals 20,833.3 feet or 3.946 miles

Bar Scale for the Raisz Landscape Maps

1 inch on the map equals approximately 82 miles on the ground

FIGURE 1.2   A comparison of map scales.

THE FACE OF THE UNITED STATES

A *landform* is an area of any size that can be separated from surrounding land on the basis of its shape (e.g., its topography). The terms *landscape, terrain,* and *physiographic province* all broadly refer to the same thing: an area of land characterized by a similar set of landforms. A physiographic region is a larger area that groups together similar physiographic provinces.

A landscape, or even a particular landform, can be classified as a plain, plateau, or mountain based on its elevation and relief in relation to adjacent land. A *plain* is a wide area of little relief (<500 feet) and relatively low elevation. A *plateau* is a wide area at relatively high elevation bounded by steep slopes that either drop down into plains or rise upward to a mountain range. A *mountain* is a landform of high relief and high elevation that rises prominently above its surroundings with relatively steep slopes and a confined summit area. The term *mountain* can be expanded to include a mountain *range*, which is a continuous line of mountain peaks, and a mountain *belt* (or mountain *system*), which is a larger landscape consisting of a number of semicontinuous mountain ranges separated by *intermontane* (between the mountain) valleys.

Figure 1.3 is a copy of the Raisz hand-drawn map of the contiguous United States. This map can be compared directly with Figure 1.4, which is a more modern, digital,

shaded-relief image that shows topography by varying brightness from an artificial sun. A quick glance at these figures suggests that, to a first approximation, we can divide the US into a least four physiographic regions. Can you visualize the boundaries between these regions? There are two mountain systems (the Cordilleran and Appalachian), an interior plains and plateaus region, and a coastal plain. The boundaries are shown on the Raisz map in Figure 1.5. A close look at Figure 1.5 suggests that there are smaller landscape provinces within the four physiographic regions. How many distinctive landscape provinces can you recognize? Obviously the correct number is subjective depending on how specific a set of landforms one chooses to define for a particular landscape. In this book, we recognize 26 separate physiographic provinces, each of which is shown in Figure 1.6 and listed in Table 1.1. They include five plains, six plateaus, and 15 mountain areas. Each is grouped into one of the four larger physiographic regions shown in Figure 1.5. A brief overview of each region is presented next.

The Interior Plains and Plateaus region encompasses the entire central part of the US from the Rocky Mountain front in the west to the Appalachian Mountains in the east. Much of the area drains into the Mississippi River valley which flows southward through the central part of the Interior

FIGURE 1.3 The Raisz landform outline map of the United States.

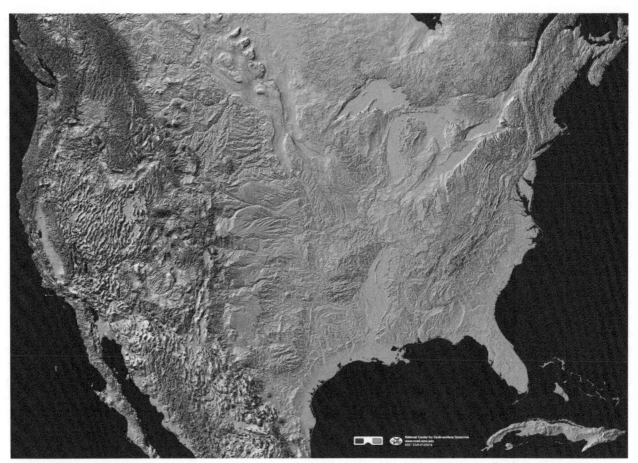

FIGURE 1.4   A digital shaded-relief image of the United States. From Thelin and Pike (1991) and downloaded from the National Center for Earth-Surface Dynamics site (www.nced.umn.edu).

region. Elevations rise from about 400 feet on the Mississippi River at St. Louis to about 5,000 feet at Denver, at the foot of the Rocky Mountains, and to about 2,000 feet in central Pennsylvania, near the western margin of the Appalachians. Included within the interior region are three plains—the Superior Upland, Central Lowlands, and Great Plains—and three plateaus, the Appalachian, Interior Low, and Ozark Plateaus. A few mountainous regions are present, such as the Wichita Mountains in Oklahoma, but only three—the Adirondack Mountains, the Black Hills, and the Ouachita Mountains—are large enough to be shown in Figure 1.6 as separate physiographic provinces. The Adirondack Mountains, the Appalachian plateau, and the Interior Low plateaus provinces form the entire eastern and southeastern margin of the Interior region where topography becomes hilly and mountainous as it rises toward the Appalachian mountain system.

The Appalachian Mountain system is narrow and characterized by a strong north-northeasterly trend. It includes the Valley and Ridge, Blue Ridge, and New England Highland mountain provinces along with the Piedmont plateau. The Ouachita Mountains are geologically similar with the Valley and Ridge but are detached from the main mountain belt and, therefore, are included with the Interior Plains and Plateaus region. The most rugged part of the Appalachians

is the Blue Ridge province of Tennessee and the Carolinas where there are at least 35 peaks above 6,000 feet, many of them within the Great Smoky Mountains National Park. Elevations decrease in all directions away from the Smoky Mountains. Toward the south, the mountain belt disappears into the Coastal Plain of Alabama and Georgia. The Appalachian belt is wide in New England, stretching from coastal Maine to eastern New York, but only Mount Washington in New Hampshire reaches an elevation above 6,000 feet. The eastern part of the mountain belt in Maine disappears into the Atlantic Ocean, only to reappear again in Nova Scotia.

The Coastal Plain extends along the entire eastern and southeastern seaboard of the US, from Cape Cod Massachusetts to the Gulf coast of Texas. It is a low-lying, nearly flat region with beaches, swamps, and wide river valleys. The Coastal Plain slopes gently toward the shoreline, and this gentle slope continues out to sea for up to 250 miles as part of the continental shelf. Ocean depth on the continental shelf is generally less than 600 feet. Beyond the shelf lies the continental slope and rise where ocean depth increases to between 13,000 and 20,000 feet, reaching the abyssal floor at the bottom of the Atlantic Ocean. Figure 1.7 is a Google Earth image of the United States that shows the extent of the continental shelf. The continental slope coincides roughly

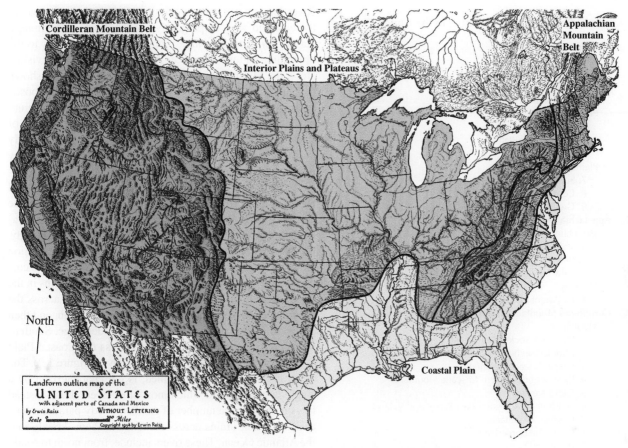

Cordilleran Mountain Belt

Interior Plains and Plateaus

Appalachian
Mountain
Belt

North

Coastal Plain

Landform outline map of the
UNITED STATES
with adjacent parts of Canada and Mexico
by Erwin Raisz   WITHOUT LETTERING
Scale ▭▭▭▭▭   ▭° Miles
Copyright 1958 by Erwin Raisz

FIGURE 1.5   The four major physiographic regions of the United States.

with the transition from continental rock to oceanic rock. It is the continental slope, and not the shoreline, that marks the geologic edge of the Coastal Plain. The entire Coastal Plain is below sea level north of Cape Cod where the Appalachian Mountains trend directly into the ocean.

The Cordillera is a complex region with three distinct mountain ranges: the Rocky Mountains in the east, the Coast Range (California Borderland) and Klamath Mountains on the west coast, and the Sierra Nevada and Cascade Mountain ranges just inland from the west coast. The highest peak is Mt. Whitney in the Sierra Nevada of California, at 14,495 feet. The US has more than 60 peaks above 14,000 feet, most of which are in the Colorado Rocky Mountains. The eastern and western mountain belts are separated by an intermontane area of plateaus, plains, and mountain blocks that include the Columbia River and Colorado Plateaus, the Snake River Plain, and the Basin and Range. There are distinct landscape changes between the Northern (Idaho-Montana) Rocky Mountains and the Middle-Southern (Wyoming-Colorado) Rocky Mountains as well as between the Northern and Central-Southern Cascade Mountains. In the Rockies, the generally linear northern part of the belt passes southward into massive mountain blocks separated by wide, flat valleys. Likewise, the North Cascades form a steep, rugged landscape of high-relief and somewhat

inconspicuous volcanoes, whereas the Central-Southern Cascades form relatively low-lying mountains with volcanoes that tower over the landscape. These differences allow us to separate each area into distinct physiographic provinces (Table 1.1; Figure 1.6).

ACROSS THE GREAT DIVIDE

A physiographic subdivision of the United States is based on the topographic expression of land in which similar landforms are grouped to form a province. An alternative subdivision is based on the distribution of major river systems with boundaries that correspond with drainage divides. A *river system* is a network of stream channels that either converge into a major river to form a drainage basin or that enter the same major body of water. River systems are separated by a *drainage divide* which is a continuous ridge of high ground where water on either side is drained into a different river system. Figure 1.8 is a schematic drawing that shows several river systems separated by divides. Only the larger divides are shown in this figure. The best-known drainage divide in North America is the Continental Divide (or Great Divide), which trends through the Rocky Mountains. This divide separates water that will eventually reach the Pacific Ocean from water that will reach either the Atlantic or Arctic Ocean.

TABLE 1.1 Physiographic Provinces of the United States

A. **Interior Plains and Plateaus**
   Mountains
      1. Black Hills
      2. Adirondack Mountains
      3. Ouachita Mountains
   Plateaus
      4. Appalachian Plateau
      5. Interior Low Plateaus
      6. Ozark Plateau
   Plains
      7. Superior Upland
      8. Central Lowlands
      9. Great Plains
B. **Appalachian Mountain System**
   Mountains
      1. Valley and Ridge
      2. Blue Ridge
      3. New England Highlands
   Plateaus
      4. Piedmont Plateau
C. **Cordilleran Mountain System**
   Mountains
      1. Northern Rocky Mountains
      2. Middle-Southern Rocky Mountains
      3. North Cascade Mountains
      4. Central and Southern Cascade Mountains
      5. Sierra Nevada
      6. Washington-Oregon Coast Range and Valleys
      7. Klamath Mountains
      8. California Borderland
      9. Basin and Range
   Plateaus
      10. Columbia River Plateau
      11. Colorado Plateau
   Plains
      12. Snake River Plain
D. **Coastal Plain**
   Plains
      1. Coastal Plain
      2. Continental Shelf (below sea level)

In this book we divide the continental US into nine major river systems as shown in Figure 1.9. Each is listed in Table 1.2. Four of the river systems drain into the Atlantic Ocean, three into the Pacific Ocean, one into the Arctic Ocean across Canada, and one does not reach an ocean.

The largest river system in the US is the Mississippi. This system extends from the Continental Divide across the entire heartland of the US to the Eastern divide, which is partly in the Appalachian Mountains. It includes the Missouri, Ohio, and Arkansas rivers as well as the Red River of Oklahoma and Texas. These tributary rivers drain directly into the Mississippi River which then empties into the Atlantic Ocean via the Gulf of Mexico. Also included are the Atchafalaya River and a few small tributaries that help to build the Mississippi River Delta region. The Mississippi

River has created a regional slope in which land areas to the east and west slope gently toward the river, forming a classic *dendritic drainage pattern* that looks similar to the branching pattern of a tree. A dendritic pattern is visible along the Mississippi River in Figure 1.9 and is also shown in Figure 1.8. This type of pattern forms where exposed rock is relatively homogeneous and where rock structure has little or no influence on the location or arrangement of streams. In this case, the underlying rock consists of fairly uniform, nearly flat-lying sedimentary layers overlain by unconsolidated sediment.

High points on the Appalachian Plateau, Valley and Ridge, Blue Ridge, and part of the Coastal Plain form the Eastern divide. This divide separates the Atlantic (or Eastern) Seaboard river system from the Mississippi. Interestingly, the divide does not follow the trend of highest elevation along the crest of the Appalachian Mountains. Instead, it cuts diagonally across the Appalachians. The divide begins in the Green Mountains of Vermont, extends westward across the Adirondack Mountains to the Appalachian Plateau, and then eastward, eventually to the crest of the Blue Ridge. South of the Blue Ridge, the divide extends across the Coastal Plain to an area just east of the Mississippi River (Figure 1.9). The Atlantic river system, as defined, is not characterized by a single large river into which most other rivers drain. Instead it is composed of a number of smaller, nearly parallel rivers, each of which drains across the Coastal Plain directly into the Atlantic Ocean. These rivers include, from north to south, the Connecticut, Hudson, Delaware, Susquehanna, Potomac, James, Cape Fear, Savannah, Suwannee, Apalachicola, Alabama, and Pearl rivers.

The natural flow of the Great Lakes is northeastward via the St. Lawrence River system to the Gulf of St. Lawrence and the Atlantic Ocean. In addition to the Great Lakes, the St. Lawrence River system also drains southeastern Canada and part of the Lake Region of northeastern Minnesota. The divide that separates the southern parts of Lake Superior, Lake Michigan, and Lake Erie consists of a series of low-lying hills, some of which are glacial ridges known as moraines. The separation is so slight that several manmade diversions into and out of the St. Lawrence drainage basin have been created in the past 200 years. One of the largest is a canal that was completed in 1900 by the city of Chicago that linked Lake Michigan with the Mississippi River system. The canal had the effect of reversing the flow of the Chicago River away from Lake Michigan (into which it previously emptied) and into the Mississippi River system. The canal was built to protect Lake Michigan, a source of drinking water, from the city's sewage and to open a shipping lane between the two waterways. Today there is serious talk of closing this connection in order to protect the Great Lakes from the dreaded Asian carp that has invaded the Mississippi River and will eventually invade the Great Lakes if given an opportunity. These fish are overcrowding and destroying native species.

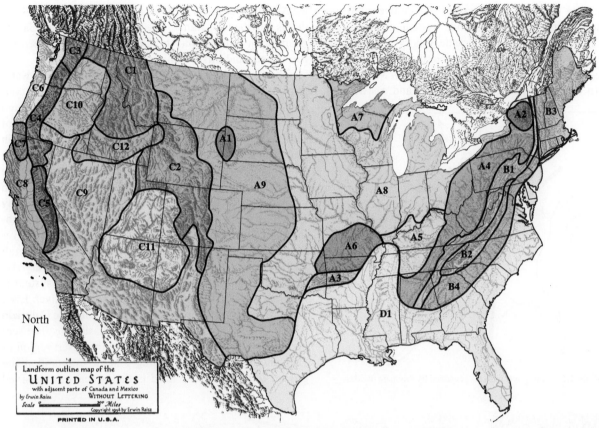

FIGURE 1.6 The 26 physiographic provinces of the United States.

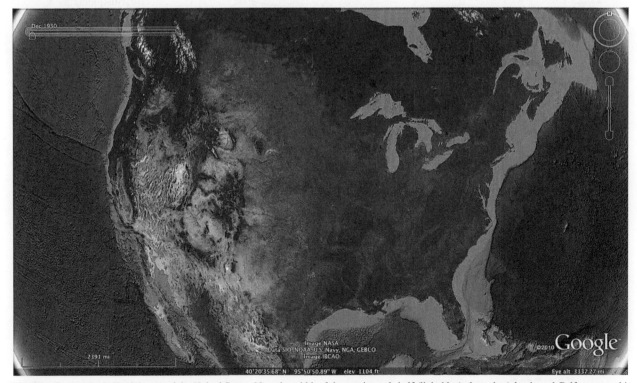

FIGURE 1.7 A Google Earth image of the United States. Note the width of the continental shelf (light blue) along the Atlantic and Gulf coasts and its abrupt transition to the continental slope and abyssal floor (dark blue). Note also the narrow shelf along the Pacific Coast.

The only other US river system that empties into the Atlantic Ocean is the Rio Grande system of southwest Texas. In addition to the Rio Grande, this system includes the Pecos, Nueces, Trinity, and Sabine rivers. The drainage area begins along the continental divide in New Mexico and southern Colorado and drains southward and eastward to

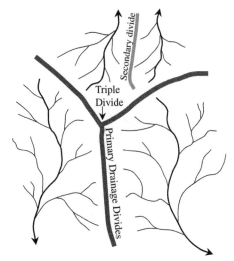

FIGURE 1.8   Three river systems separated by drainage divides.

the Atlantic Ocean via the Gulf of Mexico. The Pecos River drains into the Rio Grande. The Nueces, Trinity, and Sabine rivers drain directly into the Gulf of Mexico between the Rio Grande and Mississippi rivers.

The Colorado River system receives most of its water from the Rocky Mountains along the west side of the continental divide and flows southward and westward through the Grand Canyon to the Pacific Ocean via the Gulf of California. This system also includes the Green, San Juan, Little Colorado, and Gila rivers, all of which empty into the Colorado River. An unfortunate feature of the Colorado River is that the volume of water becomes less in a downstream direction and, during part of the year, may actually dry up near its mouth in Mexico before reaching the Pacific Ocean. By contrast, the Mississippi River becomes larger in a downstream direction due to rainfall and the addition of water from downstream tributary rivers. There are three reasons why the Colorado system is different. The first is the size of the Colorado River. Even where it contains its greatest volume of water, the Colorado is at least 10 times smaller than the Mississippi. The second is that the Colorado flows across the desert region of Utah and Arizona thereby losing water to evaporation and ground infiltration. The third is that seven water-starved

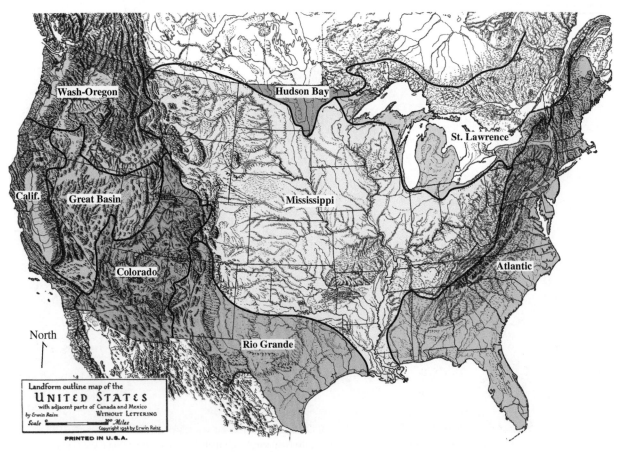

FIGURE 1.9   Major river systems of the United States.

states (Wyoming, Colorado, Utah, Arizona, New Mexico, Nevada, and California) all draw water from the river and its tributaries. The Colorado River is the lifeblood of these states. A similar situation affects the Rio Grande River, which also becomes smaller or dries up at its mouth.

The Pacific (or Western) Seaboard river system occupies the western side of the continental divide north of the Colorado River system. This system extends from Alaska southward to Baja California and, like the Atlantic Seaboard river system, is a composite of several major rivers, each of which drain directly to the Pacific Ocean. It is informally separated into a northern (Washington-Oregon) section that includes the Snake and Columbia rivers, and a

TABLE 1.2 Major River Systems of North America

1. **Mississippi River System**
   Mississippi River
   Missouri River
   Arkansas River
   Ohio River
   Drains to Gulf of Mexico
2. **Atlantic (or Eastern) Seaboard River System**
   Many separate river basins
   Drains to Atlantic Ocean, Labrador Sea, Gulf of Mexico
3. **St. Lawrence River System**
   Lake Superior (the largest lake in the world)
   Lake Huron
   Lake Michigan
   Lake Erie
   Lake Ontario
   St. Lawrence River
   Drains to Gulf of St. Lawrence
4. **Rio Grande-West Texas River System**
   Rio Grande River
   Pecos River
   Drains to Gulf of Mexico
5. **Colorado River System**
   Colorado River
   Green River
   Drains to Gulf of California
6. **Washington-Oregon River System**
   Columbia River
   Snake River
   Drains to Pacific Ocean
7. **California River System**
   Sacramento River
   San Joaquin River
   Klamath River
   Drains to Pacific Ocean
8. **Great Basin River System**
   Internal drainage (no outlet)
   Great Salt Lake
9. **Hudson Bay River System**
   Churchill River
   North and South Saskatchewan River
   Nelson River
   Lake Winnipeg
   Drains to Hudson Bay

southern (California) section that includes the Sacramento, San Joaquin, and Klamath rivers.

A rather unique drainage basin occupies the desert region of Nevada, Utah, and Oregon. It is known as the Great Basin and it truly is a great basin because any water that enters the area, such as during a rainstorm, is trapped, with no outlet to the sea. Water accumulates in low areas where it forms a temporary lake (a *playa*) before drying up in the arid climate. A few lakes still exist in the Great Basin close to mountain fronts where they receive water from melting snow. The largest include Lake Tahoe, Pyramid, Walker, and Utah lakes, and the Great Salt Lake. Most are remnants of much larger and more numerous lakes that existed during the last major glacial advance 18,000 years ago but have since evaporated. The divide that separates the Great Basin from the Sierra Nevada passes along the crest of the mountain range for about 400 miles, not allowing a single river to cross.

Part of the United States from Montana to Minnesota drains northward into the Hudson Bay river system, which occupies most of central Canada. The eastern part of the Hudson Bay system (north of the St. Lawrence river system) is a flatland characterized by a nonintegrated, deranged drainage pattern consisting of thousands of small lakes and swamps connected by slow-moving steams. There are no through-going rivers. The drainage pattern resulted from the flattening and scouring action of enormous glaciers that advanced from Canada into the United States over the past two million years or so. The glaciers destroyed a well-integrated, northward-draining system and left behind a flatland in which streams are confused as to which way to flow. The western part of the Hudson Bay system includes the Churchill and Saskatchewan rivers, which produces a much better drainage network. Water from these and other rivers eventually reaches the Arctic Ocean via Hudson Bay.

A glance at Figure 1.9 shows that there are several locations in the US where three drainage basins meet at a point. These points are known as *triple divides* (as shown in Figure 1.8). Perhaps the most important triple divide is the one located in northern Montana. This divide separates water that will eventually reach three oceans: the Atlantic, Arctic, and Pacific.

Divides are water boundaries; they are not physiographic boundaries (compare Figures 1.6 and 1.9.). Divides are present on all scales and between all rivers and streams, both large and small. A divide must exist between every stream valley shown in Figure 1.8. Divides change location rather frequently via headward erosion and stream piracy as shown in Figure 1.10. Water will take any available path to reach the ocean, even if it means crossing geological and landscape boundaries. Because they cannot be described as physiographic or geologic regions, we will end our discussion of river systems and return to our discussion of physiographic regions.

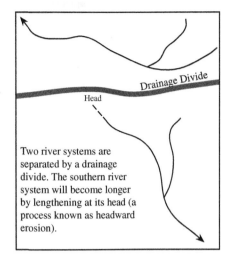 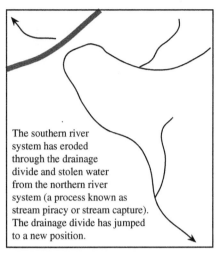

FIGURE 1.10   Headward erosion and stream piracy. North is to the top of the page.

COMPONENTS, MECHANISMS, AND VARIABLES THAT IMPART CHANGE ON A LANDSCAPE

If we assume that a landscape can completely change its look over time, then what happens to the landscape during the process of change? How does a landscape change its look? What causes some areas to change relatively quickly and other areas to change slowly? Why is the landscape in each of the 26 physiographic provinces different? One way to answer these questions is to define the components, mechanisms, and variables involved in the formation and evolution of landscapes.

Components are the constituent parts from which landscape is made. There are two major components: the *rock/ sediment type* and the *structural form* of the rock (also known as the *style of deformation* or simply the *structure*). The rock/sediment type is the substance that forms landscape; the structural form is the geometry of the substance. Together they define the geology that underlies each physiographic province.

A *mechanism* is a process or method by which something takes place. There are five mechanisms that can exact change on landscape; these are *uplift, subsidence, erosion, deposition,* and *volcanism.* These mechanisms, either singularly or in combination, can, over time, result in the destruction of a preexisting landscape and the creation of a new, different-looking landscape.

A *variable* is something that is subject to variation or change. With respect to landscape, we want to describe the variables that force change on a landscape. These are, in a sense, *forcing* variables because they represent the processes that force uplift, subsidence, erosion, deposition, and volcanism on a landscape. They are also the processes responsible for the creation, modification, and destruction of both the rock/sediment type and the

TABLE 1.3 Components, Mechanisms, and Variables
Components of Landscape
a. Rock/sediment type b. Structural form (style of rock deformation or structure of rock)
Mechanisms of Landscape Change
a. Uplift b. Subsidence c. Erosion d. Deposition e. Volcanism
Variables that Force Landscape Change

Primary
   a. Tectonic activity
   b. Climate
Secondary
   a. Sea-level change
   b. Isostatic adjustment

structural form. They can be considered variables because the magnitude and nature of the force and its effect on landscape vary over time. The two most important variables are climate and tectonic activity. *Climate* refers to the long-term condition of the atmosphere and is the principal driving force for erosion and deposition. *Tectonic activity* refers to internal stresses within the Earth that drive uplift, subsidence, and volcanism. The components, mechanisms, and variables are summarized in Table 1.3 and illustrated in Figure 1.11.

Climate and tectonic activity are the primary forcing variables. They constantly interact and compete with each other to shape the landscape and, in doing so, produce secondary variables that also affect landscape. The most

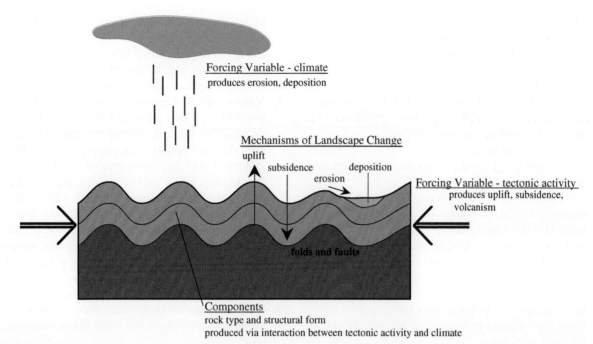

FIGURE 1.11 A sketch showing components, mechanisms (volcanism not shown), and forcing variables (sea-level change and isostatic adjustment not shown).

important secondary variables are sea-level change and isostatic adjustment. The raising or the lowering of sea level influences erosion and deposition rates, especially along coastlines and major river valleys. *Isostatic adjustment* is a process that causes vertical uplift and subsidence of land areas. The four variables are both independent and interdependent. A change in one variable will force changes in other variables, which, in turn, will cause changes in the rate at which each of the five mechanisms act upon the landscape. A simple example is a change in tectonic forces that results in rapid uplift of a land area. The rising landmass could block wind patterns, which, in turn, alters the climate of surrounding land areas. Rates of uplift, subsidence, erosion, deposition, and possibly even volcanism could all change, which, in turn, would impart change on the landscape. With these ideas in mind we can summarize landscape development with the following statement:

The landscape that characterizes a particular area is not random but is a direct result of the interaction of rock/sediment type and structural form with tectonic activity, climate, sea level change, and isostatic adjustment. These processes interact over time to drive uplift, subsidence, erosion, deposition, and volcanism, which are the mechanisms that maintain and change landscape.

Chapter 2 describes the first of the two landscape components, the rock/sediment type. Chapter 3 describes the second component, the structural form. Subsequent chapters in Part 1 of this book discuss the mechanisms and variables of landscape evolution and how they interact to produce, modify, and change landscape.

QUESTIONS

1. One inch is equal to_____ centimeters.
2. One centimeter is equal to _____ inches.
3. One mile is equal to _____ kilometers.
4. One kilometer is equal to _____ miles.
5. One meter is equal to _____ feet.
6. One foot is equal to _____ meter.
7. One mile is equal to _____ feet.
8. What major climatic event occurred between 2.5 Ma and 18,000 years ago?
9. Landscape is ephemeral. What does this statement mean?
10. What does a scale of 1:40,000 mean?
11. The Raisz maps shown in many of the figures are at an approximate scale of 1 inch equals 72.5 miles. Calculate the fractional scale of these maps.
12. Define the following terms:
    - Topography
    - Elevation
    - Relief
    - Landform
    - Landscape (terrain, physiographic province, physiographic region)
    - Physiographic region
    - Plain
    - Plateau

- Mountain
- Mountain range (mountainous region)
- Mountain belt (mountain system)

13. Name the four major physiographic regions in the contiguous United States.
14. What is the relief between Mt. Whitney in the Sierra Nevada and Owens Lake (elevation 3,563 feet) 17.2 miles to the southeast in Owens Valley?
15. What is a river system?
16. What is a drainage divide? What does the continental divide (the Great Divide) separate?
17. There are at least seven major river systems in the contiguous United States. Name the four river systems that empty into the Atlantic Ocean.
18. Draw an example of a dendritic drainage pattern.
19. In what type of rock and rock structure do dendritic drainage patterns form?
20. Name the only major river system in the contiguous United States that does not empty into an ocean.
21. Name three plateau provinces that are entirely east of the Mississippi River.
22. In what part of the country is the Coastal Plain currently entirely below sea level? Your choices are Tennessee, North Carolina, Florida, Georgia, Maine, Indiana, Texas, Louisiana, or Mississippi.
23. What physiographic province is a vast area centered in Nevada that is composed of small isolated mountain blocks and intervening valleys?
24. In what state are the Sierra Nevada?

25. Make a copy of Appendix Figure A.2. Use Figure 1.6 and Table 1.1 as a guide to color and label each individual physiographic province. Draw the boundaries of the four physiographic regions with a thick heavy line.
26. In which physiographic province do you live?
27. Make a copy of Appendix Figure A.1. Use Figure 1.9 and Table 1.2 as a guide and draw the boundaries of all the major US river systems. Be careful not to draw a divide across any river or stream. Label all the river systems. Label the continental divide. Label the five Great Lakes. Use a colored pencil to highlight the following rivers: Mississippi, Missouri, Ohio, Arkansas, Colorado, Columbia, Snake, Rio Grande, Pecos, St. Lawrence, Green, Sacramento, San Joaquin, Hudson, Susquehanna, James, Cape Fear, Savannah, Suwannee, and Pearl. You may have to consult additional references or the Internet to complete this task.
28. In what mountain range is the continental divide located?
29. What is special about the triple divide in Montana?
30. What is special about the Great Basin drainage system?
31. Name the two components of landscape evolution that define the geology that underlies each physiographic province.
32. Name the four variables of landscape evolution.
33. The Appendix contains uncolored versions of Figures 1.6 and 1.9. Color each of the 26 provinces in Figure 1.6 and provide a one- or two-sentence description of each. Do the same for the river systems shown in Figure 1.9.

Component: The Rock/Sediment Type

Three rock types are important in landscape evolution: sedimentary, crystalline, and volcanic. Each has a strong influence on landscape wherever they are exposed at the surface. Rocks not exposed at the surface are, instead, buried beneath an apron of unconsolidated (loose) sediment and soil. If the apron of unconsolidated sediment is thick enough to completely cover any trace of the underlying rock type, then the sediment will impart its own unique look upon the land thus creating a fourth landscape-forming rock/sediment type. Each type is listed in Table 2.1 along with common examples. The critical aspect of rock type with respect to landscape is its resistance to weathering and erosion relative to surrounding rock. This chapter provides an overview of the four rock/sediment types and their distribution across the United States.

WEATHERING, EROSION, AND DEPOSITION

The weathering process includes all changes that result from exposure of rock material (or any material) to the atmosphere. If you leave your bicycle outside for a year or two, it will begin to rust, which is a form of weathering. The weathering process includes physical changes that break the rock into smaller pieces, and chemical changes by which the rock reacts with water, air, and organic acids and partly or wholly dissolves. Physical weathering is equivalent to hitting a rock with a hammer. Chemical weathering is equivalent to pouring acid on a rock. The residual product of weathering is *unconsolidated sediment*.

An important form of physical weathering is frost cracking (also known as *ice wedging*). Upon freezing, water expands by about 9%. Frost cracking occurs where water collects in a small crack in a rock and then freezes. The expansion causes the crack to propagate (become wider and longer) such that the rock eventually breaks in half. Frost cracking is particularly active along mountaintops in the west where temperature is above freezing during the day and below freezing at night for much of the year. Piles of broken rock litter summit areas above tree line in areas such as the Rocky Mountains as shown in Figure 2.1.

Chemical weathering is the partial or complete dissolution of rock. Water is naturally slightly acidic and becomes more acidic when in contact with dead and decaying plant matter. Many minerals only partially dissolve and, in doing so, leave behind a clay residue. An example is the partial dissolution of potassium feldspar, which is a major mineral in granitic rock and which is also common in sandstone and some types of metamorphic rock. The following chemical equation describes the process:

$$2KAlSi_3O_8 + (2H^+ + 9H_2O) = Al_2Si_2O_5(OH)_4 + (2K^+ + 4H_4SiO_4)$$

Potassium feldspar + Acidic water = Clay
+ Components in solution in water

TABLE 2.1 Landscape-Forming Rock/Sediment Types

1. Sedimentary (sandstone, limestone, shale)
2. Crystalline (Plutonic includes granite, gabbro; metamorphic includes slate, schist, gneiss, marble)
3. Volcanic (nonexplosive basalt; explosive silicic andesite-rhyolite-tuff)
4. Unconsolidated sediment (alluvium, glacial, coastal)

FIGURE 2.1   Looking along the ridge of Mount Audubon, Front Range, Colorado. The large rock pile that litters the ridge formed via frost cracking.

The reaction removes potassium (K) and some of the silica (SiO_2) from feldspar. Both go into solution in water. The potassium then becomes available for plants to absorb. The dissolved silica (silicic acid; H_4SiO_4) may enter the groundwater system and precipitate around sand grains to form sandstone.

Quartz, which is crystalline silica, is the only common mineral that is not strongly affected by chemical weathering. It does not dissolve in water. All other common minerals are either dissolved completely or are partially dissolved and reduced to clay. For this reason quartz and clay are the two most abundant minerals in sedimentary rock. Chemical weathering is most effective when water is present. Therefore, the amount of chemical weathering in an area is controlled largely by the amount of available water.

The weathering process disaggregates rock, but it does not remove rock material from its original location. *Erosion* is the physical removal of rock and sediment from its original location by an agent such as water, ice, air, gravity, or animal/human interference. If erosion does not occur, unconsolidated sediment will remain in place, mix with organic material, and become soil. The breakdown of rock through the weathering process facilitates erosion.

Deposition is the accumulation of weathered and eroded sediment such as in a lake or subsiding basin. Deposition can occur in one of three environments: marine, nonmarine, or transitional. The term *marine* refers to something found in or produced by the sea. Marine environments include shallow marine, deep marine, and reef. Nonmarine environments are those that form on land, such as a desert, lake, river, soil, or glacier. Transitional environments form along coastlines where there is a component of both land and ocean. Such environments include beach, delta, and estuary.

ROCKS AND UNCONSOLIDATED SEDIMENT

Sedimentary rock consists of unconsolidated sediment and organic material originally deposited at the Earth's surface and subsequently buried, compressed, and cemented into rock. Sedimentary rocks are distinctive because they are almost always *stratified*, meaning they are layered. An individual layer is referred to as a *bed;* each bed has a composition, color, or texture different from underlying and overlying beds. Each bed was originally deposited as a flat-lying (horizontal) or nearly flat-lying layer. A succession of beds stacked on top of one another is referred to as a *stratigraphic sequence*, or simply as *stratigraphy*. Each bed within a stratigraphic sequence can vary in thickness from less than one inch to more than two hundred feet. Large areas of the US consist of interlayered beds of sandstone, shale, and limestone. Figure 2.2 shows a tilted sequence of sandstone and shale. The rocks were tilted after they were deposited.

Sandstone is composed of sand-sized rock particles and minerals (mostly quartz) mixed with variable amounts of silt and clay, all cemented together to form rock. The sand on the beach that you might visit this summer will eventually become sandstone. If you walk from the beach into the water, the sandy beach quickly turns to mud. *Shale* is compressed and hardened mud. It is the most abundant of all sedimentary rocks. *Limestone* is composed mostly of the mineral calcite ($CaCO_3$). It forms primarily in warm, clear marine water from shell, skeletal, and fecal debris

FIGURE 2.2   Photograph of a thick sandstone layer above thinner beds of sandstone and shale, Sunset Bay State Park, Oregon. The beds are tilted slightly to the left. The shale layer is about one foot thick.

that accumulates at the bottom of a shallow ocean. Consequently, this rock tends to be highly fossiliferous. Coral reefs, including the Florida Keys, form limestone.

In addition to sandstone, shale, and limestone, other less common sedimentary rocks include conglomerate, dolostone, and coal. A *conglomerate* contains rounded pebbles and cobbles of rock fragments in addition to sand-sized material and is often associated with sandstone. *Dolostone* is closely associated with limestone but with a slightly different composition. It is composed almost entirely of the mineral dolomite ($CaMg(CO_3)_2$). Together, limestone and dolostone are referred to as *carbonate rock* because their chemical composition includes CO_3. *Coal* is a sedimentary rock that consists almost entirely of the decayed remains of trees and other vegetation. It forms in swampy environments and is often associated with shale.

Conglomerate, sandstone, and shale are known as *clastic rocks* because they are composed of broken pieces of preexisting rock. The term *clastic* is from the Greek *klastos*, which means broken. The material that forms a clastic rock, such as individual grains of sand in sandstone, are known as *detritus*, which in Latin means to wear away (*detrital* is the adjective). As noted earlier, limestone forms primarily in shallow marine water through the action of sea animals. Limestone, dolostone, and coal are not clastic rocks. Rather than form from broken pieces of preexisting rock, they form through chemical and biochemical precipitation and the accumulation of organic material.

Igneous plutonic rocks and metamorphic rocks are grouped together as *crystalline rock* because they often occur together and because they have similar landscape

significance. Both form at depths well below the Earth's surface. The two rock types are shown in Figure 2.3.

An igneous plutonic rock crystallizes from molten magma (liquid rock) below the Earth's surface, usually at depths between 1 and 40 miles (1–65 km). For this reason plutonic rocks are referred to as *intrusive*. A *pluton* is a generic term for any body of intrusive rock, whereas a *batholith* is a large composite intrusive body that can cover nearly an entire mountain range. The most familiar plutonic and batholithic rock is granite, which is commonly pink and composed of glassy quartz, white plagioclase, pink K-feldspar, and black specks of biotite. Individual minerals in granite are large enough to be seen with the naked eye. True granite contains roughly equal amounts of quartz, plagioclase, and K-feldspar. Other granite-like plutons, such as granodiorite, syenite, and diorite, contain variable amounts of these minerals or, in some cases, may lack one or two of the minerals, or they may contain other minerals such as hornblende. We do not need to make sharp distinctions between these various rock types because all are relatively light-colored and have a similar plutonic mode of origin. We will refer to all of them as *granitic* or *granitoid*.

A second, far less common type of intrusive rock is *gabbro*. In contrast to granitic rock, gabbro is dark-colored, without K-feldspar or biotite and with very little quartz. It is composed primarily of the minerals plagioclase and pyroxene with or without olivine.

Metamorphic rocks form at depths below the Earth's surface, and for this reason they are often associated with plutonic rocks. The major difference is that metamorphic rocks do not crystallize from magma. Instead they form by solid-state recrystallization of preexisting rock in response

FIGURE 2.3 Photograph of metamorphic gneiss (dark rock) intruded by granitic rock (white), Sawatch Range, Colorado. Together these rocks form the crystalline rock type.

to changes in temperature and pressure during burial. If sedimentary or volcanic rock is buried, or if an intrusive rock is moved to a new depth (and therefore to new pressure-temperature conditions), the minerals in the rock may no longer be stable. The minerals recrystallize, thereby forming a new texture and possibly a new set of minerals.

The process is similar to baking bread. Before it is baked, the dough consists of flour, yeast, sugar, salt, and water. Let's pretend that this mixture is a sedimentary rock composed of the minerals flour, yeast, sugar, salt, and water. Nothing happens to this rock during the time it sits on the countertop. However, as soon as the rock is placed in a hot oven, the minerals begin to react to form a "metamorphic rock" that we refer to as bread. Obviously, bread has a very different texture and taste than the original dough. However, if we were to analyze both the dough and the bread we would find that both have the same chemical composition. The only difference is the loss of water during the heating process. The original ingredients never melted; they simply reacted with each other to form bread. The change occurred in the solid state and resulted in a change in both texture and mineralogy. The reaction occurred because bread (not dough) is the stable "rock" at the elevated temperature in the oven.

Real rocks behave the same way. The original chemical composition of a rock does not change during metamorphism except for the loss of water and other fluids. It is the minerals in the rock that change. The original minerals react in the solid state to produce a new set of minerals and a new texture that is stable at the new pressure and temperature.

There are many types of metamorphic rocks. *Slate* is derived from metamorphism of shale and looks very much like shale, except it is harder. The best pool tables and the

best roofing tiles in the world are made of slate. *Schist* is a shiny, biotite-muscovite-rich rock derived from the metamorphism of shale or sandstone at higher temperature-pressure conditions than those of slate. *Gneiss* is an even higher temperature-pressure rock with well-developed light and dark bands. Gneiss can be derived from metamorphism of shale, sandstone, or granite. *Quartzite* is composed primarily of crystalline quartz and is derived from the metamorphism of quartz-rich sandstone. *Marble* is derived from the metamorphism of limestone or dolostone. Finally, *greenstone, blueschist,* and *amphibolite* are derived from the metamorphism of *basalt.* As the names imply, a greenstone tends to be green and a blueschist tends to be metallic blue. Amphibolite is dark green or black. Again, we need not make sharp distinctions between all the various metamorphic rocks. They can be grouped with granitoid rocks and referred to collectively as *crystalline* for the purpose of landscape analysis.

Volcanic rocks are the final major rock type in landscape development. These are igneous rocks that crystallize from molten magma (lava) at the Earth's surface rather than below the surface like plutonic rocks. For this reason volcanic rocks are referred to as *extrusive. Basalt* (or basaltic rock) is the most common volcanic rock. It is a dark, drab rock composed of the minerals plagioclase and pyroxene with or without green olivine. A large basalt flow is shown in Figure 2.4. Here the mineralogy hardly matters because individual minerals are too small to see with the naked eye. Basaltic rock forms a vast majority of the world's ocean bottom, and for this reason it is the most common rock close to the Earth's surface. It is the rock that you would see if you visited the volcanoes of Hawaii. When basalt erupts to the surface, it tends to be nonexplosive

FIGURE 2.4   Photograph overlooking a large basalt flow at Lava Butte, Oregon. The basalt flow is approximately 7,000 years old.

and fluid. It will flow easily across great distances. Most eruptions, including those in Hawaii, are safe enough to observe. If basaltic magma does not reach the Earth's surface it will crystallize at depth to form gabbro.

Another important volcanic rock is the andesite-rhyolite-tuff family, which, collectively, will be referred to as *silicic rock*. These rocks represent the volcanic equivalent of granitic rock. They have the same composition as granitic rock; the only difference is that they crystallize at the Earth's surface rather than at depths below the surface. Unlike basalt, these volcanic rocks tend to be explosive. You do not want to be too close to these volcanoes when they erupt. Both *andesite* and *rhyolite* are lighter in color, more quartz-rich, and do not flow as easily as basalt. *Tuff* is a name used for explosive volcanic ash that has solidified into rock. It is these rocks, and not the more fluid basaltic rocks, that form the classic, pyramid-shaped volcanic cones known as *strato* or *composite volcanoes*. Examples include Mount St. Helens and Mt. Rainier in the Central and Southern Cascades.

Unconsolidated sediment is not rock but rather is the weathered product of preexisting rock. Much of the unconsolidated sediment in the United States consists of alluvium (sediment reworked, moved, and deposited by rivers), glacial drift (sediment moved and deposited by glaciers), coastal deposits (sediment moved by ocean currents and waves), or soil (material weathered in place). Unconsolidated sediment, regardless of its origin, represents sedimentary rock in the process of forming.

The three rock types, along with unconsolidated sediment, are related to each other by the rock cycle shown in Figure 2.5. All rocks ultimately begin as magma (liquid rock) at some depth within the Earth, which can vary from a few miles to about 1,800 miles. Magma is less dense than solid rock; therefore, once formed, it begins to rise. As it rises, it either cools at depth as a plutonic (intrusive) rock such as granite, or it reaches the Earth's surface as volcanic rock. If it crystallizes at depth, the plutonic rock could eventually reach the Earth's surface through erosion of overlying rock. Once rock is at the surface it will be weathered, eroded, and eventually deposited as sediment. As more and more sediment is deposited, the bottom of the pile becomes compacted and cemented to form sedimentary rock. Sedimentary rock can then be driven deep into the Earth by tectonic processes where it is metamorphosed. As temperature increases, the rock will eventually melt and become magma, thus beginning the rock cycle anew.

There are shortcuts in the rock cycle. Rather than continuing deep into the Earth and melting, sedimentary and metamorphic rocks could instead be exhumed to the Earth's surface where they would be weathered and eroded to form a second-generation sedimentary rock. Volcanic rock could be buried and metamorphosed rather than weathered and eroded. Plutonic rock could be metamorphosed rather than brought to the surface.

THE INFLUENCE OF BEDROCK ON A LANDSCAPE

Bedrock is a solid mass of rock that is physically connected to the interior Earth. When bedrock is exposed at the Earth's surface, it is known as *outcrop*. Of primary

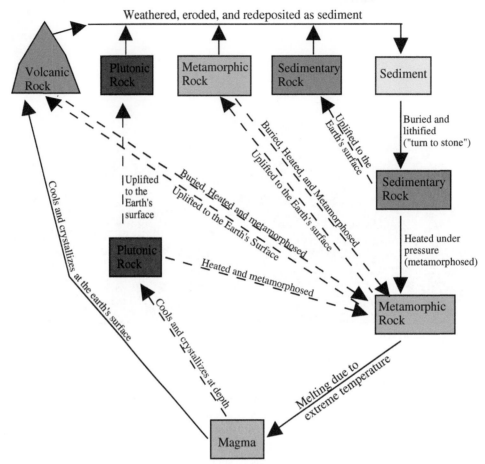

FIGURE 2.5   The rock cycle.

importance in landscape development is the way that an outcropping of bedrock reacts to weathering and erosion. Rocks that resist weathering and erosion are said to be *hard, strong,* or *resistant.* We could say that the rock has low *erodability,* meaning that it is not easily weathered and eroded. Rock that is easily weathered and eroded is said to be *soft, weak,* or *nonresistant.* Crystalline rocks, in general, are the most resistant. Volcanic rocks tend to be less resistant and sedimentary rocks the least resistant, although there will always be exceptions. More importantly sedimentary rocks are layered (stratified), and some layers are more resistant to erosion than others. Layers of shale are the least resistant. Thick layers of sandstone and limestone tend to be far more resistant than shale. The interlayering of limestone, shale, and sandstone creates a landscape of *differential erosion* where adjacent rocks weather and erode differently such that the more resistant rock protrudes above the weaker rock. Crystalline rocks, and particularly volcanic rocks, can also show layering. However, in this instance the layers typically are of similar composition. They do not show strong contrasts in erodability, and as a result they do not create a landscape of strong differential erosion. In the United States, there are two common situations where differential erosion shapes the landscape. The first is where resistant and nonresistant sedimentary rocks are interlayered. The second is where sedimentary rocks are underlain by resistant crystalline rock. Schematic examples of both are shown in Figure 2.6.

It is important to emphasize the difference between a landscape composed of sedimentary rock and one composed of crystalline rock. The tilted, eroded, edge of a layered sequence of sedimentary rocks imparts a strong grain to the landscape. Sedimentary rocks are also relatively soft. With few exceptions they tend not to form high mountains. Crystalline rock is resistant to erosion and therefore tends to form highland regions that include some of the most rugged landscapes on Earth. A majority of the mountainous regions of the United States are underlain by crystalline rock, including the North Cascade region of Washington state, the Sierra Nevada of eastern California, most of the Colorado Rocky Mountains, the Teton and Wind River ranges of Wyoming, the Blue Ridge Mountains in the Appalachians, the Adirondack Mountains, and most of the New England Highlands.

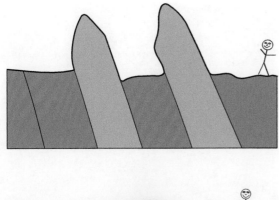

FIGURE 2.6 Differential erosion. (a) Two resistant layers of sedimentary rock protruding above nonresistant layers. (b) Resistant crystalline rock protruding above less resistant sedimentary layers.

KARST LANDSCAPE

Limestone is a common sedimentary rock which, in arid environments, tends to be as resistant to weathering and erosion as sandstone. Limestone, however, has another property unlike sandstone or shale. Limestone is composed mostly of calcite (calcium carbonate) that will dissolve in acidic water. Dolostone (calcium-magnesium carbonate) has similar properties, although it does not dissolve quite as readily. Rainwater is naturally slightly acidic. If limestone is present at or close to the Earth's surface, water will seep into cracks and bedding planes and dissolve the rock to produce *voids*, which we call *caves*. As the voids get larger, part of a cave might collapse to produce a sinkhole, which is either an open hole or a pond at the Earth's surface. In this type of landscape it is common for streams at the surface to disappear into an open sinkhole, flow underground through a cave system, and return to the surface as a spring. A landscape with a lot of caves, sinkholes, disappearing streams, and springs is referred to as a *karst landscape*.

A classic karst landscape develops where limestone is exposed at the Earth's surface. In this case, sinkholes dominate due to constant collapse of near-surface caves. The Lime Sink region of Florida, between Tallahassee and Orlando, is one example. The karst limestone in this region is famous for its natural springs and for the development of sinkholes that have swallowed roads, houses, and anything else that just happened to be built above them. Figure 2.7 is a Google Earth image that shows the Lime Sink area surrounding Orlando. The many ponds shown in the image are sinkholes.

Sinkholes are far less common in areas where limestone is covered with an overlying surface layer of insoluble rock

such sandstone or shale. The caves that form in the underlying limestone are, in this situation, protected from collapse by the overlying insoluble surface layer. Very large caves can form under these conditions. Mammoth Cave is an example. The difference between a landscape with limestone at the surface and one in which limestone underlies a surface layer of insoluble rock can be striking. Figure 2.8 is a Google Earth image of an area near Bedford, Indiana, on the Interior Low Plateaus that shows an abrupt change from a sinkhole-filled terrain underlain by limestone on the left to a landscape of dissected hills underlain by sandstone and shale on the right.

Some of the best-developed karst landscapes in the world occur in the humid climate of the Valley and Ridge, Ozark Plateau, and Interior Low Plateaus. Karst landscape is also well developed in many areas of the west, including southwest Texas, southeast New Mexico, Idaho, Montana, and the Black Hills. In most of these areas the limestone is oriented in nearly horizontal (flat-lying) layers. This is not the case in the Valley and Ridge, where layers in some areas are inclined at steep to vertical angles.

DISTRIBUTION OF ROCK/SEDIMENT TYPE AMONG THE US PHYSIOGRAPHIC PROVINCES

Figure 2.9 shows the general distribution of the three major rock types, without regard for physiographic boundaries. Here we can see that sedimentary rocks are widespread, especially in the central and southeastern parts of the United States. Volcanic rocks are mostly restricted to the Cordillera, and crystalline rocks are distributed almost randomly across the US. If we compare this figure with a figure showing the distribution of all 26 physiographic provinces (Figure 1.6), we see that certain patterns emerge. In the eastern US the boundaries shown in the two figures closely coincide. An agreement such as this implies that the type of rock strongly influences the character of landscape. Boundaries separating rock types in the Cordillera, on the other hand, tend to cut directly across physiographic boundaries. Here the tectonics are more recent and, in some cases, active. Tectonics produces the structural form (geometry) of the rock and, as a result, the structural form plays a greater role in defining boundaries of individual provinces. Figure 2.10 is a more general figure that combines rock type (Figure 2.9) with physiographic province (Figure 1.6) to show the dominant rock or rocks within each province. This figure and Table 2.2 are included to better facilitate a correlation between rock type and physiographic province.

Unconsolidated sediment is widespread in all 26 physiographic provinces. It forms a thin mantle in arid climates but can be more than 300 feet thick in humid climates. In

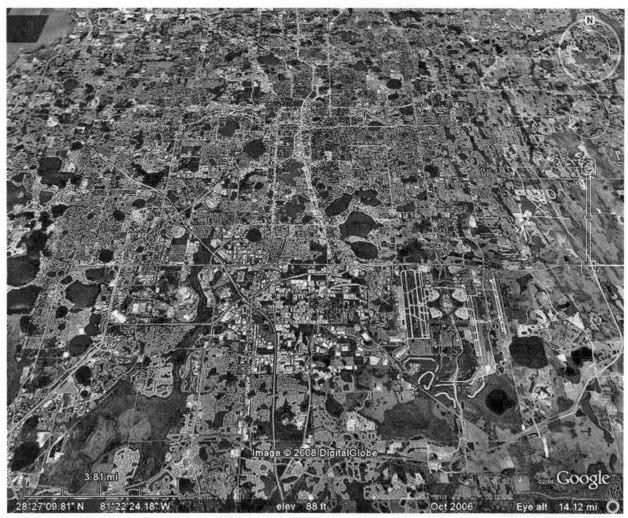

FIGURE 2.7   Google Earth image looking north at the many sinkholes in the Lime Sink area surrounding Orlando, Florida. Orlando is at upper center.

FIGURE 2.8   Google Earth image looking south at two contrasting terrains: a sinkhole-filled karst terrain underlain by limestone at left, and a hilly, dissected upland terrain underlain by sandstone and shale at right. The location is a few miles west of Bedford, Indiana, in the Interior Low Plateaus. The White River flows across the karst limestone at the top of the image.

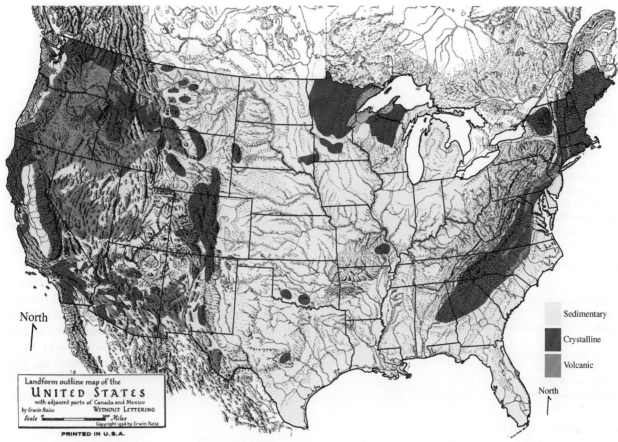

FIGURE 2.9 Landscape map showing distribution of rock types.

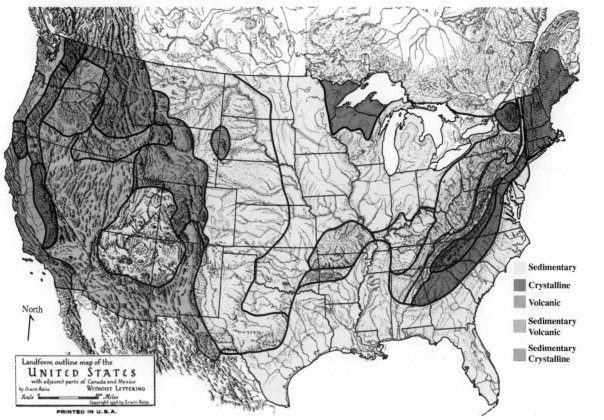

FIGURE 2.10 Landscape map showing major rock types within each physiographic province.

TABLE 2.2 Major Rock Type Within Each Physiographic Province

Dominantly Sedimentary Rock

Interior Plains and Plateaus
   Ouachita Mountains
   Appalachian Plateau
   Interior Low Plateaus
   Ozark Plateau
   Central Lowlands
   Great Plains
Appalachian Mountain System
   Valley and Ridge
Cordilleran Mountain System
   Colorado Plateau
Coastal Plain
   Coastal Plain

Dominantly Crystalline Rock

Interior Plains and Plateaus
   Adirondack Mountains
   Superior Upland
Appalachian Mountain System
   Blue Ridge
   New England Highlands
   Piedmont Plateau
Cordilleran Mountain System
   North Cascade Mountains
   Sierra Nevada

Dominantly Volcanic Rock

Cordilleran Mountain System
   Central and Southern Cascade Mountains
   Columbia River Plateau
   Snake River Plain

Sedimentary and Crystalline Rock

Interior Plains and Plateaus
   Black Hills
Cordilleran Mountain System
   Northern Rocky Mountains
   Middle and Southern Rocky Mountains
   California Borderland
   Klamath Mountains

Sedimentary and Volcanic Rock

Cordilleran Mountain System
   Basin and Range
   Washington-Oregon Coast Range and Valleys

a) Thin layer of unconsolidated sediment draped over bedrock. Bedrock controls the landscape.

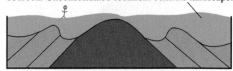

b) Thick layer of unconsolidated sediment covers bedrock. Unconsolidated sediment controls landscape.

FIGURE 2.11   Sketch showing control of unconsolidated sediment. (a) A thin layer of unconsolidated sediment covers bedrock but does not control landscape. (b) A thick layer of unconsolidated sediment completely covers bedrock and controls landscape.

underlying rock such that the rock exerts no influence on landscape. This concept is illustrated in Figure 2.11. Examples include thick glacial deposits in the northern Central Lowlands as shown in Figure 2.12, alluvial (river) deposits in the lower Mississippi River valley as shown in Figure 2.13, coastal deposits along the eastern seaboard and Gulf Coast, and thick soil in the Piedmont Plateau and Coastal Plain. None of these areas constitute an entire physiographic province.

QUESTIONS

1. Describe crystalline rocks by answering the following questions:
   a. What two rock types form crystalline rocks?
   b. What is their general resistance to erosion relative to other rock types?
   c. In what part of the Earth do crystalline rocks form?
   d. Name a common crystalline rock.
2. Define sedimentary rock and answer the following:
   a. What is their general resistance to erosion relative to other rock types?
   b. What does it mean that the rocks are stratified?
   c. Name the three most common sedimentary rocks.
   d. In what part of the Earth do sedimentary rocks form?
3. Of the three most common sedimentary rocks:
   a. Which is least resistant?
   b. Which is composed dominantly of calcite?
   c. Which forms karst topography?
   d. Which two are resistant in arid climates?
   e. Which is likely to be most resistant in humid climates?
4. Describe volcanic rocks by answering the following:
   a. Are these rocks deposited on the Earth's surface or at depths below the surface?
   b. Name a common nonexplosive volcanic rock.
   c. Name a common explosive volcanic rock.

most areas, unconsolidated sediment is draped over the bedrock such that underlying bedrock still controls the shape of the land, even though it is not exposed. In this case, it is like throwing a sheet over a table. The table controls the exterior shape even though the table cannot be seen. If, on the other hand, we were to completely bury the table in sand, then the sand, and not the table, would control the shape. Sediment-controlled landscapes form only in areas where the sediment is thick enough to completely bury the

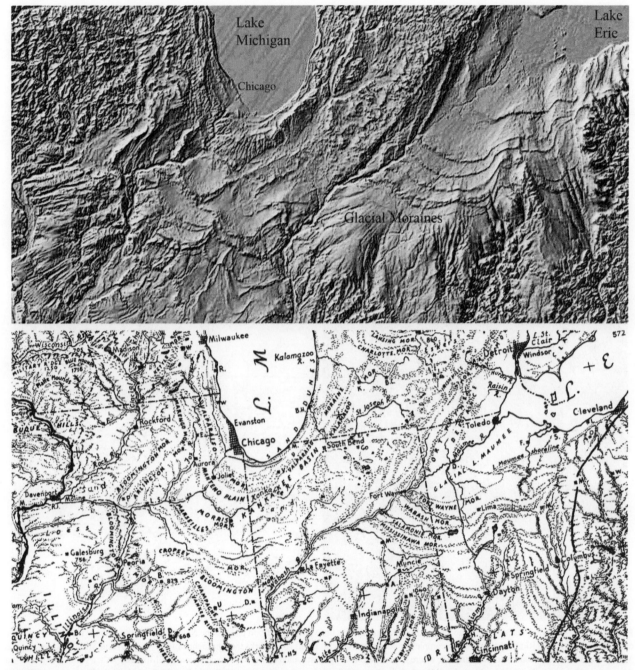

FIGURE 2.12   A digital shaded-relief map and landscape map showing a smooth glacial landscape with conspicuous ridges (moraines) in the area south of Lake Michigan and Lake Erie. Bedrock topography is completely buried in part of this area. The highly dissected area at upper left is the Driftless area, which was not glaciated. It is likely that the entire region resembled the Driftless area prior to glaciation. Indiana state boundaries parallel the north direction. The shaded-relief image is from Thelin and Pike (1991) and downloaded from the National Center for Earth-Surface Dynamics (www.nced.umn.edu).

5. Explain the following terms:
   a. Bedrock
   b. Outcrop
   c. Resistant
   d. Nonresistant
   e. Differential erosion
6. An individual layer of sedimentary rock is referred to as a _____.
7. What is a plutonic rock and where does it crystallize?

8. What is a batholith, where do batholiths form, and what type of rock characterizes batholiths?
9. Name a common intrusive (plutonic) igneous rock.
10. How does a metamorphic rock form?
11. What is the difference between a volcanic rock and a plutonic rock?
12. What is the most common volcanic rock? Name three physiographic provinces where volcanic rocks dominate.
13. Name three rock types that form silicic volcanic rock.

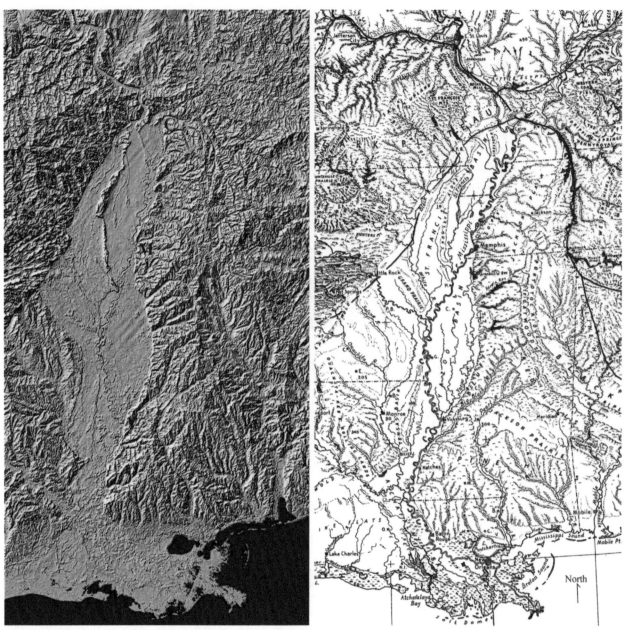

FIGURE 2.13   A digital shaded-relief map and landscape map showing the Mississippi flood plain in the Mississippi Embayment. With the conspicuous exception of Crowley's Ridge, the flood plain completely buries bedrock. St. Louis (SL) marks the confluence of the Missouri and Mississippi rivers. The Mississippi River flows southward to Cairo (C), where it enters the Embayment joined by the Tennessee River. The Mississippi then flows along bluffs at the eastern edge of the floodplain to Memphis (M), where it turns westward, forming an arc into the flood plain that loops back to the bluffs at Vicksburg (V). Southward, the river maintains its course along the eastern edge of the flood plain until it reaches the Mississippi Delta and turns eastward. Note also the large sediment accumulation along the shoreline west of the Mississippi Delta, with sparse accumulation to the east. The shaded-relief image is from Thelin and Pike (1991) and downloaded from the National Center for Earth-Surface Dynamics (www.nced.umn.edu).

14. In which physiographic province are silicic volcanic rocks abundant?

15. Of the three major rock groups (excluding unconsolidated sediment), which is most widespread in the United States?

16. Under what conditions will the type of rock (sedimentary, crystalline, volcanic) have no influence on the shape of landscape?

17. Describe the weathering process.

18. What is alluvium?

19. If weathered material remains in place, it will form what?

20. Name a physiographic province where glacial landforms are well developed.

21. Name a physiographic province where alluvial landforms are well developed.

22. Name six physiographic provinces in the Interior Plains and Plateaus region that are composed mostly of sedimentary rock. Name three additional provinces composed of sedimentary rock.
23. Name three physiographic provinces in the Appalachian Mountain system composed mostly of crystalline rock. Name four additional provinces composed of crystalline rock.
24. Name four physiographic provinces in the Cordilleran Mountain system composed mostly of a combination of sedimentary and crystalline rock. What other physiographic province fits this category?
25. Name two physiographic provinces composed mostly of a combination of sedimentary and volcanic rock. In which physiographic region are they located?
26. The three rock groups, along with unconsolidated sediment, are related to each other by the _____.
27. Regarding Figure 2.12, what evidence is there that Lake Erie was once larger than it is today?
28. Regarding Figure 2.13, explain why it is advantageous for Memphis and Vicksburg to be located where they are.
29. What do the terms *marine, carbonate, clastic,* and *stratigraphy* mean?
30. Name some typical marine, nonmarine, and transitional depositional environments.
31. The Appendix contains an uncolored version of Figure 2.9. Make a photocopy of this figure and color each of the three rock types. You may ignore the rather busy area in the southwestern US. Write a paragraph or two on the distribution of each rock type.

Chapter 3

Component: The Structural Form

Chapter Outline

Deformation is a combination of distortion (a change in shape) and translation (bodily movement) of rock by dominantly horizontal and vertical stresses that were active at the time of deformation (the stresses are not necessarily still active). *Stress* is force per unit area and can be imposed on rock by plate tectonic movements. A *compressional stress* squeezes rock, an *extensional stress* pulls rock apart, and a *shear stress* forces one rock to slide horizontally past another. The application of the three types of stress on a square object is shown in Figure 3.1. All three types are capable of bending and breaking rock to produce folds, faults, and fractures. The *structural form* (or style of rock deformation) is the geometry that the rock body assumes following deformation. The structural form of a layered sequence of rock imparts a strong influence on landscape. Table 3.1 lists the common types of structures. We will introduce each of these and discuss their landscape significance in this chapter.

STYLE OF ROCK DEFORMATION (STRUCTURE)

Folds and faults are best displayed in sedimentary rocks because the layering of rock accentuates the structure. Additionally, sedimentary rocks make up 70 to 80% of all exposed rock in the United States. Sedimentary rocks are originally deposited in flat- or nearly flat-laying layers. When deformed, these layers become inclined (no longer horizontal). The inclination is described by the strike and dip of the layer. If you were to hold a pencil perfectly horizontal on the surface of an inclined plane, the compass direction of the pencil would define the *strike* of the rock. The *dip* is the angle of inclination of the plane measured perpendicular to the strike. In other words, the dip is the maximum slope of

the layer. A dip of 1° is nearly horizontal. A dip of 90° is vertical. For our purposes, a dip of less than 10° is gentle; a dip between 70° and 90° is steep; and a moderate dip lies between 10° and 70°. The strike and dip of inclined layers are shown in Figure 3.2.

A *fold* is a bend in an originally flat sedimentary layer. There are three categories of folds as outlined in Table 3.1. An *anticline* is formed when originally flat sedimentary layers are bent into an upward arch such that the beds dip moderately to steeply away from a center hinge line. A *syncline* is a downward bend such that beds dip moderately to steeply toward the center hinge line. These folds resemble the crinkling of a thin stack of paper and are shown in Figure 3.3a. A *dome* is an upward bulge in which layers are bent into the shape of a dome. Grab your shirt with two fingers and pull the shirt away from your body; the shirt will form a dome. A *structural* (or *geologic*) *basin* is a downward bowl-shaped indentation (or sag) such that sedimentary layers dip toward a low point near the center of the fold. A mattress would form a structural basin around a person sitting on a bed.

Anticlines and synclines come in all sizes. Those that form landscape are typically between a quarter mile and 50 miles wide. It is not uncommon for normal or reverse faults to be present at the margin of these folds. The Middle-Southern Rocky Mountains consists of large faulted anticlines and synclines. For simplicity we will refer to this type of folding as *anticlinal*, with the understanding that the term refers to anticlines, synclines, domes, and basins.

A *monocline* is a single, sharp, moderate to steep dip in otherwise flat-lying layers. These folds are less common than anticlines and synclines but are well developed on the

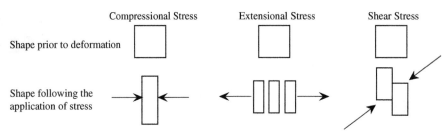

FIGURE 3.1 Types of stress and resulting deformation. The square in each figure represents the rock prior to deformation. Compressional stress squeezes rock. The most common structures are folds, reverse faults, and thrust faults. Extensional stress pulls rock apart. Common structures are joints and normal faults. Shear stress forces one rock to slide horizontally past another. Common structures are strike-slip faults.

TABLE 3.1 Style of Rock Deformation (Structure)

1. Folds
   a. Anticlinal (including anticlines, synclines, domes, and basins)
   b. Monoclines
   c. Nearly flat-lying layers (including flat-lying rocks, broad anticlines, synclines, domes, basins, and gently tilted rock)
2. Joints
3. Faults
   a. Normal faults (also known as block faults, gravity faults, or extensional faults)
   b. Reverse faults (these are high-angle compressional faults)
   c. Thrust faults (these are low-angle compressional faults)
   d. Strike-slip faults (also known as shear faults)
4. Metamorphic Complex
5. Batholiths

Colorado Plateau. It is common for a monocline to develop above a blind (hidden) fault as depicted in Figure 3.3b.

Sedimentary layers in many parts of the US are flat-lying or with such a gentle dip (<10°) that they appear nearly flat-lying. They can be *homoclinal* in the sense that all rock layers dip in the same direction, as on the Coastal Plain, or they can be in the form of a broad anticline (referred to as an *arch*), syncline, dome, or basin that can span a distance of more than 200 miles as shown in Figure 3.3c. These structures are transitional with the previously described steeper-dipping anticlinal fold structures. They are widespread across the Interior Plains and Plateaus region. For example, the entire state of Michigan is down-folded into a large structural basin known as the Michigan Basin.

A *fracture* is a plane along which rock is broken. A *joint* is an extensional fracture with movement perpendicular to the fracture surface such that rock on either side separates as shown in Figure 3.3d. Separation is on the order of an inch or two, but the space is widened by weathering and erosion. Joints form under conditions of extensional stress and are particularly well developed in massive crystalline rock and in thick layers of nearly flat-lying sandstone. Joints typically occur in sets of parallel, evenly spaced, near-vertical clusters

as shown in Figures 3.3d and 3.4. Although joints influence and modify many areas, they are not the primary landscape-controlling structure within any physiographic province. Joints, however, form the dominant structure in several of our most famous national parks, including Bryce Canyon and Arches National Park, as discussed in Chapter 10.

A *fault* is a fracture along which there has been movement of rock parallel with the fracture surface. As suggested in Table 3.1, there are several different types of faults that influence landscape. A *normal fault* forms where extension pulls rock apart. It is different from a joint in the sense that there is enough extension for the rock to slide down the dip of the fracture surface under the force of gravity. Thus, a normal fault is "normal" in the sense that the major displacement force is gravity. Although steeply dipping (>60°) near the Earth's surface, many normal faults flatten to a near-horizontal dip at depth. Normal faults occur along the margins of folds, below monoclines, and wherever rocks are pulled apart. This type of faulting is also referred to as *block faulting* because when one side drops down under the influence of gravity, the rock on the other side is left standing as a high, isolated mountain block. Normal faults commonly occur in semiparallel sets that form alternating valleys (basins) and mountain blocks. The down-thrown block is known geologically as a *graben*; the up-thrown mountain block is a *horst*. A typical horst-and-graben structure is illustrated in Figure 3.3e. Normal faults have a major effect on landscape and are the primary structure in the Basin and Range Province.

Reverse faults are the opposite (i.e., the reverse) of normal faults in the sense that they work against gravity as shown in Figure 3.3f. This type of fault forms where rocks undergo compression. In this case, the rocks are pushed upward along the dip of the fracture plane against the force of gravity. Reverse faults are defined by their moderate to steep dip. They occur along the margins of folds, below monoclines, and wherever rocks are compressed. Moderate to steeply dipping reverse faults rarely create enough displacement to have a significant effect on landscape. They are significant in a few areas that include the California Coast Ranges and the Middle-Southern Rocky Mountains.

A *thrust fault* is a special type of reverse fault in which the dip alternates between a nearly flat-lying portion with near horizontal dip and a ramp portion with dips between

FIGURE 3.2   Strike and dip of layers. Strike is the compass direction of the horizontal line in the inclined plane. Dip is the angle of inclination of the plane measured perpendicular to strike. The Google Earth image is of inclined layers along the northern flank of the Uinta Mountains in northeastern Utah, just west of the southern end of Flaming Gorge National Recreation area. The symbol gives the approximate strike and dip of the layers.

30° and 45°. The flat-ramp-flat geometry results in the stacking of older rock layers above younger layers and the development of anticlines above ramp areas as shown in Figure 3.3g. Displacement on thrust faults can be on the order of 50 miles or more; therefore, this type of fault is an important contributor to landscape. It is the major landscape-forming structure in the Valley and Ridge as well as in the eastern part of the Northern Rocky Mountains.

A *shear stress* produces strike-slip faults. In this case, the rock breaks along a vertical fracture surface and the two sides slide horizontally past each other with little associated upward (reverse fault) or downward (normal fault) movement as shown in Figure 3.3h. Strike-slip faults can have hundreds of miles of displacement; therefore they are major contributors to landscape. A right-lateral strike-slip fault displaces rock to your right as you look across the fault as depicted in Figure 3.3i. A left-lateral strike-slip fault displaces rock to your left as depicted in Figure 3.3j. The famous San Andreas fault is a landscape-forming right-lateral strike-slip fault. Right-lateral offset of Wallace Creek is clearly visible in Figure 3.5.

Crystalline rocks are either not particularly well layered, or they contain complex structural forms that do not have a consistent trend and do not fit well in any of the preceding structural categories. These rocks fall into one of two special categories: those composed dominantly of massive (nonlayered) intrusive granitic rock, known as *batholiths*, and those composed dominantly of layered metamorphic rocks, referred to here as *metamorphic complexes*.

Batholiths are large masses of granitic rock that cover or nearly cover an entire mountain or mountain range. Many form originally as underground feeder chambers below active volcanoes. They are brought to the surface following extinction and erosion of the volcano. Metamorphic rocks are present but are of secondary importance. Granitic

batholiths tend to be homogeneous in the sense that there is no layering, no obvious folds, and few faults. With the common exception of joint sets, these rocks are structureless. Areas underlain by batholithic granitic rock tend to produce a somewhat random, unpredictable landscape without consistent structural trend as suggested in Figure 3.3k. The major batholiths in the US include the Sierra Nevada in California and the Idaho batholith that forms part of the Northern Rocky Mountains. Figure 3.6 is a photograph of the Sierra Nevada batholith near Whitney Portal, California. Note that the granitic rock is homogeneous and nonlayered and that the most conspicuous structures are vertical fractures.

Layering in a metamorphic complex can produce a regional grain to the landscape. An obvious example is the regional north-northeast landscape grain of the Blue Ridge, Piedmont, and New England Appalachians, which follows the strike of inclined layers, folds, and faults (Figure 2.10). However, the landscape, at the scale of an individual mountain, is influenced less by these structures than in sedimentary rock because the process of metamorphism tends to lower contrasts in erodability between layers. Adjacent layers of similar erodability will behave (from a landscape point of view) as if they are a single homogeneous layer as suggested in Figure 3.3l. Under these circumstances the rocks are capable of producing a somewhat random landscape similar to that of batholiths. The rugged crystalline landscape of the North Cascade Mountains, shown in Figure 3.7, is an example.

INFLUENCE OF GEOLOGIC STRUCTURE ON LANDSCAPE

Regardless of whether a landscape is underlain by sedimentary, crystalline, or volcanic rock, the strike and dip of rock layers or fractures (if present) are capable of imparting

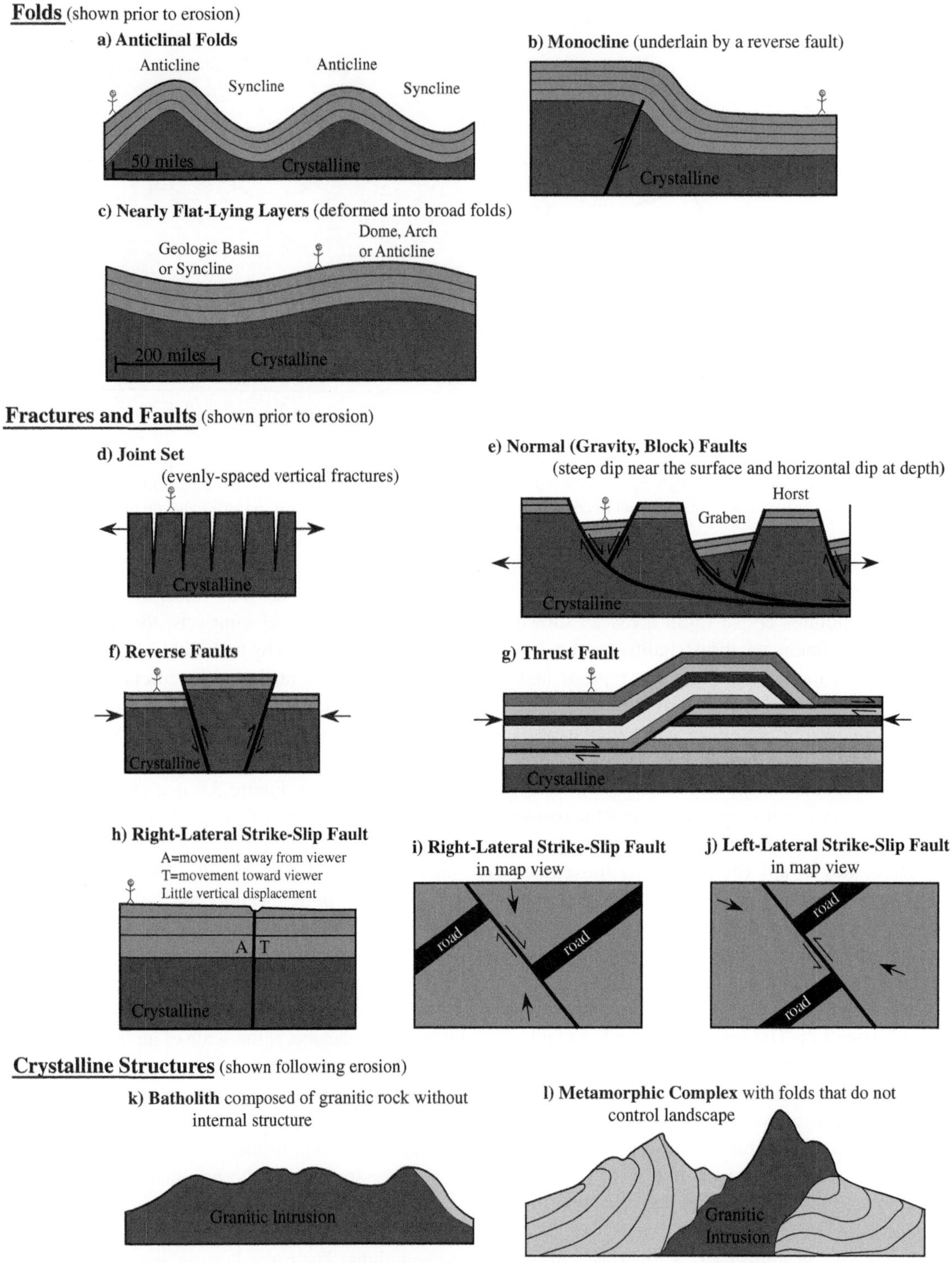

Folds (shown prior to erosion)

a) Anticlinal Folds

Anticline   Anticline

Syncline   Syncline

50 miles   Crystalline

b) Monocline (underlain by a reverse fault)

Crystalline

c) Nearly Flat-Lying Layers (deformed into broad folds)

Dome, Arch
or Anticline

Geologic Basin
or Syncline

200 miles   Crystalline

Fractures and Faults (shown prior to erosion)

d) Joint Set

(evenly-spaced vertical fractures)

Crystalline

e) Normal (Gravity, Block) Faults

(steep dip near the surface and horizontal dip at depth)

Horst

Graben

Crystalline

f) Reverse Faults

Crystalline

g) Thrust Fault

Crystalline

h) Right-Lateral Strike-Slip Fault

A=movement away from viewer
T=movement toward viewer
Little vertical displacement

A   T

Crystalline

i) Right-Lateral Strike-Slip Fault
in map view

road   road

j) Left-Lateral Strike-Slip Fault
in map view

road

road

Crystalline Structures (shown following erosion)

k) Batholith composed of granitic rock without
internal structure

Granitic Intrusion

l) Metamorphic Complex with folds that do not
control landscape

Granitic
Intrusion

FIGURE 3.3   Landscape-forming structures.

a strong imprint on the land. In order to better understand how dipping structures interact with landscape, we will look at two possibilities: nearly flat-lying rock layers, and moderately inclined to vertical rock layers and fractures.

A thick pile of nearly flat-lying rock layers often, but not always, produces bench-and-slope landscape in which a series of flat or gently tilted plains step up or down across narrow sloping areas like a staircase. The geometry is shown in Figure 3.8. Bench-and-slope landscape is well developed on the Colorado Plateau and Interior Low Plateaus, among other provinces, as discussed in Chapter 10.

FIGURE 3.4   A photograph of a closely spaced joint set in massive granitic rock at City of Rocks National Reserve, Idaho.

Moderately inclined and vertical rock layers and fractures characteristically produce a linear landscape that follows the strike of inclined beds and structures. Figure 3.9 shows a steep, narrow, linear, ridge known as a *hogback* along the Colorado Front Range south of Denver; it was created by a particularly resistant and nearly vertical layer of sedimentary rock. Faults can also produce a linear landscape. Faulting produces narrow planar zones of crushed rock that are weaker than surrounding rock. They regularly erode into lowland valleys. California's Tomales Bay, as shown in Figure 3.10, is a straight, narrow valley produced by erosion of the nearly vertical San Andreas fault. A linear valley is also shown in Figure 3.5. Thrust faults follow the strike of dipping rock layers in both the Valley and Ridge and the Northern Rocky Mountains thus contributing to landscape linearity in both provinces. Normal faults too will produce a linear landscape of range-basin-range (horst-graben-horst) that parallels the strike of faults. Examples include the Basin and Range of Nevada and The Grabens area on the Colorado Plateau as shown in Figure 3.11. In some instances, thick horizontal layers of rock will fracture into a vertical joint pattern that creates a strong linear aspect to the landscape. Such a pattern is seen in horizontal sandstone layers near Moab, Utah (Figure 3.12).

It should be pointed out that landscape linearity, on the scale of an entire mountain range, can be created along the margins of crystalline rocks where they are in contact with weaker sedimentary or volcanic rocks. Examples include the western side of the Blue Ridge Mountains and along both sides of the Sierra Nevada.

THE RESPONSE OF DIPPING LAYERS TO EROSIONAL LOWERING

In addition to creating a linearity to the landscape, the dip of a rock layer also controls the way the landscape will respond to erosional lowering. A resistant rock layer dipping at a moderate to gentle angle will form a cuesta at the surface that will migrate (retreat) in the direction of the dip during erosional lowering. A vertical resistant rock layer, on the other hand, forms a hogback at the surface that does not retreat. Its position during erosional lowering does not change. To put this another way, an inclined layer changes position with elevation by retreating in the direction of the dip; a vertical layer maintains its position. These relationships are shown in Figure 3.13. Geometrically, this implies that a vertical layer or structure will cut a straight path across a landscape regardless of topography, whereas a gently inclined layer will cut a sinuous

FIGURE 3.5   A Google Earth image looking ENE at the right-lateral offset of Wallace Creek along the San Andreas fault at Carrizo Plain National Monument, between Bakersfield and Santa Maria, California. Relative displacement along the fault is shown with half-arrows. Wallace Creek is to the left of the half-arrows.

FIGURE 3.6   A photograph of Mt. Muir (14,015 feet) on the eastern face of the Sierra Nevada batholith as seen from near Whitney Portal, California. The granitic rock is massive, homogeneous, nonlayered, and highly fractured.

FIGURE 3.7 A Google Earth image looking west across the North Cascade Mountains. The complex structure in the metamorphic and plutonic rock complex does not produce strong landscape linearity of mountain ranges. Mt. Baker (a Cascade volcano) is the high peak at distant center. Mt. Shuksan is the high peak at front and slightly to the right of Mt. Baker.

FIGURE 3.8 The bottom figure is a cross-section of bench and slope landscape in which hard sandstone layers form the bench. Slope areas expose weak shale layers, which undercut the sandstone. The top figure is a Google Earth image looking southward at Monument Valley on the Colorado Plateau, showing a nicely formed bench at front and a second bench in the far upper right. The spire at upper right is Agathia Peak.

FIGURE 3.9   A Google Earth image looking north at a resistant hogback of steeply dipping layers along the Colorado Front Range between Castle Rock and Littleton, Colorado.

FIGURE 3.10   A Google Earth image looking SSE at Tomales Bay, California. The straight, narrow valley is a product of erosion of the nearly vertical San Andreas fault. San Francisco is at distant center. Point Reyes National Seashore is at right.

FIGURE 3.11 A Google Earth image looking NNE at the Grabens, located on the Colorado Plateau in Canyonlands National Park just south of the confluence of the Green and Colorado rivers, Utah. The valleys form down-dropped grabens bordered on one or both sides by normal faults.

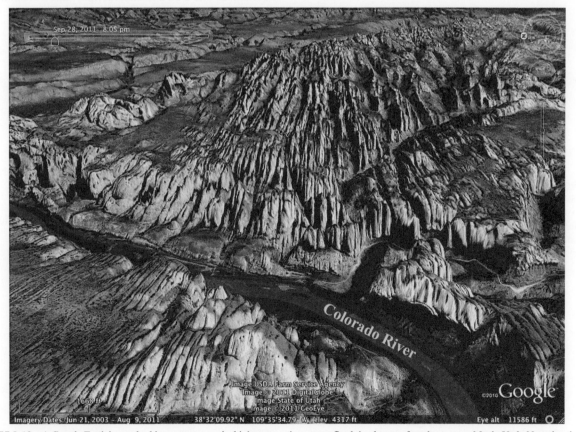

FIGURE 3.12 A Google Earth image looking east at a vertical joint set cutting across flat-lying layers of sandstone near Moab, Utah. Note that the joint set cuts across the Colorado River without deflection.

path across a topography. Note in Figure 3.12 how the vertical joint set cuts directly across the Colorado River without deflection and in Figures 3.5 and 3.10 how erosion of the vertical San Andreas fault creates a very linear valley. Conversely, note how erosion of the gently dipping layer in Figure 3.8 produces a sinuous path. As a rule of thumb, dipping layers are deflected in the direction of dip as they cross low areas such as river valleys. Figure 3.14 shows this relationship.

The terms *cuesta* and *hogback* are used repeatedly throughout this book, so you need to understand exactly what they refer to. Both are landforms. *Cuesta* is Spanish for *slope*. The term is used most commonly in the US for an escarpment or steep slope that separates two nearly flat areas of land at different elevations. Used in this sense, a cuesta is equivalent with an escarpment; both refer to a steep slope or cliff that separates land of different height. Cuestas typically form along the eroded edge of nearly flat-lying sedimentary or volcanic rock layers. The eroded edges of the benches shown in Figure 3.8 are cuestas. A hogback is a steep, narrow ridge that forms where a steeply dipping resistant layer protrudes above the surrounding land. The ridge in Figure 3.9 is a hogback. Two hogbacks are shown in Figure 2.6.

THE SHAPE OF LAND VS. THE SHAPE OF ROCK STRUCTURE

It is important to clearly distinguish landscape (the topography of land) from the rock structure (the shape of rock) that underlies the landscape. These are two very different and independent concepts. Landscape, for example, can be described as a mountain, a plain, or a topographic basin, regardless of the structure of rock. Structure can be described as an anticline, a syncline, or a structural basin, regardless of the landscape. Figure 3.15 shows how any of four structural types (flat-lying layers, anticline, syncline, crystalline nonlayered) can produce any of three landscape forms (mountain, valley, flatland).

A *topographic high* refers specifically to the highest elevation of land, such as the top of a mountain or hill. A *structural high* refers to the highest part of a particular rock layer within the structure, even if the rock layer has been removed by erosion. The crest of an anticline is a structural high regardless of the level of erosion. Landscape and structure can be described in combination. An anticlinal mountain is one in which topographic and structural highs coincide. An anticlinal valley is one in which the topographic low is pared with a structural high. Split Mountain, near the Colorado-Utah border, is an anticlinal mountain in one area and an anticlinal valley in another as shown in Figure 3.16.

QUESTIONS

1. Describe the three types of stress:
   a. Compressional
   b. Extensional
   c. Shear (or transform)
2. What is the strike of a rock?
3. What is the dip of a rock?
4. What type of fault results from compressional forces in which rocks are pushed toward each other?
5. What is another name for a "gravity" fault?
6. What type of fault results from extension in which the rocks are pulled apart?
7. What type of fault typically forms from shear stresses?

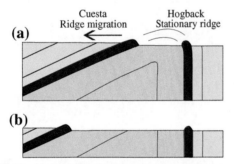

FIGURE 3.13   (a) A cross-section of a thin resistant layer between thick nonresistant layers. The rocks are deformed into an anticline with a moderate dip on the left side and a vertical dip on right side. The resistant rock layer forms a cuesta on the left side and a hogback on the right. (b) The same cross-section following erosional lowering. The cuesta retreats in the direction of dip. The hogback remains stationary.

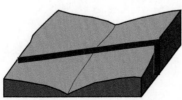

A vertical layer cuts straight across the river valley.

An inclined layer deflects in the direction of dip as it crosses the river valley.

FIGURE 3.14   Dipping layers and topography.

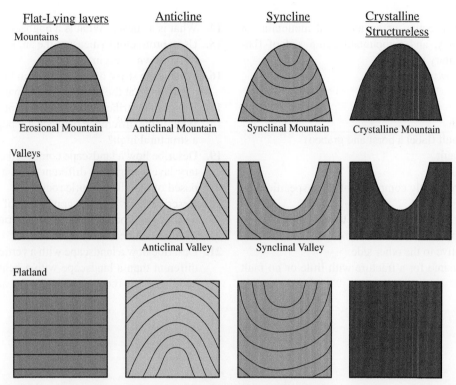

FIGURE 3.15 A series of cross-sections that show how any of four structural types (flat-lying layers, anticline, syncline, crystalline nonlayered) can produce any of three landscape forms (mountain, valley, flatland). A few of the landscape/structure types are named.

FIGURE 3.16 A Google Earth image looking west at the Split Mountain anticline and Ladore Canyon, located in the Unita Mountains near Flaming Gorge and Dinosaur National Monument east of Vernal, Utah. The Green River cuts through the mountain, forming an anticlinal valley. Eroded layers surrounding the mountain form cuestas.

8. Sketch a cross-section of an anticlinal mountain, an anticlinal valley, and a mountain composed of flat-lying sedimentary rock.

9. Draw the following structures:
   a. Syncline
   b. Anticline
   c. Monocline
   d. Normal fault (label a horst and graben)
   e. Reverse fault
   f. Thrust fault

10. Batholiths are mostly composed of what specific type of rock?

11. What type of fault typically has a vertical dip?

12. Name the type of fault in which one side has moved to the right relative to the other side.

13. What is the name for a fracture with little or no fault movement?

14. What is a cuesta? What is a hogback?

15. Use a protractor to measure the angle of dip of the hogback shown in Figure 3.9.

16. What is the strike of the hogback in Figure 3.9? Use the north arrow at the upper right of image.

17. What is the strike of the vertical joints in Figure 3.12?

18. What is the difference between a topographic high and a structural high?

19. Describe how a landscape composed of tilted sedimentary layers might be different from a landscape composed of massive granitic rock.

20. Describe how a ridge composed of steeply dipping beds might erode differently than a ridge composed of gently tilted beds.

21. Describe how a landscape with a vertical fault might be different than a landscape with a gently tilted fault or bedding plane (compare Figures 3.8 and 3.10).

Mechanisms That Impart Change to Landscapes

Before we look at the variables that force change on landscape, let us first look at the processes, the mechanisms, by which change occurs. There are five mechanisms that can exact change on landscape: uplift, subsidence, erosion, deposition, and volcanism. This chapter explains each of these and provides some insight as to how they are measured.

UPLIFT AND SUBSIDENCE

The terms *uplift* and *subsidence* refer to the raising or lowering of part of the Earth's surface relative to mean sea level. Uplift and subsidence can be regional in the sense that they encompass an entire mountain range, plateau, or valley, or they can be local in which case they refer to a small area or point that has been displaced such as along a fault where one side drops down relative to the other (and relative to mean sea level). In the United States, mean sea level is defined as the mean height of the surface of the sea as recorded at hourly intervals over a 19-year period. The frequency of measurement is required in order to average out the tidal highs and lows caused by gravitational forces from the moon and sun.

Present-day uplift and subsidence rates are easily measured using space-based satellites and ground-based leveling techniques. These rates, however, may not be indicative of long-term rates measured over thousands or millions of years. One reason for this is that uplift and subsidence are not necessarily steady. Instead, they tend to occur in cycles or pulses of rapid uplift (or subsidence) that can last several thousand or several million years, separated by long periods with little or no uplift. A land area, for example, can uplift or subside by more than 20 feet in less than a minute during an earthquake and then remain relatively stable for hundreds or thousands of years. It is also possible for periods of uplift to be separated by periods of subsidence. This phenomenon has occurred in Yellowstone National Park in the past 100 years where bulging is attributed to the movement of magma at depth below an area, and subsidence to movement away from an area.

Figure 4.1 illustrates average rates of present-day regional uplift and subsidence across the United States based on pre-1972 leveling techniques. It shows that much of the Cordillera, the Great Lakes region, and part of the southeastern US are currently experiencing broad uplift at rates between 4 and 60 inches per 100 years. The Central Valley of California, and most of the Atlantic and Gulf coast areas, are subsiding at rates between 4 and 40 inches per 100 years. Not shown on this diagram are small areas of local uplift and subsidence such as along part of the California coastline. This is likely due to the regional scale of the data in which uplift of a small ridge is canceled by subsidence of an adjacent valley. Taken together, this creates a larger area with zero net elevation change.

The highest uplift rates in the United States are in the Lake Superior region. Much of this is due to isostasy, which will be explained in greater detail in Chapter 7. For now we need only understand that, if a weight is placed on the Earth's surface, the land below the weight will subside. If the weight is removed, the land will rebound (uplift) to its original elevation but no more. Isostatic subsidence and uplift occur only until the original isostatic equilibrium is restored. Less than 18,000 years ago continental glaciers covered the Great Lakes region, causing land below the ice to subside. The land is currently isostatically rebounding to its original elevation following removal of the glacier. Ice thickness was presumably greater in the Lake Superior region; therefore, this area was depressed to a greater degree and is now rebounding at a higher rate relative to

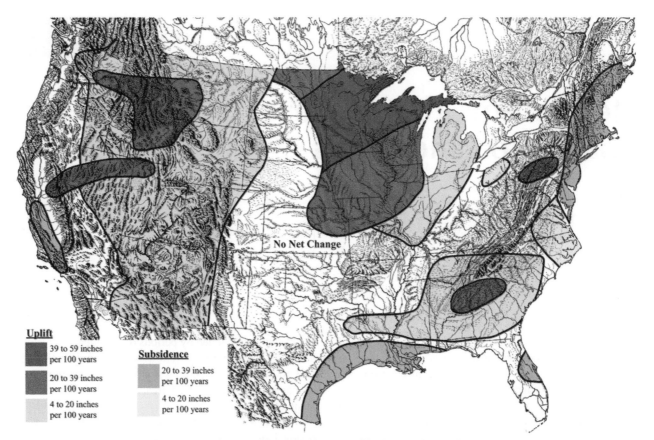

FIGURE 4.1   Areas of active uplift and subsidence. Modified from Keller and Pinter, 2002, p. 93.

surrounding areas in the US. These relatively high uplift rates extend into Canada such that the area north of the Great Lakes is uplifting faster than the area to the south. The result is an overall southward tilt of the Great Lakes region. If this continues, Lake Michigan may eventually spill into the Mississippi River Valley.

Isostatic rebound of the Great Lakes area is an example of rapid present-day uplift rates that cannot be sustained over the long term. The land will likely rise, perhaps a few hundred feet in the next thousand or several thousand years, but uplift will slow and eventually cease as isostatic equilibrium is restored. This type of uplift is not an indication of mountain building, which, instead, is driven by tectonic forces that can be sustained for millions of years (see Chapter 5). The point here is that measured present-day rates could merely be the result of short-term or transient causes. Present-day uplift/subsidence rates, measured over the past 100 years or so, may not be the same as long-term rates if measured over thousands or millions of years.

Although present-day uplift rates can be measured, there are no unequivocal methods by which to measure ancient uplift rates or to measure maximum elevations attained by ancient mountain belts such as the Appalachian Mountains. A perceived indicator of uplift, such as a change in fossil population from warm-temperature species

to cold-temperature species, could instead result from a change in climate without any change in elevation.

Present-day uplift rates, coupled with circumstantial geological evidence, suggests that long-term tectonic uplift of a mountain system (or subsidence of a basin) has occurred at rates of less than 25 inches per 100 yrs with typical rates on the order of 4 to 10 inches per 100 years. However, it should be noted that maximum uplift rates experienced within an actively deforming and uplifting mountain range (or subsiding basin) can exceed 100 inches per 100 years, especially for short time periods of several thousand or several tens of thousands of years. With limited erosion and an uplift rate of 10 inches per 100 years, a 5,000-foot mountain range can emerge from sea level in 600,000 years.

EROSION AND DEPOSITION

As noted in Chapter 2, erosion is the removal of rock and sediment from its original location on the Earth's surface by an agent such as water, ice, air, gravity, or animal/human interference. Deposition is the accumulation of eroded sediment, as in a lake or subsiding basin.

It is possible to measure present-day erosion and deposition rates; however, as with uplift and subsidence, these rates may not accurately reflect long-term rates that have occurred over the past 100,000 or 1,000,000 years. However, unlike uplift/subsidence rates, it is possible to estimate ancient erosion/deposition rates with reasonable accuracy. We will explore how to do this later in the chapter. For now, let us look at ways to measure recent and present-day rates of erosion.

From a purely qualitative perspective, the existence of high mountains suggests that rates of erosion are slower than rates of uplift. If the opposite were true, then mountains would not exist and most of the Earth's land surface would be close to sea level. We will discover, however, that there is a limit to the height of a mountain.

One way to measure present-day erosion rates is to measure the volume of sediment carried away by rivers within a specified drainage basin. This amount can be divided into the surface area of the drainage basin to arrive at an average *denudation rate*, which is the rate of overall vertical (erosional) lowering of land. The procedure is illustrated in Figure 4.2. The present-day average denudation rate across the United States is about 0.24 inches per 100 years. What this means is that, on average, a 0.24-inch-thick layer of rock and sediment has been removed from all land areas in the US over the past 100 years. If we assume zero uplift, it means that the average elevation of the US has been lowered by 0.24 inches. I stress that this is an average rate. Some areas undoubtedly were eroded at much higher rates; other areas were not eroded at all or experienced a gain in elevation through deposition. It has been suggested that this rate is twice as high as it was only a few hundred years ago. Apparently the disruption of land surface by agricultural practices has substantially increased the rate of erosion.

The *rate of bedrock river incision* (that is, the rate at which a river erodes downward into rock) is a measure of erosion on a smaller scale than regional denudation. It measures erosion on the scale of an individual river or stream. As rivers cut downward, they meander slightly, thereby leaving old river cuts elevated above the present-day river channel (known as *terraces* or *strath lines*). If the radiometric age of the elevated surface can be determined along with the elevation of the present-day channel, the average rate of river incision can be measured as shown in Figure 4.3. Incision rates can be extrapolated into the recent past if the radiometric age of two abandoned river surfaces at different elevations can be measured. Using this method, we can extrapolate as far back as the age of the oldest river terrace, which can be several million years or more. Dating methods include fossils or lava flows if present, or a relatively new but complicated technique known as *cosmogenic radionuclide dating*. Cosmic rays emanating from the center of the Milky Way galaxy are used to date how long a surface has been free of water and subject to long-term cosmic ray bombardment. An age is obtained by measuring daughter products produced by the collision of cosmic rays with the dry rock surface. The method dates how long ago the surface last

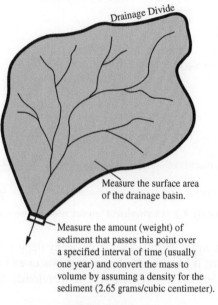

Calculation of Denudation Rate

Denudation rate is the rate of overall vertical (erosional) lowering of land.

Drainage Divide

Measure the surface area of the drainage basin.

Measure the amount (weight) of sediment that passes this point over a specified interval of time (usually one year) and convert the mass to volume by assuming a density for the sediment (2.65 grams/cubic centimeter).

$$\text{Volume} = \frac{\text{weight}}{\text{gravity} \times \text{density}}$$

$$\text{Denudation rate} = \frac{\text{volume of sediment}}{\text{area of drainage basin}}$$

FIGURE 4.2 Calculation of denudation rate. The calculation is equivalent to taking all the sediment removed over the course of a year and spreading it evenly over the entire drainage basin. The amount of erosional lowering (the denudation rate per year) would be equivalent to the thickness of the sediment layer.

emerged from below water (river) level. Typical nuclides used in this dating technique include ^3He, ^{10}Be, and ^{26}Al.

River incision can occur due to any number of factors. An obvious cause is uplift of surrounding land. Under these conditions the river could potentially maintain a constant elevation by downcutting into the surrounding elevated land area. River incision, however, does not imply uplift. Incision can occur if the surrounding land is stationary or even subsiding. Downcutting into a stationary or subsiding land area could occur, for example, if a river empties into an ocean or lake and then sea level (or lake level) drops faster than the subsiding land. Under these conditions, the river will be forced to cut downward to create a smooth path to the ocean or to the lake.

Similar to uplift/subsidence, the rate of incision is not necessarily steady. Instead, there may have been periods of rapid incision, periods of little or no incision, and periods of deposition during the measured time interval. A good indication that a river is currently incising and eroding its channel is when it flows over bare bedrock. In this case, the river must be removing material rather than depositing. If sediment is present at the bottom of a river channel, then it

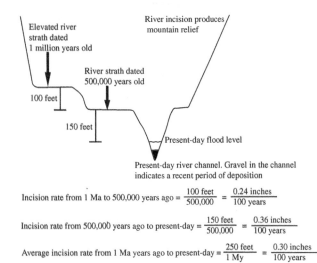

Incision rate from 1 Ma to 500,000 years ago = $\dfrac{100\ \text{feet}}{500,000} = \dfrac{0.24\ \text{inches}}{100\ \text{years}}$

Incision rate from 500,000 years ago to present-day = $\dfrac{150\ \text{feet}}{500,000} = \dfrac{0.36\ \text{inches}}{100\ \text{years}}$

Average incision rate from 1 Ma years ago to present-day = $\dfrac{250\ \text{feet}}{1\ \text{My}} = \dfrac{0.30\ \text{inches}}{100\ \text{years}}$

FIGURE 4.3   Calculation of bedrock river incision rates.

indicates either that the river is currently depositing material or that it has recently begun to downcut and erode its channel but has not yet removed all previously deposited material.

Typical incision rates vary from less than one inch to several inches per 100 years and thus can be much faster than regional denudation rates to the point where they approach or exceed uplift rates. As with uplift, maximum incision in mountain belts can be several tens of inches or more per 100 years, especially at high elevation where powerful glacial incision occurs. Because river incision is restricted to the river channel itself, it results in a carving effect that shapes the mountain and produces spectacular relief.

We have discussed two very different methods of measuring erosion rates: one on the scale of an entire drainage basin, the other on the local scale of an individual river. We have also seen that rates can vary from very low to those approaching rates of uplift. Now, we can ask ourselves, what causes overall rates of erosion to be slow in some instances and rapid in others? Based on observation and logic, we can conclude that rates of erosion are dependent on climate, rock hardness, elevation, relief, and vegetation. Obviously, soft rock will erode at a higher rate than hard rock. Abundant vegetation tends to slow the rate of erosion because roots hold soil in place. Water, on the other hand—especially running water—will increase erosion rates simply because it is capable of moving loose soil. Thus, other factors being equal, an ideal situation that allows for rapid erosion is a semihumid climate where there is water but limited vegetation.

In any climate, the rate of erosion increases with elevation and particularly with relief. High elevation forces air to rise, which creates precipitation. Cold air at high elevation limits the amount of vegetation and creates the potential for glaciation. Glaciers are very powerful erosional agents. On balance, they are orders of magnitude more effective than running water. High elevation alone, however, does not necessarily

cause an increase in erosion rates, especially without glaciers. A flat plateau at high elevation could conceivably erode at about the same rate as a flat plain at low elevation simply because there is little relief and therefore little, if any, moving water. It is relief (the change in elevation between two points) that allows gravity to work in favor of erosion. High relief implies steep slopes, which allow for less vegetation, faster-moving streams, and greater potential for landslides. The highest erosion rates are typically found in mountain belts because there we have a combination of elevation and relief as well as the potential for glaciers. Still, the fact that we have high mountains with glaciers implies that rates of erosion are, in general, slower than rates of uplift.

Is there an erosional limit to how high a mountain can grow? The answer apparently is yes because the rate of erosion seems to increase as a mountain grows higher and steeper. At some point, the rate of erosion will equal the rate of uplift. At this point, mountain growth has reached what is known as a *steady-state condition* with erosion. A steady-state mountain implies that any additional surface uplift (mountain growth) will be compensated by an equal amount of erosion. The mean surface elevation and the basic shape of the mountain remain constant as long as uplift continues and the climate and the rocks do not change. If the rate of uplift increases, the mountain will gain elevation. If the rate of uplift begins to slow or cease, or some other variable changes such that erosion becomes the dominant force, the mountain will be lowered at a rapid rate at first and then at progressively slower rates as elevation and relief (and therefore the rate of erosion) are reduced.

An interesting aspect of this process is that most of the erosion in a mountain range is likely to occur in stream valleys between mountain peaks via river incision and the clearing of landslides. The focused removal of rock from a stream valley will lessen the overall weight of the land, which, in turn, triggers isostatic uplift of the entire mountain range. During the initial stage, when erosion first becomes the dominant force, an individual uneroded mountain peak can potentially be isostatically uplifted to a higher elevation, even though the mountain range as a whole is eroded to a lower elevation. This is a short-lived phenomenon because erosion will, eventually, begin to wear away at the mountain peak.

Periods of rapid river incision, when extrapolated into the ancient past, can be used to indirectly determine periods of rapid uplift and therefore periods of mountain building. This procedure must account for climate and other factors such as sea-level change because, as we have already noted, such changes could potentially trigger the same effect without uplift. However, if these factors can be taken into account, and a period of rapid river incision can be isolated between two periods of relatively slow river incision, then we can reasonably conclude that incision is due to uplift and mountain building.

Deposition is the opposite of erosion in the sense that a land area is raised relative to surrounding land as a result of sediment accumulation. Average rates of deposition for both the present day and the geological past are estimated in a fairly simple manner by measuring the thickness of a sequence of rock (or sediment) and dividing by the time interval over which the rocks were deposited. In this calculation, the amount of compaction that occurs when sediment is turned to rock must be taken into account. When this is done, it is found that rates of deposition are roughly equal to rates of erosion. This is not surprising given that any eroded material must be deposited somewhere. However, it should be realized that a single mountain range, or perhaps even a single mountain peak, can erode into several separate depositional basins. Alternatively, a single depositional basin can receive sediment from multiple sources.

As a land area subsides, it will form a low-lying basin that will fill with water and receive sediment from erosion of surrounding highlands. The actual amount of elevation decrease due to subsidence is a function of the amount of basin infilling by sediment. If the amount of sediment infilling equals the amount of surface subsidence, elevation of the land will not change (in other words, surface subsidence is equal to zero). This is equivalent to a mountain in a steady state. If the rate of deposition is faster than subsidence or, occurs without subsidence, the basin will eventually fill with sediment and any body of water, such as a lake, will disappear to form dry land. Slow subsidence and simultaneous deposition over the course of millions of years will produce huge thicknesses of sediment and, eventually, sedimentary rock. An example of deposition during slow subsidence can be seen along the eastern North American coastline where tens of thousands of feet of sedimentary rocks have accumulated, all in shallow water. Measuring the time interval over which these rocks were deposited gives an indication of long-term subsidence rates.

Subsidence without deposition will result in a loss of elevation. Such a situation can occur in desert areas where rates of erosion (and, therefore, rates of deposition) are extremely slow. In this case, it is possible for a land area to sink below sea level. An example is Death Valley.

EXHUMATION

Exhumation is the process by which once deeply buried rocks are brought to the surface. Ancient long-term erosion rates are estimated based on the rate of exhumation. Uplift and simultaneous erosion have the effect of transporting deeply buried rock to the surface. In other words, as uplift occurs, the surface of a mountain is eroded thereby exposing previously buried rock underneath. Note that if uplift occurs but there is no erosion, buried rock is transported to a higher elevation but remains as deeply buried as it was prior to uplift. Note also that a rock becomes unburied (exhumed)

at the same rate as erosion irrespective of the rate of uplift. The relationship between uplift and erosion is illustrated in Figure 4.4.

It is important to understand that Figure 4.4 does not show a mountain in a steady state. The figure is actually somewhat unrealistic in that it shows all uplift occurring first, followed by erosion, which lowers the mountain and exhumes deeply buried rock. We would expect any mountain that is undergoing active uplift to also undergo simultaneous erosion. This is true if the mountain is in a steady state, in which case there is no surface uplift (that is, mountain elevation remains constant). It would also be true if the rate of erosion were less than the rate of uplift. In this case, the mountain would continue to grow higher until it reached a steady state with the rate of erosion. On this basis we can make a distinction between *surface uplift*, which is a topographic increase in surface elevation, and *rock uplift*, which is the vertical displacement of rock with or without a corresponding increase in surface elevation. Any form of vertical uplift will result in rock uplift. Surface uplift is equal to rock uplift minus erosion. In a steady-state mountain, there is rock uplift but no surface uplift. In the case of Figure 4.4, the second panel from the top shows 400 feet of rock uplift and 400 feet of surface uplift (there is no erosion). This uplift was followed by erosional lowering of the mountain, without any additional rock uplift. Thus, after 400 feet of erosion, rock R3 is exposed at the surface (the bottom panel of Figure 4.4). Total surface uplift equals rock uplift (400 feet) minus erosion (400 feet). Total surface uplift, therefore, is equal to zero, even though rock R3, which was once deeply buried, is now exposed at the surface.

As previously discussed, both uplift and erosion (that is, exhumation) are strongest in mountain belts. Thus it is common to find crystalline rock (rock that originally formed deep within the earth) exposed at the top of a mountain. Some lowland areas, such as the Superior Upland and the Piedmont Plateau, are composed of crystalline rock. The presence of these rocks at the Earth's surface suggests that at one time long ago, a mountain range was present.

For many rocks it is possible to accurately estimate the pressure at which the rock formed as well as its age. Pressure is a function of rock density, depth, and the force of gravity. Rock density and the force of gravity are well-known variables that do not change very much, at least within the top 40 miles of Earth. They can be treated as constants. If pressure is also known, the depth (d) of burial is easily calculated as pressure (P) divided by the product of rock density (r) and gravity (g).

$$P / (r \times g) = d$$

For example, if P = 2.47 kbar (or 247,000,000 Pascal), r = 2,750 kg/m^3, and g = 9.8 m/s^2, depth is equal to 9,165 meters (30,061 feet; note that Pascal, and not kbar, must

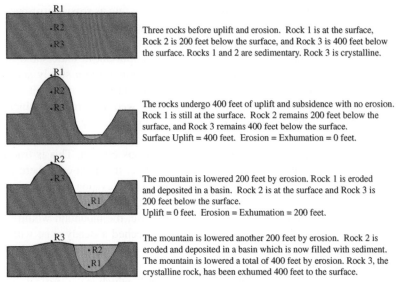

Three rocks before uplift and erosion. Rock 1 is at the surface, Rock 2 is 200 feet below the surface, and Rock 3 is 400 feet below the surface. Rocks 1 and 2 are sedimentary. Rock 3 is crystalline.

The rocks undergo 400 feet of uplift and subsidence with no erosion. Rock 1 is still at the surface. Rock 2 remains 200 feet below the surface, and Rock 3 remains 400 feet below the surface. Surface Uplift = 400 feet. Erosion = Exhumation = 0 feet.

The mountain is lowered 200 feet by erosion. Rock 1 is eroded and deposited in a basin. Rock 2 is at the surface and Rock 3 is 200 feet below the surface. Uplift = 0 feet. Erosion = Exhumation = 200 feet.

The mountain is lowered another 200 feet by erosion. Rock 2 is eroded and deposited in a basin which is now filled with sediment. The mountain is lowered a total of 400 feet by erosion. Rock 3, the crystalline rock, has been exhumed 400 feet to the surface.

FIGURE 4.4   Example of exhumation.

be used in the equation in order to keep units consistent). Because the rock is currently at the Earth's surface, the average rate of exhumation is simply depth divided by age. Although crude, this simple relationship allows geologists to estimate average exhumation rates deep into the geological past. For example, if a rock now at the Earth's surface originally formed 56 million years ago at a depth of 30,061 feet, the average rate of exhumation is 30,061 feet per 56 million years (or 0.64 inches/100 years). This, of course, is exhumation averaged over the entire 56-million-year interval. It assumes that exhumation began soon after the rock formed, which is not necessarily true. Exhumation could have started at any time after the rock formed. If, for example, exhumation occurred entirely during the last 14 million years, the actual rate of exhumation would be four times the previously calculated rate over the shorter time period. Fortunately, sophisticated dating techniques allow geologists to estimate changes in exhumation rates over time periods that are shorter than the age of the rock. Thus it is possible to determine when, during the 56-million-year interval, most of the exhumation may have occurred.

Common methods used to estimate rates of exhumation over specific time intervals include argon 40-argon 39 ($^{40}Ar/^{39}Ar$) uranium-thorium/helium (U-Th/He), and fission track dating techniques. Argon is a product of the radioactive decay of potassium, which is a common element in a variety of minerals. Because it is an inert (nonreactive) noble gas, all argon will diffuse (move) out of a mineral when that mineral is at high temperature. The *closure temperature* of a mineral is the temperature at which most of the argon is retained in the mineral. It is at this time and temperature that the radiometric clock begins to record an age. In the simplest case, an argon age is the amount of time that has passed since the sample cooled below the closure temperature during

exhumation to surface. The mineral hornblende has a relatively high closure temperature in the argon-argon system of around 500±25°C; muscovite and biotite have intermediate closure temperatures of about 425±25°C and 330±25°C, respectively; and K-feldspar has a low but variable closure temperature between 150°C and 300°C.

The same principle is employed with the (U-Th)/He method except that the mineral apatite is used and the inert gas is helium rather than argon. Helium is a product of the radioactive decay of uranium and thorium. It is expelled from apatite at temperatures above 70°C and is completely retained within the mineral at temperatures below 30°C. Helium is partially retained between these temperatures.

Fission tracks are lines (tracks) of near-constant length that represent damage within the minerals apatite and zircon, caused by radioactive decay of small amounts of enclosed uranium. The tracks are completely erased in apatite at temperatures above about 110°C, partially retained between 110° and 50°C (track lengths are shortened), and fully retained below 50°C. Fission tracks recorded in zircon are erased above about 310°C, retained below about 210°C, and shortened between these two end-member temperatures. The temperature range across which track lengths are shortened is referred to as the *partial annealing zone*.

Pressure cannot be calculated with these methods; therefore, the exhumation rate is determined by inferring a geothermal gradient (which is the increase in temperature with depth in the Earth). Once established, a geothermal gradient is used to infer depth of burial at the time the closure temperature was reached. Normal geothermal gradients are between 20°C and 30°C per km of depth. Thus, if several dating methods are employed within an area, average exhumation rates can be determined at various stages of exhumation. For example, if we infer a geothermal gradient of 25°C/km and determine an ^{40}Ar/^{39}Ar date on biotite with a closure temperature of 330°C, and a (U-Th)/He date on apatite with a closure temperature of 30°C, we can determine the rate at which the rock was exhumed from a depth of 13.2 km (the closure depth of biotite) to 1.2 km (the closure depth of apatite) and from a depth of 1.2 km to the surface. Note here that we are making another assumption that the average surface temperature is 0°C. If the biotite age is 17 My and the apatite age is 8 My, the exhumation rate from 13.2 to 1.2 km depth is 12 km/9 My (or 5.25 in/100 years). From 1.2 km depth to the surface, the exhumation rate is 1.2 km/8 My (or 0.59 in/100 years).

There are inherent problems and assumptions in using these techniques. The closure temperature can vary by 25°C or more depending on the size of the mineral grain, the speed at which ions move through the mineral (a process known as *diffusion*), the mineral purity, and the rate of cooling. Furthermore, it is possible for some of the inert gas to be retained prior to reaching the closure temperature or for inert gas to be expelled for a short time after reaching the closure temperature. Additionally, there is no straightforward correlation between the temperature of a rock and the depth at which the rock is buried. Thus, barring independent pressure data, the geothermal gradient (and, therefore, the depth) at the time of closure must be estimated. These problems create potentially large uncertainty in the age of the sample and, therefore, in the absolute rate of exhumation. Fortunately, geologists have developed techniques and models to account for these and other uncertainties to the point where they allow for tight constraints on what the isotopic age means in terms of an exhumation rate. Regardless of the uncertainties, the relative rates between the different minerals, using the same assumptions, are precise enough to show trends. Many geologists correlate periods of rapid exhumation with periods of rapid surface uplift, and thus with mountain building. In the preceding example, mountain uplift likely occurred between 17 Ma and 8 Ma (resulting in a 5.25 in/100 years erosion rate) and then slowed during the past 8 million years (resulting in a 0.59 in/100 years erosion rate).

It is important to realize that what is being measured is the rate of erosion averaged over a certain time period. The actual rate of surface uplift and the elevation achieved cannot be precisely calculated. The underlying assumption here is that rapid uplift is responsible for rapid erosion. This assumption is valid for narrow mountain ranges where uplift is expressed by an increase in surface relief and slope. It is not necessarily valid for plateau uplift of a wide area where surface relief and slope are not substantially increased.

In the preceding example, we equated measured rates of rock exhumation with the rate at which the ancient mountain belt was eroded. This is an example of erosional exhumation in which all exhumation (the transport of once deeply buried rock to the surface) was accomplished by erosion (as in Figure 4.4).

Erosional exhumation is probably valid for many mountain belts, including the Appalachians, and is the most important and most common process by which exhumation occurs. Erosion, however, is not the only process by which exhumation can occur. Faulting, particularly normal faulting, can also result in the exhumation of rock; in later chapters we will refer to this as *tectonic exhumation*.

VOLCANISM

Volcanism is the process by which magma and associated gases rise to the surface to form lava flows, volcanoes, and highly explosive discharges. This is a special type of landform formation and modification that can shape or reshape a landscape in a matter of days or months rather than the thousands or hundreds of thousands of years required for uplift and erosion. The consequences of volcanism are discussed in more detail with respect to specific physiographic provinces later in the book.

QUESTIONS

1. Define the following:
   a. Uplift
   b. Subsidence
   c. Erosion
   d. Deposition
   e. Denudation
   f. Exhumation
2. Sea level has risen at an approximate rate of 0.20 inches per year for the past 18,000 years. How many feet has sea level risen in the past 18,000 years?
3. What has caused the rise in worldwide sea level for the past 18,000 years?
4. What is meant by the statement "Mountain elevation is in a steady state"?
5. If rocks that originally formed deep in the Earth are exposed on a flat plain close to sea level, what type of landscape likely was present at that location a long, long time ago? (*Hint:* In what type of landscape are deeply buried rocks brought to the surface?)
6. Explain the difference between rock uplift and surface uplift.
7. What are the inferred typical rates of uplift for the geological past (per 100 years)?
8. Is it possible to determine ancient uplift/subsidence rates or ancient elevations? Explain.
9. Which is typically faster at low elevation; rates of uplift or erosion?
10. Explain how the rate of erosion varies with:
    a. Climate
    b. Hardness of rock
    c. Elevation
    d. Relief
    e. Vegetation

11. If relief between river bottom and a river terrace is 120 feet and the age of the terrace is 420,000 years, what is the average rate of river incision in inches per 100 years?

12. Does river incision necessarily imply uplift of land? Explain.

13. What is the exhumation rate for a rock buried 4 miles 9 million years ago (in inches per 100 years)?

14. What is the rate of deposition for a 5-mile-thick sequence of rocks deposited over a time interval of 270 million years (in inches per 100 years)?

15. True or false: The combination of uplift and erosion results in exhumation of deeply buried rock.

16. The chapter states that typical river incision rates are much faster than regional denudation rates. Explain why this would logically make sense.

17. As explained in the chapter, when does isostatic uplift (or subsidence) occur?

18. How is isostatic uplift different from tectonic uplift in terms of its longevity?

19. What is erosional exhumation?

20. What can be inferred regarding the history of a mountain if exhumation rates are found to be very slow over a period of time?

21. What can be inferred regarding the history of a mountain if exhumation rates are found to be very fast over a period of time?

Forcing Variable: The Tectonic System

The Earth's landscape is in a constant state of change due to the interaction of internal forces as exemplified by the tectonic system and external forces as exemplified by the climatic system. These are the primary variables that force landscape change. Together they create the components that form the landscape, and they control the rate, intensity, and longevity of the mechanisms that exact change on landscape. Without these forces of nature, the Earth would be an unchanging, stagnant planet. Because the strength and character of these forces varies over time, we can refer to them as *forcing variables*. This chapter provides a short introduction to the tectonic system. A critical aspect of the tectonic system with respect to landscape is the distinction between areas of active or recent tectonics, and areas where tectonic activity has been dormant for 100 million years or more.

FIRE AND ICE

The *tectonic system* involves the constant large-scale horizontal movement of tectonic plates. Energy for the tectonic system is provided by the outward flow of heat from the hot center of the Earth and by heat-producing radioactive decay of elements within the Earth. The tectonic system is primarily responsible for three of the five mechanisms of landscape formation and change: uplift, subsidence, and volcanism. It is also responsible for the creation of crystalline and volcanic rock and for the structure of rock. In other words, it is, in large part, responsible for the components that form landscape. Because the surface expression of the tectonic system is the creation (uplift) of a landform, we can think of this system as the fire that builds landforms. It is the tectonic system that helps to build elevation.

The *climatic system* is driven by the heat of the sun. Disproportionate heating of one area relative to another, along with the Earth's rotation, drives atmospheric and oceanic circulation, which, along with gravity, drives the remaining mechanisms of landscape formation and change: erosion and deposition. The surface expression of erosion and deposition is the reduction of an uplifted land area to a flat plain close to sea level. As such, the climatic system produces the weathered unconsolidated sediment that eventually forms sedimentary rock. If the tectonic system represents the fire that builds a landform, then the climatic system is the ice that modifies and eventually destroys a landform.

It is the competition and interplay between the tectonic and climatic systems that, over time, build and modify landscape. If the tectonic system dominates, mountains are formed. If the climatic system dominates, a highland is reduced to level ground. Sea-level change and isostatic adjustment are a consequence of this interplay. Both cause vertical displacement (uplift/subsidence) of the landscape.

THE TECTONIC SYSTEM

Plate tectonic theory states that the outer layer of Earth is broken into brittle, rigid tectonic plates that move slowly over a weak, partially melted layer of the mantle known as the *asthenosphere*. The theory evolved over several decades and was finally accepted by most of the geological community in the late 1960s on the basis of newly discovered data on the structure and composition of ocean basins. Today, with the advent of satellites and global positioning systems, present-day rates of tectonic plate movement are easily measured. The interaction of these plates as they move toward, away, or laterally past each other produces most of the world's volcanoes and earthquakes and is ultimately responsible for the rock type and the structural form that underlies individual landscape provinces. The displacement of tectonic plates plays a key role in the development of present-day landscape.

The tectonic system results from the interaction of the top three layers of Earth: the crust, the solid upper mantle, and the asthenosphere. The uppermost layer, which includes the land surface, is the Earth's crust. The crust that forms continental areas is different from the crust that forms ocean basins. In continental areas the crust averages between 20 and 30 miles thick (32–48 km) and is granitic in composition. Oceanic crust averages between 2 and 4.5 miles thick (3–7 km) and is basaltic in composition. Basaltic (oceanic) crust is dense and heavy compared to an equal volume of granitic (continental) crust. The solid upper mantle underlies both continental and oceanic crust, and in both areas it is composed of peridotite, which is a heavy, magnesium-rich rock that is rare at the Earth's surface. Like the Earth's crust, the solid upper mantle is thicker below continents than below ocean basins. Continental and oceanic crust are imbedded into the solid upper mantle and together these two layers form the rigid, brittle lithosphere. In plate tectonic theory it is the lithosphere that forms the tectonic plate.

Continental lithosphere is, on average, between 90 and 100 miles thick and can be as much as 150 miles thick. Oceanic lithosphere typically is less than 50 miles thick. A single tectonic plate can consist of both oceanic and continental lithosphere. The transition from continental to oceanic lithosphere within a single tectonic plate coincides topographically with the continental slope, which is an off-shore region where the depth of the ocean basin increases relatively abruptly from a few hundred feet to several thousand feet. The eastern United States and the western Atlantic Ocean are both part of the same North American tectonic plate. A cross-section that shows the transition from the North American continent to the Atlantic Ocean is shown in Figure 5.1. The transition is also visible in Figure 1.7. Continental and oceanic lithosphere are both underlain by the asthenosphere, which is a partially melted part of the mantle (1–2% partial melt) that acts like a lubricated surface on which the lithosphere (i.e., the tectonic plate) can slide.

Worldwide there are seven major tectonic plates and many additional smaller ones. As each plate moves, it interacts with adjacent plates such that two plates diverge (move away from each other), converge (move toward each other), or slide horizontally past each other (known as *shear* or *transform*). When two plates diverge, they create a void that is filled with magma that cools and attaches itself to the two diverging plates thereby enlarging the size of both plates. In Figure 5.2, Plates B and C are diverging from each other. The Mid-Atlantic ridge, as shown in Figure 5.1, is a divergent plate boundary that separates the North American plate from the Eurasian plate.

Convergence of two plates often results in the sinking (*subduction*) of one plate beneath the other. The subducting

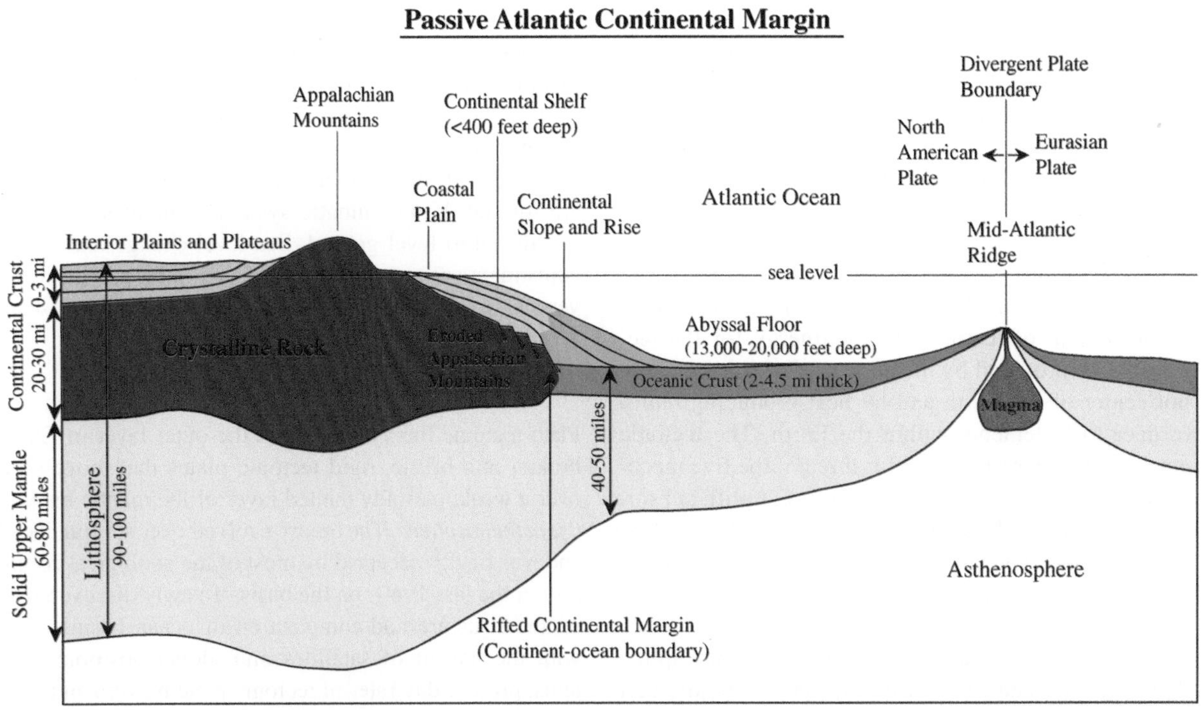

Passive Atlantic Continental Margin

FIGURE 5.1   A cross-section of the US passive Atlantic continental margin. Note that the North American continental crust and the western Atlantic oceanic crust reside on the same (North American) tectonic plate.

plate is consumed and, therefore, becomes smaller. Interestingly, this process can enlarge the overriding plate if material is scraped off the subducting plate and attached (accreted) to the overriding plate. The concept of tectonic accretion is discussed in more detail later in this chapter. In Figure 5.2, Plate B is converging with, and subducting below, Plate A.

In its purest form, a transform boundary conserves mass because nothing is added or removed as one plate slides past another. However, in reality, in many cases there is a certain amount of divergence or convergence between the plates. Divergence could potentially create subsidence, normal faulting, or volcanism. Where volcanism occurs, the transform is sometimes referred to as a *leaky transform*. Convergence between plates could create folds, thrust faults, and tectonic uplift. Transform faults are shown between Plates B and C and between Plates B and A in Figure 5.2.

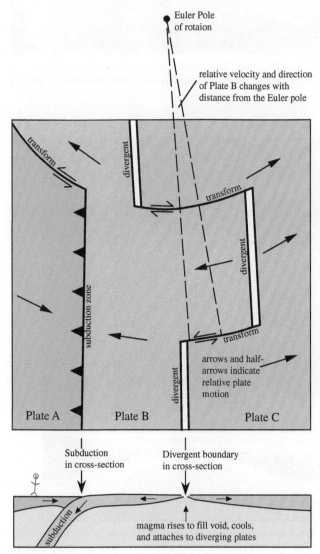

FIGURE 5.2 A sketch showing interaction of the three plate tectonic boundaries.

When tectonic plates move, they typically do not move across the Earth in a line as if pushing a book across a table. The Earth is spherical, which requires that plates rotate about a pole of rotation that emanates upward from the center of Earth and intersects the Earth's surface. The location of a pole at the Earth's surface is referred to as an *Euler pole* (pronounced *oiler* pole). The Euler pole for Plate B is shown in Figure 5.2. An Euler pole can be used to determine both the direction and velocity of movement of one plate relative to another. Because the plate is rotating about an axis, the relative velocity, and therefore the rate of divergence or convergence, increases with distance from the Euler pole. This phenomenon is shown in Figure 5.2 with dashed lines emanating from the Euler pole. Because a transform plate boundary is, by definition, a location where two plates are sliding past each other, the orientation of a transform is approximately parallel with the movement direction of the two plates on either side. Note in Figure 5.2 that arrows and half-arrows showing direction of plate movement are approximately parallel with the orientation of transforms. If one were to draw a series of lines emanating perpendicular from a transform fault, those lines should intersect at a point which is the Euler pole. For example, the two dashed lines drawn in Figure 5.2 both intersect the Euler pole.

Tectonic plates move at rates that vary from less than 6 feet per 100 years to 66 feet per 100 years; and these rates may have been faster in the ancient past. At an average rate of 33 feet per 100 years, a tectonic plate can move 62 miles in one million years. Such rates seem slow, but over the course of several million years, a tectonic plate can move into an entirely different climatic region. If two plates are colliding, then 62 miles of Earth must be destroyed every 1 million years. Because each plate is rigid and moves as a single unit, most of the world's earthquakes, deformation, plutonic activity, metamorphism, and volcanism occur at or near plate boundaries. Plate collision is a major driving force in mountain building.

THE ATLANTIC PASSIVE CONTINENTAL MARGIN

A *passive continental margin* is one in which the continent and the adjacent ocean basin reside on the same tectonic plate. The eastern North American seaboard is a passive continental margin because, as shown in Figures 5.1 and 5.3, the plate boundary (the divergent Mid-Atlantic ridge) is located in the middle of the Atlantic Ocean. The absence of a plate boundary along the continental margin results in few earthquakes, no volcanoes, and no active mountain building. Slow subsidence and wave erosion over many millions of years have flattened and beveled the eastern continental margin to produce the Coastal Plain and the wide continental shelf. Currently, the North American plate is moving southwestward relative to the Pacific plate at an approximate rate of 8.5 feet per 100 years.

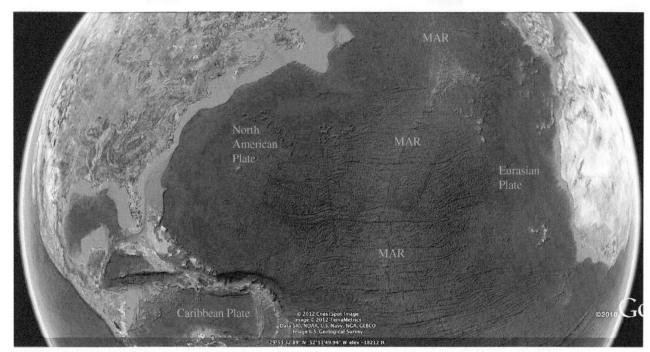

FIGURE 5.3   A Google Earth image looking north that shows the location of the Mid-Atlantic ridge (MAR) relative to the eastern North American seaboard. Notice that the MAR is offset along transform faults that are oriented approximately perpendicular to the ridge system. This offset is especially apparent in the southern part of the image.

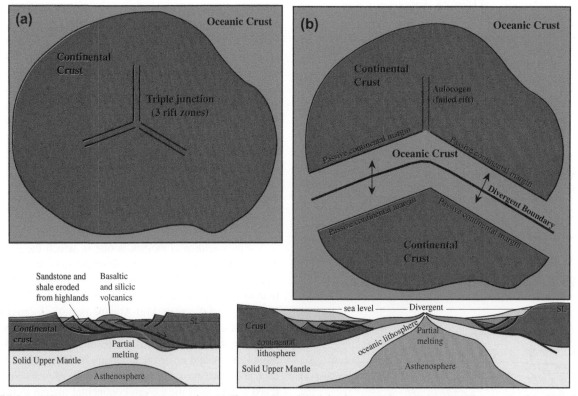

FIGURE 5.4   Rifting in map view and in cross-section. (a) The map shows initial development of a triple junction. The cross-section shows bulging and stretching of the crust and development of normal faults and volcanism. (b) The map shows development of a divergent plate boundary between two continental fragments as well as a failed rift. The cross-section shows development of a passive continental margin on both sides of the divergent plate boundary. Based on Van der Pluijm and Marshak (2004, p. 398).

A passive continental margin forms by *rifting*, which is the splitting apart of a single continent into two or more continents separated by a divergent plate boundary. The process is illustrated in Figure 5.4a which shows a continent stretching and breaking initially in three directions, forming what is

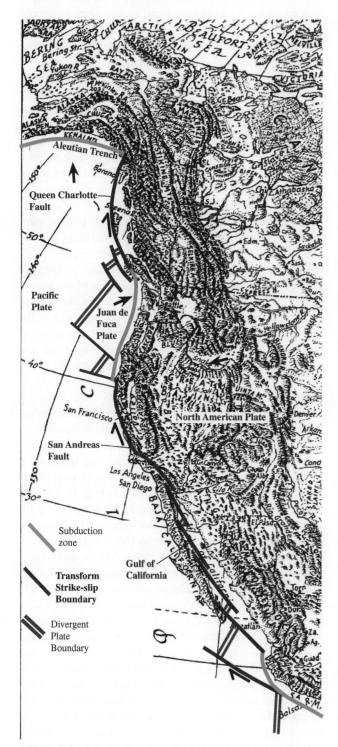

FIGURE 5.5 A map of western North America showing the plate tectonic configuration.

known as a *triple junction*. In many cases, only two of the three arms of the triple junction develop into a divergent plate boundary (Figure 5.4b). Extension and normal faulting allow warm mantle rocks to rise and partially melt, creating volcanic rocks that attach themselves to the diverging plates. A passive continental margin is developed on both sides of the divergent plate boundary. The ocean basin will widen as the two continental fragments drift apart.

A rift zone that does not go to completion is known as a *failed rift* or *aulacogen* (Figure 5.4b). These structures create a weak zone in the middle of a continent that often develops into a major river valley. In the United States, the Mississippi River follows an ancient rift zone that was created initially about 800 million years ago. Faults associated with this rift zone periodically reactivate (that is, they come back to life). The most recent and significant reactivation produced the highly destructive New Madrid earthquakes of 1811 and 1812.

THE PACIFIC ACTIVE CONTINENTAL MARGIN

The western North American seaboard is an active (as opposed to passive) continental margin, meaning that the transition from continental lithosphere to oceanic lithosphere coincides with a plate boundary. Because individual plates are sliding past each other or colliding, an active continental margin is characterized by volcanism, earthquakes, intense deformation, and areas of rapid tectonic uplift and subsidence.

Active tectonics along the western seaboard involves three tectonic plates as shown in Figure 5.5: the North American plate, the Pacific plate, and the Juan de Fuca plate. The boundary between the Pacific and North American plates in California is a transform plate boundary marked by the San Andreas strike-slip fault. Along this fault, the Pacific plate is moving northward relative to the North American plate at an approximate rate of 16.5 feet per 100 years. A small piece of California, including Los Angeles, is west of the San Andreas fault. San Francisco is east of the fault. If present-day plate motions continue, Los Angeles will slide northward and will reach the city of San Francisco in about 12 million years.

The San Andreas fault extends from the vicinity of Cape Mendocino southward to the Salton Sea. Figure 5.6 shows the southern termination of the San Andreas fault. Farther south, Figure 5.6 shows a series of strike-slip (transform) fault boundaries within the Gulf of California that connect small divergent plate segments. The transforms have collectively forced the rifting of Baja California from Mexico resulting in the opening of the Gulf of California. The Gulf began to open only 5 to 6 million years ago and is the most recent rifting event to affect North America. Prior to 5 or 6 million years ago, Baja California was part of Mexico and the Gulf of California did not exist. Over time, the Gulf will continue to open and Baja

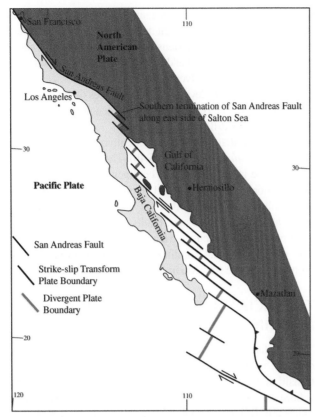

FIGURE 5.6 A map showing the tectonic setting of the Gulf of California. Based on Wallace (1990, p.76).

California will slide northward with Los Angeles, eventually reaching San Francisco and beyond.

A small tectonic plate, referred to as the *Juan de Fuca plate*, is present north of the San Andreas fault off the northern California-Oregon-Washington coastline. This plate is moving northeastward relative to North America along a convergent plate boundary such that the Juan de Fuca plate is sliding below (subducting beneath) the North American plate at an approximate rate of 13.3 feet per 100 years. Subduction of the Juan de Fuca plate causes melting and magma generation in the mantle which rises to the surface to create the Cascade volcanoes.

The Juan de Fuca plate extends a short distance into Canada where it is replaced by the Queen Charlotte strike-slip fault (Figure 5.5). The Queen Charlotte fault is a transform boundary between the North American and Pacific plates, much like the San Andreas. It extends to Alaska, where the Pacific-North American plate boundary makes nearly a 90° turn and the Pacific plate subducts beneath North America at the Aleutian trench.

TECTONIC ACCRETION, UNDERPLATING, AND SUTURE ZONES

The subduction of one tectonic plate beneath another produces several geologic landforms and phenomena that include earthquakes, an oceanic trench, an accretionary prism, a forearc basin, and a mountain range composed of explosive volcanoes. Figure 5.7a is a cross-section that shows the connection between tectonics and landscape. A narrow oceanic trench, usually less than 50 miles wide and up to 36,000 feet deep (more than twice the normal depth of an ocean basin), marks the location where one plate subducts beneath another. Friction between the two plates at the trench causes part of the subducting oceanic lithosphere to be scraped off and added (accreted) to the underside of the overriding plate. These rocks mix with sediment eroded from surrounding highlands, and are deformed, producing a jumbled, chaotic, thrust-faulted, *mélange* of accreted oceanic rock known as an *accretionary prism* (or *accretionary wedge*). Most of the accretionary prism material remains below sea level; however, the underplating process produces a wedging affect that causes uplift and tilting that can elevate part of the accretionary prism above sea level. The process of underplating is shown in Figure 5.8.

As the subducting oceanic lithosphere sinks into the mantle, it begins to fracture and break apart producing a string of earthquakes known as the *Benioff zone* (Figure 5.7a). The subducting slab also begins to heat up and, in doing so, releases fluids into the mantle that cause part of the mantle to melt. The rising magma eventually reaches the surface to produce an arc of explosive silicic volcanoes such as the Cascade volcanoes. Underplating and uplift at the accretionary prism, coupled with inland development of a volcanic arc, produce a lowland between the two topographic highs. This lowland is referred to as a *forearc basin* (that is, a basin in front of the volcanic arc).

The rocks and sediment that form the accretionary prism are literally pasted (accreted) onto the edge of the overriding plate. This process, in effect, enlarges the overriding plate. The process of adding nonsubducting material to the overriding plate is known as *tectonic accretion* and is essentially the opposite of continental rifting. In addition to small fragments of oceanic lithosphere, it is possible for a large accreted terrane to be added (accreted) to the edge of the overriding plate. This may include a volcanic island, a volcanic seamount chain, an oceanic plateau, a divergent spreading center, or even a small continental mass known as a *microcontinent*. These large, coherent blocks accrete to the edge of the overriding plate because they are too buoyant to subduct. Their accretion forces active subduction to either jump to the oceanward side of the accreted terrane as shown in Figure 5.7b, or to flip direction as schematically in Figure 5.9. In either case, the accreted terrane is welded onto an enlarged overriding plate and the old intervening accretionary prism becomes a suture zone. As such, we can define a *suture zone* as an ancient, inactive, accretionary prism that marks the location where a preexisting oceanic basin has been completely subducted (destroyed). The entire North American continent consists of accreted

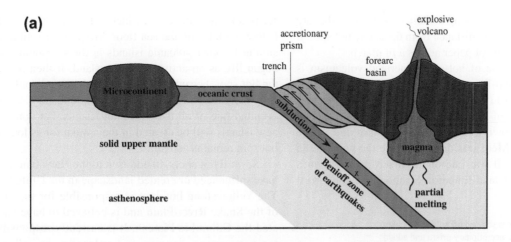

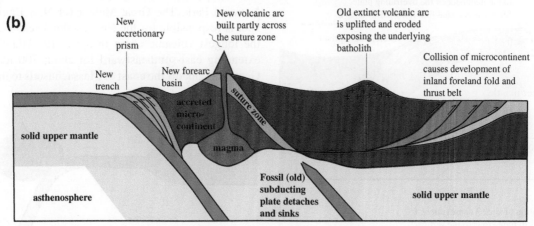

FIGURE 5.7 Cross-sections that show subduction characteristics and accretion. (a) Characteristics prior to accretion of a microcontinent. (b) Characteristics following accretion.

terranes. About two thirds of North America was amalgamated prior to 1 billion years ago and the rest during Appalachian and Cordilleran mountain building events that began about 470 million years ago. The accretion of large lithospheric fragments to an existing continent is a major driving force of mountain building and will be discussed in more detail in Part 3 of this book.

Note the spelling of the word *terrane,* which is different from *terrain. Terrane* is used in geology to signify an area of similar geological history that is distinct from the geological history of surrounding areas. *Terrain* is used to signify a physical stretch of land. It has geographic meaning but no unique geologic meaning.

THERMAL PLUMES AND HOT SPOTS

A *thermal plume* is a column of hot (but not liquid) rock that rises vertically from the base of the mantle. The rising plume takes the shape of a mushroom with a large head and a long tail. As the plume encounters the rigid lithosphere, upward motion is temporarily halted, causing the hot rock to flatten against the base of the lithosphere. As more and

more hot rock accumulates, it begins to melt which stretches and weakens the overlying lithosphere eventually allowing the magma to reach the surface. Because of the mushroom shape of the plume, the initial outpouring of lava is tremendous, producing what are known as *flood basalts.* The sequence of events is depicted in Figure 5.10. The mile-plus thickness of basalt that forms most of the Columbia River Plateau is considered by many to be an example of hot spot flood basalt. Once the plume head is depleted, the tail of the plume will continue to create volcanism at the surface over the course of millions of years, although the total amount of volcanic material is far less voluminous than the initial outpouring. This constant, thermal plume-fed volcanism at the Earth's surface creates what is known as a *hot spot.*

It is important to understand that the thermal plume, the hot spot, is stationary, but that the lithosphere (the tectonic plate) is in motion. Active volcanism is restricted only to that part of the tectonic plate that is directly above the hot spot. As the tectonic plate moves, the area of active volcanism will move off the hot spot and become inactive. At the same time, the adjacent area, which is now above the hot spot, will become the active volcanic site. In this manner, a hot spot is capable of creating a chain of volcanoes that are progressively older with distance from the active hot spot. It is like moving a piece of paper across a lit match.

A classic example of hot-spot-generated volcanism is the Hawaiian Island-Emperor Seamount Chain, which, as shown in Figure 5.11, stretches from the active volcanoes on the big island of Hawaii directly above the hot spot, to a seamount known as Meiji at the western end of the Aleutian Islands. Meiji was directly above the hot spot 85 million years ago. It was one of the first volcanic islands in the Hawaiian Island-Emperor Seamount Chain. Most

seamounts are ancient, extinct volcanoes that, today, sit below sea level on the sea floor. Meiji underwent a history similar to other volcanic islands in the seamount chain. It began life as an active volcanic island. It then moved off the hot spot, became inactive, eroded, cooled, and isostatically sank below sea level. This simple history will be the eventual fate of all the Hawaiian Islands, but don't worry, new islands will be created in their wake for as long as the hot spot remains active.

Several hot spots, in addition to the Hawaiian hot spot, have influenced or created landscape in the United States. The Yellowstone hot spot is responsible for the creation of the Snake River Plain and is believed to have also created the Columbia Plateau. It is currently responsible for the hot springs, geysers, and volcanism in Yellowstone National Park. The Great Meteor (or New England) hot spot is responsible for the New England seamount chain, the longest volcanic chain in the North Atlantic Ocean extending east-southeastward for about 700 miles from Georges Bank off the coast of Massachusetts to the middle

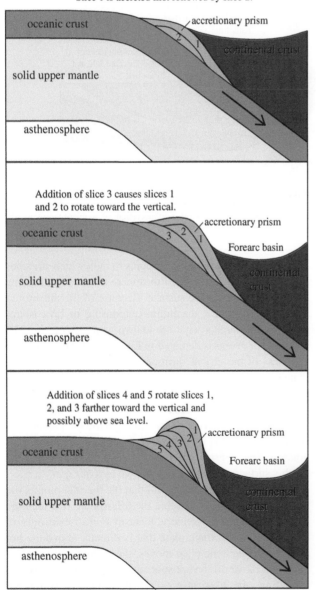

FIGURE 5.8   A series of cross-sections that show the process of underplating.

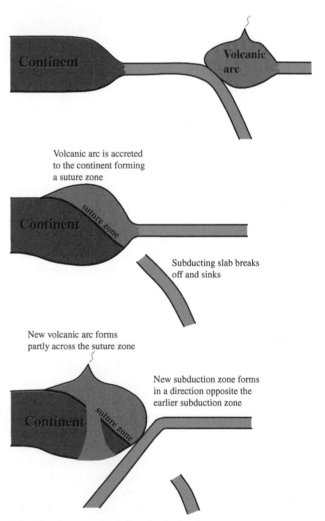

FIGURE 5.9   A cross-section sketch showing a switch in subduction direction following accretion.

of the Atlantic Ocean. The western part of the chain at Georges Bank began to form about 100 million years ago; the eastern part, in the North Atlantic, about 80 million years ago. The seamount chain is shown in Figure 5.3 located just above the words North American plate. Prior to creating volcanism in the Atlantic Ocean, the New England hot spot began life on the North American continent in Canada possibly as early as 200 million

years ago. It is believed to be responsible for the younger (130–100 million-year) phase of magmatism in the White Mountains of central New Hampshire and possibly for uplift of the Adirondack Mountains. Another somewhat obscure, poorly understood, and debated thermal plume, known as the Bermuda hot spot, may be responsible for some of the landscape in the lower (southern) Mississippi Valley.

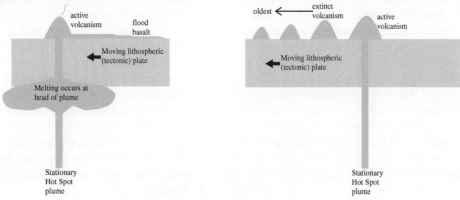

FIGURE 5.10   The origin of hot-spot volcanism.

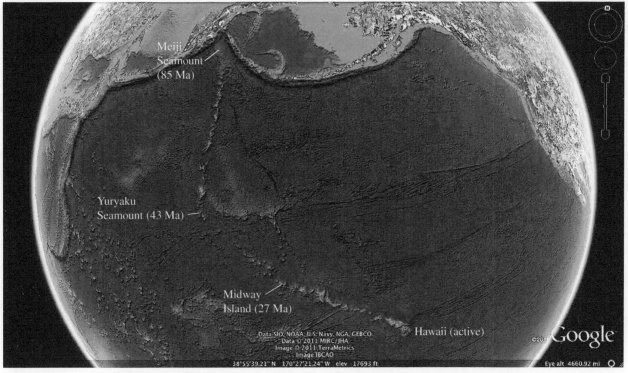

FIGURE 5.11   A Google Earth image looking north at the Hawaiian Island-Emperor Seamount chain.

TEKTON: THE CARPENTER, THE BUILDER

Tectonics plays a pivotal role in landscape formation and evolution. Without an active tectonic system, the climatic system would have no competition and, through erosion and deposition, would have long ago worn the landscape down to a flat surface. There would be no mountains and few, if any, hills. Much of the land area would be at sea level or below a shallow ocean perhaps only a few feet deep. There would be no crystalline rock exposed at the surface, no volcanic rock, and probably very little, if any, sedimentary rock—only unconsolidated sediment. The Earth would be rather bland.

Tectonics is the builder, the carpenter, of Earth. It is the tectonic system that is largely responsible for putting the rock cycle in motion. The interior of Earth, to a depth of 1,800 miles, is a nicely layered sequence of primarily solid rock. If left undisturbed, there would be no impetus for the transfer of heat, the melting of rock, and ultimately the uplift and subsidence of land areas. The Earth would literally be a rock in space covered with water. It is the tectonic system, the movement of plates, that perturbs the interior Earth and, in so doing, creates conditions favorable for land modification. Subduction pushes cold rock deep into the mantle; rifting causes warm rock to rise toward the surface. Both disturb the perfect layering of Earth creating magma and metamorphic rocks that make their way to the surface either through erosion as crystalline rock or directly as volcanic rock. Once at the surface, the rocks are immediately attacked by the climatic system which is the other force that sets the rock cycle in motion. The tectonic system, in a sense, feeds the climatic system. Without the tectonic system to drive rock toward the surface, the climatic system would have nothing to do on a flat Earth except to (literally) storm around frustrated at the absence of exposed bedrock to erode.

The tectonic system not only creates rock, it also produces the structure in rock. The interaction of two adjacent tectonic plates moving in different directions creates stresses strong enough to fold and fault large masses of rock. Thus, it is the tectonic system that is primarily responsible for the formation of the two components of landscape evolution (the rock/sediment type and the rock structure) and for three of the five mechanisms of landscape change (uplift, subsidence, and volcanism). Tectonic forces are the only forces on Earth that can be sustained long enough and rapidly enough to create a mountain range. In Chapter 8, we will make a distinction between landscape shaped by recent and active tectonic activity, and landscape in which tectonic activity is old and long since inactive.

QUESTIONS

1. Make an enlarged copy of the west side of Figure 1.3 with plenty of open space on the left side of the sheet. Draw the following and provide a short tectonic explanation of each feature: Cascadia (Juan de Fuca) subduction zone, Juan de Fuca plate, Juan de Fuca plate boundary with the Pacific plate, San Andreas fault, and the divergent plate boundary within the Gulf of California.

2. Speculate as to why the Hawaiian Island–Emperor Seamount chain bends at Yuryaku. Is it due to movement of the hot spot? All the islands and seamounts are part of the Pacific plate.

3. Use Google Earth to find the last island in the Hawaiian Island–Emperor Seamount chain. Once you have located this island, determine or estimate its age. The big island of Hawaii, although still active, is currently moving off the Hawaiian hot spot. If we presume that volcanism on the island will end within a million years, how many years can we expect Hawaii to exist as an island once it has moved off the hot spot?

4. Using Google Earth, locate the New England hot-spot seamount chain off the coast of Massachusetts. Measure the visible length from one end to the other. Assuming that the North American plate has moved at a rate of 8.5 feet per 100 years, how long would it have taken to create the seamount chain? In what direction would the North American plate have moved? Which side of the seamount chain is oldest?

5. Describe the difference between a passive continental margin and an active margin. Which is the most appropriate name for the Texas Gulf Coast?

6. What is the tectonic system? Where does it obtain its energy?

7. What is the climatic system? Where does it obtain its energy?

8. Describe the competition between the tectonic system and the climate system.

9. What are typical rates of plate movement in inches per year, inches per 100 years, millimeters per year, centimeters per year, centimeters per 100 years, miles per million years, and kilometers per million years? What is the correlation between millimeters per year and kilometers per million years?

10. Describe, with a drawing, the process of tectonic accretion.

11. Describe the process of underplating.

12. What is an accretionary prism?

13. What is a suture zone?

14. Describe the three major types of plate boundaries. Which are responsible for major volcanism? Which are responsible for major, destructive earthquakes?

15. What is a thermal plume?

16. Where are some of the major active hot spots in the United States?

17. What are flood basalts and why do they occur?

18. How does the movement of tectonic plates build landscape?

Forcing Variable: The Climatic System

Climate controls temperature and seasonal precipitation in a given area, which, in turn, controls the type of vegetation, the type and rate of weathering, and the depth of soil. Climate also influences other Earth-surface phenomena such as erosion and mass wasting processes, the development of river systems, groundwater, karst, and glacial systems, and short-term sea-level changes. The result of these processes is the wearing down of landscape to form a nearly flat surface.

A critical aspect of climate with respect to landscape is climate change. In the absence of active tectonic uplift or subsidence, landscape tends toward an equilibrium with climate in which rates of erosion remain relatively steady and landscape changes slowly and predictably. In Chapter 8 we refer to this as a landscape that slowly grows old. Climate change throws landscape out of equilibrium which can result in increased rates of erosion and deposition and lead to landscape reincarnation. The most profound climate change in the past 2.4 million years has been glaciation. Glaciers have advanced across the north-central United States, New England, and the mountainous areas of the Cordillera resulting in strong modification and, in some cases, complete reincarnation of the landscape.

PRESENT-DAY CLIMATE ZONES

Figure 6.1 shows present-day climate zones in the United States, each of which are listed in Table 6.1.

The United States is approximately 3,000 miles east to west and 2,000 miles north to south. Climate changes dramatically across this large area. Climate is warm and humid in the southeast (wet-dry savanna, humid subtropical) and becomes cooler and somewhat less humid in the

northeast (humid continental). Both areas are wet enough to support a thick soil and a lush hardwood deciduous forest of maple, oak, ash, beech, and hickory. Much of the farmland region of Ohio, Indiana, Illinois, Iowa, and Missouri was once covered by forest. Farther north, the deciduous forest mixes and finally gives way, in eastern Canada, to a coniferous forest of spruce, fur, pine, hemlock, and cedar. This region is cooler and drier than the northeastern US producing a subpolar climate. As we head west to the Great Plains, the climate becomes markedly drier (steppe) and the lush forests of the east give way to open grasslands and a much thinner soil. By the time we reach the mountainous states of New Mexico, Colorado, Wyoming, and Montana, semi-arid to arid conditions prevail in lowland areas producing steppe and desert climates, whereas the surrounding mountains produce their own (mountain) climate which tends to be cooler and wetter than lowland areas. Moist air coming off the Pacific Ocean keeps the coastal areas of Washington, Oregon, and northern California very wet (moist coastal) and southern California seasonably wet (Mediterranean).

Weathering (but not erosion) tends to be most intense (and therefore most rapid) in areas across the eastern half of the US and along the Pacific Coast where climate is wet for most of the year. A wet climate promotes lush vegetation, which in turn adds acids to the water, causing chemical weathering. Roots hold weathered material in place, inhibiting erosion and promoting development of a deep, rich soil. Weathering tends to be less intense across much of the Great Plains and parts of the Cordillera where the climate is dry and often alternates between hot and cold. In this situation, frost cracking dominates, vegetation is sparse, chemical weathering is curtailed, and only a thin soil develops such that underlying rocks are widely exposed.

FIGURE 6.1 A landscape map showing present-day climate regions of the US.

TABLE 6.1   Climate Zones in the United States
Wet-dry savanna
Humid subtropical
Humid continental
Steppe
Desert
Mediterranean
Moist coastal
Mountain
Subpolar (northeastern Canada)

CONTROLS ON CLIMATE

The climates outlined in this brief synopsis of the United States are controlled largely by latitude, proximity to large water bodies (including the ocean), global wind patterns, and mountains. The effect of latitude is obvious. Areas close to the equator receive more direct rays of the sun and therefore are warmer. The sun is lower on the horizon with distance from the equator, and therefore the northern part of the US tends to be cooler.

Proximity to a large water body has several effects, one of which is lake-effect snow for which Buffalo, New York, is famous. More importantly, large bodies of water have the effect of moderating air temperature. Even if two cities are about the same distance from the equator, a coastal city such as Seattle is cooler in summer and warmer in winter relative to an inland city such as Minneapolis. Mid-August average high temperature for Seattle is 76°F versus 82°F for Minneapolis. Mid-January high temperatures are 45°F versus 26°F. The reason for these differences is that, relative to land, water takes longer to heat and to cool. This is why, when the temperature outside is 95°F, the water in your pool remains relatively cool. If hot air passes over a body of cool water, the air itself will become cooler. Seattle benefits from its proximity to the cool Pacific Ocean during summer months. Minneapolis receives only hot inland continental air. Conversely, the ocean warms frigid air during winter months, which keeps Seattle warm relative to the blustery cold air that blows across Minneapolis.

The moderating influence of the ocean is felt most strongly on the West Coast because air masses move across the Pacific Ocean onto land. The effect in southern California in particular is to produce relatively moderate,

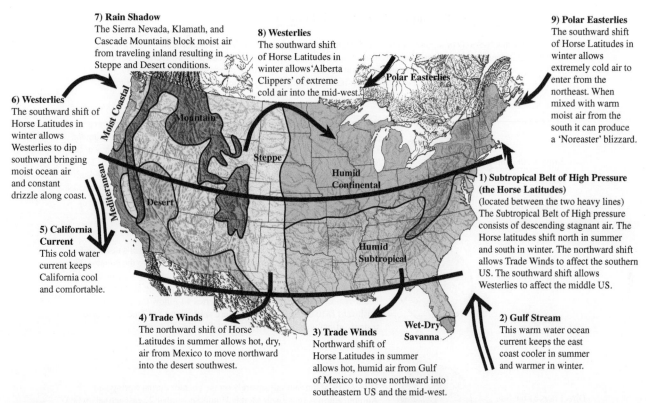

7) Rain Shadow
The Sierra Nevada, Klamath, and Cascade Mountains block moist air from traveling inland resulting in Steppe and Desert conditions.

8) Westerlies
The southward shift of Horse Latitudes in winter allows 'Alberta Clippers' of extreme cold air into the mid-west.

9) Polar Easterlies
The southward shift of Horse Latitudes in winter allows extremely cold air to enter from the northeast. When mixed with warm moist air from the south it can produce a 'Noreaster' blizzard.

6) Westerlies
The southward shift of Horse Latitudes in winter allows Westerlies to dip southward bringing moist ocean air and constant drizzle along coast.

5) California Current
This cold water current keeps California cool and comfortable.

4) Trade Winds
The northward shift of Horse Latitudes in summer allows hot, dry, air from Mexico to move northward into the desert southwest.

3) Trade Winds
Northward shift of Horse Latitudes in summer allows hot, humid air from Gulf of Mexico to move northward into southeastern US and the mid-west.

1) Subtropical Belt of High Pressure (the Horse Latitudes)
(located between the two heavy lines) The Subtropical Belt of High pressure consists of descending stagnant air. The Horse latitudes shift north in summer and south in winter. The northward shift allows Trade Winds to affect the southern US. The southward shift allows Westerlies to affect the middle US.

2) Gulf Stream
This warm water ocean current keeps the east coast cooler in summer and warmer in winter.

FIGURE 6.2 A landscape map showing present-day wind and ocean current patterns in the US. The figure should be read clockwise beginning with the subtropical belt of high pressure.

pleasant year-round temperatures. Temperatures on the eastern seaboard are also moderated, but because air typically moves from land areas toward the Atlantic Ocean, the effects are not as pronounced. In New England, several inches of snow could fall in Vermont while at the same time coastal areas receive only rain. Similarly, coastal areas remain cool relative to inland areas during summer months. The effect of summer cooling is less obvious along the Gulf Coast because perennially warm water in the Gulf of Mexico does little to moderate the hot, humid, summer air.

Global wind patterns have a huge influence on climate. The southern part of the United States is dominated by the subtropical belt of high pressure (also known as the *Horse Latitudes*). This is a belt of relatively stagnant, heavy, descending air. As the air descends, it warms and dries. Many of the world's deserts, including the Mohave Desert, lie within the dry, descending air of the Horse Latitudes. The movement of air is vertical (downward); therefore there is very little wind associated with the Horse Latitudes except along the margins as air moves horizontally toward low-pressure areas. Wind is the movement of air from high-pressure areas to low-pressure areas. Global wind currents develop at the margins of the high-pressure Horse Latitudes. The Westerlies form the dominant wind current in the northern part of the US, and the Trade Winds affect the very southern part of the country. A third global wind pattern, the

Polar Easterlies, emanates from high-pressure areas near the Arctic Circle. The Easterlies primarily remain north of the US, affecting Canada, but on occasion they dip southward into New England and the north-central part of the country. All northern hemisphere winds rotate clockwise due to the counterclockwise (eastward) rotation of Earth. This results in the Polar Easterlies and Trade Winds blowing mainly out of the east and the Westerlies out of the west. The Westerlies produce the dominant west-to-east movement of weather systems in the northern part of the US, and are responsible for jet streams with wind speeds up to 250 mph.

All northern hemisphere wind currents shift northward 5 to 10 degrees during the summer months, and shift southward 5 to 10 degrees during the winter months as explained below. A synopsis of global wind patterns and their affect on the US is presented in Figure 6.2. Take a moment to read though this figure beginning with the subtropical belt of high pressure.

The summer warmth and winter cold that we all experience is due to the fact that the Earth's axis is tilted 23.5 degrees to the plane of its orbit around the sun. As shown in Figure 6.3, the orientation of the axis does not change position with respect to the sun such that, as the Earth orbits the sun, the north axis will point directly away, and directly toward, the sun one day per year. On these days, the sun's direct rays are 23.5 degrees below (south of)

The Seasons and The Shift in Global Wind Patterns

The Earth's N-S axis of rotation is tilted 23.5 degrees to the plane of its orbit around the Sun. The axis does not change position with respect to the Sun such that the north axis points directly away, and directly toward, the Sun once per year during the Winter and Summer solstice respectively. The north-south axis is oriented exactly perpendicular to the Sun's rays twice per year during the Spring and Autumn equinox.

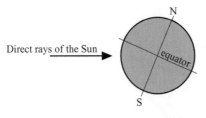

Winter arrives in the Northern Hemisphere when the Earth's north axis points away from the Sun and direct rays of the Sun are below the equator. The Winter solstice occurs on or near December 22 when direct rays of the Sun are at their southern-most latitude, the Tropic of Capricorn. The Horse Latitudes, Trade Winds, Westerlies, and Polar Easteries all follow the Sun and shift southwad 5-10 degrees. The southward shift allows the Westerlies and, to a lesser extent, the Polar Easterlies, to strongly affect weather patterns in the US during Winter.

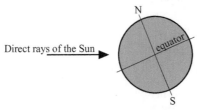

Summer arrives in the Northern Hemisphere when the Earth's north axis points toward the Sun and direct rays of the Sun are above the equator. The Summer solstice occurs on or near June 22 when direct rays of the Sun are at their northern-most latitude, the Tropic of Cancer. The Horse Latitudes, Trade Winds, Westerlies, and Polar Easteries all follow the Sun and shift northward 5-10 degrees. The northward shift allows the Trade Winds to strongly affect weather patterns in the US during Summer.

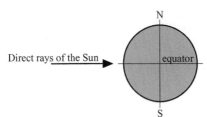

Direct rays of the Sun fall squarely on the equator during Spring and Autumn equinox on or near March 21 and September 22 respectively when the Earth's north-south axis is oriented perpendicular to the Suns rays.

FIGURE 6.3 The seasons and the shift in global wind patterns.

the equator, exactly on the Tropic of Capricorn, and 23.5° above (north of) the equator, exactly on the Tropic of Cancer. We refer to these times as the winter and summer solstice, respectively. They are the shortest and longest days of the year. Also, for two days of the year, the Earth's axis is oriented exactly perpendicular to the sun's rays. During these days the sun's direct rays fall exactly on the equator. This is the spring and autumn equinox, when the day is 12 hours long everywhere.

In a matter of speaking, all global wind patterns follow the sun. As the vertical rays of the sun shift northward during the northern hemisphere summer, the global wind patterns follow, and vice versa in winter. The shift, however, is delayed somewhat from that of the sun and is generally only 5 to 10 degrees rather than 23.5 degrees. The shift in global wind systems produces major climatic effects as described in Figure 6.2. The summer northward shift of dense, descending air associated with the subtropical belt of high pressure has the effect of blocking the cool Westerlies from reaching the southern half of the US. At the same time, the warm Trade Winds rotate northward. In the southeastern US, the Trade Winds move over the Gulf of Mexico where they pick up moisture producing hot,

humid conditions (#3, Figure 6.2). In the western US, these same winds move over the dry desert region of Mexico, resulting in hot, dry air over Arizona and New Mexico (#4, Figure 6.2).

The winter southward shift of the Horse Latitudes allows the Westerlies to rotate into the heartland of the US. On the West Coast they bring moist air from the Pacific Ocean, resulting in long, cool (but not freezing) days of constant drizzle (#6, Figure 6.2). In the central US the Westerlies bring frigid Alberta Clippers of extreme cold, windy air (#8, Figure 6.2). The Polar Easterlies also shift southward during winter months, carrying with them frigid temperatures. In New England, Polar Easterlies can, on occasion, mix with moist coastal air emanating from the south. Mixing of these two air currents produces what are known as *nor'easters* in New England. Elsewhere they are known as major blizzards and cold snaps (#9, Figure 6.2).

Mountains affect climate because they block air from moving across a region. Mountains force air to rise in order to pass over them. As air rises, it cools to its saturation point which first produces large cumulus clouds and then thunderstorms. This is why clouds tend to always be present in the vicinity of a mountain and why a passing afternoon

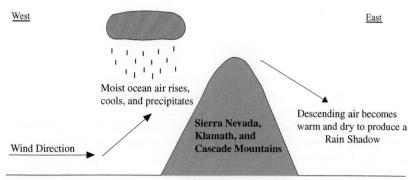

West   East

Moist ocean air rises,
cools, and precipitates

Sierra Nevada,
Klamath, and
Cascade Mountains

Descending air becomes
warm and dry to produce a
Rain Shadow

Wind Direction

FIGURE 6.4   Development of a rain shadow east of the Cascade-Klamath-Sierra Nevada.

mountain thunderstorm is commonplace. Temperature decreases by about 3.57°F for every 1,000 feet of elevation gain (6.5°C/1000 m). Vegetation is temperature and precipitation dependent and as a result, vegetation changes dramatically in a vertical direction up a mountain.

The forcing of precipitation as air moves over a mountain range produces a climatic effect known as a *rain shadow*. The effect is shown schematically in Figure 6.4. The Cascade-Klamath-Sierra Nevada ranges extend nearly the width of the Cordillera producing a strong rain shadow which partly explains the arid to semi-arid climate throughout the west. As air moves inland from the Pacific Ocean, it rises, cools, and precipitates. By the time the air reaches the eastern side of the mountain ranges, it is dry. To see the effect of a rain shadow, one could contrast the thick, lush, forest region west of the Cascade Mountains in Washington and Oregon shown in Figure 6.5a, with the arid grasslands, wheat fields, and vineyards east of the mountain range shown in Figure 6.5b. The Appalachian Mountains, by contrast, do not produce a pronounced rain shadow because prevailing westerly winds blow from west to east across dry land.

The situation just described represents only the present-day climate, which has been fairly stable for the past 7,000 to 10,000 years. Before that, for the past 2.4 million years, we have experienced a series of glaciations, during which time our climate has sometimes been warmer but more commonly cooler. At certain times during the past, and most recently only 18,000 years ago, nearly all of Canada and a large part of the northern US were covered in ice (glaciers) more than a mile thick. The glaciations have had a dramatic and lasting effect on our landscape.

A DAUGHTER OF THE SNOWS: THE CONTINENTAL GLACIATION

Fifty million years ago, North America and the rest of the world was much warmer than today. At that time palm trees, alligators, and turtles could have sunned themselves close to the Arctic Circle. The Earth has cooled since then, reaching a low point during the Ice Age beginning about 2.4 million years ago and, as some would argue, continuing today.

The Ice Age did not consist of a single glaciation. Instead, there were multiple glacial advances and retreats, and during some of the glacial retreats the Earth was warmer than today. ^{26}Al-^{10}Be cosmogenic radionuclide dating of till in the central part of the US suggests that a major glacial advance took place 2.4 million years ago, but there is no evidence of a second advance south of 45 degrees north latitude (the approximate latitude of Minneapolis) until about 1.3 million years ago. Additional glacial advances became common only in the past 800,000 years. The last prolonged interglacial warm stage occurred between about 130,000 and 115,000 years ago. This warm period was followed by the last glacial stage to affect the central US. Ice advanced into the central part of the US and then retreated several times during the past 80,000 years, with the final major advance beginning about 35,000 years ago and ending 18,000 years ago. At the height of the final glacial advance 18,000 years ago, glaciers covered most of Indiana and Ohio as well as New York and New England. By 10,000 years ago these glaciers had retreated to Canada and northern New England. By 6,000 years ago, large areas of glacial ice had all but disappeared from Canada. This final glacial advance did not push as far south as earlier glaciations, which reached Kansas, Missouri, and (just barely) Kentucky.

This last glacial advance, as well as the ones that came before, have shaped and reshaped all of Canada and large parts of the United States. Glaciers are thick ice masses that originate on land and show evidence of flow. A glacier originates in the *zone of accumulation*. This is the area where, over the course of a year, the amount of snow accumulation exceeds the amount of melting. If snow accumulates at an average rate of only 6 inches per year, a snow pack 1 mile thick will have formed in less than 11,000 years. As snow accumulates and turns to ice, the weight of the accumulating snow causes ice at the bottom of the pile to flow outward from under the snow pack into warmer areas that otherwise would not be covered by ice. Once a snowpack of this type begins to flow outward in this manner, a continental glacier is born. As long as snow continues to accumulate at the top of the pile, ice will continue to flow outward from below the pile and move into warmer areas until it eventually melts. Normally the movement of ice is in the form of lobes that follow lowland regions (stream valleys). The warmer region into which a glacier flows and melts is known as the *zone of ablation* (*ablation* means to remove by melting, evaporation, and vaporization). Figure 6.6 shows some of the landscape characteristics typical of the zone of accumulation and zone of ablation.

Figure 6.7 shows areas of the US that have been affected by glaciation. For the purpose of our discussion of landscape development we will distinguish among areas dominated by continental glacial erosion, continental glacial deposition, alpine glaciation, and areas primarily too far south to have been glaciated. These regions are listed in Table 6.2. Alpine (or Valley) glaciers form on mountaintops in a manner similar to continental glaciers. As snow accumulates, a glacial lobe will begin to flow downward into the valley, primarily under the force of gravity, where it eventually melts. This type of glacier is generally confined to individual mountain

FIGURE 6.5   The rain shadow effect in Oregon. (a) Deep forest in the Coast Range west of Corvallis, Oregon. (b) Sparse vegetation east of the Cascade Mountains near Bend, Oregon.

valleys; however, in some instances, several alpine glaciers will coalesce such that they completely cover the valley

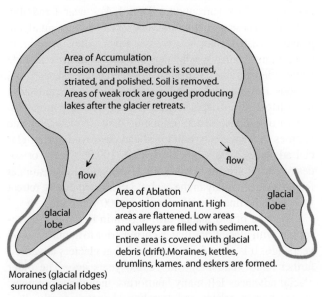

FIGURE 6.6   A representation of continental glaciation.

floor and most of the surrounding mountain tops. Such was the case with the ice sheet that covered the Canadian Cordillera. This continuous mass of ice was separate from ice sheets that covered the central US and is known as the *Cordilleran Ice Sheet*. The Cordilleran Ice Sheet reached as far south as the northern fringe of the US Cordillera as shown in Figure 6.7. Alpine glaciers south of the Cordilleran Ice Sheet are confined to mountaintops and a few valleys.

All glaciers transport and deposit sediment. *Drift* is a general term for any sediment of glacial origin. Drift can be

TABLE 6.2 Glacial Zones

1. Areas of continental glacial erosion
2. Areas of continental glacial deposition
   a. Young glacial drift (Wisconsinan)
   b. Old glacial drift (pre-Wisconsinan)
3. The driftless area
4. The Cordilleran Ice Sheet
5. Areas of Alpine glaciation (all ages)
6. Not glaciated

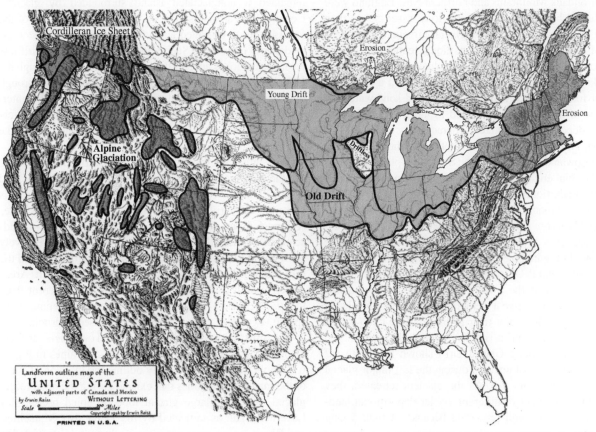

FIGURE 6.7   A landscape map showing areas that were glaciated in the past 2.4 million years. Areas of erosion, young drift, old drift, the driftless area, the Cordilleran ice sheet, and alpine glaciation are shown. Uncolored areas were not glaciated.

deposited directly via melting of ice, or the deposited material can be reworked by running water and redeposited in layers. When it is dropped directly by ice it is known as *till* or *unstratified drift*. When reworked by running water it is known as *stratified drift*.

As suggested in Figure 6.6 and shown in Figure 6.7, the area dominated by continental glacial erosion is in central Canada where it coincides roughly with the zone of accumulation. During the last glacial advance 18,000 years ago, glaciers accumulated to a thickness between 1 and 3 miles (5,000–16,000 feet). The outward flow of ice in the zone of accumulation resulted in scouring of the underlying land surface, which, in this area, consists of crystalline rock. Soil was removed and high spots were flattened. The rocks were scraped clean. What was left after the glaciers retreated was a nearly flat, shapeless landscape of bare bedrock. The pre-existing river network was destroyed in favor of a deranged network consisting of slow-moving streams that connect thousands of lakes and ponds now located where weak rock was gouged by the glacier. This type of landscape characterizes part of the Superior Upland, but mostly it characterizes the region north of the Great Lakes. Figures 6.8a, 6.8b, and 6.8c show schematic before and after cross-sections of glacial erosion in a crystalline terrain and a Google Earth image of present-day landscape. As suggested in Figure 6.7, the Adirondack Mountains and northern New England were also shaped by glacial erosion but to a lesser extent than central Canada. In addition to glacial erosion, these areas were also shaped by glacial depositional features.

The removal of soil and the plucking and scraping of rock in the zone of accumulation produced a huge volume of loose material that was transported southward into the zone of ablation and deposited as the ice melted. Thus glacial depositional features characterize the zone of ablation. This region includes most of the northern part of the Central Lowlands and much of the Superior Upland. As in the area of glacial erosion, high spots were flattened by moving ice, but in addition low spots, including river valleys, were filled with sediment. This created a wide, nearly flat region of thick, richly fertile glacial soil that covers nearly all bedrock. Before and after cross-sections and a Google Earth image are shown in Figures 6.8a, 6.8d, and 6.8e. One of the largest rivers in the central US prior to glaciation, the Teays River, once flowed across the zone of ablation in central Indiana. This river and others like it were destroyed during the glacial advance. Today the major river in Indiana (the Ohio River) has established itself primarily just south of glacial drift.

The area of Young Glacial Drift shown in Figure 6.7 represents the zone of ablation during the last major glacial advance 18,000 years ago. As the glaciers retreated, they left behind a number of different glacier depositional landforms that remain well preserved because of their young age. The most common landforms are moraines. A *moraine* is a general term for any landform composed of unstratified drift (till). The most recognizable moraines are long, arcuate

ridges that develop along the margins of glacial lobes. They are shown schematically in Figure 6.6. Moraines are also a characteristic feature of alpine glaciation. Figure 6.9 shows glacial moraines surrounding Twin Lakes near Leadville, Colorado. These particular moraines formed when an alpine glacier retreated from the Arkansas River valley. In addition to moraines, other glacial depositional landforms include kettles (depressions), kames (steep-sided hills), eskers (long, sinuous ridges), and drumlins (streamlined, asymmetrical hills). Most glacial deposition landforms, including moraines, consist of unconsolidated sediment. They are rather delicate features that are easily bulldozed and destroyed during glacial advances, or are quickly (within 100,000 years or so) destroyed by erosion. Nearly all of the glacial depositional features we see today in the central US formed during retreat following the final glacial advance 18,000 years ago.

The area of old glacial drift shown in Figure 6.7 corresponds with glaciations that occurred prior to about 130,000 years ago. It is clear from Figure 6.7 that glaciers advanced farther south than they did 18,000 years ago. The older glacial advances left many landforms, including moraines, kettles, kames, eskers, and drumlins. However, nearly all of these have already been destroyed by erosion, leaving only flat plains composed of glacial drift.

The driftless area, located in southwestern Wisconsin, remained a nonglaciated island surrounded by glaciers. Certainly this area was covered in snow during much of the Ice Age, but to be called a glacier, the snowpack must show evidence of movement. Such evidence is lacking in the driftless area. Glacial lobes apparently advanced southward though the Lake Superior lowland into Iowa and southwestward through the Lake Michigan lowland to southern Indiana, but the intervening highland was bypassed. This relatively hilly area offers a glimpse at what the Central Lowlands may have looked like prior to glaciation.

The area south of glacial deposition includes the Southern Appalachians, the Interior Low Plateaus, and Ozark Plateau. Much of this area, like the driftless area, is relatively hilly. Even at the scale of the entire United States (Figure 6.7), one can easily see the abrupt change in landscape in southern Illinois and Indiana between the relatively flat Central Lowlands, which are covered with drift, and the hilly Interior Low Plateaus south of the glacial limit. Areas directly south of the glaciation looked very different from today. They were likely barren in the cold climate, with sparse vegetation and wide areas of exposed bedrock.

Contrasts between areas of young and old glacial drift are displayed in more detail in Figure 6.10. Moraines in Figure 6.10 (abbreviated *Mor.*) clearly outline the location of some of the major glacial lobes that existed during the most recent glacial retreat. A large glacial lobe occupied present-day Lake Michigan as indicated by the Valparaiso, Marseilles, and Bloomington moraines among others. A lobe existed in present-day Saginaw Bay as indicated by the Lansing and Charlotte moraines; and another occupied present-day Lake

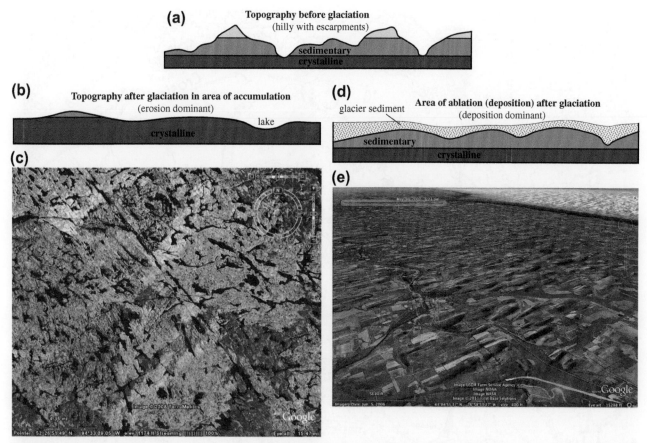

FIGURE 6.8 Glacial landscapes. (a) Cross-section depicting landscape before glaciation. (b) Cross-section that shows scoured and flattened crystalline rock following glaciation within the zone of accumulation. (c) Google Earth image looking north across western Ontario, Canada, of a scoured and flattened landscape composed of deformed crystalline rock. (d) Cross-section showing depositional landscape in the zone of ablation following glaciation. Sedimentary rocks are buried beneath a layer of glacial sediment. (e) Google Earth image looking westward at glacial depositional features just north of the Finger Lakes, near Rochester, New York. The streamlined topography consists of drumlins, which are linear hills composed of unstratified drift. The drumlins are oriented in the direction of glacial flow.

FIGURE 6.9 A photograph looking southeastward from near the summit of Mt. Elbert overlooking Twin Lakes, the Arkansas River valley, and the Mosquito Range. The lakes are surrounded by glacial moraines.

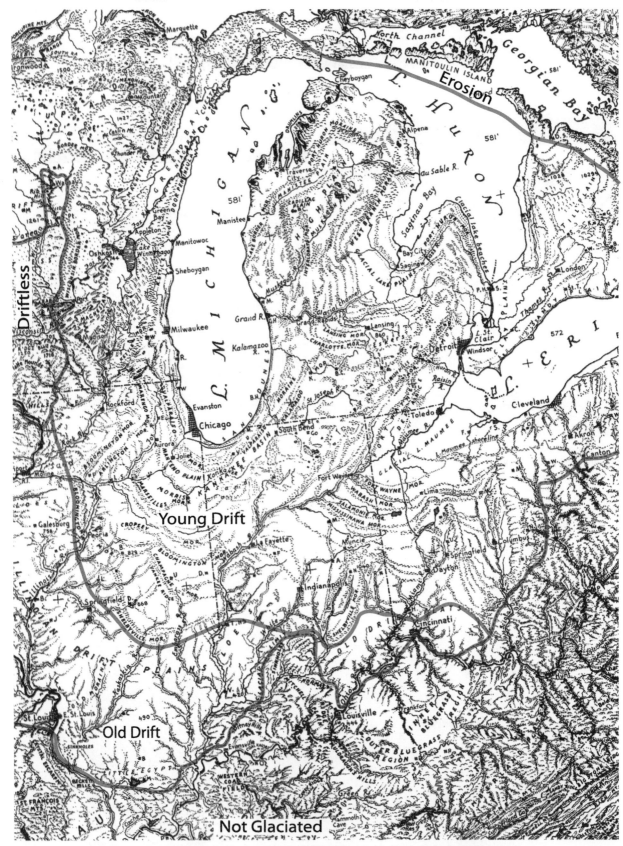

FIGURE 6.10   A detailed landscape map showing glacial features in the Lake Michigan region.

Erie, as indicated by the Fort Wayne, Wabash, Salamonie, and Mississinawa moraines. The presence of these moraines suggests that the Great Lakes were stream valleys 2.4 million years ago and that they were gouged presumably many times during the glaciation. This implies that the present-day Great Lakes are young landforms less than 2.4 million years old. Note also in Figure 6.10 how abruptly the glacial moraines end at the boundary between young and old drift. The area of old drift is truly a flat landscape occupied primarily by the Illinoian Drift Plains with only a few glacial landforms. This area was last glaciated between about 310,000 and 130,000 years ago and likely also experienced earlier glaciations.

The northeast corner of Figure 6.10 shows a transition from glacial erosion features north of Lake Huron, where streamlined topography and many small lakes are evident, to glacier depositional features in Michigan, where moraines and drumlins dominate the landscape. Finally, in the southern part of Figure 6.10, note the distinct and abrupt change in landscape between the Illinoian Drift Plains (the area of old drift) and the hilly, unglaciated region to the south. In Indiana, this boundary corresponds with the boundary between the Central Lowlands and the Interior Low Plateaus. A Google Earth image of this abrupt boundary is shown in Figure 6.11. It is amazing to think that the

landscape north of this line is no older than about 310,000 years, whereas the landscape to the south probably has not changed appreciably in more than 200 million years.

The retreat and melting of glaciers generates large volumes of water, which, because of the slope of the land and the presence of moraines, tends to pond along the margins of the melting glacier. Lakes of this type are known as *proglacial lakes* and they were widespread along the margins of both young drift and old drift. The Great Lakes formed originally as proglacial lakes. Many of the lakes were destroyed and buried during subsequent glacial advances. Others eventually drained into a river system, dried up, or were filled with sediment. Many would leave behind a characteristically flat landscape of level land that represents what was once the bottom of the lake.

One of the largest lakes to have ever existed in North America was glacial Lake Agassiz. The lake formed initially about 12,000 years ago and expanded as glaciers retreated into Canada. Throughout its life the lake would vary in size depending on the movement of glaciers and the availability of drainage outlets. At its maximum size about 10,000 years ago, the lake was 700 miles long and 250 miles wide extending from South Dakota well into Canada. Most of the glacial ice had melted by 6,000 to

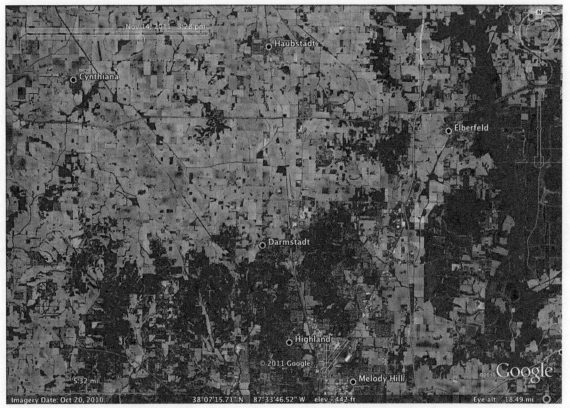

FIGURE 6.11   A Google Earth image looking north at the transition between hilly, forested, unglaciated Interior Low Plateaus in the south, and flat, glaciated farm fields of the Illinoian Drift Plains in the north. Location is just north of Evansville, Indiana. Much of the lowland area was periodically occupied by lakes.

7,000 years ago, and the giant lake drained northward into the newly opened Hudson Bay. Today the area formally occupied by Lake Agassiz is a flat landscape of "10,000 lakes," abandoned beach ridges, swamps, and deltas. Lake Winnipeg, Lake Winnipegosis, Lake Manitoba, and Lake of the Woods (all mostly in Canada) are all remnants of giant Lake Agassiz. The most prominent remnant in the US is in the Red River Valley which separates North Dakota from Minnesota. Here, an arm of Lake Agassiz extended all the way to the Mississippi River system divide south of Lake Traverse near the border with South Dakota. Figure 6.12 shows the extent of

FIGURE 6.12 A detailed landscape map showing glacial features in the Minnesota region.

Lake Agassiz in the United States including its southward extent to Lake Traverse. The lake left behind a flat landscape with deltas and sandy beaches now occupied by the undersized, north-flowing, Red River, which meanders incessantly through the middle of the valley as shown in Figure 6.13.

Note the area of dissected loess in the southwestern part of Figure 6.12 east of Omaha. This is an area of old (pre-310,000-year-old) glacial drift where all glacial landforms, and nearly all glacial deposits, are eroded or covered by younger sediment, which, in this case, includes loess. *Loess* is a deposit of fine-grained, wind-blown sediment. The rubbing of rocks at the bottom of glaciers produces a large amount of fine-grained material that is left behind when the glacier retreats. Over time, the loess dries out and is carried downwind to be deposited elsewhere. Loess east of Omaha was deposited above older glacial drift following the most recent glacial advance 18,000 years ago. Loess is also present in the nonglaciated southwestern part of the map west of Omaha. The southern extent of old drift is not shown in Figure 6.12, but, as suggested in Figure 6.7, it extends southward to Manhattan, Kansas, and then turns eastward following the Kansas and Missouri rivers to Topeka, Kansas City, Jefferson City, and St. Louis where its eastern continuation is shown in Figure 6.10.

Less distinctive is the transition from primarily glacial erosion to glacial deposition features in the New York-New England area. This area is shown in Figure 6.15. The reason for the absence of a distinct transition is the preservation of glacial deposition landforms across the entire area and the overall mountainous landscape which produced minor alpine glacier features. Nevertheless, there is a subtle change from a more rugged erosional landscape in the north to a more subdued depositional landscape in the south. This is especially true along the coastline of New England, which changes from rocky beaches with many inlets in the north, to sandy beaches farther south. The beaches of Cape Cod, Rhode Island, and Long Island result from sediment (moraines) dropped by retreating glaciers and then reworked by ocean currents. Notice also that there is little in the way of an obvious landscape change across the boundary that separates glaciated areas in Pennsylvania from nonglaciated areas. This is again due to the hilly and mountainous landscape that masks changes. The boundary can be located on the ground by the presence of glacial deposits and a few moraines not shown in Figure 6.14.

ALPINE GLACIATION

Alpine glaciation has strongly affected mountainous areas of the Cordillera. High peaks in the Rocky Mountains, Cascades, Sierra Nevada, Olympic Mountains, and Klamath Mountains were all glaciated, as were peaks and high areas in the Basin and Range and Colorado Plateau. These areas are identified in Figure 6.7. Prior to glaciation, many of the mountains in the western US were somewhat

FIGURE 6.13   A Google Earth image looking north at the Red River Valley south of Grand Forks, North Dakota. Note the flat landscape, the highly meandering Red River (highlighted in blue), and the long beach ridges along the eastern side.

FIGURE 6.14   A detailed landscape map showing glacial features in the New York-New England region.

rounded in a manner similar to the present-day Southern Appalachians. They were, to put it bluntly, far less spectacular in their appearance. Glacial erosion sculpted and eroded the mountains into sharp jagged peaks and ridges (horns and arêtes), rock amphitheaters (cirques), and alpine lakes (tarns). Narrow stream valleys were scoured, flattened, and gouged into steep-sided, flat-floored, U-shaped valleys such as Yosemite Valley in the Sierra Nevada. Figure 6.15 is a schematic illustration that shows the difference between a nonglaciated mountain and one that is glaciated. Figure 6.16 shows a snowless cliff face on Mt. Rainier carved by the active Carbon glacier, which can be seen extending down the mountain. Notice in this figure the clear distinction between fresh snow in the zone of accumulation, and the dark, dirty snow in the zone of ablation.

As an analogy to what an alpine glacier can do to a mountain, imagine a five-pound stick of butter. Stream erosion would be analogous to cutting into the butter with a hot knife, thereby producing a deep V-shaped incision. Moving ice, on the other hand, would be like gouging the butter with a hot spoon, thereby producing a deep, wide, U-shaped chasm. Naturalists, photographers, hikers, and rock climbers all owe a debt of gratitude to the glaciation that produced some of the most spectacular mountain scenery on Earth.

As you might expect from all the melting of ice, the US was wetter during the last glaciation than it is today. An area that hosted a large number of lakes was the Great Basin

centered in Nevada and Utah where perhaps more than 1,000 lakes may have existed between about 25,000 and 10,000 years ago. A map that shows the location of some of the larger lakes is presented in Figure 6.17. Most of these

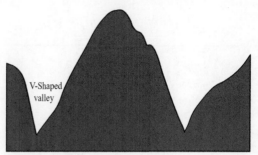

Mountain before glaciation
Mountain is round with v-shaped valleys.

Mountain following glaciation
Mountain is sculpted with sharp jagged peaks and ridges. Valleys are u-shaped with flat bottoms and steep sides.

FIGURE 6.15   A sketch showing how alpine glaciers carve mountains.

FIGURE 6.16   A Google Earth image looking south at Mt. Rainier. Carbon glacier is on the right side and Winthrop glacier on the left.

FIGURE 6.17   A detailed landscape map showing glacial lakes in the Great Basin, Nevada-Utah.

lakes have simply evaporated in today's arid climate. A remnant of one particularly large lake is the Great Salt Lake, which today covers about 2,000 square miles to an average depth of 14 feet. Twenty thousand years ago the lake was more than 1,000 feet deep and covered more than 20,000 square miles, extending from Idaho and Nevada almost to Arizona. Known as ancient Lake Bonneville, the dried

remains of this lake form the famous Bonneville Salt Flats. Evidence of the size and depth of Lake Bonneville is seen in the many terraces and wave-cut beaches along hillsides that surround Salt Lake City. A second exceptionally large lake was Lake Lahontan, which existed in northwestern Nevada. Remnants of Lake Lahontan include Walker Lake, Carson Sink, Pyramid Lake, and Honey Lake.

QUESTIONS

1. Describe the global wind patterns and the resulting climate in the area where you live.

2. At what latitude are the Tropics of Cancer and Capricorn located, and what is their significance with respect to climate?

3. Describe how the climate where you live would be different if (a) Earth's axis were oriented exactly perpendicular to the sun's rays during the entire year, and (b) the Cascade-Klamath-Sierra Nevada ranges did not exist, but a large mountain range did exist extending from Minnesota to Louisiana.

4. Notice in Figure 6.12 that deltas are present along the western side of Lake Agassiz in the Red Valley, whereas beaches characterize the eastern side. Suggest several scenarios that could explain this relationship.

5. Notice in Figure 6.12 that beaches are present in the middle of Lake Agassiz in the area southwest of the Lake of the Woods. Suggest several scenarios that could explain this relationship.

6. Notice the area of glacial erosion in the northeast corner of Figure 6.12. Describe how this area is different from areas of glacial deposition to the immediate south.

7. What evidence is there in Figure 6.12 that a large glacier occupied Lake Superior? How far south did this glacier advance?

8. Study the color patterns of the rocks in Figure 6.8c as well as the pattern of the lakes. Describe what you see. Are there dominant and secondary trends? What can you say about the style of deformation?

9. Do you live in an area that was glaciated? Describe the landscape and how it may have been modified during the glaciation. Are there glacial deposits or landforms near where you live? Describe them.

10. Can you find any glacial landforms in the area of old drift in Figure 6.10?

11. Can you name the three volcanoes seen in the distance in Figure 6.16?

12. Which of the following were glaciated?
    a. New England Appalachians
    b. Southern Appalachians
    c. Shawnee National Forest
    d. Klamath Mountains
    e. Sierra Nevada
    f. Colorado Rockies

13. Research topic: Nine climate zones were mentioned briefly at the beginning of this chapter. Research two of these and determine on what basis they are defined and how one is different from the other.

14. Research topic: Find the average temperature and precipitation for mid-August and mid-January in (or near) the town where you live and compare them with those in Seattle, Minneapolis, and Boston. Speculate as to why they are similar or different.

15. Research topic: What would you expect unstratified drift to look like? How would it be different from stratified drift?

16. Research topic: Describe how moraines, kettles, kames, eskers, and drumlins form.

17. Research topic: Describe how cirques, tarns, arêtes, and horns form.

18. Research topic: Research the history of Lake Lahontan or Lake Bonneville. When did they exist? What happened to them? Did they cause floods?

19. Research topic: A large lake once existed in Death Valley. What was the name of this lake? When did it exist? What happened to it? Did it cause any floods?

20. The Appendix contains uncolored versions of Figures 6.1 (the climate map) and 6.7 (the glacial map). Color each of these and provide a one- or two-sentence description of each climate and glacial feature.

Forcing Variables: Sea Level and Isostasy

The final two forcing variables are sea level change and isostasy. Both result in broad vertical displacement of land, and both occur in response to tectonic activity and climate-induced surface processes. This chapter discusses the cause of sea level change and isostasy and their effect on landscape.

SEA-LEVEL CHANGES

Global sea level is dependent on the total volume (area and depth) of the Earth's ocean basins, the amount of water available on Earth, and the temperature of the ocean. All have varied throughout geologic time. Sea level, therefore, has also varied. Ocean basins open via rifting and close via subduction. Old ocean basins are typically deeper than young ones and therefore hold more water. If tectonic movement results in a decrease in the volume of ocean basins, seawater will be displaced onto continents and global sea level will rise. The opposite occurs if ocean basins become larger or deeper over time. Tectonic plates move at rates that vary between about 11 and 125 miles per million years (6 and 66 feet per 100 years). This rate of movement has had huge consequences on sea level over the entire history of Earth; however, because the movement is relatively slow, the changing size of ocean basins has had only a minor effect on today's landscape. Climate, and particularly the size and thickness of glaciers, strongly influences the amount of available water on Earth. Generally, we can say that when glaciers are few, sea level is high. When glaciers are plenty, sea level is low.

Measuring changes in sea level is not straightforward because, in addition to actual changes in sea level, one must also account for vertical changes in land (uplift/subsidence). Relative sea-level change measured at any one location over a period of time is equivalent to the sum effect of sea-level change plus the vertical displacement of land. One way to mitigate the effect of vertical land displacement is to map the distribution and age of marine sediments (e.g., rocks deposited from ocean water). By studying and correlating the distribution of marine sedimentary sequences worldwide, it is possible, to a first approximation, to remove the effects of vertical land displacement and study long-term eustatic (global) sea-level changes. Such measurements extend back hundreds of millions of years. However, to understand the effect of sea-level change on landscape, we need more precise estimates, especially over the past several million years.

Estimates of sea-level change over the past several hundred thousand to several tens of millions of years can be accomplished using a variety of relatively new dating techniques developed over the past 30 years or so, some of which have made it possible to directly date marine sediments. One relatively new technique employs radioactive uranium (U). Uranium is dissolved in seawater, allowing marine animals, particularly coral, to extract U and incorporate it in their shell. By analyzing the parent/daughter ratios of certain decay products of U, it is possible to date marine sediments as far back as 500,000 years. Another method employs oxygen isotopes to determine the growth and melting history of continental glaciers. Glacial ice is enriched in ^{16}O, which is a light oxygen isotope. As glaciers increase in volume, ocean water becomes enriched in the heavy isotope (^{18}O). These values are preserved in the shells of foraminifera, which are microscopic sea organisms that drift in the ocean and subsequently are deposited in marine sediment. This technique can be used as far back in time as unaltered foraminifera are found which can be several tens of millions of years. These same oxygen isotope measurements are also garnered directly from ice cores but only as far back as drills can reach, which is less than 750,000 years.

Based on the location, age, and continuity of marine sedimentary rocks, we can say with a fair degree of certainty that sea level has been higher than present-day levels throughout much of the past 500 million years. Unfortunately, measurement of the exact magnitude of sea-level rise and fall remains elusive. Recent studies suggest that between 100 and 34 million years ago, sea level varied from about 165 feet to more than 720 feet above present-day levels, with many oscillations. Climate was generally hotter throughout this period, with few, if any, glaciers. Sea level was high enough to drown part or nearly all of the Coastal Plain. Climate began to cool beginning about 34 million years ago resulting in glaciers becoming more common and better established worldwide, particularly on Antarctica. With the appearance of glaciers, sea level began to fall to its present level, but, again, with many oscillations. Some of the largest oscillations in sea level have occurred in the past 2.4 million years in conjunction with the glacial advances and retreats during the Ice Age.

Figure 7.1 shows areas of the US that are inferred to have been below sea level during three high sea-level stands approximately 80, 50, and 15 million years ago based on the distribution of marine sedimentary rock. Do not infer from this sequence that sea level has steadily lowered over this time span; that is definitely not what happened. Note the presence, 80 million years ago, of a large inland sea that covered the most of the Great Plains, the Colorado Plateau, and the Middle and Southern Rocky Mountains. This was

the last inland sea to have inundated the US. The Coastal Plain, on the other hand, has been a lowland throughout its entire existence and has been frequently inundated during high sea-level stands. Much of Florida was below sea level just 3 million years ago. Sea-level changes on the West Coast are not shown in Figure 7.1 because of ongoing mountain building and tectonic modification. The West Coast consists of accreted terranes, and some areas of California, Oregon, and Washington did not exist as a part of the North American continent prior to 50 million years ago.

The rock record indicates that sea level was high circa 120,000 years ago during a rather warm interglacial stage that lasted approximately from 130,000 to 115,000 years ago. During this time interval, oxygen isotope and sediment studies indicate that mean global surface temperatures were at least 2°C warmer than present, and sea level was 13 to 20 feet (4 to 6 m) higher than today covering part of the Coastal Plain including all of southern Florida south of Palm Beach.

The circa 120,000-year-old rise in sea level can be contrasted directly with a time, only 18,000 years ago, when glaciers covered three times as much land area as they do today and sea level was about 400 feet lower than it is today, exposing nearly all of the Atlantic continental shelf (refer back to Figure 1.7). It was at this time that Florida was about

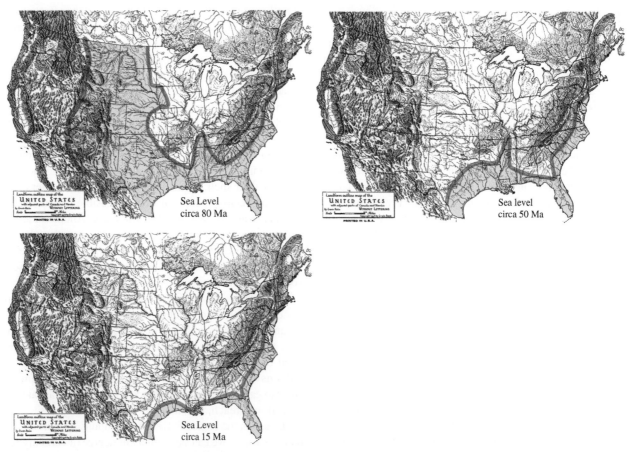

FIGURE 7.1   Sea level 80, 50, and 15 million years ago.

twice the size it is today and New York City was 100 miles from the coastline. Sea level has been on the rise for the past 18,000 years due, not only to glacial melting, but also to thermal expansion of ocean water as the climate warms.

The amount of continental drowning during a rise in sea level is dependent on the mean elevation of continents, the size of low-lying coastal areas, and the magnitude of uplift or subsidence along the coast. The Atlantic and Gulf coastal areas are low-lying, wide, and generally subsiding. The result is substantial drowning. Much of the Pacific seaboard, by contrast, is rising faster than the worldwide rise of sea level and, as a consequence, is currently rising out of the sea. The Pacific continental shelf was deeply incised by rivers during the height of glaciation, but because the shelf is so narrow, the location of the coastline did not change appreciably.

A total sea level rise of 400 feet implies an average rise of more than 26 inches per 100 years over the entire 18,000-year period. The highest rates of more than 3 feet per 100 years occurred between about 15,000 and 6,000 years ago when Earth temperatures periodically rose above present-day temperatures and when there was considerably more glacial ice on continents to melt than there is today. Rates, at times, could have approached 15 feet per 100 years. It is lucky for us that sea level is not rising at such a high rate today.

Nearly the entire North American continental ice sheet had melted by 6,000 years ago, and since that time climate has been rather steady and sea level has risen an average of 2 to 4 inches per 100 years. However, the rise in sea level has accelerated over the past 150 years. Sea level is estimated to have risen 7 to 8 inches during the 20th century. Available climate data suggest that the past three decades have been the warmest in at least the last 1,000 years. Warming will accelerate glacial melting and thermal expansion, and as a result, sea level is expected to rise a total of 15 to 20 inches during the 21st century, with extreme predictions suggesting 3 feet or more. Sea level is currently rising at a rate of 11 to 13 inches per 100 yrs. If the Greenland Ice Sheet were to melt completely, sea level would rise by about 23 feet. This is more than enough to drown all the world's coastal cities and cause severe inland flooding. About 90% of the world's ice is contained in the Antarctic Ice Sheet. Sea level would rise another 200 feet or so if this ice sheet were to melt. Such a catastrophic rise would drown nearly all of the Coastal Plain.

RIVER RESPONSE TO SEA-LEVEL CHANGES

An important characteristic of a river is its ability to cut downward (incise) into bedrock and form a narrow V-shaped valley. As downcutting occurs, the river channel progressively reaches lower elevations. The lowest elevation that a river channel can erode is referred to as *ultimate base level*. For rivers that empty into the ocean, this elevation is equal to sea level. For most rivers, including those that empty into an ocean, there are local base levels along the river where

the gradient (slope) decreases to nearly zero. *Local base level* represents the lowest elevation that the river upstream from that point can erode. Situations that produce local base level include an upstream lake or dam; a landslide; a location where the river flows over especially hard, flat rock; or anyplace where the steepness of the river gradient becomes markedly lower. Local base levels are temporary, and once they are removed by erosion, a stream can cut downward as far as the next local base level downstream. Figure 7.2 is a longitudinal profile of a river with both local and ultimate base level. A longitudinal profile shows the elevation of a river as if you are walking along the river bottom beginning at its head (where the river first forms) and ending at its mouth (where the river empties into the ocean).

Rivers also have the ability to deposit sediment in their channel and thereby raise their elevation. Whether a river erodes its channel or deposits sediment depends primarily on three factors: base level, the amount of water passing a certain point in the river (its discharge), and the amount of sediment supplied to the river. In general, an increase in water discharge favors erosion whereas an increase in sediment supply favors deposition. A river high above base level would have significant downcutting power, which would favor erosion. A sudden drop in base level, such as the sudden draining of the lake shown in Figure 7.2, would significantly increase downcutting power upstream from the lake which would favor erosion. Because ultimate base level coincides with sea level, any change in sea level will result in a change in the dynamics of all river systems that empty directly into the ocean. Thus, changes in sea level not only profoundly affect coastal landscapes, they also have a major effect on inland landscapes.

In theory, a river will develop a relatively smooth concave-upward longitudinal profile that is steep near its head and flat near its mouth so that it merges gently with the ocean. Under such perfect conditions there are no local base levels. A smooth concave-upward longitudinal profile of a river is shown in Figure 7.3a. Because zero elevation begins at mean sea level, a rapid drop in sea level is equivalent (from the perspective of the river) with a global rise of all land area. Such a possibility would perturb the smooth concave-up profile of a river that empties into an ocean and

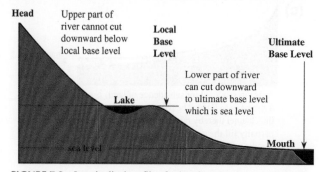

FIGURE 7.2 Longitudinal profile of a river from head to mouth showing local and ultimate base level.

likely result in downcutting and erosion of the stream channel. Much of the downcutting would occur near its mouth so that the river could again enter the sea smoothly at the lower elevation. In this scenario, the landscape could potentially evolve from a relatively flat plain, to a dissected, canyon-filled, plateau. This situation is shown in Figure 7.3b.

A rise in global sea level is equivalent with worldwide subsidence of land. This would have the opposite effect and likely result in deposition and meandering as stream channels attempt to remain at or above sea level. As the river channel fills with sediment, the channel itself becomes smaller and the amount of water the channel can hold becomes less. The displaced river water will cause larger and more frequent inland floods. Such a situation could materialize in the US if worldwide glaciers continue to melt. This situation is shown in Figure 7.3c.

We know that sea level has varied by as much as 400 feet during the past 2.4 million years as a result of glacial advances and retreats. Some of this variation can be seen along the lower (southern) Mississippi River in the form of poorly preserved terraces above the

present-day floodplain. A *river terrace* is a fragment of a former river floodplain that now stands above the level of the present-day floodplain. Low sea-level stands during glacial advances have resulted in downcutting, entrenchment of the river, and the development of river terraces. Conversely, high sea-level stands during interglacial times have caused the Mississippi River to back up, which has resulted in channel filling, flooding, and deposition, thus drowning and destroying elevated river terraces. Our most recent history (the past 18,000 years) has been one of rising sea level, channel filling, and presumably more frequent flooding. The rise, however, has not been high enough to drown all the previously formed river terraces. In other words, river level was once higher than it is today.

ISOSTASY AND ISOSTATIC EQUILIBRIUM

The term *isostasy* is from the Greek *isos,* which means *equal,* and *stasis,* which means *standing.* It is the rising or settling of a portion of the Earth's lithosphere that occurs when weight is removed or added in order to maintain equilibrium between buoyancy forces that push the lithosphere upward and gravity forces that pull the lithosphere downward. When these two forces balance, the lithosphere is said to be at *isostatic equilibrium,* which implies that it is in gravitational equilibrium such that the lithosphere literally floats on the underlying weak, malleable asthenosphere. The concept is similar to the way a boat behaves in water (Figure 7.4). An empty boat will weigh less and be less dense than the same boat loaded with cargo. It will float high on the water. However, as the boat is loaded with cargo it will become dense and heavy. It sinks slightly into the water to compensate. Unload the boat and it rises again. The boat maintains isostatic equilibrium with the water. In the same manner, most of the Earth's surface is at or close to isostatic equilibrium with the underlying asthenosphere.

Isostatic compensation of the Earth's lithosphere does not occur instantaneously as it does with a boat in water, but it does occur rather quickly, within a year or so. For example, measurable subsidence occurred around Lake Mead within a few years following the building of Hoover Dam.

The weight of a glacier sitting on a continent is enough to cause several hundred to several thousand feet of isostatic subsidence over the course of a few thousand years. The Antarctic Ice Sheet has depressed the crust an estimated 3,000 feet. The lithosphere below the Hudson Bay region in central Canada is understood to have been depressed more than 1,300 feet during the height of the glaciation 18,000 years ago. The removal of glaciers since about 8,000 years ago has resulted in rebound (isostatic uplift), which, in turn, has caused the shoreline of Hudson Bay to retreat, leaving a series of elevated shoreline terraces that can be dated to determine the

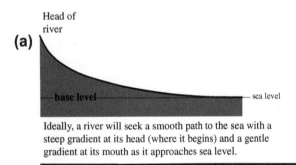

Ideally, a river will seek a smooth path to the sea with a steep gradient at its head (where it begins) and a gentle gradient at its mouth as it approaches sea level.

A drop in sea level (or uplift of land) is equivalent with a drop in base level. The river reacts by eroding downward possibly creating a canyon.

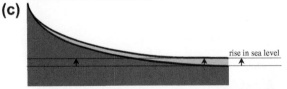

A rise in sea level (or subsidence of land) is equivalent with a rise of base level. The river reacts by depositing material in the channel thereby raising river elevation. Deposition will partially fill the river channel creating the possibility of flooding. The mouth of the river may be inundated by the ocean.

FIGURE 7.3　(a) Longitudinal profile of a river that shows, (b) a drop in sea level (ultimate base level), and (c) a rise in sea level.

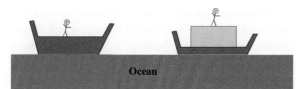

A boat with no cargo floats high on the water. The same boat loaded with cargo is heavy and sinks deeper in the water. Both boats are in isostatic equilibrium with the water.

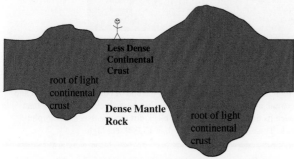

A deep root of continental crust is more buoyant because it displaces a greater amount of dense mantle rock. As a result, the continental crust floats higher on the asthenosphere to form high mountains. As the mountain is eroded, the crustal root isostatically uplifts to compensate for the loss of weight. For every 5 feet of erosion, the crustal root (and the mountain) will isostatically uplift approximately 4 feet. Isostatic uplift continues until the crustal root is gone and the mountain is reduced to a nearly level plain at an elevation similar to the surrounding area. Isostatic equilibrium is maintained throughout the erosional process.

FIGURE 7.4 Examples of isostatic equilibrium.

rate of uplift in a manner similar to dating strandline terraces along a river (refer back to Figure 4.3). The analysis suggests that at least 935 feet of isostatic uplift has occurred along the Hudson Bay shoreline over the past 8,000 years. This is an amazingly high overall average uplift rate of almost 12 feet per 100 years. Even more amazing is that the rate must have been much higher when isostatic uplift first began because rates will slow as the land area approaches equilibrium. It is estimated that uplift rates were initially as high as 39 feet per 100 years. Today, the Hudson Bay shoreline continues to rise at rates of up to 4.3 feet per 100 years. In this case, we see that several thousand years are required for the lithosphere to completely rebound to its original equilibrium elevation. We noted earlier that, because the glacier was thicker in the north, the Great Lakes are rising faster in the north than in the south and that, in time, some of the Great Lakes may spill into the Mississippi River valley.

A large delta such as the Mississippi River delta in Louisiana can deposit such an enormous amount of sediment that the weight causes the continental margin to slowly subside. This keeps the delta below sea level such that more sediment can be deposited. Deposition coupled with slow isostatic subsidence will result in the accumulation of many thousands of feet of sediment in shallow offshore waters. This is the current situation along the Atlantic and Gulf coast shorelines.

Isostatic compensation also occurs below volcanoes. The island of Hawaii is a massive volcanic pile. Figure 7.5 shows how the weight of the pile isostatically depresses the sea floor below it, creating a moat around the island. When measured from the bottom of the ocean to the summit of Hawaii's highest point (Mauna Kea), the Hawaiian island is about 33,476 feet high. On this basis, Mauna Kea is the tallest structure on Earth.

If we were to remove all ocean water from the surface of Earth, we would see an obvious difference in height between continental and oceanic crust. Isostasy explains this topographic difference. Continental crust is primarily granitic in composition and less dense than oceanic crust which is primarily basaltic in composition. Similar to a boat in water, the lighter continental crust reaches isostatic equilibrium higher on the asthenosphere than the more dense oceanic crust. A graph that shows the distribution of elevation on the Earth's surface is presented in Figure 7.6. The graph clearly shows two distinct highs. Continental crust is mostly between sea level and 1 kilometer above sea level (0 and 3,280 feet). Oceanic crust is mostly between 4 and 5 kilometers below sea level (13,120 and 16,400 feet). This bimodal distribution of the Earth's surface elevation is a

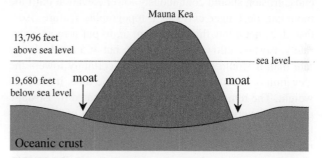

FIGURE 7.5 Isostatic depression of oceanic crust surrounding Hawaii.

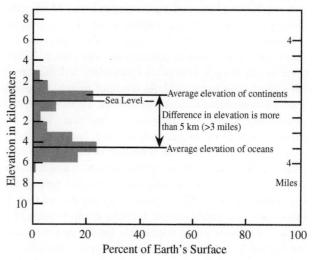

FIGURE 7.6 A graph that shows the distribution of elevation at the Earth's surface. Based on Moores and Twiss (1995, p. 8).

direct result of the density difference between oceanic and continental lithosphere. Oceanic lithosphere can become so heavy that it will literally sink (subduct) into the asthenosphere, producing a subduction zone. Sinking is facilitated where two tectonic plates are colliding. In this case, buoyancy contrasts dictate that the heavy oceanic lithosphere will subduct below the lighter continental lithosphere.

The existence of high mountains is broadly attributed to the convergence of tectonic plates, a process that drives tectonic uplift. Isostasy also causes uplift and is responsible, in many respects, for the preservation of high mountains long after tectonic uplift has ceased. The normal thickness of continental crust is about 25 miles. During tectonic plate convergence the continental crust is thickened to between 30 and 55 miles. Crustal thickening below the mountain occurs in the form of a root of relatively light continental crust that sticks downward into the mantle, much like the root of an iceberg and similar to what is shown in Figure 7.4. The crustal root replaces surrounding heavier mantle, and as a result, the land rises to maintain isostatic equilibrium. It is this isostatic rise that, in part, creates the mountain in the first place even during tectonic collision. The area will obviously remain mountainous for as long as the collisional process continues. Once the collision process ends, erosion should continuously lower elevation until the mountain no longer exists as a topographic feature. Isostasy does not allow this to happen; or, to put it more accurately, isostasy allows this to happen but at a much slower rate than it would otherwise. Erosion certainly lowers the elevation of the mountain but, at the same time, it removes weight. The removal of weight is compensated by isostatic uplift of the crustal root, which causes uplift of the mountain. Isostatic uplift will continue until the crustal root is reduced to a normal crustal thickness, at which point the mountain would likely be reduced to a nearly flat plain at an elevation similar with surrounding areas. The situation is very much like a large iceberg floating in water. Most of the iceberg forms a root below surface water. As ice (weight) is removed from the top of the iceberg, the root below the water surface will rise to compensate. The root will continue to rise until it disappears completely and the top of the iceberg is at water level. As a rule of thumb, for every five feet of elevation removed by erosion, the mountain will isostatically uplift by approximately four feet. Thus, the overall height of the mountain is lowered, but the lowering is at a much slower rate than if erosion alone were the only factor. An important outcome of this process is that it results in the exhumation and surface exposure of once deeply buried crystalline rock. This explains why crystalline rock is often found on mountaintops. Isostatic compensation of a thickened crust is partly responsible for the persistence of the Appalachian Mountains 265 million years after tectonic collision had ended, and why once deeply buried crystalline rock now makes up most of the mountain belt. The removal

of the lithospheric root of a mountain during erosion and isostatic uplift, and the erosional exhumation of deeply buried rock, are shown in Figure 7.7. The entire process

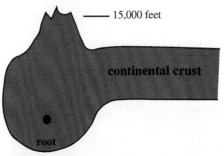

Mountain range immediately after the end of continental collision and tectonic uplift. The continetal crust has been thickened to more than twice its normal thickness. The black dot represents a metamorphic rock. The mountains are more than 15,000 feet high.

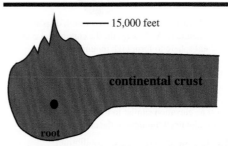

Erosion without tectonic uplift reduces the weight of continental crust in the mountain range which results in isostatic uplift. For every five feet of erosion, the mountain isostatically uplifts about four feet. Erosion occurs primarily in valleys so, if isolated peaks are not eroded, it is possible for them to rise higher than they were prior to cessation of tectonic uplift. This is a short-term phenomenon that lasts only until the mountain top is eroded. The metamorphic rock (the black dot) has exhumed toward the surface as a result of erosion and uplift.

The mountain root has almost disappeared and mountain elevations have been reduced to low-lying hills. The metamorphic rock has risen closer to the surface.

The moutain root is completely removed and the mountain has been reduced to a flat plain at about the same elevation as adjacent land. The rate of erosion becomes very slow as does any isostatic adjustment. The metamorphic rock has been exhumed to the surface.

FIGURE 7.7   Isostatic compensation following a mountain-building event.

is possible because isostatic adjustment is rapid enough to maintain isostatic equilibrium during erosion.

Another example of isostasy has to do with river incision and the focused removal of rock from a river canyon. Such a situation could potentially cause the lithosphere directly below the canyon to isostatically rise to a greater extent than areas surrounding the canyon where bedrock has not been removed. The result could be the formation of a broad anticline directly below the canyon. In this case, the anticline is a product of focused erosion in which formation of the river valley precedes development of the anticline. Here we have a situation in which erosion causes isostatic uplift and deformation.

A superb example of an anticline that follows the path of a river canyon is the Meander Anticline, which precisely follows the Colorado River for about 25 miles in the vicinity of the Needles at the confluence of the Colorado and Green Rivers in Utah. In this case, the river flows across an area where low-density evaporite (salt) beds are present at depth. The unloading of overlying rock has caused the lighter evaporite rock to rise toward the surface and, in doing so, has arched overlying beds at the rim of the canyon into an anticline. Part of the Meander Anticline is shown in Figure 7.8. Note that the layers are flexed upward ever so slightly as they cross the river.

Even the presence of a rain shadow can influence isostasy. A rain shadow will produce abundant precipitation on one side of a mountain relative to the other which could result in differences in rates of erosion. A greater amount of erosion on the wet side of a mountain will result in higher rates of isostatic uplift and steeper slopes relative to the dry side.

Thus we see that isostasy can affect landscapes on time scales of a few years to millions of years and that the rate of isostatic uplift or subsidence can be remarkably high over a short interval of time. The accumulation of ice, volcanic rock, or sediment will cause isostatic subsidence, whereas the melting of a glacier, the removal of rock by erosion, and the replacement of heavy mantle rock with a root of light continental crustal rock will cause land areas to isostatically rise. We have also seen that focused erosion along a deeply incising river valley can cause not only isostatic uplift, but also deformation processes such as the folding of sedimentary layers.

Differences in density between two adjacent rock types, and the addition or removal of weight, are two factors that affect isostatic equilibrium. Another factor that affects isostasy is the thermal condition of the lithosphere (the crust and the solid upper mantle). Elevated temperature in the

FIGURE 7.8 A Google Earth image looking NNE along the Colorado River just north of the confluence with the Green River on the Colorado Plateau. The sedimentary layers are flexed upward slightly where they cross the river as part of the Meander Anticline. The La Sal Mountains form the skyline at upper right.

lithosphere causes thermal expansion which lowers density. The lithosphere responds by uplifting. Conversely, slow cooling of the lithosphere will result in thermal contraction, higher density, and slow subsidence. We can refer to temperature-controlled isostasy as *thermal isostasy* or *thermal uplift* (subsidence). Thermal isostastic uplift explains the abnormally high average elevation of the Basin and Range (4,600 feet) even though the region is underlain by thin crust. Thermal isotactic buoyancy also helps explain the height of the Colorado Rocky Mountains whose crustal thickness is less than the adjacent, and much lower, Great Plains. Thermal isostasy explains relatively high elevation at Yellowstone National Park and even the presence of the Mid-Atlantic ridge, which is a broad upwarp more than 1,500 miles wide and 8,000 feet above the surrounding ocean bottom (Figure 5.3). In both cases, warm, rising magma has resulted in thermal isostatic uplift.

TECTONIC VERSUS ISOSTATIC UPLIFT/SUBSIDENCE

Because uplift and subsidence are functions of both tectonic compression and isostatic compensation, it is important to distinguish between the two. We can refer to one as tectonic uplift (or subsidence) and the other as isostatic uplift (or if necessary, thermal isostatic uplift). It is sometimes difficult to distinguish tectonic from isostatic uplift, especially when both are occurring at the same time. There are, however, significant differences in their mode of uplift, their sustainability, and their final outcome.

Tectonic uplift (or subsidence) is often focused along fault zones or fold crests and can be sustained episodically over the course of hundreds of thousands to tens of millions of years. Tectonic uplift is associated with strong deformation of rock, and because uplift can be an order of magnitude faster than erosion, it can result in the creation of high mountain ranges. Sustained uplift rates of 4 to 10 inches per 100 years will occur in pulses associated with earthquakes. There could conceivably be zero tectonic uplift between earthquakes. Uplift rates, therefore, must be averaged over many thousands to millions of years.

Isostatic uplift (or subsidence) is primarily regional in the sense that it is not normally focused or localized along fault zones but instead results in broad warping of a wide area. For example, focused erosion within a river valley will result in broad isostatic uplift of the entire area that surrounds the river valley. This form of uplift is not capable of producing strong deformation in rock.

Isostatic uplift can occur at very high rates, on the order of feet per 100 years, but such rates cannot be sustained long enough to create high mountains. Once an isostatic imbalance presents itself, isostatic compensation will occur on a time scale of years and will continue for only as long as the isostatic imbalance is present.

In contrast to periodic, earthquake-generated pulses of tectonic uplift, isostatic compensation is a continuous process. Continuous erosion of a mountain will drive continuous isostatic uplift. In the absence of tectonic uplift, the end result of coupled erosion and isostatic uplift is an overall lowering of mountain elevation because isostatic uplift compensates for only four-fifths of erosional lowering. Isostasy can sustain a high mountain range built via tectonic processes for a much longer time period than otherwise might be expected. However, the end result of coupled erosion and isostatic uplift will be a flatland in which crustal thickness is equal to the crustal thickness of surrounding areas (Figure 7.7). In other words, the end result will be the same as if erosion were acting alone. The major difference is the time involved in the overall lowering process.

In the case of thermal isostatic compensation, it is possible to create (by added heat) and maintain (by sustained heat) areas of high elevation, such as the Basin and Range, for as long as the crust and upper mantle remain abnormally warm, even if the crust is relatively thin. This process, like density-driven isostatic compensation, results in broad regional uplift of a land area and, in and of itself, is not capable of strong deformation or the production of a high mountain range.

QUESTIONS

1. How is climate different from weather?
2. Describe how the melting of the Greenland glacier might affect worldwide sea level. How might this melting affect isostatic equilibrium on Greenland?
3. Based on what you have read in this chapter, how can we be reasonably sure that the Colorado Rockies did not exist 80 million years ago?
4. How has sea level changed over the past 120 million years?
5. What are some of the highest inferred rates of sea-level rise over the past 18,000 years?
6. How might a major river near where you live be affected by rising sea level? How might the effect be different if the river emptied directly into the ocean versus directly into another river?
7. What are the significant differences between tectonic uplift and isostatic uplift in terms of their mode of uplift, sustainability, and final outcome?
8. Explain why ocean basins are, on average, more than 13,000 feet below sea level and continents are above sea level.
9. If two oceanic lithospheres were to collide, what would you have to know in order to predict which lithosphere would subduct beneath the other?
10. How soon after applying a weight to the Earth's surface would you expect isostasy to respond?

11. What is thermal isostasy? Why might the introduction of heat into the lithosphere cause uplift?

12. How could a mountain peak increase its elevation in a mountain range that is undergoing erosion but no tectonic uplift?

13. Research the Meander Anticline in Utah. What is it? Where is it exactly? How did it form?

14. Given the fact that sea level has risen about 400 feet in the past 18,000 years, how much higher could it rise if all of the world's glaciers were to melt?

15. Explain why isostasy allows mountains to maintain elevation over a much longer time period than might otherwise be expected.

16. Go to http://geology.com/sea-level-rise/ and describe how a sea-level rise of between 1 and 100 feet might affect Florida and the Chesapeake Bay area. Do the same for other areas of your choice.

17. Provide some examples of isostasy.

18. What variables must be taken into account when studying global sea level change over geologic time?

19. Why might you expect old ocean basins to be deeper than young ones?

20. What is the difference between local and ultimate base level?

21. What situations produce local base levels?

22. What role does isostatic adjustment play in mountain building?

23. How can a rain shadow influence isostasy?

24. Explain thermal isostatic uplift.

Chapter 8

Interaction of Tectonics, Climate, and Time

Chapter Outline

Time plays a major role in the ultimate shape of land. An important consideration is the amount of time during which each forcing variable acts at a steady rate. If all four variables remain at a steady rate over a long period of time, or if the rate of change varies slowly over time, then the landscape tends toward equilibrium with these conditions such that modification occurs in a predictable way. This is not to say that no changes occur. Only that change is slow enough or predicable enough so that the landscape remains recognizable. For example, if the rate of uplift of a mountain equals the rate of erosion, the mountain is in a steady state and will neither grow larger nor diminish in size because, on average, for every inch of uplift there will be an inch of erosion. The mountain will maintain its youthful look.

In another example, a tectonically quiet area could, in a sense, grow old over time because, without active tectonic forcing, there will be no significant uplift, subsidence, or volcanism (only minor changes due to isostasy and sea-level change). We are left with climate-controlled erosion and deposition as the primary mechanisms of change. Under these conditions, the landscape will lose relief over time as high areas are eroded and low areas undergo deposition. The change would be slow and predicable. The Southern Appalachians have undergone erosion and deposition without strong tectonic activity over the past 265 million years. During that time, and despite periods of isostatic uplift, sea-level oscillation, and climate change, the only major change to the landscape has been lower and less dramatic relief. The mountains are evolving slowly and predictably. They are growing old.

If, on the other hand, there is rapid change in one or more of the variables, such as a shift to active tectonism, a change in climate, or the erosional uncovering of a different rock type or rock structure, then rates of uplift, subsidence, erosion, deposition, and even volcanism could all change. The landscape, which had adjusted to the previous equilibrium condition, is now out of equilibrium and will begin a relatively rapid and profound change toward a new equilibrium shape that potentially is unrecognizable relative to its previous state. This, in effect, is reincarnation of land. A period of sustained uplift driven by renewed tectonic activity could, for example, cause a low-lying plain with lazy, meandering rivers to evolve into a high plateau with deep river canyons. Rivers on the Colorado Plateau are an example.

Figure 8.1 compares lazy river meanders on the Mississippi River in Missouri with deeply entrenched meanders on the Green River in Utah. It is likely that meanders on the Green River were established at a time of low relief on the Colorado Plateau and were entrenched during renewed uplift or base-level lowering. Entrenchment of meanders is an example of landscape reincarnation. There are many areas in the Cordillera that have completely reincarnated themselves in the past 20 million years due primarily to renewed tectonic activity and volcanism. Much of this activity is so young that relicts of earlier landscapes are preserved.

In this final chapter of Part I, we discuss how tectonics, climate, and time interact to shape two end-member types of landscape: a structure-controlled landscape, and an erosion-controlled landscape. We then divide the US into nine structural provinces, each of which is discussed in Part II.

STRUCTURE-CONTROLLED VERSUS EROSION-CONTROLLED LANDSCAPES

Active tectonics refers to any form of folding, faulting, earthquake, plutonic intrusion, volcanism, metamorphism, uplift, or subsidence that is occurring today as a result of horizontal stress imparted on rock by the interaction of moving tectonic plates. Because tectonic activity produces most of the folds and faults evident in rock,

FIGURE 8.1 Comparison between river meanders on the Coastal Plain with those on the tectonically active Colorado Plateau. (a) A Google Earth image looking east at meander bends on the Mississippi River at New Madrid, Missouri. River elevation is between 265 and 290 feet which is slightly lower than the surrounding farmland which is at 275 to 300 feet. Reelfoot Lake was formed or enlarged during a series of three major (and now famous) earthquakes that occurred between December 16, 1811, and February 7, 1812. The earthquakes were responsible for as much as 20 feet of vertical displacement of surrounding land areas. (b) A Google Earth image looking southwestward at entrenched canyon meanders on the Green River just north of Canyonlands National Park. The river is at about 3,958 feet. The abandoned meander at the center of the photo is between 4,050 and 4,200 feet. The canyon rim is between 4,750 and 5,000 feet. Spring Canyon is at lower right, Hell Roaring Canyon is at lower left, and Horseshoe Canyon is at upper center where Barrier Creek occupies part of the abandoned meander.

(a)
<u>Structure-Controlled Landscape (no erosion)</u>

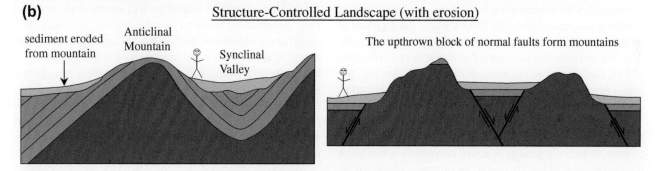

Anticlinal Mountain

Synclinal Valley

The upthrown block of normal faults form mountains

exposed fault surface

(b)
<u>Structure-Controlled Landscape (with erosion)</u>

sediment eroded from mountain

Anticlinal Mountain

Synclinal Valley

The upthrown block of normal faults form mountains

(c)
<u>Erosion-Controlled Landscape</u>

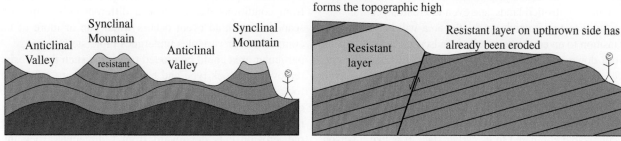

Synclinal Mountain

Anticlinal Valley

Anticlinal Valley

Synclinal Mountain

resistant

Resistant layer on the down-thrown side of fault forms the topographic high

Resistant layer

Resistant layer on upthrown side has already been eroded

FIGURE 8.2   Structure-controlled and erosion-controlled landscapes.

and because tectonic uplift and volcanic accumulation can initially be more than 10 times faster than erosion, any form of active tectonics can, relatively quickly (within a few million years), destroy an existing landscape and create an entirely new one—a process referred to in this book as *reincarnation*. Volcanism, in particular, can reincarnate a landscape possibly within a human lifetime.

Because the rate of erosion is initially low relative to the rate of uplift, the newly developed active tectonic landscape will closely mimic the underlying structure that is developing in the rock. If the structure consists of folds, the landscape will consist of anticlinal mountains and synclinal valleys; if the structure consists of faults, mountains will coincide with uplifted sections along the fault,

and valleys with down-dropped sections. We will refer to a landscape that mimics the structure of underlying rock as a *structure-controlled landscape*, even if the correlation between structure and topography is not an exact match. Figure 8.2a shows ideal structure-controlled landscapes composed of folds (left) and normal faults (right) without any erosion. In this instance the landscape exactly replicates the underlying structure of the rock. Figure 8.2b shows a structure-controlled landscape where, in spite of erosion, the landscape continues to closely mimic the shape of the underlying structure.

A major difference between a tectonically induced mechanism such as uplift and a climatically induced mechanism such as erosion is that the tectonic force will, at some point in time, end. Erosion, on the other hand, will continue although the rates of erosion may change. If we allow erosion to proceed without tectonic activity, we would expect the effect of erosion to be most intense on steep slopes at high elevation. Erosion would remove material from high areas and deposit material in low areas such that the overall effect would be for relief to decrease. The end result of erosion is the development of a relatively flat surface. However, the intensity of erosion at any one location is dependent on the hardness of rock. Therefore, an equally possible result would be a landscape that maintains some relief in which topographically high areas correspond with hard rock, and low areas with soft rock. This will occur regardless of the structure of the rock potentially creating topographic highs that do not correspond with structural highs such as anticlinal valleys, synclinal mountains, and fault blocks where the highland is on the downthrown side. Such topography is shown in Figure 8.2c. We can refer to a landscape that is primarily a reflection of rock hardness as an *erosion-controlled landscape* (or *erosional landscape*) even if the structure continues to impart at least partial influence on the landscape.

Structure- and erosion-controlled landscapes are the result of the competition between tectonic activity and climate. If tectonic activity is active or recently active, then the landscape will tend to mimic the structure produced by the tectonics simply because erosion has not yet had enough time to modify the newly emerging landscape. The result of active tectonics is to continually reshape the land surface to resemble the underlying structure thereby maintaining a structure-controlled landscape even as erosion occurs.

Once tectonic activity ends, the area will begin its long transition to an erosion-controlled landscape. As landscape is worn down, the rate of erosion begins to slow, and the rate of landscape modification also slows. Landscape change becomes predictable in the sense that rates of weathering and erosion will have adjusted to the prevailing climate and rock. The landscape evolves but at the same time remains recognizable compared to its earlier state. It is like a person growing old. If you knew that person when he was young, chances are you would recognize him 10 or 20 years later. We all know that some people appear to grow old faster than others. The same is true of landscape. Some areas are worn down faster than other areas. The rate of erosion depends on a number of factors already discussed, including climate, vegetation, hardness of rock, elevation, and relief. Thus, a landscape of soft rock in a humid climate would likely grow old faster than one of hard rock in a dry climate. Similarly, an area located close to a river would likely evolve differently from one just a few miles away. We must recognize that growing old is a relative term.

It is tempting to consider structure-controlled landscapes as young, and erosion-controlled landscapes as old, but this is not always correct. These two terms should be used only to describe the physical relationship between structure and landscape without the implication of time. The presence of tectonic activity favors development of a structure-controlled landscape. The absence of tectonic activity favors development of an erosion-controlled landscape. But neither is guaranteed. There are many exceptions to the dominantly structure-controlled landscape in the tectonically young Cordillera (Figure 3.16 is an example) as well as to the dominantly erosion-controlled landscape in the tectonically old Appalachians. As we shall see, it is possible for an old, erosion-controlled landscape to slowly, and predictably, evolve into what would appear to be a structure-controlled landscape.

THRESHOLDS AND REINCARNATION

An erosion-controlled landscape will continue to grow old until disrupted by renewed tectonic activity, at which time rapid reincarnation will destroy the old landscape and create a new structure-controlled landscape. Even if the previous landscape was young and structure-controlled, any change in tectonic activity that results in changes in uplift/subsidence rates or an outpouring of volcanic lava, could result in profound landscape reincarnation. In all instances relicts of the previous landscape would likely be preserved during the reincarnation process.

Although renewed tectonic activity is an obvious method by which to reincarnate landscape, it is not the only method. An erosion-controlled landscape could be in the process of growing old when a threshold is reached. A threshold separates conditions that produce a given effect from conditions that produce a different effect. In landscape, a threshold event occurs when one or more of the components or variables undergoes rapid steady change. A particular area could be eroded over time such that erosion removes one rock type and begins eroding into a different rock type or into a different style of deformation. In this case, simple erosion could reincarnate a landscape without renewed tectonic activity, given enough time. If, for example, erosion cuts downward from sedimentary rock into crystalline rock, an anticlinal valley could evolve into an anticlinal mountain. Figure 8.3 shows a scenario in which an anticlinal mountain of sedimentary rock erodes first to a flatland, then to an anticlinal valley, then back to a flatland, then to an anticlinal mountain, and finally back to a flatland, this time composed of crystalline rock. Rather than active tectonics, this entire process simply reflects long, continuous differential erosion of soft versus hard rock in an ancient erosion-controlled landscape. A nontectonic scenario like this would likely require hundreds of millions of years to complete. The St. Francois Mountains of Missouri and the Adirondack Mountains may have progressed through the first five stages of this sequence. The process appears to have been completed in the northern crystalline part of North America which includes the Superior Upland province and areas north of the Great Lakes.

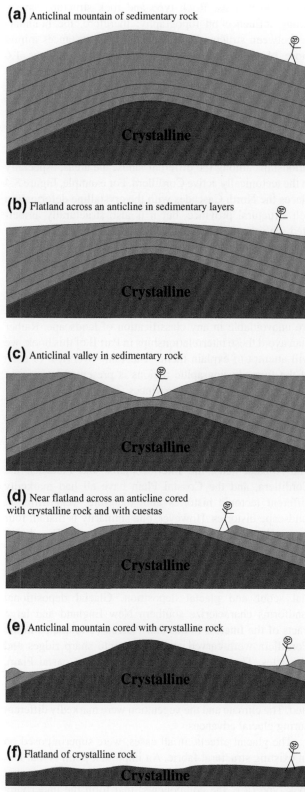

(a) Anticlinal mountain of sedimentary rock

(b) Flatland across an anticline in sedimentary layers

(c) Anticlinal valley in sedimentary rock

(d) Near flatland across an anticline cored with crystalline rock and with cuestas

(e) Anticlinal mountain cored with crystalline rock

(f) Flatland of crystalline rock

FIGURE 8.3 Evolution of an anticlinal mountain via erosion. Given a small amount of compressional stress, an additional step (not shown in the figure) would be for the crystalline flatland (bottom panel) to rise to form a domal mountain.

In another example, a rapid change in climate could change erosion/deposition rates to such an extent that landscape, which may have reached equilibrium with the earlier climate, reincarnates under new climatic conditions. The glaciation during the past 2.4 million years was a climatic threshold event that both modified and reincarnated landscape. We have already noted that the glacial landscape of the Superior Upland and Central Lowlands is some of the youngest landscape in the country. These areas have been completely reincarnated by continental glaciation creating a glacially scoured and flattened glacial erosion landscape in the north (primarily Canada), and a glacial deposition landscape to the south. Alpine glaciation has modified and reincarnated mountainous areas of the Cordillera creating many of the jagged peaks we see today. New England was modified by both continental and alpine glaciation.

One part of the country that is constantly changing and reincarnating itself as a result of nontectonic forces is our nation's shorelines. Waves and tides combine to produce one of the most powerful and relentless natural forces on Earth. The constant pounding of wave action can reshape a shoreline in one person's lifetime, particularly in areas where there is an abundance of unconsolidated material or soft sedimentary rock. Hurricanes can reshape a shoreline in a matter of hours. The landscape of the Coastal Plain, in particular, has undergone tremendous change in only the past 18,000 years as a result of glacial melting and a 400-foot rise in sea level.

To summarize, the term *active tectonics* refers to the presence of an active horizontal stress in the rock large enough to cause folding, faulting, earthquake, plutonic intrusion, volcanism, metamorphism, uplift, or subsidence at rates that are faster than erosion. This condition results in the reincarnation of any existing landscape into a structure-controlled landscape that mimics the developing structure of rock. A structure-controlled landscape, once established, will persist for as long as the structure-producing tectonic activity is active. As the landscape grows higher, it will eventually reach a topographic steady-state with erosion. Under these conditions, the landscape will remain structure-controlled and, in a sense, forever young.

Although modified by erosion, a structure-controlled landscape will persist for a certain amount of time even after tectonic activity ends. The amount of time could be more than 50 million years in some instances, or it could be only a few thousand years depending on climate and the distribution of hard rock. Eventually, however, we would expect the land area to evolve into an erosion-controlled landscape of relatively low relief with high areas composed exclusively of hard rock. In this case, the landscape will evolve toward equilibrium with climate such that the transition from a structure-controlled landscape to an erosion-controlled landscape will be gradual and predictable. The landscape will, in a sense, grow old. The process of growing old may involve punctuated periods of rejuvenation (gains in elevation or relief) or erosional flattening (losses in elevation or relief) depending on a number of factors that include the strength and frequency of minor tectonic pulses, changes in climate-driven erosion rates, rapid sea-level changes, and possible thermal isostatic adjustment. These changes, particularly sea-level changes, could be responsible for 1,000 feet or more of elevation change particularly in landscapes south of the glaciation such as the Southern Appalachians, the Ozark Plateau, and the Interior Low Plateaus where there has been no strong tectonic activity for at least the past 265 million years. In these areas, landscape modification and reincarnation have been slow and primarily the result of rock thresholds where erosion uncovers rock of different hardness or different structure (Figure 8.3).

A CLASSIFICATION OF STRUCTURAL PROVINCES

A physiographic subdivision is one that separates land areas based on their topographic expression. In essence, a physiographic province is one that simply looks different from surrounding areas. In this book we have delineated 26 physiographic provinces each with a different combination of rock/sediment type, rock structure, and forcing variables. This approach allows us to look in great detail at individual provinces particularly with respect to what each province is composed of and why each looks different from surrounding provinces. It is a descriptive approach specific to each province. However, this type of approach does not lend itself very easily to a more general understanding of landscape. For example, given a different set of variables, such as young versus old tectonics, it is clear that two provinces with generally the same rock/sediment type and rock structure can produce two very different-looking landscapes. Rather than using topography as the basis for subdivision, we can instead use landscape components and tectonic setting. We can define *structural provinces* based on similar or related rock/sediment type and rock structure without regard for topography. This type of subdivision has an advantage in that it allows us to see how two landscape areas with the same rock and structural characteristics evolve when subjected to a different set of forcing variables (i. e. tectonic activity and climate).

It is possible to divide the contiguous United States into eight structural provinces as shown in Figure 8.4 and listed in Table 8.1. Unconsolidated sediment is thick enough in a few areas to completely control landscape and thus, can be considered a ninth structural province. Areas of thick unconsolidated sediment are not shown in Figure 8.4 but they include the Mississippi Embayment region, areas of glacial drift and loess in the Central Lowlands, Superior Upland, and Great Plains, and coastal areas particularly along the Atlantic Coast. In Part 2 of this book we will use

the classification of structural provinces for discussion of landscape regions. Rock type and rock structure have a strong influence on topography; therefore, the boundaries between structural provinces in some instances mimic those of physiographic provinces (compare Figures 1.6 and 8.4). Both classifications are useful in discussing individual areas; therefore, we will incorporate physiographic provinces and physiographic regions into our discussion of structural provinces where appropriate.

We must also realize that even a structural division as presented in Figure 8.4 is incomplete in describing interrelationships among the different landscape areas, especially in the tectonically active Cordillera. For example, Figure 8.4 places the North Cascades with the crystalline deformation belt structural province, but it is also structurally associated with the Cascadia volcanic arc system. Similarly, the Colorado Plateau is shown as part of the nearly flat-lying sedimentary layers structural province, but it also has strong structural ties to the Middle-Southern Rocky Mountains which are grouped with the crystalline-cored mid-continent anticlines and domes structural province. Such ambiguities are unavoidable in any classification of landscape. Rather than avoid these interrelationships in Part II of this book, we will attempt to explain them. A tectonic overview of each of the four physiographic regions is presented here as an introduction to structural provinces.

A STRUCTURAL OVERVIEW OF THE FOUR PHYSIOGRAPHIC REGIONS

The Appalachians, the Interior Plains and Plateaus, the Cordillera, and the Coastal Plain have all had markedly different tectonic histories and, therefore, very different landscape histories. However, one similarity is that all four were affected by glacial advances and retreats. The glaciation is so young that the effects it imparted on the landscape have hardly been modified. The New England Appalachians and northern areas of the US show the effects of both glacial scour and glacial deposition. Glacial depositional landforms characterize southern New England and large tracts of the Interior Plains and Plateaus. Mountains in the Cordillera were carved and shaped into sharp ridges and pinnacles. The Southern Appalachians, the Coastal Plain, and other areas south of the glacial limit were incised at times by rivers, and drowned during other times as sea level fluctuated. The climate and the vegetation were markedly different during glacial advances.

The glacial effects, in all cases, were superimposed on an existing structural fabric. An important point is that it is the rock type and the most recent structure imposed on that rock type (i. e. the components) that form the underlying character of a given landscape. Clearly, the age of landscape can be no older than the age of the rocks that form each

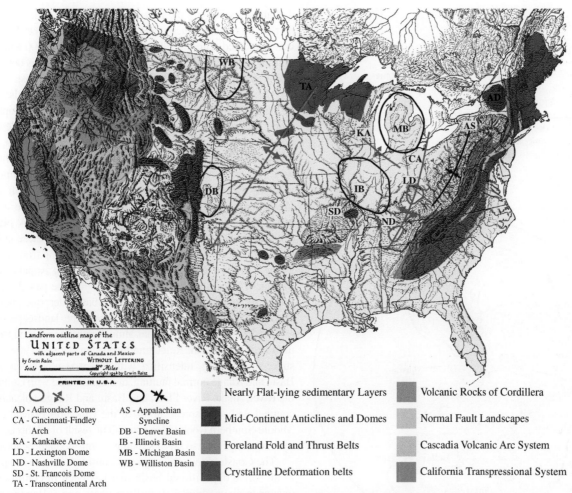

AD - Adirondack Dome
CA - Cincinnati-Findley Arch
KA - Kankakee Arch
LD - Lexington Dome
ND - Nashville Dome
SD - St. Francois Dome
TA - Transcontinental Arch

AS - Appalachian Syncline
DB - Denver Basin
IB - Illinois Basin
MB - Michigan Basin
WB - Williston Basin

Nearly Flat-lying sedimentary Layers

Mid-Continent Anticlines and Domes

Foreland Fold and Thrust Belts

Crystalline Deformation belts

Volcanic Rocks of Cordillera

Normal Fault Landscapes

Cascadia Volcanic Arc System

California Transpressional System

FIGURE 8.4 A landscape map showing structural provinces.

landscape or, with the exception of relict landscapes, the age of the rock structure that shapes each landscape. Here we present a short overview of the rocks and the structure that underlie each physiographic region.

The Appalachians are an ancient mountain belt that has not undergone reincarnation since the final phases of its formation some 265 million years ago. The mountain still retains its original rock type and rock structure from that time period. Most of the faults and folds strike NNW producing the overall topographic trend and grain of the mountain belt. Details about the formation of the Appalachian mountain belt are reserved for Part 3 of this book. For now, we can say that both the Appalachian and Ouachita belts developed over the course of several hundred million years ending 265 million years ago following collision of eastern and southeastern North America with Europe, Africa, and South America. The collision produced a supercontinent known as Pangea in which all of the world's major continents amalgamated to form a single large supercontinent. Figure 8.5 is a sketch of the Pangean supercontinent circa 265 million years ago. Note that North America lay close to the equator and was rotated clockwise from its present position. The Pangean supercontinent began to separate (rift) beginning about 230 million years ago which allowed North America to drift to its current position away from other continents, thus producing the Atlantic Ocean. Since that time the Appalachian-Ouachita mountain belt has undergone erosion with punctuated periods of isostatic and thermal uplift, but has not undergone tectonic or structural reincarnation. The landscape has grown old but, for the most part, hasn't changed its general look or features in all that time. A large part of the mountain belt was already eroded and covered with younger rock of the Coastal Plain by about 130 million years ago. Today the remaining mountain is an old erosional landscape where crystalline rock is widely exposed at the surface and where topographically high areas are composed of hard rock

TABLE 8.1 Structural Provinces
Unconsolidated sediment
Nearly flat-lying sedimentary layers
Crystalline-cored mid-continent anticlines and domes
Foreland fold and thrust belts
Crystalline deformation belts
Young volcanic rocks of the Cordillera
Normal fault-dominated landscapes
Cascadia volcanic arc system
California transpressional system

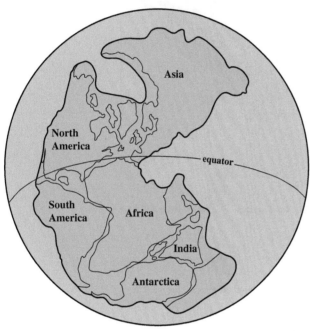

FIGURE 8.5 A sketch that shows the supercontinent Pangea circa 265 million years ago. Note that the southern part of the US was on the equator and rotated clockwise at this time. Based on Chernicoff and Whitney (2007, p. 29).

regardless of the structure. Some areas in the Blue Ridge and possibly New England may have existed continuously as a land area for more than 450 million years.

The Interior Plains and Plateaus region as a whole was not subject to strong Appalachian, Ouachita, or Cordilleran mountain-building episodes. The sedimentary rocks that form most of the province still retain a nearly flat-lying geometry. The only real structural change has been the slow development of broad domes and basins, some of which have formed over the course of several hundred million years and were actively forming during deposition of sedimentary rock. The eastern half of the region, in particular, reflects hundreds of millions of years of erosion such that, like the Appalachians, hard rock forms highlands. Much of the landscape has been above sea level and slowly growing old for at least the past 290 million years. An important exception is the glaciation that reincarnated part of

the landscape in only the past 2.4 million years. The western part of the region, which includes primarily the Great Plains, was inundated with a shallow sea between about 100 and 75 million years ago (Figure 7.1). This area, therefore, is considerably younger than the nonglacial landscape in the east.

The Cordillera is a recent to active tectonic landscape in which structurally high areas tend to be topographic highs regardless of rock type. The mountain system developed over the course of several hundred million years by collision of numerous small lithospheric islands. There was no continent-continent collision on a scale similar with the Appalachians. An ocean basin has existed off the West Coast since before the time of Pangea (Figure 8.5). Most of the collisions that formed the Cordillera took place prior to about 50 million years ago. Collectively these collisions produced a series of mountain belts extending from Mexico to Canada that probably resembled a young Appalachian Mountain belt. More recently, particularly in the past 20 million years, almost the entire Cordillera has undergone intense tectonic reincarnation to the point where the landscape that existed 50 million years ago has almost entirely vanished from the face of the Earth. Reincarnation began with San Andreas strike-slip faulting, intensified with Cascade volcanism and Basin and Range normal faulting, and culminated with Columbia River, Snake River Plain, and Basin and Range volcanism coupled with widespread uplift of the Colorado Plateau and Middle-Southern Rocky Mountains. The result of all this activity is a young, diverse set of structure-controlled landscapes.

The Coastal Plain and adjacent continental shelf are part of an active depositional basin that has been in continuous existence since about 230 million years ago following the breakup of Pangea, the opening of the Atlantic Ocean, and erosion of part of the Appalachian chain to sea level. The sedimentary rocks that form the Coastal Plain were deposited directly on top of crystalline rock that at one time formed the inner core of the Appalachian-Ouachita Mountain belt. Nearly all of the sedimentary rocks exposed today on the Coastal Plain are younger than about 130 million years. These rocks were deposited during slow subsidence of the continental shelf such that they tilt gently toward the ocean. Most of the oldest exposed rocks are inland, adjacent to the Piedmont Plateau. The youngest are adjacent to the ocean. The present-day landscape can be no older than these rocks. Sea level, at about 50 million years ago, was much higher than it is today (Figure 7.1). At that time, ocean water covered most of the Coastal Plain which implies that much of the landscape is less than 50 million years old. Parts of the Coastal Plain near the ocean last emerged from below sea level less than 3 million years ago implying a very young landscape. Sea level has certainly varied over the past 50 million years but perhaps not as much as during the glaciation of the past 2.4 million years. The most recent variation was 18,000 years ago during the height of the last glacial advance when sea level was 400 feet below present-day conditions. The rapid sea-level rise associated with the melting of glaciers has resulted in complete reincarnation of the shoreline during this short time interval.

QUESTIONS

1. What is a structure-controlled landscape?
2. What is an erosion-controlled landscape?
3. What is implied by the term *reincarnation* as used in this book?
4. What is implied by the term *growing old* as used in this book?
5. What are some expected differences between a landscape dominated by active or recent tectonics and one where tectonics have been inactive for a long time (other factors being equal)?
6. What is a meant by a landscape at equilibrium with climate?
7. What is a threshold event? Name several possible changes that could result in a threshold event.
8. As described in this book, how is a structural province different from a physiographic province?
9. What was Pangea?
10. In terms of tectonic activity, how are the Appalachian Mountains different from the Cordilleran Mountains?
11. An uncolored version of Figure 8.4 can be found in the Appendix. Photocopy and color this map. Describe each of the structural provinces and their distribution across the US.
12. How is the Appalachian physiographic region different from the Cordilleran region in terms of their tectonic and erosional history?
13. Why is the landscape in the western Interior Plains and Plateaus younger than in the east?
14. Describe the geologic history of the Coastal Plain.
15. Compare Figures 1.6 (physiographic provinces) and 8.4 (structural provinces). Which provinces from each map appear to most closely mimic each other? How is the Basin and Range physiographic province different from the Normal Fault Landscapes structural province? How is the Middle-Southern Rocky Mountain province portrayed on the structural provinces map? Which physiographic provinces contain crystalline-cored mid-continent anticlines and domes? How are the Appalachian physiographic provinces portrayed on the structural provinces map? How is the California Borderlands-Sierra Nevada physiographic region different from the California transpressional structural province? Which structural provinces form the Northern Rocky Mountain physiographic province? In terms of their structural provinces, how is the Northern Rocky Mountains-North Cascades region similar to the Valley and Ridge-Blue Ridge/Piedmont region? How is the Central-Southern Cascade physiographic province different from the Cascadia Volcanic arc structural province? What does the Colorado Plateau have in common with the Appalachian and Interior Low Plateaus?
16. How does a synclinal mountain form? Show with sketches.
17. Is an erosional-controlled landscape necessarily older than a structure-controlled landscape? Why or why not?
18. What are some non-tectonic means of reincarnation?

Structural Provinces

PART II

Structural Provinces

Unconsolidated Sediment

Chapter Outline

Unconsolidated sediment covers nearly all of the United States but controls landscape only in areas where it is thick enough to completely bury underlying bedrock (refer back to Figure 2.11). Under these conditions the rock type and rock structure have no direct influence on the shape of land. Areas where unconsolidated sediment is thick enough to control landscape include river valleys on the Coastal Plain particularly along the lower (southern) Mississippi River, the Atlantic and Gulf coast shorelines, areas of glacial deposition in the north-central US, and a huge area of sand dunes in Nebraska. In terms of its erodability, unconsolidated sediment is much like a thick layer of shale in which streams can meander across or down-cut with equal ease producing a low-lying dissected landscape without a great deal of relief. In this chapter we concentrate on unconsolidated sediment on the Coastal Plain but also briefly discuss the Nebraska Sand Hills and the Pacific Coast.

THE NEBRASKA SAND HILLS REGION AND THE OGALLALA AQUIFER

Some people might assume that Nebraska is a flat plain across which those of us from the east must travel in order to get to the mountains of the west. However, from a landscape point of view, about 30% of Nebraska is unique. The entire north-central part of the state, as much as 23,600 square miles, is covered with the largest sand dune formation in the Western Hemisphere. As shown in Figure 9.1a, the Sand Hills region lies east of the Wyoming Rocky Mountains between the Niabrara and Platte Rivers. Some of the dunes grew to be almost 400 feet high and 20 miles long before being stabilized by vegetation. The dunes are not entirely stable today. During past droughts, enough of the vegetation died to allow the dunes to shift. Today, special measures are in place to prevent excessive shifting so that the region does not revert back to a desert-like landscape.

The sand dunes are part of the Ogallala formation which consists of unconsolidated sand, silt, and gravel transported eastward from the Rocky Mountains via rivers and wind between 17.5 and 2 million years ago and deposited to a thickness of nearly 900 feet. The Ogallala formation forms the flat, sloping surface layer across the entire High Plains region of the Great Plains from the Sand Hills southward to Texas. The entire Ogallala formation is unconsolidated except for the top surface layer where the sand has been cemented by calcium carbonate into a hard, nearly impermeable rock layer known as *caliche*. The layer of caliche that forms the top of the Ogallala formation is 10 to 30 feet thick and is referred to as *caprock*. It is the caprock layer that directly underlies nearly all of the High Plains serving to protect the underlying unconsolidated sediment from erosion. This protective caprock layer has been removed (eroded) in the Sand Hills region thus exposing the loose, highly permeable, unconsolidated sand that is present in the middle part of the formation. Once the sand was exposed, wind pushed and shaped the sand into dunes and, at the same time, picked up the finer-grained silt and blew it eastward to accumulate on the loess plains of western Nebraska and Iowa. Some of these loess plains are visible in Figure 9.1a in the vicinity of Hastings.

The Sand Hills lie in the recharge area of the Ogallala aquifer. An *aquifer* is a body of permeable rock or sediment that contains groundwater. The top surface of an aquifer is known as the *water table*. Rainwater and snowmelt enter the Ogallala aquifer in the Sand Hills region and flows underground southward, below the caprock layer, at a rate of about one foot per day all the way to Texas. The thickness of the water-saturated zone varies from about 900 feet in Nebraska to less than 100 feet in Texas. The aquifer forms a giant underground water storage area nearly the size of Lake Huron. This makes the Ogallala aquifer one of the largest in the world. It is the primary source of water for much of the High Plains region, and it supports as much as one-fifth of the wheat, corn, and cattle produced in the United States. Its significance and importance cannot be overstated; without this source of water to provide

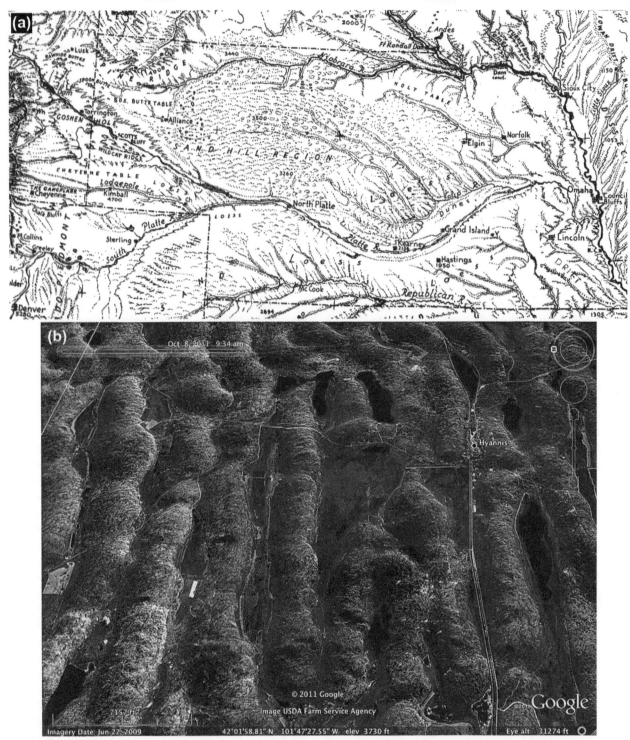

FIGURE 9.1   (a) A landscape map of the Nebraska Sand Hills. (b) A Google Earth image looking east across the Sand Hills at Hyannis, northern Grant County, Nebraska. Lakes and isolated farming plots are located between the dunes. The white line is the boundary with Cherry County.

irrigation, much of the land on the Great Plains would not be suitable for farming.

Some interesting features of the Sand Hills region are the hundreds of small ponds and water holes that form in low areas between the dunes. A few are obvious in Figure 9.1b. The reason for this is shown in Figure 9.2. The water table (the top of the aquifer) lies so close to the surface in the Sand Hills region that it intersects the Earth's surface in low areas between the dunes thus forming the many lakes. In Texas, the water table is as deep as 500 feet.

THE ATLANTIC AND GULF COAST SHORELINE

The Coastal Plain extends along the Atlantic coastline from Cape Cod to Texas. It is a low-lying, nearly flat plain less than 500 feet above sea level that slopes gently toward the ocean. Weak sedimentary rock and a humid climate have produced a thick cover of unconsolidated sediment that controls most (but not all) of the landscape features. Sand in the form of ancient sandbars, barrier islands, and beaches covers large areas of the Coastal Plain. These coastal features were left when sea level was higher than it is today. Inland continental deposits left by marshes, rivers, floodplains, and deltas are also present, as are thick soil horizons. Relief is rarely more than a few tens of feet. The Coastal Plain itself extends offshore below sea level as the continental shelf for distances that vary from a few miles to over 200 miles (Figure 1.6). Much of the shelf region is less than 400 feet deep. This is important because the shallow water, combined with the warm Gulf Stream, provides some of the best fishing grounds in the world. The inland edge of the Coastal Plain is marked by the Fall Line which extends semicontinuously from Cape Cod at least as far south as Montgomery, Alabama. The Fall Line is an escarpment 100 to 300 feet high that marks the boundary between resistant crystalline rock of the Piedmont Plateau and soft sedimentary rock of the Coastal Plain. In this section we discuss the Atlantic-Gulf coast beginning in New England and ending in Texas.

Most of the Atlantic and Gulf coast shoreline is currently losing ground to the ocean through subsidence, sea-level rise, or a combination of both. Such coastlines tend to be highly irregular because the rising ocean claims coastal low areas, particularly river valleys, producing deep, inland bays, coves, marshes, and near-shore islands that were once areas of high ground.

Evidence of recent drowning is best seen along the north Atlantic Coast. The entire New England Coastal Plain north of Cape Cod is below sea level. Ocean waves in Maine lap directly against crystalline rock of the Appalachian Mountains producing a highly irregular rocky coastline with long inlets that represent drowned glacial and river valleys, and small coastal mountains and islands that represent highland areas not yet covered by water. In Figure 9.3 we see a strong Appalachian structural fabric that trends at an angle to the coastline. Mt. Desert Island, in Acadia National Park, is shown in Figures 9.3 and 9.4. The high point on this island,

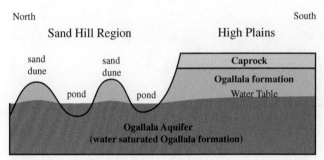

FIGURE 9.2   A cross-section that shows the caprock layer below the High Plains and the intersection of the Ogallala aquifer with the ground surface in the Nebraska Sand Hills region.

FIGURE 9.3   An image of the drowned Maine coastline from Portland to Acadia National Park. Image downloaded from USGS, EROS image gallery, Landsat State Mosaics, http://eros.usgs.gov/imagegallery/collection.php?type=landsat_states#18.

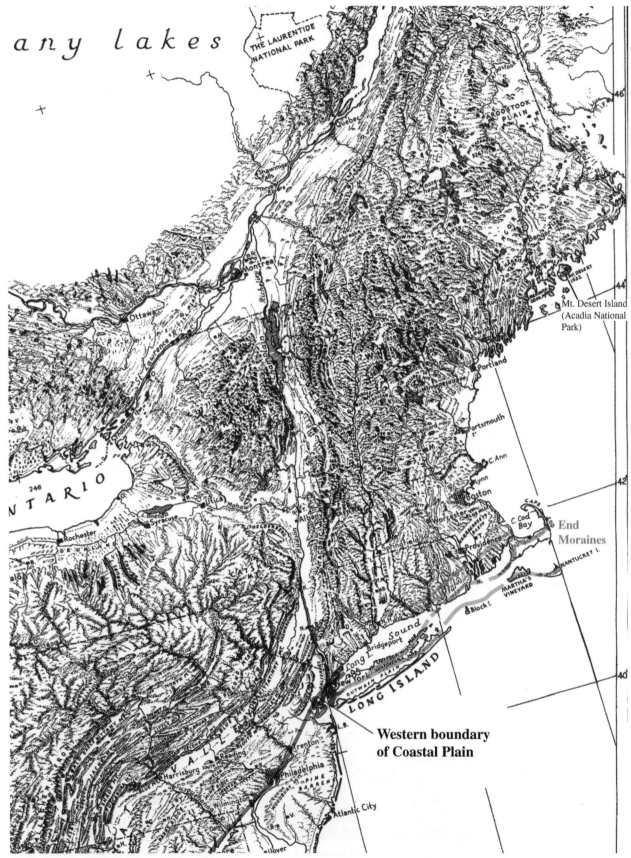

FIGURE 9.4 A landscape map of New England. Note the rocky drowned coastline north of Portland, Maine.

Cadillac Mountain, tops out at an elevation of 1,528 feet to form the highest point along the Atlantic seaboard. As they say in New England, this is where the mountains meet the sea.

The Coastal Plain emerges from below sea level at Cape Cod, and the coastal landscape changes to one dominated by unconsolidated sand and silt deposited initially by glaciers and more recently by rivers. Once deposited, the sediment is shaped, transported, and reshaped by waves and currents via longshore drift. The process of longshore drift is illustrated in Figure 9.5. The Fall Line extends through Long Island Sound where it marks the boundary between crystalline rock of the New England Highlands and sedimentary rock of the Coastal Plain. Long Island Sound itself is a product of glaciation. Weak sedimentary rock at the sedimentary-crystalline rock contact was gouged out to create the ocean inlet; at the same time, it left an inward-facing cuesta of sedimentary rock on the north side of Long Island as shown in Figure 9.6. This area marks the southernmost extent of continental glaciation on the East Coast. Glaciers

terminated at the present location of Cape Cod, Long Island, and Martha's Vineyard producing thick mounds of sediment in the form of end moraines that have helped shape and build the islands. The location of these moraines is traced on the landscape map of New England in Figure 9.4.

South of the glacial limit, the Delaware Bay, Chesapeake Bay, and Pamlico Sound are drowned river valleys protected in part by barrier islands that include Cape Hatteras. This area is shown in Figure 9.7. Based on tide-gauge records, the combination of sea-level rise and land subsidence in the Chesapeake Bay area has resulted in a present-day drowning rate of 12 to 16 inches per 100 years.

Farther south, the smooth scalloped shape of the coastline between Cape Lookout and Cape Romain shown in Figure 9.7 is rather unique for the Atlantic seaboard and is more indicative of uplift than of subsidence. According to Figure 4.1, this area is located at the northeastern margin of an area of active uplift. It is also an active earthquake area, and there is evidence, discussed later in this chapter, that terraces as young as 110,000 years have been tilted. The rocks themselves show evidence of long-lived broad uplift. Rocks exposed along part of the Cape Fear River are older than surrounding rocks. These rocks form a broad anticlinal structure shown as the Cape Fear arch in Figure 9.7. Uplift and erosion associated with this arch is most likely responsible for the oceanward protrusion of Cape Hatteras and for the exposure of relatively old rock in the Cape Fear river valley. Given the evidence, the scalloped shoreline is also likely the result of active or recently active uplift associated with development of the Cape Fear arch.

The South Carolina-Georgia coastline near Savannah again shows evidence of recent drowning although not to the same extent as at Chesapeake Bay. A glance at Figure 9.8 would help you understand why this part of the coastline is known as the Sea Island section. The drowned river valleys have left a series of islands just offshore that include

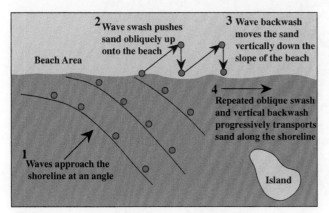

FIGURE 9.5 A sketch map of ocean water breaking on a beach, creating longshore drift. The sequence is numbered. Longshore drift can move sand along the shoreline at rates that exceed 3,000 feet per day.

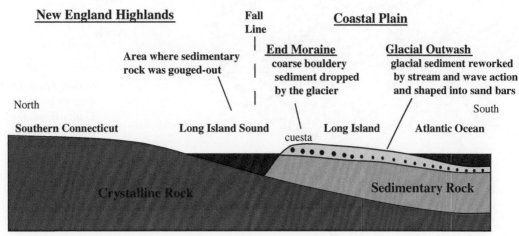

FIGURE 9.6 A cross-section from Connecticut to Long Island showing a cuesta built on an end moraine and glacial outwash that covers most of the island.

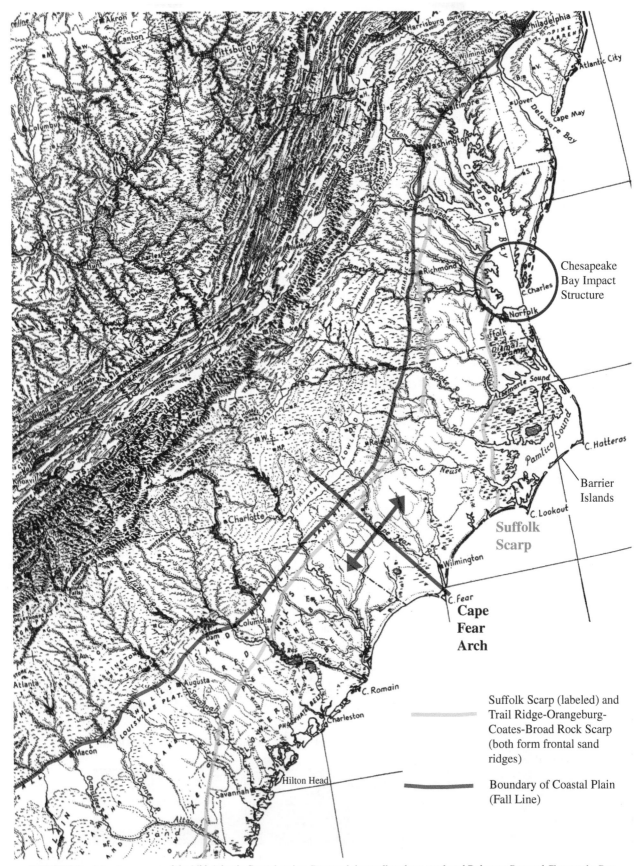

Chesapeake
Bay Impact
Structure

Barrier
Islands

**Suffolk
Scarp**

**Cape
Fear
Arch**

Suffolk Scarp (labeled) and
Trail Ridge-Orangeburg-
Coates-Broad Rock Scarp
(both form frontal sand
ridges)

Boundary of Coastal Plain
(Fall Line)

FIGURE 9.7   A landscape map of the Mid-Atlantic Coastal region. Drowned river valleys have produced Delaware Bay and Chesapeake Bay.

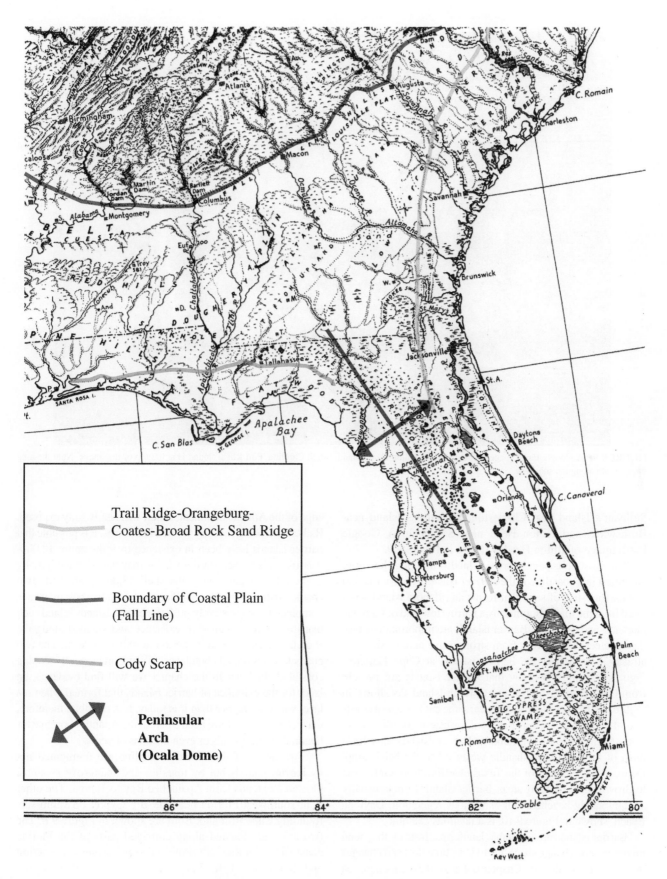

FIGURE 9.8 A landscape map of the Georgia-Florida area.

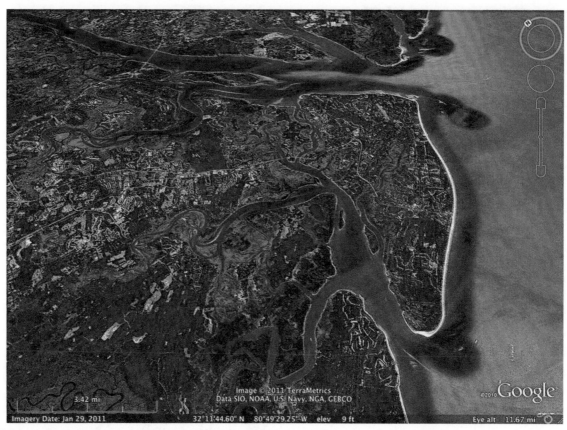

FIGURE 9.9   A Google Earth image looking NNE at Hilton Head Island, South Carolina. Port Royal Sound is at the top of the image. Note the long sandy beach along the shoreline.

Sullivan's Island near Charleston, St. Simons Island near Brunswick, and Hilton Head near Savannah. A Google Earth image of Hilton Head is shown in Figure 9.9.

The most consistently present landform along the entire Atlantic-Gulf coastline is the *barrier* island. A barrier island is a narrow strip of sand that forms an offshore island separated from the main shoreline by a narrow lowland of lagoons, marshes, and estuaries. Barrier islands are important because they protect inland areas from strong wave action. We have already pointed out the barrier islands at Cape Hatteras. Figures 9.7 and 9.4 show that barrier islands are present from Cape Hatteras northward to Long Island. As shown in Figure 9.8, barrier islands form nearly the entire eastern coastline of Florida as well as part of the western coastline from Cape Romano to St. Petersburg. Barrier islands also form most of the Florida panhandle westward to the Mississippi Delta, as well as part of the Texas coastline from Galveston to Brownsville. In many areas, barrier islands form unusually straight sandy beaches with or without vegetation. Daytona Beach, shown in Figure 9.10, is a fabulous example.

Barrier islands are not stable landforms. Instead, they tend to migrate and change shape due to longshore drift and changes in sea level. Recall from Chapter 6 that 18,000 years ago, sea level was about 400 feet lower than it is today and that the outer

edge of the Atlantic continental shelf was nearly fully exposed. Rather than being recently formed landforms, it is possible that barrier islands have been in existence over the entire 18,000-year rise of sea level. Barrier islands may have formed initially at the edge of the continental shelf 18,000 years ago. Two characteristics of the Coastal Plain and Atlantic shelf may have combined to progressively push the barrier islands inland such that they have remained in existence and situated along the shoreline during the entire period of rising sea level. The two characteristics are an abundance of sand, and a gently sloping coastal shelf. Later in the chapter we will find evidence, on land, for the existence of barrier islands that formed when sea level was even higher than it is today. It is possible, therefore, that barrier islands have been a part of the Atlantic shoreline for hundreds of thousands or even millions of years.

Two areas of the Florida coastline are dominated not by barrier islands but by marshes and mangrove swamps. One area extends from Apalachee Bay to Tampa. The other is in southwest Florida near the Big Cypress Swamp surrounding Cape Romano. Both are shown in Figure 9.8. The two areas are located along embayed parts of the Florida coast such that they are protected from strong wave action and longshore drift. Large accumulations of sand are absent. The marshland at Apalachee Bay sits on a platform

FIGURE 9.10 A Google Earth image looking northward along Daytona Beach, Florida. Note how the sandy barrier island protects the inland tidal marsh.

of karsified limestone less than 50 feet above sea level that tilts gently westward due to its location on the west side of the Ocala dome. Sinkholes are clearly evident in the Google Earth image of the area shown in Figure 9.11, as are tree-filled marsh islands (known as *marsh hammocks*) which form on isolated limestone knobs that are high enough above the salt water to support trees. Tallahassee, the largest city in the area, is built across Cody scarp at the northern margin of the limestone platform. Cody scarp is an escarpment that separates the karst platform to the west from a decidedly more hilly and river-dissected karst terrane to the east. Cody scarp is one of the more prominent topographic features in Florida. In a few areas it has as much as 75 feet of relief, which is enough to raise Tallahassee to elevations between 100 and 200 feet above sea level. The scarp likely developed as a shoreline feature (a cliff) during a high sea-level stand when waves rolled across the karst platform.

The Cape Romano area is known as the Ten Thousand Islands. Limestone in this area is covered with 10 to 15 feet of mangrove plant matter and thus does not produce karst topography. A mangrove is a type of plant that can grow in warm, muddy, salty water. In contrast to other parts of Florida that are losing ground to rising sea level, the proliferating mangrove forest is building seaward. The plants have an interlocking root system that traps river mud

before it can reach the open ocean. The trapped mud eventually builds above sea level, thus adding to the coastline. Inland, the southern part of Florida is characterized by freshwater wetlands that include the Everglades and the Big Cypress Swamp (Figure 9.8). These areas result from a combination of flat topography, an elevation within 30 feet of sea level, abundant rainfall, and poor surface drainage.

The Mississippi Embayment is the region of Coastal Plain that swings northward into the midsection of the United States. It is a superb example of an alluvial (that is, river) landscape shaped by the meandering Mississippi River. Figure 9.12 shows a wide, flat, swampy floodplain (the St. Francis-Yazoo-Atchafalaya Basin) covered with abandoned stream channels and oxbow lakes. Crowley's Ridge is a remnant highland along the west side of the river within the floodplain that has not yet been worn down. It is underlain by young (<50 million years) sedimentary rocks and partly covered with glacial-derived, wind-blown silt (loess). Crowley's Ridge is 150 miles long, up to 12 miles wide, and 250 to 550 feet above the floodplain. Other ridges on the floodplain consist of old river deposits. One shown in Figure 9.12 lies just west of Natchez from Sicily Island northward.

Escarpments define both the eastern and western margins of the Mississippi River floodplain. The eastern

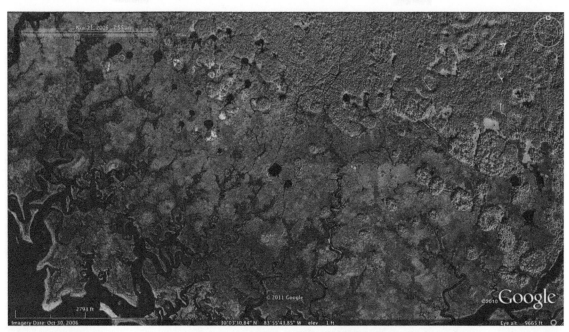

FIGURE 9.11   A Google Earth image looking vertically downward at marshlands in Apalachee Bay southeast of Tallahassee, Florida. Sinkholes on the limestone platform are evident at upper left-center. Marsh hammocks are at lower right and at upper left-center, intermingled with sinkholes. The Econfina River is at lower right. The shoreline is at lower left. The Econfina River is not labeled in Figure 9.8. but is drawn where it crosses the first *e* in *Tallahassee*.

escarpment is a bluff that extends almost continuously from Mississippi to Tennessee. In places, the bluff rises more than 200 feet above the valley floor. Baton Rouge, Natchez, Vicksburg, and Memphis are all located where the river swings eastward against the bluff, offering access to the river as well as relative safety above the floodplain. The bluff consists of young sedimentary rocks and is known as Bluff Hills at Vicksburg, where it too is capped with loess. The well-developed bluff is shown in the vicinity of Vicksburg in Figure 9.13. The escarpment at the western edge of the floodplain forms a less prominent bluff that does not overlook the present course of the Mississippi River. Nevertheless, several cities, including Little Rock and Poplar Bluff, are built on the escarpment.

The muddy Mississippi River carries an enormous amount of sediment into the Atlantic Ocean which has the effect of building the Mississippi Delta outward into the Gulf of Mexico in the shape of a bird's foot. The delta continues to build outward because more sediment accumulates than can be removed by wave action. The location of the delta, however, is not permanent; it has shifted several times over the past 5,000 years. The current bird's-foot delta has been building for only 400 to 500 years. Prior to its present location, the delta emptied into what is today Breton Sound. Prior to that, the river emptied into the area between New Orleans and Atchafalaya Bay. Today these areas are mostly swampland (Figure 9.12).

Beyond the swampland of the Mississippi embayment lies another coastal area of nearly continuous barrier islands extending from Galveston to Brownsville, Texas. This area

is shown on the landscape map depicted in Figure 9.14. The barrier islands protect a highly irregular drowned coastline consisting of numerous bays similar to, but smaller than, Chesapeake Bay. Tide gauges along the Texas coast suggest that subsidence (sea-level rise plus vertical land movement) is occurring at rates between about 8 and 32 inches per 100 years, with rates as high as 7.5 feet per 100 years in some areas. Very high subsidence rates are certainly the result of excessive ground subsidence brought on by a combination of groundwater withdrawal, oil and gas production, and natural subsidence.

ANCIENT SHORELINES OF THE COASTAL PLAIN

The inland Coastal Plain, particularly from Georgia to Virginia and along the Texas Gulf Coast, steps downward toward the coast across a series of as many as seven scarps of unconsolidated sediment that represent shorelines that existed less than 6 million years ago when sea level was higher than it is today. The physiography of the Coastal Plain changes slightly across each of these scarps. Two scarps are particularly well developed and are shown in Figures 9.7 and 9.8. The Trail Ridge-Orangeburg-Coates-Broad Rock scarp extends discontinuously from northern Florida to Richmond, Virginia close to the Fall Line. Sections of this scarp have been recognized as far north as New Jersey and as far south as southern Florida. The scarp represents the frontal edge (the face) of sand ridges and sand dunes that developed along a beach between about 6 million and 1 million years

FIGURE 9.12 A landscape map of the Mississippi Embayment area. The boundary of the Coastal Plain is drawn.

FIGURE 9.13 A Google Earth image looking NNE at Bluff Hills, Vicksburg, Mississippi. The Hills extend continuously for about 330 miles, from Natchez, Mississippi, to Gilt Edge, Tennessee, where they form the eastern edge of the Mississippi River floodplain.

ago. In the Carolinas, this scarp forms a terrace that separates the Carolina Sand Hills section of the Coastal Plain from lower, flatter sections closer to the ocean. The transition across the scarp is best seen in the vicinity of Columbia, South Carolina (Figure 9.7). The Sand Hills themselves are a strip of beach sand dunes that formed against the Fall Line, possibly as early as 23 million years ago.

The Suffolk scarp extends along the eastern side of the Coastal Plain across Virginia and part of North Carolina. This scarp represents the ancient front of barrier island sand dunes, much like present-day Daytona Beach (Figure 9.10). Correlative sediments extend as far south as southern Florida where they have been dated at about 110,000 years. The scarp represents a shoreline that existed during a short-lived high-sea-level stand circa 120,000 years ago as mentioned previously in Chapter 7. Apparently the present-day coastal landscape of barrier islands has not changed greatly in the past 110,000 years even though the shoreline itself has migrated with changing sea level. The Suffolk scarp forms a terrace that separates the present-day, low-lying, back-barrier tidal flats and marshes to the east, from circa 110,000-year-old tidal flats and marshes to the west that are now elevated 40 to 50 feet above sea level. The scarp is easily recognized along the western side of Dismal Swamp,

Virginia as shown in Figure 9.15. The city of Suffolk is built on the scarp.

The scarp lines provide evidence of differential uplift across the Coastal Plain. Differential uplift implies greater uplift in one area than another. Scarp lines represent ancient shorelines and, as such, individual scarp lines of the same age should have formed at the same elevation (at sea level). Regardless of sea-level change, an individual scarp line that has not undergone differential uplift should remain at the same elevation along its length. If a scarp line varies in elevation along its length, then we can assume that the higher elevations correspond with areas of greater uplift. The Trail Ridge-Orangeburg-Coates-Broad Rock scarp varies in elevation by as much as 125 feet. The only explanation is that the land surface must have uplifted or subsided in different areas along the scarp line to produce this relative difference in elevation. The evidence is consistent with deformation (broad folding) across the central part of the Coastal Plain within the past 6 million years.

Differences in elevation of 10 to 15 feet along the Suffolk scarp suggest that broad deformation of the Coastal Plain has been active during the past 110,000 years. The deformation can be in the form of differential uplift, subsidence, or both. Apparent recent uplift is consistent with

FIGURE 9.14 A landscape map of the Texas coast. The boundary of the Coastal Plain is drawn.

FIGURE 9.15   A Google Earth image looking approximately north along the Suffolk scarp. The scarp represents the front of an 110,000-year-old barrier island. The low-lying tidal flat at Dismal Swamp is to the east; the higher, circa 110,000-year-old tidal flat is to the west.

formation of the scalloped coastline between Cape Lookout and Cape Romain described earlier. A maximum of 15 feet of uplift over the past 110,000 years amounts to an average uplift rate of only 0.16 inches per 100 years. This is clearly very slow, broad deformation probably associated with continued development of the Cape Fear arch.

THE PACIFIC COAST

The Pacific coastline is nearly the opposite of the Atlantic coastline. Whereas most of the Atlantic and Gulf coasts are subsiding relative to sea level, the Pacific coast, for the most part, is uplifting at a higher rate than rising sea level. It is emerging from the ocean and, as such, it is shaped more by erosion than deposition. Rather than an irregular, embayed, Atlantic-type coastline, constant uplift will produce a straight, rocky coastline with numerous sea stacks and narrow beaches that are often inundated during high tide. The coastline migrates oceanward leaving only a narrow continental shelf. Such are the characteristics of the Pacific coastline (Figures 1.3, 1.4, 1.7).

Figure 9.16 shows how wave action can erode and undercut a coastal highland to create a smooth, gently sloping bench at sea level that terminates against the steep, eroded face of the uplifted highland, thus forming a relatively straight coastline. A narrow bench of this type is known as a wave-cut platform. Once cut, the bench collects sand through erosion and longshore drift to form a narrow beach. Figure 9.17 shows a wave-cut platform that developed across tilted bedrock on Sunset Beach, Oregon. Sea stacks situated on the platform represent the erosional remnants of resistant high ground not yet removed by wave action. If an earthquake occurs, a platform can be elevated above sea level and left high and dry to become a wave-cut (marine) terrace. A younger platform would then develop along the newly emergent shoreline. Repetition of this process creates a series of marine terraces that stair-step up a straight coastline as depicted in Figure 9.16.

California is replete with wave-cut terraces, many of which result from earthquake-generated land displacement. Figure 9.18 shows at least five terraces beveled into young (less than 23 million years old) mudstone and fine-grained sandstone in the Santa Cruz Mountains just north of Santa Cruz. The lowest terrace is about 100 feet, and the highest, 750 feet above sea level. Cosmogenic ^{10}Be and ^{26}Al dating indicates that the lowest terrace is about 65,000 years old, and the highest is about 226,000 years old, although there are errors on the order of 10,000 to 25,000 years associated with these ages. Given uncertainties, the emergence rate (uplift plus sea-level change) has been between about 1.8 and 4.4 inches per 100 years for the past 225,000 years. These rates agree with geodetic studies that suggest present-day emergence rates on the order of 3.2 inches per 100 years. We would expect emergence to occur episodically in pulses associated with earthquakes rather than gradually as might be expected from sea-level changes alone.

QUESTIONS

1. Under what conditions does unconsolidated sediment control landscape?

2. Name some areas where landscape is controlled largely by unconsolidated sediment.

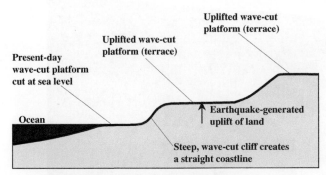

FIGURE 9.16   A sketch of a wave-cut platform and terrace.

3. Draw the Adirondack, New England, Piedmont, Blue Ridge, Valley and Ridge, and Appalachian Plateau physiographic boundaries on Figure 9.4.

4. With reference to Figure 9.6, explain the origin of Long Island Sound. What is glacial drift?

5. Draw the Piedmont, Blue Ridge, Valley and Ridge, and Appalachian Plateau physiographic boundaries on Figure 9.7.

6. What are the conditions that favor development of barrier islands?

7. Describe the different landscape areas east and west of the Suffolk scarp in Figure 9.7. Describe the different landscape areas east and west of the Trail Ridge-Orangeburg-Coates-Broad Rock scarp. Why are landscape differences across the Suffolk scarp more distinct?

8. Examine the area around the Cape Fear arch in Figure 9.7; is there anything in the landscape as depicted on this map to suggest the presence of an arch? Speculate as to why the arch is or is not reflected in the landscape.

9. The symbol for swamp/marshlands on the Raisz landscape maps is shown surrounding the Big Cypress

FIGURE 9.17   A photograph of a wave-cut platform, Sunset Beach, Oregon.

FIGURE 9.18 A Google Earth image looking NNW at wave-cut terraces just north of Santa Cruz, California. The present-day wave-cut platform and at least five terraces can be seen.

Swamp in Figure 9.8. Color all swamp areas in Figure 9.8 yellow. What can you infer about the elevation of these areas? Where are swamp/marshlands concentrated?

10. Examine Figure 9.9. Describe the landscape and the network of rivers. What types of landforms are present in addition to rivers? Is this a drowned landscape or an emergent one? Explain your reasoning. In what way are ocean-facing headlands protected from wave action? How would this landscape change during a steady rise in sea level?

11. Try to guess the angle that the waves are breaking relative to the shoreline in Figure 9.10. Based on your conclusion, is sand in this area undergoing longshore drift? Why or why not?

12. Why are there few homes directly behind the beach area in Figure 9.10?

13. The entire area shown in Figure 9.11 is less than 4 feet above sea level. What type of rock underlies this area? Why are there no trees along the coast? Is this an erosional or a depositional landscape?

14. Study the landscape map of the Mississippi embayment in Figure 9.12. What is the dominant land type surrounding New Orleans? What major structural landform is indicated on the map southeast of Atchafalaya Bay?

15. The southwestern edge of the Mississippi flood plain in Figure 9.12 can be followed southward from Little Rock to Pine Bluff, Arkansas (*P.B.* on map), to Monroe, Alexandria (*A.*), and finally to Opelousas, Louisiana (*O.*). The edge of the floodplain is visible on a Google Earth image, especially if you fly to each of these cities. Color the entire Mississippi floodplain in Figure 9.12 yellow. What do these cities have in common? What do Memphis, Vicksburg, Natchez, and Baton Rouge have in common? How is the landscape of the floodplain different from surrounding areas?

16. Note the sharp turn in the Mississippi River south of Natchez in Figure 9.12. Describe what might happen if the Mississippi were to jump its banks and flow into the Atchafalaya River.

17. Why do the strong rock trends seen in the Ouachita Mountains west of Little Rock in Figure 9.12 abruptly terminate at the Mississippi floodplain?

18. Draw the boundaries to the Ouachita, Ozark, Interior Low, and Central Lowland physiographic provinces in Figure 9.12.

19. Why is there sand on the left (west) bank of the Mississippi River in Figure 9.13?

20. In Figure 9.14, speculate on the origin of Sandy Hills east of Dallas.

21. In Figure 9.14, speculate why the Balcones escarpment faces east (steps down from west to east) and the Austin Chalk cuesta faces west.

22. Why are there no farms east (right) of the ridge in Figure 9.15?

23. Look closely at Figure 9.17. Is there any evidence for slope erosion and retreat along the hillside? Speculate as to why there is no sand.

24. In Figure 9.18, draw lines along the bench-slope intersection for as many terraces as you can. In Google Earth, fly to Santa Cruz. How far northward can you follow some of these terraces?

25. In terms of their mode of origin, in what way are Pacific wave-cut terraces different from the Atlantic sandbar terraces? Which mode of origin appears best suited to explain the limestone platform and Cody scarp at Apalachee Bay, Florida? Explain your answer.

26. In Figure 9.1, use a blue pencil to color areas underlain by sand and a yellow pencil for areas underlain by loess. Note that the area east of the Missouri River at Council Bluffs consists of dissected loess. Speculate on the pattern. Does the pattern suggest a prevailing wind direction? Speculate on what may control the distribution of sand. Could the sand have been derived from the Rocky Mountains, or could it have formed in place? What is loess? Where did the loess come from? Was it blown in with the sand, or could it have some other origin? Compare this figure with Figure 10.4. Why do the Sand Hills end in the vicinity of Norfolk?

27. *Research topic:* Describe the types of unconsolidated sediments that might be produced in the following environments: river, lake, delta, beach, glacial, shallow marine, tidal flats, swamp, deep marine, volcanic, reef, desert.

Nearly Flat-Lying Sedimentary Layers

The dominant structural style across a vast area of the United States is that of nearly flat-lying sedimentary layers. This style is evident across the Great Plains; the Central Lowlands; the Appalachian, Interior Low, and Ozark Plateaus; the Coastal Plain; the Colorado Plateau; and the Wyoming Basin which forms a large part of the Middle Rocky Mountain province in Wyoming (Figure 8.4). The rocks are not perfectly flat; instead they are tilted primarily into a series of broad folds in which the dip is so gentle (<10°) that the layers appear nearly flat lying at outcrop. Some of the layers form monoclines in which nearly horizontal dips are punctuated with local areas of steep (>45°) dip. Beautifully developed monoclines are present on the Colorado Plateau. Rock layers on the Coastal Plain tilt gently mostly in one direction which is toward the ocean.

Nearly flat-lying layers erode to form one of two dominant landscape types, erosional mountains or bench and slope. Figure 10.1a shows a stack of gently folded sedimentary layers prior to erosion. *Erosional mountains* form where erosion cuts consistently downward, leaving only isolated high areas as depicted in Figure 10.1b. *Bench-and-slope landscape* forms where erosion is dominantly horizontal, resulting in the stripping, or the peeling off, of layers as shown in Figure 10.1c. In this situation the landscape consists of a series of flat or nearly flat benches that step up or down across a steep, narrow, sloping area (an escarpment or cuesta) to another bench. Recall from Chapter 3 that the terms *cuesta* and *escarpment* both refer to a steep slope or cliff that separates land of different heights. Monoclines produce a similar bench-and-slope landscape with the addition of a hogback in areas of steep dip. Figure 10.1d shows a monocline prior to erosion. In Figure 10.1e the

monocline is eroded to bench-and-slope landscape with cuestas and a hogback. Other erosional patterns are possible. In Figure 10.1f the monocline is eroded to a flat surface. Figure 10.1g shows bench-and-slope landscape on the right side of the panel and exposure of crystalline rock on the left.

The sedimentary sequence across the central US and across the Colorado Plateau is, for the most part, less than 3 miles thick, less than 542 million years old, and underlain by a slab of ancient crystalline rock nearly all of which is more than 1 billion years old. This is an important point that must be repeated. The oldest sedimentary rocks across the central United States are, for the most part, less than 542 million years old. The crystalline rock that directly underlies these sedimentary layers is more than 1 billion years old. At a minimum, there is nearly half a billion years of time that cannot be accounted for in the rock record. A contact across which not all of geologic time can be accounted for is known in geology as an *unconformity*. Because there is no record of any rock being deposited over this time span, we can presume instead that the area was above sea level and that it underwent erosion during at least part of the time interval. Many of the sedimentary rocks that unconformably overlie the crystalline slab are marine in character meaning that they were deposited in a shallow ocean. The geologic story that we can infer is that the ancient crystalline rock was deformed more than a billion years ago probably into a mountain range. The mountain was eroded to a flat plain at or below sea level during the next half-billion years or more (thus creating the unconformity) and, by about 500 million years ago, was receiving deposition of the overlying sedimentary layers. We will describe unconformities in more detail in Chapter 20. For now, we can simply think of an unconformity as an ancient erosion surface.

Landscape Evolution in the United States. http://dx.doi.org/10.1016/B978-0-12-397799-1.00010-5

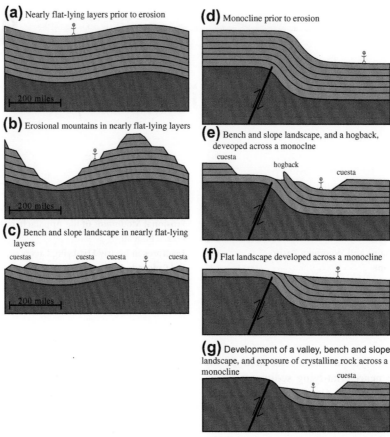

FIGURE 10.1   Schematic cross-sections showing typical erosion patterns of broad folds and monoclines.

The slab of ancient crystalline rock that underlies the central United States is 15 to 30 miles thick and extends continuously into both the Cordilleran and Appalachian Mountain belts. This slab represents the hardened crystalline core, the nucleus, of North America. It is the North America that existed prior to Cordilleran and Appalachian mountain building. We will refer to it as the *ancient North American crystalline shield* or, more succinctly, as the *crystalline shield*.

There is no sharp distinction between the nearly flat-lying sedimentary layer structural province, and the crystalline-cored mid-continent anticlines and domes (Chapter 11). The two structural styles are transitional. The reason for this is the presence of the crystalline shield across the central US directly below a thin sequence of sedimentary rocks. Crystalline rocks reach the surface in areas where a fold becomes tight enough, and where uplift and erosion are adequate enough, to remove the overlying sedimentary layers. The Great Plains, Ozark Plateau, Wyoming Basin, and Colorado Plateau are all dominated by nearly flat-lying sedimentary layers where crystalline-cored domes locally pierce the surface. The most notable areas where rocks of the crystalline shield crop out from below its sedimentary cover are the Superior Upland, the Middle and Southern Rocky Mountains, the Black Hills, along the southern margin of the Colorado Plateau, at the bottom

of the Grand Canyon, in a few localities in the Basin and Range, and in the Adirondack Mountains. Shield rocks are also present in the Blue Ridge Mountains and in the Green Mountains of western Vermont. In this chapter we will concentrate on landscape produced in nearly flat-lying sedimentary layers and reserve discussion of crystalline rocks for later chapters.

We begin our survey with a short overview of the Interior Plains and Plateaus region and the Coastal Plain. This short overview is followed by a more in-depth overview of the Great Plains and the Wyoming Basin. We then discuss bench-and-slope landscape and erosional mountains with examples from the Colorado Plateau, Interior Low Plateaus, Appalachian Plateau, and Ozark Plateau. We take a quick look at fractures on the Colorado Plateau and finish with a discussion of Coastal Plain landscape.

OVERVIEW OF THE INTERIOR PLAINS AND PLATEAUS REGION AND THE COASTAL PLAIN

Broad fold structures cored with sedimentary rock are especially characteristic of the Interior Plains and Plateaus region. Anticlinal structures include the Cincinnati Arch and the Transcontinental Arch. Broad down-warps include the Appalachian syncline and the Michigan, Illinois, and Williston Basins. These features are located in Figure 10.2a. Also shown are two cross-sections. Figure 10.2b extends from the Colorado Plateau eastward to Lake Superior, and Figure 10.2c from Lake Superior southward to the Valley and Ridge. Both cross-sections show the broad fold structure of the sedimentary rocks and the underlying ancient crystalline shield.

Note in Figure 10.2b that the youngest sedimentary layers underlying the Great Plains are wedge-shaped. The sediment is thicker and more coarse-grained in the west reflecting its derivation from erosion of the Rocky Mountains. The wedge shape produces a topographic slope of more than 5 feet per mile that persists all the way to the Mississippi River. This is why many rivers traverse the Great Plains, and why, when driving from St. Louis to Denver, the road seems flat but, in reality, carries you steadily upward toward the Mile-High City. A similar but older sedimentary wedge is developed on the Appalachian Plateau via erosion of the Appalachian Mountains.

The cross-section in Figure 10.2c crosses the Michigan Basin, Cincinnati Arch, and Appalachian Syncline. The Lower Peninsula of Michigan is between 140 and 200 miles wide from east to west. Notice in Figure 10.2c that this entire region is part of the Michigan Basin in which the oldest sedimentary layers crop out along the shorelines of the Great Lakes, and the youngest crop out in the center of the state. Notice also in Figure 10.2a the curvilinear form of Michigan's Upper Peninsula and the continuation of this form across Lake Huron into Canada such that the curvilinear form surrounds the Lower Peninsula. This area is an extension of the bowl-shaped Michigan Basin. Progressively older rock layers are exposed with distance from the center of the structural basin culminating with exposure of rocks of the ancient crystalline shield in the northwest corner of the Upper Peninsula south of Lake Superior. This type of broad fold represents the typical structure within the Interior Plains and Plateaus region. The Illinois Basin, Williston Basin, Appalachian syncline, Transcontinental Arch, and Cincinnati Arch are all of similar dimensions. An important point here is that, in many cases the landscape does not mimic the geologic structure (these are erosion-controlled landscapes). For example, the highest elevations in the Lower Peninsula of Michigan are in young sedimentary rocks near the center of the structural basin in north-central Michigan.

Sedimentary layers on the Coastal Plain are younger (mostly <130 million years old), thicker, and generally softer than those in the continental interior. With the exception of a few broad domes, they dip gently and consistently toward the ocean producing an overall gentle, seaward topographic slope. The geometry of the rock succession is that of a wedge that thickens from zero at the boundary with the Piedmont Plateau to more than 40,000 feet in some areas below the continental shelf. Crystalline rocks are not exposed on the Coastal Plain. However, drill and seismic data indicate that crystalline rocks are present at depth and

that they are younger than 1 billion years old. They are too young to be part of the ancient North American crystalline shield. The rocks are similar in age to those found in the Blue Ridge and Piedmont (600 to 260 million years old), which suggests that they were once part of the Appalachian Mountain belt. Apparently this part of the Appalachian Mountain belt had already eroded and subsided to sea level prior to deposition of the Coastal Plain sedimentary sequence.

OVERVIEW OF THE GREAT PLAINS AND WYOMING BASIN

Nearly flat-lying sedimentary layers dominate the Great Plains, Wyoming Basin, and Colorado Plateau; however, these areas are more complex than other regions of flat-lying sedimentary rock in that other structures, including crystalline-cored domal mountains, synclinal down-folds, and volcanic landforms, are also present. In fact, the landscape of this region is closely associated with the anticlinal mountains that form the Middle and Southern Rocky Mountains. We have already noted that all of these areas were inundated 80 million years ago by the last inland sea to have covered the United States (Figure 7.1). We also know, based on the age of some of the deformed layers, that much of the deformation in the Great Plains and Wyoming Basin occurred between 75 and 40 million years ago.

The presence of an inland sea 80 million years ago, and the presence of a deformational phase between 75 and 40 million years ago, suggest that the landscape is less than 75 million years old. In Chapter 11, following discussion of the Middle and Southern Rocky Mountains, we will provide evidence to suggest that, prior to about 10 million years ago, the Middle-Southern Rocky Mountain landscape did not exist in its present form. Instead, it was part of an expanded Great Plains along with the Colorado Plateau and Wyoming Basin. In other words, 10 million years ago it would have been possible to drive from Denver to Salt Lake City without even seeing a large mountain range. Beginning less than 10 million years ago, the entire region was elevated between 1,640 and 3,280 feet into a broad dome structure centered among the high mountain peaks of Colorado. It was this uplift, and its associated erosion, that created the present-day landscape.

A number of maps are provided to facilitate both a geographic and a landscape understanding of the Great Plains-Wyoming Basin region. Figure 10.3 is a Google Earth image of the region with place names. Figure 10.4 is a regional map with structural divisions. Figures 10.5, 10.6, 10.7, and 10.8 are landscape maps that show most of the region. These maps will be referred to both in this chapter and in Chapter 11. Note in Figures 10.3 and 10.4 that nearly flat-lying sedimentary layers are continuous from the Great Plains into the Wyoming Basin through a gap between the Laramie and Big Horn Mountains, then into the Colorado Plateau through a gap east of the Uinta Range.

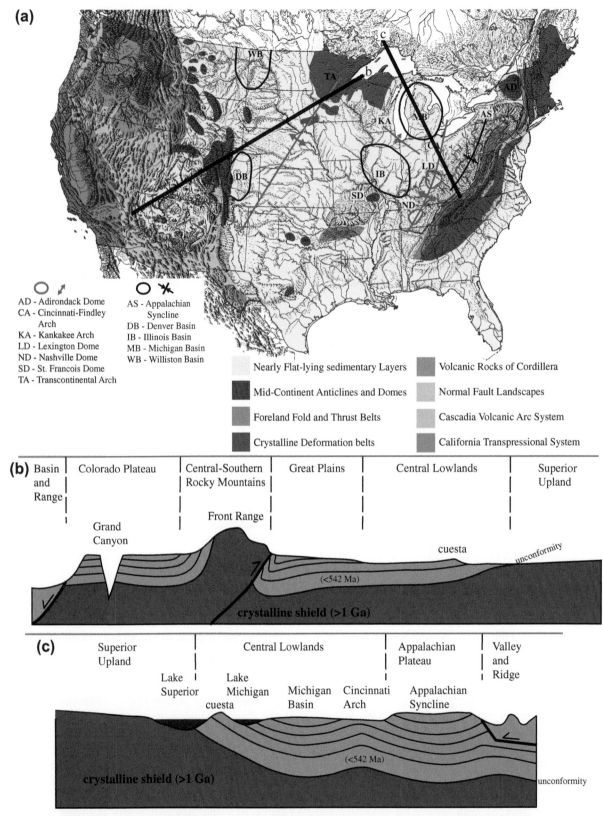

FIGURE 10.2   (a) A structural province map showing location of cross-sections. (b) A cross-section from the eastern edge of the Basin and Range to the Superior Upland. (c) A cross-section from the Superior Upland to the western margin of the Valley and Ridge. A thin layer of nearly flat-lying sedimentary rock unconformably overlies the North American crystalline shield in both areas.

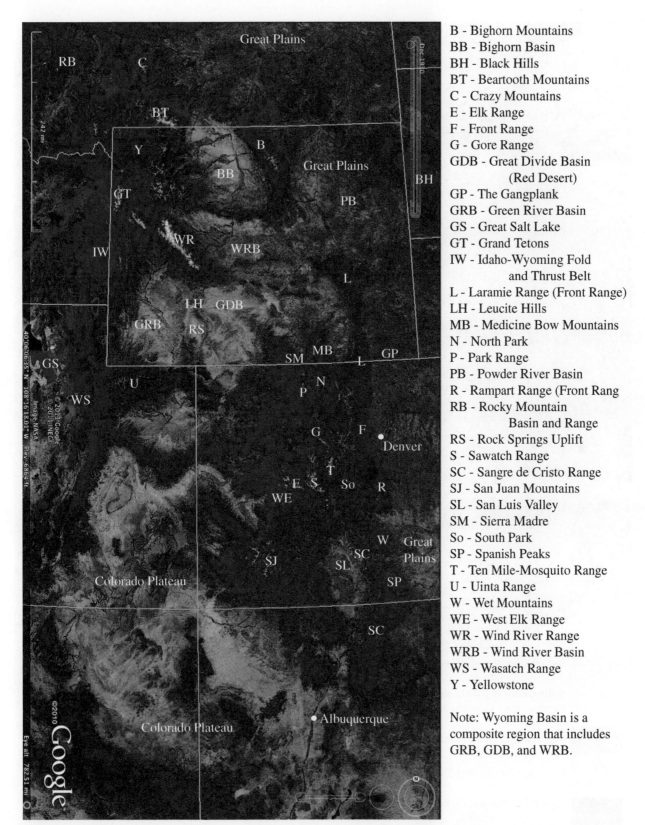

B - Bighorn Mountains
BB - Bighorn Basin
BH - Black Hills
BT - Beartooth Mountains
C - Crazy Mountains
E - Elk Range
F - Front Range
G - Gore Range
GDB - Great Divide Basin
        (Red Desert)
GP - The Gangplank
GRB - Green River Basin
GS - Great Salt Lake
GT - Grand Tetons
IW - Idaho-Wyoming Fold
        and Thrust Belt
L - Laramie Range (Front Range)
LH - Leucite Hills
MB - Medicine Bow Mountains
N - North Park
P - Park Range
PB - Powder River Basin
R - Rampart Range (Front Rang
RB - Rocky Mountain
        Basin and Range
RS - Rock Springs Uplift
S - Sawatch Range
SC - Sangre de Cristo Range
SJ - San Juan Mountains
SL - San Luis Valley
SM - Sierra Madre
So - South Park
SP - Spanish Peaks
T - Ten Mile-Mosquito Range
U - Uinta Range
W - Wet Mountains
WE - West Elk Range
WR - Wind River Range
WRB - Wind River Basin
WS - Wasatch Range
Y - Yellowstone

Note: Wyoming Basin is a
composite region that includes
GRB, GDB, and WRB.

FIGURE 10.3 A Google Earth image that shows the landscape continuity among the Great Plains, Wyoming Basin, and Colorado Plateau. North oriented to top of page. State boundaries are shown for reference.

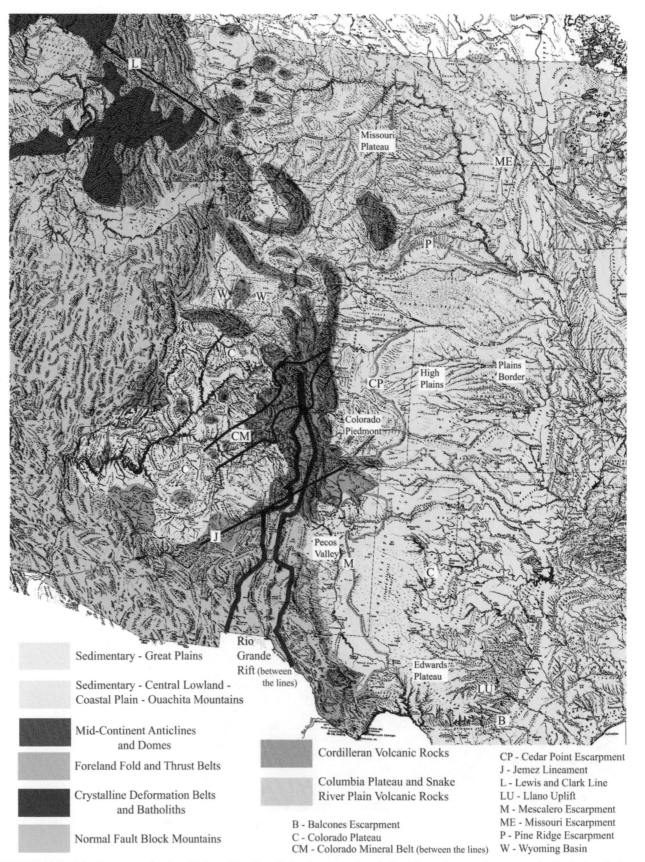

Sedimentary - Great Plains

Sedimentary - Central Lowland -
Coastal Plain - Ouachita Mountains

Mid-Continent Anticlines
and Domes

Foreland Fold and Thrust Belts

Crystalline Deformation Belts
and Batholiths

Normal Fault Block Mountains

Cordilleran Volcanic Rocks

Columbia Plateau and Snake
River Plain Volcanic Rocks

B - Balcones Escarpment
C - Colorado Plateau
CM - Colorado Mineral Belt (between the lines)

CP - Cedar Point Escarpment
J - Jemez Lineament
L - Lewis and Clark Line
LU - Llano Uplift
M - Mescalero Escarpment
ME - Missouri Escarpment
P - Pine Ridge Escarpment
W - Wyoming Basin

FIGURE 10.4 A landscape map showing divisions of the Great Plains, the extent of nearly flat-lying sedimentary layers, and the distribution of other structural provinces and lineaments.

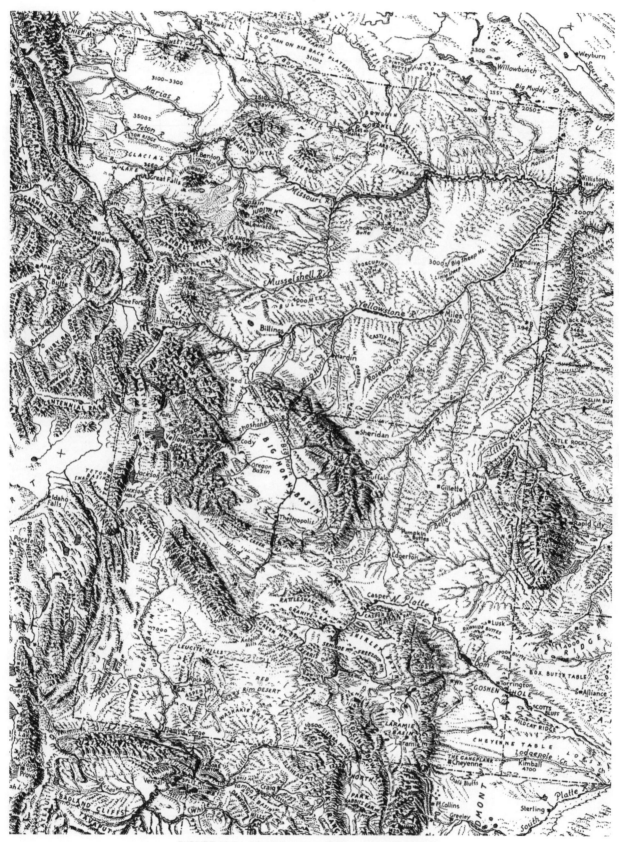

FIGURE 10.5 A landscape map of the northwest Great Plains.

FIGURE 10.6　A landscape map of the northeast Great Plains.

FIGURE 10.7 A landscape map of the central Great Plains.

The rocks that form the surface of the Great Plains are, in general, younger than those in the Central Lowlands. They were derived chiefly from erosion of the Rocky Mountains, and they form a wedge that tapers toward the Central Lowlands. Figure 10.4 shows that the Great Plains are divided into two large areas (the High Plains and Missouri Plateau) and several smaller areas that include a few crystalline-cored domes and volcanic areas (Colorado Piedmont, Pecos Valley, Edwards Plateau, Plains Border).

The High Plains region comes closest to being the expansive flat surface that everybody dreams about when they think of the Great Plains. It is actually a young

(<12 million years old), little dissected, gently east-tilted, flat, plateau surface, underlain, in most areas, by the hard caliche caprock layer that forms the top of the Ogallala formation (mentioned briefly in Chapter 9). The High Plains are surrounded on most sides by escarpments (cuestas) that formed via erosion of the caliche caprock layer. The most prominent of these are the Mescalero escarpment (abbreviated *M* in Figure 10.4), which forms the border with the Pecos Valley; the escarpment at Cedar Point (*CP*) that forms the border with the Colorado Piedmont; and the Pine Ridge escarpment (*P*) north of the Sand Hills region, which forms the border with the Missouri Plateau. These escarpments

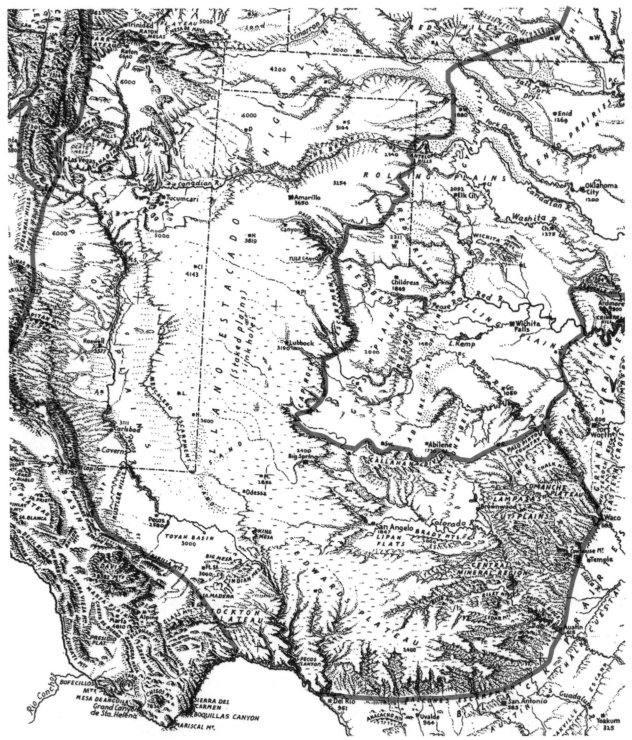

FIGURE 10.8   A landscape map of the southern Great Plains with physiographic divisions.

are labeled and traced with a heavy line in Figure 10.4. There is no obvious physiographic border between the High Plains and Edwards Plateau, or between the High Plains and the Plains Border region. The distinction between these areas is based on the presence of the

Ogallala caprock layer in the High Plains, and its absence on the Edwards Plateau and Plains Border. The Ogallala caprock layer is eroded back (westward) along river valleys in the Plains Border region such that the High Plains drops down to the Central Lowlands across a series of small escarpments rather than across a single high escarpment (Figure 10.7). Several other prominent escarpments form the eastern border of the Great Plains. The Balcones escarpment (*B*, Figure 10.4) separates the Edwards Plateau from the Coastal Plain, the Caprock escarpment (*C*) separates the High Plains from the Central Lowlands, and the Missouri escarpment (*ME*) separates the Missouri Plateau from the Central Lowlands.

The fact that the Ogallala formation is eroded along its margins implies that, prior to erosion, both the Ogallala formation and the High Plains once extended farther east, west, and north of its present extent. There is little doubt that the High Plains once extended across both the Colorado Piedmont and Pecos Valley to the foot of the Rocky Mountains. The evidence is seen west of the Sand Hills region where the Ogallala formation and High Plains today extend to the Rocky Mountain front in Wyoming (Figure 10.4). Denver is the famous Mile-High City, but it is at a lower elevation than Cedar Point due to erosion of the Ogallala formation to form the Colorado Piedmont (Figure 10.7). Erosion of the Pecos Valley has exposed highly soluble halite (rock salt) and limestone, resulting in karst topography that includes the massive Carlsbad Caverns (Figure 10.8).

North of the Pine Ridge escarpment, the Missouri Plateau is divided by the Missouri River into two distinctive landscape areas (Figure 10.6). The area west of the Missouri River either was not glaciated or the glacial landforms have largely been destroyed by erosion. This part of the Great Plains, unlike the High Plains to the south, has been so thoroughly dissected by rivers that most of the younger rock has been removed. The area east and north of the Missouri River preserves a long, continuous glacial moraine known as the Coteau Du Missouri. It forms a hummocky landscape west of Minot, Jamestown, and Aberdeen, where small indentations have resulted in thousands of small lakes and ponds (Figure 10.6). Rather than forming a bedrock cuesta, the 500- to 600-foot-high Missouri escarpment forms the eastern edge of the Coteau Du Missouri overlooking the Central Lowlands.

The geology and landscape of the Great Plains continue into the Wyoming Basin. In fact, if you were to travel westward through the Wyoming Basin hoping to see the Rocky Mountains, you might be disappointed because it is possible to miss the mountains completely. The ease with which one could bypass the Rocky Mountains has made the Wyoming Basin a major thoroughfare. The Oregon Trail, the first transcontinental railroad, and Interstate 80 all pass through this region.

The Wyoming Basin is truly a transitional area between the mid-continent anticlines and domes that dominate the surrounding landscape, and the nearly flat-lying sedimentary layers of the Great Plains. The average elevation is between 6,000 and 8,000 feet, nearly as high the Colorado Plateau. It is not a single basin but a conglomerate of several basins including the Green River (or Bridger), the Red Desert (or

Great Divide), and the Wind River basins (Figures 10.3 and 10.5). Each of these basins are separated by anticlinal mountains some of which are crystalline-cored. The Bighorn Basin lies to the north of the Wyoming Basin on the west side of the Bighorn Mountains, and the Power River Basin lies at the edge of the Great Plains between the Black Hills and the Bighorn Mountains (Figure 10.3). All of the basins are synclinal in form. The young surface rocks are weakly folded and form a continuous surface with the Great Plains and the Colorado Plateau. However, unlike sedimentary layers on the Great Plains, older buried rocks are strongly tilted and down-folded. These rocks crop out only along the margins of surrounding anticlinal uplifts. The thickness of the sedimentary sequence within the synclinal down-folds is between 3 and 6 miles, much thicker than sedimentary rock across most of the Great Plains and Central Lowlands.

The largest of the Wyoming basins is the Green River (or Bridger) Basin (Figures 10.3 and 10.5). It is separated from the Red Desert (or Great Divide Basin) by the Rock Springs anticlinal uplift and the Leucite Hills. The Rock Springs anticline is small and does not expose crystalline rock in its core. The core of the anticline is a topographic basin known as Baxter Basin that was eroded into nonresistant rock. One can view the tilted sedimentary layers along the margins of the fold while traveling along I-80. The Leucite Hills are a series of volcanic flows mostly between 940,000 and 890,000 years old that are now eroded to isolated mesas and buttes. The rocks contain a variety of relatively unusual minerals, including gem minerals. To the east of the Leucite Hills, the Continental Divide extends along the crest of the Wind River Range and then passes southward through the Red Desert and along the crest of the Sierra Madre. The divide bifurcates into two branches as it passes through the Red Desert enclosing the Great Divide Basin and creating a situation similar to that of the Great Basin in Nevada where water is trapped with no outlet to the ocean. The Green River and Great Divide basins are separated from the Wind River Basin by the Wind River and Green Mountains. Farther north lies the Bighorn Basin, which connects with the Great Plains at its north end. Beautifully developed folds are exposed along the northeastern margin of the Bighorn Basin in the vicinity of Sheep Mountain on the western flank of the Bighorn Mountains as shown in Figure 10.9.

BENCH-AND-SLOPE LANDSCAPE

In areas of flat or nearly flat-lying sedimentary rock, it is common for resistant layers such as sandstone and limestone to alternate with nonresistant layers such as shale. This initial situation is depicted in Figure 10.10a with shale at the surface prior to erosion. Because shale is weak, thick layers will not last very long at the Earth's surface. Where present, they produce a landscape of gentle, low-lying hills covered with sediment, with wide meandering rivers such as

FIGURE 10.9 A Google Earth image looking NNW at folds and hogbacks along the northwest margin of the Bighorn Mountains north of Greybull, Wyoming. The northern part of the Sheep Mountain anticline is shown at lower right. Note the anticlinal basin at upper left-center.

what is shown in Figure 10.10b. As more shale is removed, an underlying resistant layer of sandstone or limestone is uncovered. The underlying resistant layer will impede downward erosion, forcing rivers and streams to meander and sweep away any remaining shale. The result is a nearly flat plateau or plain underlain by a hard caprock layer of resistant rock (Figure 10.10c).

In order to maintain slope, the stream will cut into the resistant caprock layer where it will encounter strong walls that impede meandering. This forces the stream to cut downward, producing a deep valley. Many of the famous gorges in the United States, including Zion Canyon in Utah and the Grand Canyon, result from down-cutting through resistant, nearly horizontal rock layers. In some cases down-cutting is facilitated by the presence of vertical joints or faults that form narrow, weak zones.

As the stream cuts downward through a caprock layer, it may encounter an underlying layer of shale. In this situation, the weak walls of the shale allow the stream to once again meander. This action undercuts the sandstone, producing cliffs that break off and retreat (Figure 10.10d). The result of this process is a series of flat benches underlain by caprock layers that step up (or down) across a series of slopes that form where the caprock layer has been removed. A mature bench-and-slope landscape is depicted in Figure 10.10e.

The bench-and-slope landscape is methodically lowered by sequential removal (horizontal stripping) of higher caprock layers via undercutting, wearing away, and slope retreat. In this way, a plateau can evolve into a plain. The geology is predictable in the sense that if you step down from one caprock surface onto a lower surface, you are also stepping down across an escarpment (cuesta) onto a lower layer of sedimentary rock.

Bench-and-slope landscape is well developed across the entire Interior Plains and Plateaus region, on the Colorado Plateau, and, to a lesser extent, on the Coastal Plain. The actual shape of bench-and-slope landscape is dependent on tectonics, climate, the density of through-going rivers, and the erodability of rock from one layer to the next. We will compare bench-and-slope landscape on the Colorado Plateau with that on the Interior Low Plateaus. Although these areas possess similar landscape components (rock type and rock structure), they are different in terms of climate and tectonic activity.

THE COLORADO PLATEAU

The rock and landscape of the Wyoming Basin continue into the Colorado Plateau across the area east of the Uinta Mountains. Here we enter one of the most spectacular regions in the world. The Colorado Plateau is home to over two dozen national parks, national monuments, state parks, and scenic wonders, including Grand Canyon, Bryce Canyon, Zion, Arches, Canyonlands, and Capital Reef National Parks; the Painted Desert; Monument Valley; Shiprock, and Meteor Crater. Most of these areas are labeled on the landscape map shown in Figure 10.11. In addition to numerous broad folds and monoclines, there are crystalline-cored anticlines, fault and fracture zones, and young volcanic rocks. At an average elevation of about 6,000 feet, it is the highest plateau in the country. The general shape of the Colorado Plateau is that of a southwesterly tilted topographic bowl with high elevations along its rim that drain toward the center of the plateau where the only through-going river, the Colorado River, carries water and sediment southwestward to the Pacific Ocean via the Gulf of California. Physiographically, the Colorado Plateau can be

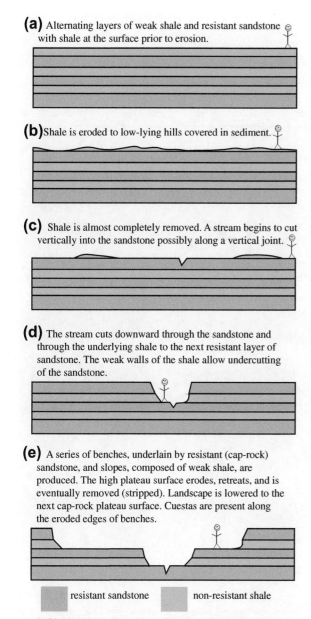

(a) Alternating layers of weak shale and resistant sandstone with shale at the surface prior to erosion.

(b) Shale is eroded to low-lying hills covered in sediment.

(c) Shale is almost completely removed. A stream begins to cut vertically into the sandstone possibly along a vertical joint.

(d) The stream cuts downward through the sandstone and through the underlying shale to the next resistant layer of sandstone. The weak walls of the shale allow undercutting of the sandstone.

(e) A series of benches, underlain by resistant (cap-rock) sandstone, and slopes, composed of weak shale, are produced. The high plateau surface erodes, retreats, and is eventually removed (stripped). Landscape is lowered to the next cap-rock plateau surface. Cuestas are present along the eroded edges of benches.

resistant sandstone   non-resistant shale

FIGURE 10.10   Formation of bench-and-slope landscape.

divided into a least four areas: the High Plateaus in the northwest, which includes the Pink Cliffs and volcanic rocks on the Paunsaugunt and Aquarius plateau; the canyon region along the Colorado and Green Rivers; the Navaho region, which encompasses the entire southeastern part of the plateau including Monument Valley, the Painted Desert, Black Mesa, and Chacra Mesa; and volcanic regions in the south which include the San Francisco Mountains near Flagstaff, and volcanic fields surrounding Mt. Taylor near Albuquerque (Figure 10.11). Figure 10.4 locates the larger lava fields on the Colorado Plateau. Similar to the Wyoming Basin, the climate is arid particularly in the Navaho region.

Bench-and-slope landscape is well displayed in the nearly flat-lying and monoclinal layers northeast of the

Grand Canyon. Many of the folds and monoclines are located in a Google Earth image shown in Figure 10.12. Note in this figure that some of the broad anticlinal folds are referred to variably as *uplifts, upwarps*, and *swells*. The term *reef* is used for a long, continuous rocky cliff that poses a barrier to travel. The classic reef is a hogback that forms along the eroded steep dip of a monocline as shown in Figure 10.1e. The Echo Cliffs in Arizona, between The Gap and Bitter Springs, are an excellent example as shown in Figure 10.13.

The lack of vegetation and absence of a thick soil in the dry climate result in widespread exposure of rock. Rather than forming vegetation-covered slopes, the edges of benches form an angular landscape of vertical cliffs. Figure 10.14 is a schematic cross-section that shows this characteristic feature. A famous bench and slope landscape on the Colorado Plateau is the Grand Staircase, located north of the Grand Canyon (see Figure 10.12 for location). Figure 10.15 is a north-to-south cross-section from the Paunsaugunt Plateau in the High Plateaus region, southward across the Grand Staircase to the San Francisco Mountains. Note in this cross-section that the Grand Canyon cuts directly across the Kaibab Uplift (anticline) and that a complementary syncline underlies the High Plateaus area to the north. The rim of the Grand Canyon sits at an elevation of about 8,000 feet. Elevation increases northward toward the High Plateaus and up the Grand Staircase in a series of benches that preserve progressively younger rock layers toward the center of the syncline. Slopes along the Grand Staircase correspond with eroded cliff faces (cuestas) that include the Chocolate, Vermilion, White, Gray, and Pink cliffs. The Vermilion, White, and Pink cliffs are clearly visible in a Google Earth image of the Grand Staircase in Figure 10.16. Zion National Park is carved into the White Cliffs. The Pink Cliffs host Bryce Canyon National Park.

Along the western margin of the Colorado Plateau, the High Plateaus region drops precipitously some 5,000 feet across an escarpment marked by the Hurricane normal fault onto the Basin and Range (Figure 10.15). This escarpment, although not bound continuously by a fault, extends for several hundred miles along the western and southern margins of the plateau. In the south, the escarpment is known as the Mogollon Rim. As shown in Figure 10.17, the Mogollon Rim clearly separates the Colorado Plateau physiographic province from the Basin and Range. The dissected area at lower right in Figure 10.17, between the Mogollon Rim and the basin to the south, is known as the *transition zone*. This zone becomes wider to the east and represents a part of the Mogollon Rim that has been dissected. Rocks of the ancient North American crystalline shield are exposed at the base of the transition zone. In Figure 10.11, the southern boundary of the Colorado Plateau is placed on the rim above the transition zone.

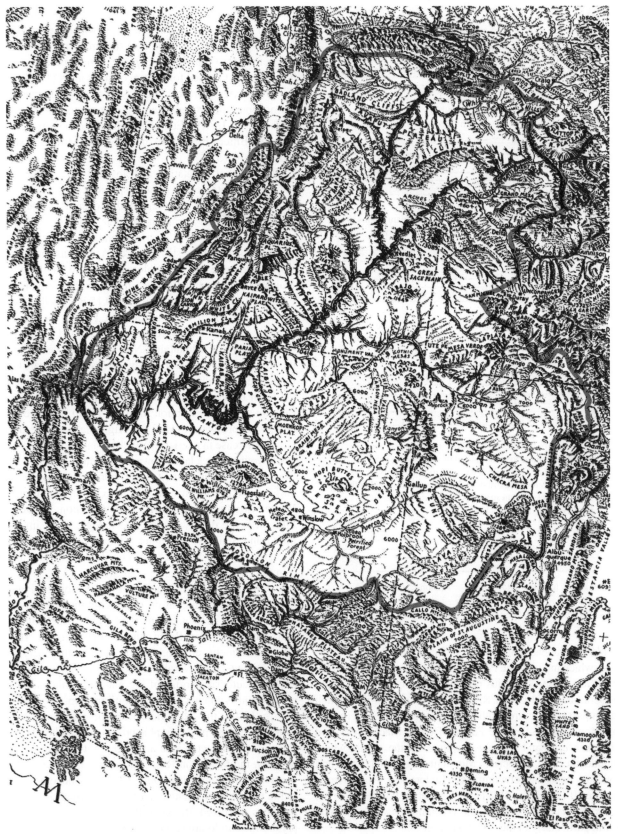

FIGURE 10.11 A landscape map that shows the physiographic boundary of the Colorado Plateau.

FIGURE 10.12   A Google Earth image looking north at a variety of structures in the southwestern part of the Colorado Plateau.

FIGURE 10.13   A Google Earth image looking northward at the Echo Cliffs reef (monocline) and Route 89. The town of Gap is at bottom-center of image. Note the truck on the road.

Slope retreat due to undercutting of the caprock layer is a dominant form of erosion. As the caprock bench undergoes erosion, sections of a bench may become isolated, producing first a mesa and then a butte before being completely removed (Figure 10.14). Figure 10.18 shows beautiful examples of mesas and buttes in the wondrous area of Monument Valley.

In addition to the Kaibab Uplift, broad anticlines on the Colorado Plateau include the San Rafael swell, the

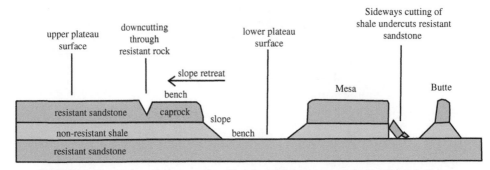

FIGURE 10.14   A schematic cross-section of bench-and-slope topography on the Colorado Plateau.

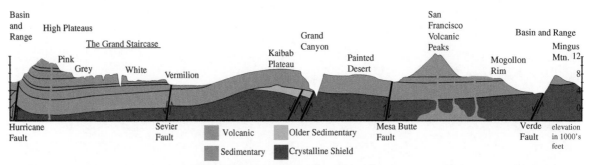

FIGURE 10.15   A north-to-south cross-section from the High Plateaus region across the Grand Staircase and Grand Canyon to the Mogollon Rim. Based on a cross-section by P. Coney and W. J. Breed, Museum of Northern Arizona and the Zion Natural History Association, Zion National Park, Utah, 1975.

FIGURE 10.16   A Google Earth image looking eastward at the Grand Staircase near Kanab, Utah. The Vermilion, White, and Pink Cliffs are visible. The Henry Mountains (including Mt. Ellen) are at center skyline.

Monument upwarp, and the Raplee anticline (Figure 10.12). The anticlines are so broad and flat-topped that the margins in many areas form monoclines. The east side of the San Rafael swell is a perfect example as shown in Figure 10.19. The east side of the Kaibab uplift, the west side of the Raplee anticline, and Comb Ridge on the east side of Monument upwarp, all form monoclinal structures. Another major monocline is the Waterpocket fold which extends continuously for about 100 miles. Part of this fold forms Capital Reef National Park.

As noted in Figure 10.1e, erosion will strip layers off the top of monoclines producing bench-and-slope landscape on one or both sides of the fold. The Waterpocket fold forms the bottom of a bench and slope staircase with the Circle Cliffs and Straight Cliffs (Figure 10.12). In some instances, as shown in Figure 10.1e, an erosion-controlled landscape with hogbacks (reefs) develops within the tilted part of the monocline. The Echo Cliffs and Comb Ridge are examples.In other instances, a structure-controlled landscape

FIGURE 10.17   A Google Earth image looking northeast along the western and southwestern rim of the Colorado Plateau. The deeply incised area at center-top is the Grand Canyon.

develops in which topography follows the structure of the fold. Figure 10.20 is a Google Earth image of the Water-pocket fold in which the landscape follows the structure of the monocline. Figure 10.21 is a Google Earth image of the East Kaibab monocline in which the southern part of the fold is eroded to form a hogback, whereas landscape in the northern part follows the structure of the fold. Note the development of benches in the distance of both folds.

A possible origin for many of the monoclines in the United States, particularly those on the Colorado Plateau, is illustrated in Figure 10.22. This figure shows that the North American crystalline shield was cut by faults more than 800 million years ago. The faults then became inactive and the crystalline shield was eroded to a flat plain prior to deposition of the overlying sedimentary sequence. Following deposition of the sedimentary layers, the old faults became active again (i.e., they were reactivated) and, in doing so, they cut upward into the overlying sedimentary layers. Some of these faults do not reach the surface but instead remain at depth to produce blind faults. The geometry of blind faults on the Colorado Plateau requires that rocks must be folded (or draped) over the fault as shown in

Figure 10.22. The resulting surface geometry is that of a monocline underlain by a (blind) fault not visible at the surface.

THE INTERIOR LOW PLATEAUS

Bench-and-slope topography on the Colorado Plateau can be compared with similar topography developed in a warm, humid climate that has not undergone recent uplift. One such province is the Interior Low Plateaus of Kentucky, Tennessee, and southern Indiana. A landscape map that outlines the plateaus region is shown in Figure 10.23. A sketch map and two cross-sections that identify major features are presented in Figure 10.24. In sharp contrast to the Colorado Plateau, elevations in the Interior Low Plateaus are mostly between 400 and 1,000 feet. It is truly a low plateau. Nevertheless, the province is sharply defined against the highland region of the Appalachian Plateau to the east, the Coastal Plain lowland to the southwest, and glacial lowlands of the Central Lowlands. Traditionally, the northern boundary of the province east of Louisville was arbitrarily placed along the Ohio River. I have instead placed the northern boundary at the inferred southern limit of glacial deposition, which

FIGURE 10.18   A Google Earth image looking southeast at mesas and buttes in Monument Valley. The ridge at top center is Comb Ridge, which forms a continuous cuesta for about 100 miles. Comb Ridge is part of an eroded east-dipping monocline that marks the southeastern margin of Monument Upwarp. The white line is the Utah-Arizona border.

roughly follows the Ohio River. The contrast between the rolling hills of this province and the flat glacial drift plains of the Central Lowlands is clearly indicated on the landscape map surrounding the town of Evansville (Figure 10.23). The Interior Low Plateaus have been tectonically inactive for at least 265 million years implying that the landscape has slowly grown old over that time span in the sense that it has evolved in a predictable, recognizable way.

A major difference with the Colorado Plateau is the absence of canyons in the Interior Low Plateaus. This is not surprising considering the general absence of recent uplift. There are, however, deeply incised valleys. This, perhaps, also is not surprising. The incision is at least partly in response to periodic drops in sea level during glacial advances, the most recent of which was the 400-foot drop that occurred 18,000 years ago. Today, with rising sea level, many of the rivers are undergoing deposition in their channels. A second major difference with the Colorado Plateau is the presence of a thick soil and plenty of vegetation that covers most everything. A glance at the cross-sections in Figure 10.24 shows that, rather than the rocky cliffs of Figure 10.14, the

edges of caprock layers form a more rounded landscape of soil-covered slopes and steep hillsides.

The primary landscape-controlling structure is the Cincinnati Arch which is a broad anticline that trends approximately north-south through the eastern part of the province. Rocks are inclined less than 10 degrees away from the center of the arch such that most of the rocks dip gently westward across the province toward the Illinois Basin (Figure 10.2a). Two domal structures are developed along the crest of the arch: the Lexington (or Jessamine) Dome and the Nashville Dome. The Nashville Dome is eroded to form a topographic basin referred to as the Nashville Basin. This is a classic example of inverted topography in an old erosion-controlled landscape where elevated rock in the core of a dome is preferentially eroded to produce the topographic basin. The anticlinal basin is clearly evident in the cross-section of Figure 10.24c. The central part of the Lexington Dome, by contrast, forms a plain known as the Inner Bluegrass or Lexington Plain. Topography of the glacial drift plains in the Central Lowlands is fairly continuous into the Lexington Plain and that is one reason why

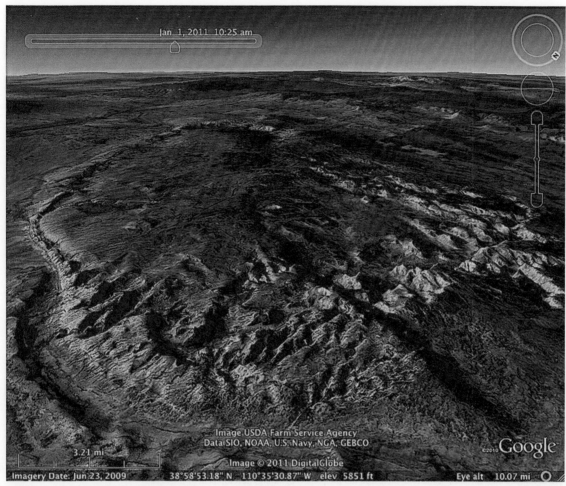

FIGURE 10.19   A Google Earth image looking southwest at the flat-topped San Rafael Swell (anticline). Note the well-developed monocline along the southeastern (left) side, which is partly eroded into hogbacks.

the boundary between the two physiographic provinces was drawn at the Ohio River. Transported glacial rocks, known as *erratics*, have been reported within the Lexington Plain as far south as Frankfort, suggesting that an early glaciation (pre-135,000 years ago) may actually have extended into western Kentucky.

Monoclines are not common on the Interior Low Plateaus; therefore, the landscape can be compared most directly with the bench-and-slope landscape of the Grand Staircase on the Colorado Plateau. There are several benches (plateau surfaces) that step upward onto younger rock toward the northwest, away from the Cincinnati Arch. The two most prominent are the Highland Rim-Pennyroyal (or Pennyrile) Plateau, and the Western Coal Field-Chester Uplands. Both are shown in the cross-section of Figure 10.24c. In contrast to the Colorado Plateau where elevation increases across benches of the Grand Staircase, elevation does not change substantially across the Interior Low Plateaus. The absence of elevation gain is due to the relatively low height of the escarpments that separate the benches, the distance between the benches, and the general northwestward dip of the rock layers which negates the elevation gain.

The Highland Rim-Pennyroyal plateau completely encircles the Nashville Basin and partly encircles the Lexington Plain. As shown in Figure 10.24a, the edge of the plateau forms a well-developed cuesta known as the Highland Rim that surrounds the Nashville Basin on all except the southern side where it is partly eroded. The equivalent cuesta that partly surrounds the Lexington Plain is known as the Knobstone-Muldrough's Hills escarpment. This particular escarpment is most conspicuous immediately west of Louisville where it rises as much as 600 feet above the Lexington Plain. Here, as shown in Figure 10.25, it forms a cuesta not unlike the cuestas that develop on the Grand Staircase except with greater vegetation and human population. Southward along this escarpment, in the vicinity of Muldraugh's Hills, the Pennyroyal Plateau becomes so thoroughly dissected that only a few isolated hills remain. Here it may be difficult to see where the Lexington Plain ends and the Pennyroyal Plateau begins. The isolated hills that form the border region are referred to as *knobs*. They are erosional remnants of the Pennyroyal Plateau, and as such they are the humid equivalent of buttes on the Colorado Plateau. Excellent examples of knobs are

FIGURE 10.20 A Google Earth image looking northward along the Waterpocket monocline. Note how topography follows the structure, and note that benches are developed on both sides of the fold. Note also the incipient (beginning) development of hogbacks.

shown in Figure 10.26; these can be compared with the buttes of Figure 10.18.

The escarpment that separates the Highland Rim-Pennyroyal Plateau from the Western Coal Field-Chester Uplands is known as the Dripping Springs-Chester (or Springville) escarpment (Figure 10.24c). This escarpment is distinctive in its southern part as shown in Figure 10.27, but becomes less distinctive farther north. Limestone forms the caprock layer across much of the Highland Rim-Pennyroyal Plateau south of the escarpment producing karst features such as sinkholes, some of which are visible in Figure 10.27. Limestone is present at depth in the Western Coal Fields-Chester Uplands region north of the escarpment below an insoluble caprock layer. The dissolving limestone in this area is protected from collapse by the overlying insoluble caprock layer which allows caves, including Mammoth Cave, to develop. The Dripping Springs escarpment is so named because some of the water moving through the limestone below the caprock layer reaches the surface along the face of the escarpment.

Mammoth Cave, the largest cave system in the world, is located in the Western Coal Field region near the boundary with the Pennyroyal Plateau and is labeled in Figure 10.23. Figure 10.28 is a cross-section of the Mammoth Cave region that shows the protective caprock layer and the caves underneath.

In Chapter 9 we defined an aquifer as a body of permeable rock or sediment that contains groundwater. The water table forms the top surface of an aquifer. The dissolution of limestone and the formation of caves typically occur at or just below the water table. The reason for this is that the dissolution of limestone has the effect of neutralizing acidic water. Once neutralized, the water can no longer dissolve limestone. The water table is the level at which fresh (that is, acidic) rainwater is added to the aquifer. Water at greater depth has been neutralized. The elevation of the water table, therefore, controls the elevation at which cave formation is active.

The water table is not a flat surface. Instead, it mimics the ground surface by rising slightly below hills and dipping slightly below valleys. Water generally flows away from high areas toward low areas such that the water table intersects the surface of major rivers and lakes. For example, in Chapter 9 we described how the water table intersects low areas in the Nebraska Sand Hills region, thus forming ponds between the dunes (Figure 9.2). An active water table

FIGURE 10.21   A Google Earth image looking north at the East Kaibab Monocline showing an erosion-controlled landscape of hogbacks along the southern part of the fold and a structure-controlled landscape along the northern part where topography follows the structure. The Paria Plateau (part of the Vermilion Cliffs) is visible at upper right. The flat-topped Kaibab uplift (anticline) is visible at left.

is what keeps many rivers flowing and many lakes full of water during summer drought.

As suggested in Figure 10.28, water in the Mammoth Cave system seeps into the ground or descends through vertical cracks (joints and faults) to the water table and then moves horizontally northwestward along nearly horizontal bedding planes, eventually to intersect the Green River. The elevation of the Green River corresponds roughly with the elevation at the top of the water table and therefore with the elevation at which cave formation occurs. If the Green River cuts downward, active cave formation will drop to a lower level, leaving dry caves at higher elevations. If the Green River rises due to increased water supply or through deposition in its channel, active cave formation will rise to a higher level and any cave below that level would be below the water table and therefore flooded.

There are 390 miles of interconnected passages in Mammoth Cave which is more than twice the length of Jewel Cave in South Dakota, the second longest cave system in the world. The uppermost passages, at elevations of 570 to 690 feet, are the oldest and largest with openings that are more than 100 feet wide (the ground surface is at an elevation of about 745 feet). These caves are probably more than 5 million years old and could have begun forming as long ago as 30 million years. This part of the cave system formed when the elevation of the Green River (and therefore the water table) remained relatively stable for a long period of time. This was also the time when most of the karst landscape on the Highland Rim-Pennyroyal Plateau developed. A wetter glacial climate during the past 2 million years caused the Ohio River to lengthen and to cut downward. The Green River is a tributary to the Ohio so it too initiated down-cutting, thus lowering the water table and therefore lowering the elevation at which caves form. Major cave passages are present at 550 and 500 feet suggesting that elevation of the Green River (and the water table) remained stable during the time these caves developed. Terraces along the Green River at these elevations are understood to have formed at about the same time as the caves. The 550-foot terrace has been dated at 700,000 to 800,000 years old, and the 500-foot terrace at more than 350,000 years old. These cave passages, therefore, are considerably younger than

Normal faulting of the ancient North American crystalline shield between 900 and 800 million years ago creates weak zones in the crust.

The faults became inactive, the crystalline shield is eroded to a flat plain, and a younger sedimentary sequence unconformably covers the faults.

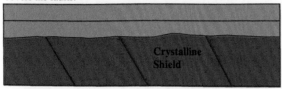

Following deposition of sedimentary layers, the old faults are reactivated and cut upward into the overlying sedimentary layers. Faults that do not reach the surface are known as blind faults. Sedimentary layers are draped over the blind faults to create a monocline.

FIGURE 10.22 A series of cross-sections that show the origin of many of the monoclines on the Colorado Plateau.

those above them. Additional cave passages extend down to and below the present elevation of the Green River at 425 feet.

Another area of interest in the Interior Low Plateaus is the Shawnee Hills region in southern Illinois (Figure 10.24). This is a somewhat transitional area between the Western Coal Field to the east and the Ozark Plateau to the west. It is a surprisingly hilly region that rises abruptly from the flat glacial plains of the Central Lowlands. Most of the layers are close to horizontal; however, steeply dipping layers occur along several fault zones that trend through the area. The abundance of faults makes this area one of the most deformed in the central US. The faults are part of the Rough Creek-Pennyrile fault zone which is an extension of an ancient rift zone known as the Reelfoot Rift. These faults have a history that is essentially identical to faults on the Colorado Plateau as depicted in Figure 10.22. Much like the Colorado Plateau, the faults formed initially between about 800 and 500 million years ago within the underlying ancient crystalline shield prior to deposition of the overlying sedimentary layers. The faults have periodically reactivated since that time, cutting into and displacing the sedimentary layers and, in some areas, forming monoclines. Stress along these buried, reactivated faults is responsible for earthquakes in the area, including the great New Madrid earthquakes of 1811 and 1812.

The Google Earth image and cross-section in Figures 10.29 and 10.30 show an abrupt boundary between the northeastern part of the Shawnee Hills and the glaciated plains of the Central Lowlands. This boundary corresponds with the inactive Shawneetown fault and with the northern limb of the Eagle Valley syncline. The highlands are due to the presence of a thick, resistant layer of sandstone that forms a cuesta several hundred feet high. The cuesta can be followed continuously around to the south side of the syncline where, at one location known as the Garden of the Gods, the rocks are eroded into a variety of odd shapes. The Shawneetown fault is present along the north and west sides of the cuesta. This fault has had a complex history of reactivation that includes both south-side-up and south-side-down displacement. This type of displacement is not unusual for reactivated faults. Such movement is shown in Figure 10.22. The cuesta, in this case, is an example of an erosion-controlled landscape in which the highland coincides with the hardest rock irrespective of which side of the fault is the up-thrown block.

EROSIONAL MOUNTAINS OF THE APPALACHIAN PLATEAU

Any highland area is subject to erosion; therefore, any mountain could be considered an erosional mountain, especially those where tectonic processes are dormant. In this section we restrict the term *erosional mountain* to an originally level or nearly level plateau surface that was subsequently so thoroughly dissected by stream erosion that only isolated parts of the original surface are left to form highlands. Dissection occurs in stages, as shown in Figure 10.31, in which streams continually invade flat areas until the entire surface is dissected. Rather than sharp, steep summit areas, erosional mountains characteristically develop broad, flat mountaintops composed of remnants of the original plateau surface. Valleys between summit areas are often sharp and steep. Roads and houses, therefore, are built on the wide summit areas rather than in the valleys, thus creating a second distinction with tectonically uplifted mountains.

Erosional mountains in flat-lying sedimentary layers represent an extreme form of bench-and-slope landscape, one in which streams continually cut downward through caprock layers rather than strip layers horizontally. The formation of erosional mountains is favored over bench-and-slope landscape in areas of regional uplift, falling sea level, or along the edge of a plateau surface where there is a steep drop to a low area such that the river on the plateau is well above base level. Erosional mountains are a characteristic feature of plateau areas underlain by sedimentary rock; however, they can form in any rock type.

Plateau areas are, by definition, elevated above surrounding areas and, because of this, are vulnerable to the down-cutting action of streams. The beginning stage in

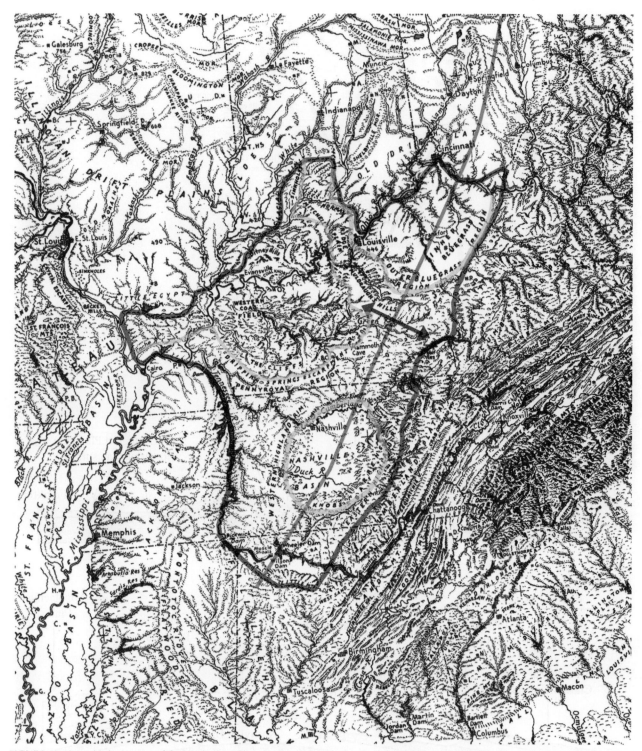

FIGURE 10.23 A landscape map of the Interior Low Plateaus showing physiographic divisions. The crest of the Cincinnati Arch is shown with a heavy gray line and double arrow.

the formation of an erosional mountain is illustrated by the Llano Estacado, an area also known as the Palisaded Plains. It is part of the High Plains region of the Great Plains along the New Mexico-Texas border underlain by the gently

east-tilted Ogallala formation. Figure 10.8 is a landscape map of the southern Great Plains that shows the Llano Estacado bordered on the north by the Canadian River, on the east by the Caprock Escarpment (which also forms the

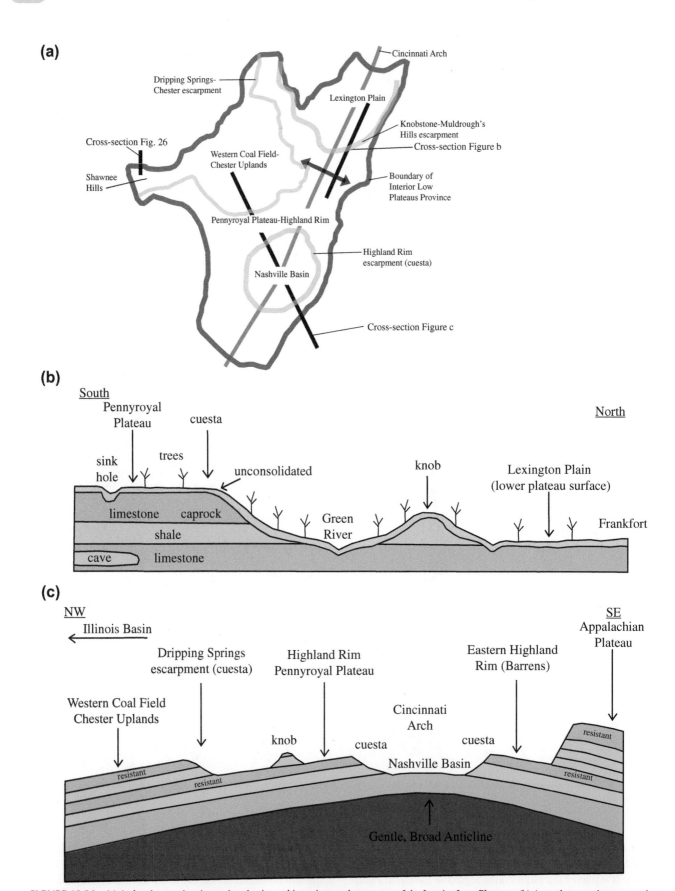

FIGURE 10.24   (a) A sketch map showing major physiographic regions and structures of the Interior Low Plateaus. (b) A north-to-south cross-section from the Lexington Plain to the Highland Rim. (c) A northwest-to-southeast cross-section of the Nashville Basin and Chester Uplands.

FIGURE 10.25 A Google Earth image looking southwest across the Knobstone Escarpment.

FIGURE 10.26 A Google Earth image looking north toward Lexington, Kentucky, at tree-covered knobs along the eroded Muldrough's Hills Escarpment, the southern margin of the Lexington Basin. The white line is the Campbellsville-Lebanon County line.

FIGURE 10.27   A Google Earth image looking NNE along the Dripping Springs Escarpment. The many ponds on the Pennyroyal Plateau are sinkholes produced by dissolution of limestone at the surface. Caves form in the Chester Uplands north of the escarpment because the limestone is protected from collapse by an overlying layer of insoluble caprock. A cross-section of the Dripping Springs Escarpment is shown in Figure 10.28.

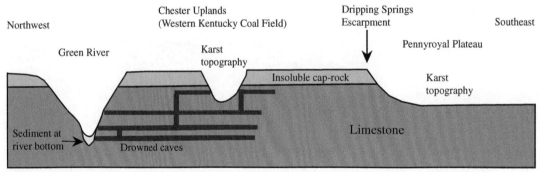

FIGURE 10.28   A cross-section of Mammoth Cave from the Green River southeastward to the Pennyroyal Plateau. Based on Harris, Tuttle, and Tuttle (1997).

border with the Central Lowlands), and on the west by the Mescalero Escarpment which overlooks the Pecos River valley. A glance at Figure 10.8 reveals that no rivers south of the Canadian River cross the Llano Estacado. The arid climate and the lack of major through-going rivers have produced one of the flattest, least dissected, treeless areas in the United States. The area rises westward at a rate of about 10 feet per mile, from less than 3,000 feet in the southeast to 5,000 feet in the northwest. Figure 10.32 is a Google Earth image of the Caprock Escarpment at the edge of a

monotonous expanse of the Llano Estacado. At one time, less than 2 million years ago, rivers did flow from the Rocky Mountains across the Llano Estacado, but rapid headward erosion of the Pecos River beheaded those rivers and captured (pirated) their water as depicted in Figure 10.33. This left the Llano Estacado high and dry. It will take many years and perhaps a change to a wet climate for this area to evolve into erosional mountains.

Many areas, including parts of the Colorado Plateau, are at intermediate stages of dissection. An excellent

FIGURE 10.29 A Google Earth image looking east across the Shawnee Hills at Equality, Illinois. A resistant sandstone layer forms a cuesta that surrounds the Eagle Valley syncline on three sides. Part of the cuesta is underlain by the Shawneetown fault. The Wildcat Hills form the north side of the syncline; Garden of the Gods is on the south side. Resistant sandstone also surrounds Hicks (structural) Dome to form a domal basin. Lakes periodically occupied nearly all of the flat farmland in the image during the glaciation. The Ohio River crosses the top of the image.

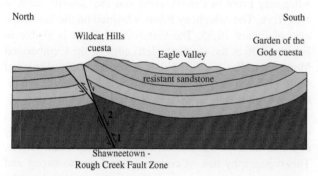

FIGURE 10.30 A north-to-south cross-section of the Shawneetown-Rough Creek fault zone and Eagle Valley syncline, southeastern Illinois. The numbers indicate reverse-fault motion along the Shawneetown fault, followed by normal fault motion.

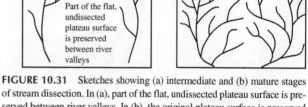

FIGURE 10.31 Sketches showing (a) intermediate and (b) mature stages of stream dissection. In (a), part of the flat, undissected plateau surface is preserved between river valleys. In (b), the original plateau surface is preserved only as isolated, flat-topped erosional mountains between river valleys.

example is the area west and south of Moab, Utah. Figure 10.34 shows several stream tributaries near Moab that are just beginning to cut into flat areas of the plateau. This area differs from the Llano Estacado by the greater presence of water and perhaps greater and more recent uplift. The result is the development of dissected stream canyons separated by flat areas where dissection has not yet commenced. On a grander scale, tributaries to the Colorado River in the Grand Canyon area show a similar pattern as shown in Figure 10.11. Streams, over time, will invade the flat areas until the entire region is dissected into erosional mountains.

A mature degree of dissection is present on the Appalachian Plateau, most of which is outlined on the landscape map shown in Figure 10.35. A quick look at Figure 10.35 shows stream valleys almost everywhere across the province. Figure 10.36 is a Google Earth image looking SSW

across the completely dissected Cumberland Plateau of southeastern Kentucky. The long ridge in the image is Pine Mountain, an outlier of the Valley and Ridge and underlain by a thrust fault. Pine Mountain separates the dissected Cumberland Plateau to the north (right) from the equally dissected, but topographically higher, Cumberland Mountains, which includes Black Mountain (4,145 feet). Pine Mountain and the Cumberland Mountains are structurally part of the Valley and Ridge but, physiographically, they are more closely connected with the Appalachian Plateau in the sense that the Cumberland Mountains are classic erosional mountains, very similar in many respects to the Alleghany and Catskill Mountains to the north, both of which are on the Appalachian Plateau. The Cumberland Plateau, Pine Mountain, and the Cumberland Mountains are shown as part of the Appalachian Plateau in the southern part of Figure 10.35 west of Knoxville.

FIGURE 10.32   A Google Earth image near Post, Texas, looking west across Caprock Escarpment and the wide expanse of the Llano Estacado (High Plains).

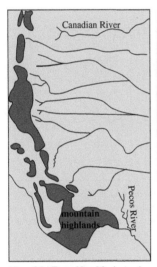

Map of the Texas-New Mexico region prior to about 2 million years ago showing east-flowing rivers crossing the Llano Estacado.

Headward erosion (lengthening) of the Pecos River resulted in the capture of rivers flowing east leaving the Llano Estacado high, dry and flat.

FIGURE 10.33   Headward erosion and stream piracy of the Pecos River.

Notice in Figure 10.35 that most of the rivers on the Appalachian Plateau drain westward to the Ohio River and the Mississippi Valley. As a result of this drainage pattern, the western half of the plateau is more thoroughly worn down, leaving dissected mountains in the eastern half that include the Allegheny and Pocono Mountains of Pennsylvania, the Catskill Mountains of New York, and the Cumberland Mountains. Mountaintops at the very eastern edge of the Appalachian Plateau form a major escarpment that overlooks the Valley and Ridge province to the east. This escarpment is referred to as a *front,* such as the

Allegheny Front in Pennsylvania and the Catskill Front in New York. The Alleghany Front is labeled on the landscape map of Figure 10.35. The Cumberland Front is visible in Figure 10.36 at the southern (left) edge of the Cumberland Mountains. The Catskill Front is spectacularly displayed south of Albany as shown in Figure 10.37. Here, the mountains rise more than 3,000 feet above the Hudson Valley.

The Grand Canyon itself could be considered an area of erosional mountains especially if you are standing at the bottom of the canyon next to the Colorado River. This topography has, of course, formed due to strong and persistent river down-cutting. Resistant layers such as the Red Wall limestone and Coconino sandstone form vertical cliffs separated by sloping benches of less resistant rock. Erosional mountains form where cliffs become isolated between river valleys.

The Appalachian Plateau, Colorado Plateau, and Llano Estacado show different stages of the same process: the slow dissection of a plateau surface by river incision to form a terrain of erosional mountains. If the process of river dissection were to go to completion without a major change in forcing variables, the mountains themselves will eventually be eroded such that the entire plateau region is lowered to a nearly flat-lying plain. The amount of time it takes to produce erosional mountains is dependent on climate, the number of through-going rivers, the hardness of rock, the magnitude of relief between the edge of the plateau and adjacent plain, and how base level varies (that is, the amount of uplift/subsidence or sea-level change). Thus, even though a landscape is in the mature stage of erosional dissection, it could be younger (in actual years) than a relatively undissected landscape. For example, the Llano Estacado is as old or older than the more dissected landscape of the Colorado Plateau.

FIGURE 10.34   A Google Earth image looking northward at tributaries to Kane Springs Canyon southwest of Moab, Utah. The areas between tributary valleys remain nearly flat and undissected. This area is in the intermediate stage of dissection.

The edge of a plateau, where it drops to a plain, is a prime location for development of erosional mountains because the steep slope provides velocity and increased cutting power to streams. Under these conditions, streams cut progressively into the plateau surface such that it may be possible to see the entire transition from an undissected flat plateau to dissected erosional mountains at one location. The Edwards Plateau along the southeastern margin of the Llano Estacado is one such location (Figure 10.8). Streams gain velocity and cutting power as they drop from the plateau surface across Balcones Escarpment onto the Black Prairies. In the east, this has resulted in nearly complete dissection of the plateau forming Postaok Ridge, Cedar Mountain, and Riley Mountain, a situation similar to the Appalachian Plateau. Streams to the west of this dissected area are just beginning to cut into the nearly flat, undissected plateau, forming an intermediate stage of deep canyons separated by flat areas (Figure 10.8). Still farther west, the Llano Estacado remains undissected. Figure 10.38 shows the Postoak Ridge area west of Austin, Texas. Note the dissected erosional mountains at the front of the image, and the undissected plateau at the rear. Also note that most of the homes and roads are built across flat summit areas. This entire erosional landscape will, over time, migrate westward via slope retreat. Postaok Ridge and other erosional mountains will be reduced

to the level of the Black Prairies and the dissected margin of the Edwards Plateau will migrate westward into the Llano Estacado.

OZARK PLATEAU

The Ozark Plateau is a large domal structure centered in the St. Francois Mountains south of St. Louis where nearly flat-lying sedimentary layers dip gently away from a crystalline core. The St. Francois crystalline core is discussed in Chapter 11. Here we discuss the surrounding sedimentary rocks.

This area, like the Interior Low Plateaus, is an old erosion-controlled landscape located south of the southernmost advance of glaciers. The dome is asymmetric and steeper on the east side which explains why the crystalline St. Francois Mountains are on the eastern side of the Ozarks. Although the overall structure is domal, the sedimentary layers are truly nearly flat lying. The dip of sedimentary layers across the dome is less than 3 degrees, even on the steep eastern side.

A landscape map of the Ozark Plateau is shown in Figure 10.39. The dome structure is such that generally younger rocks are exposed with distance from the St. Francois crystalline core. Three age groups of sedimentary rocks form three separate areas of the plateau. The first age group (between

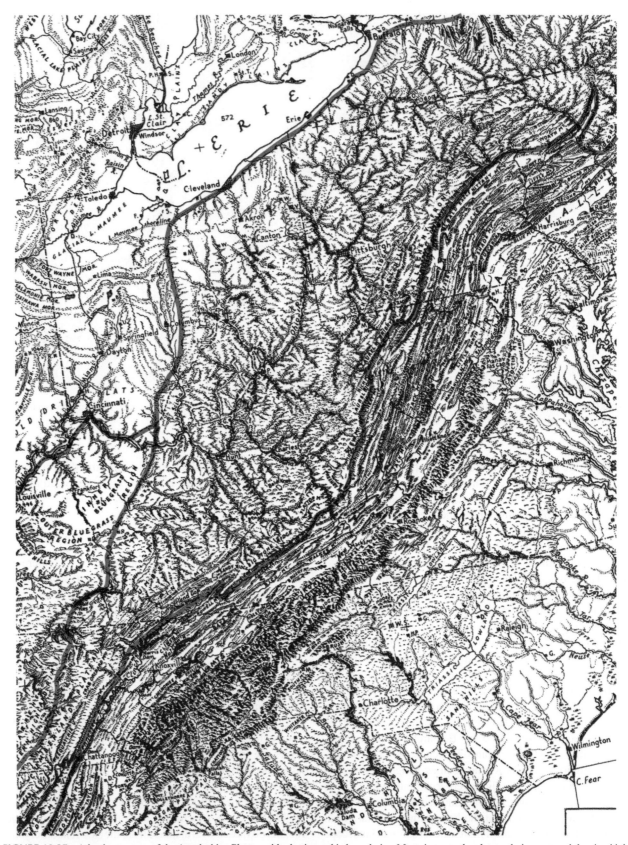

FIGURE 10.35 A landscape map of the Appalachian Plateau with physiographic boundaries. Most rivers on the plateau drain westward, leaving highlands (fronts) on the east that overlook the Valley and Ridge, such as the Alleghany Front in Pennsylvania.

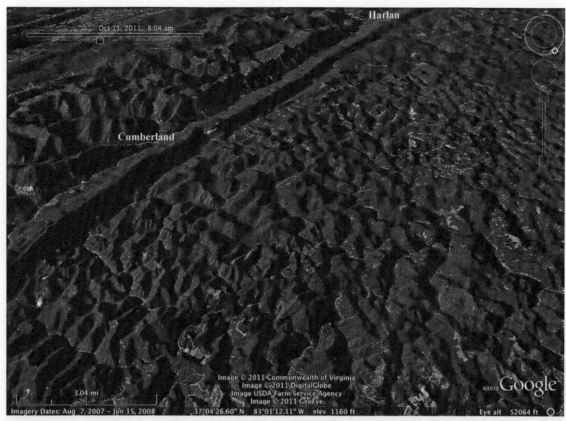

FIGURE 10.36   A Google Earth image looking SSW across the dissected Cumberland Plateau in southeastern Kentucky, near the border with Virginia and Tennessee. The long ridge between Harlan and Cumberland is Pine Mountain. The Cumberland Plateau is at right, the Cumberland Mountains at left, and the Valley and Ridge at extreme upper left.

542 and 443 million years old) forms the Salem Upland (or Salem Plateau) surrounding the St. Francois Mountains. The second (between 359 and 318 million years old) forms the Springfield Plateau. The youngest rocks (between 318 and 306 million years old) form the Boston Mountains. The addition of younger rock with distance from the crystalline core creates a series of inward-facing (that is, facing toward the crystalline core) escarpments that separate each of the three areas and creates a poorly defined bench-and-slope staircase. The Burlington (or Eureka Springs) escarpment east of Springfield, Missouri is shown in Figure 10.39. It forms a low-lying, eroded cuesta that steps up from the Salem Plateau onto younger rock of the Springfield Plateau. Similarly, the Boston Mountains escarpment forms a cuesta that steps up from the Springfield Plateau onto younger rock exposed in the Boston Mountains. The Boston Mountains escarpment is the largest and best defined with as much as 800 feet of relief. Smaller benches are developed within each of the three areas along rock layers of differing resistance. A cross-section that shows some of these features is presented in Chapter 11, Figure 11.3.

The topography of the Salem and Springfield plateaus is one of highly dissected rolling hills and incised stream valleys. Elevations are mostly between 1,000 and 1,500 feet with less than 500 feet of relief. The presence of incised meandering river channels, such as Crooked Creek in Arkansas, is reminiscent of the rivers on the Colorado Plateau. Their presence suggests a phase of recent uplift. The damming of some of these rivers along the Missouri-Arkansas border has created large sinuous lakes such as Table Rock Lake in Figure 10.40 that are ideal for canoeing, fishing, and general recreation.

Limestone and dolostone are the dominant rock types on the Salem and Springfield plateaus, and given the humid climate, karst features are abundant, including the second highest concentration of large springs in the country (slightly less than the Snake River Plain but more than the Lime Sink region of Florida). Karst features are, however, less abundant than one might expect. The reason for this is that much of the limestone is laced with chert nodules. *Chert* is microcrystalline quartz and is essentially immune to dissolution. The chert weathers out of the limestone and collects in great quantities along the banks of the many streams in the area.

The Boston Mountains are higher and more rugged than the Salem and Springfield plateaus. They are erosional mountains, a dissected plateau, with maximum elevation above 2,500 feet and relief on the order of 1,000 feet or

FIGURE 10.37   A Google Earth image looking north at the Catskill Front along the Hudson Valley in New York. The Hudson River is at far right. The Catskill Mountains form a broad synclinal fold that is part of the Appalachian syncline. The mountains at distant right are the Taconic, Berkshire, and Green Mountains. The Adirondacks are at distant center-left.

more. In contrast to other parts of the Ozark Plateau, the rocks here are dominantly sandstone and shale.

The Salem Plateau boasts a well-exposed meteor crater known as the Crooked Creek impact structure. The crater is understood to be about 80 million years old. As shown in Figure 10.41, it forms a circular landscape 4 to 7 miles in diameter located along Crooked Creek between the towns of Cook Station and Cherryville, Missouri. Evidence of meteor impact is present within strongly deformed rocks exposed along Crooked Creek in the northern part of the crater. Figure 10.41 also shows the dissected, low-lying hills typical of the Salem Plateau.

The uplift history of the Ozark Plateau is not well con-strained. One interpretation is that the plateau was uplifted to its highest elevation by 265 million years ago during the last phase of Appalachian-Ouachita mountain-build-ing events, and has been an upland ever since. Changes in climate-driven erosion rates, sea-level changes, and pos-sible thermal isostatic adjustment could be responsible for 700 feet or more of elevation change over that time span; however, the absence of strong tectonic activity since 265 million years ago suggests that there has been no major reincarnation of the land surface. Other than slow ero-sional lowering and relatively small elevation changes,

the landscape probably has not changed significantly. Based on this interpretation, we could conclude that, like the Interior Low Plateaus, the Ozark Plateau is an ancient landscape that has grown old over several hundred million years. At a minimum, we can surmise that the landscape was in existence 80 million years ago when the Crooked Creek impact crater formed.

FRACTURES IN NEARLY FLAT-LYING LAYERS ON THE COLORADO PLATEAU

Strong fracture patterns within layers of sedimentary rock, coupled with the dry climate of the Colorado Plateau, have produced some of the most unusual and picturesque land-scape in the world including the landscape in and around Arches, Zion, and Bryce Canyon National Parks.

Arches National Park is located close to the Colo-rado River just north of the confluence of the Colorado and Green Rivers near Moab, Utah. The Arches land-scape developed in a thick sandstone layer known as the Entrada Formation, which overlies a thick, buried layer of salt rock. Salt layers, at depth, tend to be unstable because they are less dense than the rock around them and because they are weak enough to flow when under pressure. The

FIGURE 10.38   A Google Earth image looking northward at erosional mountains along the edge of Edwards Plateau in the Postoak Ridge area west of Austin, Texas. Undissected Edwards Plateau is seen at far upper right.

combination of density and plasticity causes the salt to rise as a column, similar to an igneous intrusion. As the salt rises, the overlying sedimentary rocks are bent, fractured, and faulted. A combination of bending and faulting resulted in the stretching of the Entrada Formation across the Salt Valley anticline, causing it to break into a series of subparallel, vertical fractures. Figure 10.42 is a Google Earth image of these fractures along the eastern margin of the anticline. Because of the dry climate and general absence of soil, the best place for vegetation to take root is within the fractures themselves. Freeze-thaw weathering and erosion have progressively widened the fractures, and in doing so have isolated individual vertical walls of rock called *fins*. Several fins can be seen on the right side of Figure 10.42. The base of a fin tends to weather more quickly than the top due to the natural accumulation of water in shaded areas close to the base. Weathering and erosion will eventually breach the lower part of a fin but will leave the upper part intact. The result is the development of an arch. Figure 10.43 shows the resulting arch.

Fractures also helped to form Zion National Park. This park is located in the White Cliffs in the High Plateaus region of Utah north of the Grand Canyon. The primary rock formation is a 2,000+-foot-thick white sandstone known as the Navajo Formation, famous for its cross-beds

that are easily seen along cliff faces. Here again, a series of fractures extends vertically through the nearly flat-lying rock unit. The North Fork Virgin River has cut vertically downward along one of the fractures to produce the upper part of Zion Canyon. Note in Figure 10.44 that the Virgin River flows parallel with the fractures before turning southwestward across the fractures into Zion Canyon, which is located at lower left in the Google Earth image. Where the Virgin River parallels the fractures, it cuts downward so quickly that some of the tributary rivers cannot keep up and, instead, form hanging valleys (waterfalls) where they merge with the river. Here, along four miles of the river in an area known as the Narrows, the canyon is as little as a few feet wide and the river occupies the entire canyon floor. Vertical bedrock walls soar upward as much as 2,000 feet on both sides of the canyon floor. On a good day, when the river is low, one can walk along the river into the Narrows and into one of the most spectacular sights in the world.

Bryce Canyon National Park forms another fracture-controlled landscape, this one carved into the Pink Cliffs of the High Plateaus region of Utah. Two vertical fracture directions oriented almost at right angles to each other, and a nearby fault, control the landscape. The fractures cut through nearly horizontal beds of limestone, mudstone,

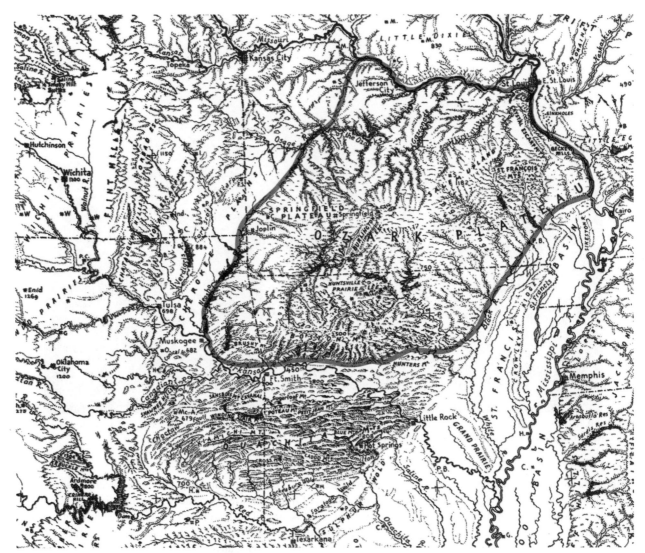

FIGURE 10.39   A landscape map of the Ozark Plateau.

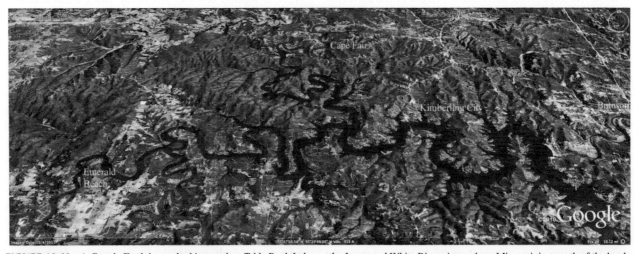

FIGURE 10.40   A Google Earth image looking north at Table Rock Lake on the James and White Rivers in southern Missouri, just north of the border with Arkansas.

FIGURE 10.41   A Google Earth image looking north at the Crooked Creek impact crater and the Salem Plateau. The area of the crater is indicated with a circle.

FIGURE 10.42   A Google Earth image looking SSE at a section of Arches National Park. The Entrada Formation dips gently NNE (left) away from the crest of an anticline where vertical fractures have eroded to form fins. Salt Valley is visible at extreme upper right.

FIGURE 10.43   A photograph of Skyline Arch eroded from a fin.

FIGURE 10.44   A Google Earth image looking north at Zion Canyon. Note the strong vertical fractures in the White Cliffs of the Navaho Formation. Zion Canyon is the wide canyon at lower left that contains the North Fork Virgin River. The Narrows are located where the upper part of the river turns parallel with the fracture pattern.

and sandstone of the Claron Formation. Erosion into the canyon wall, combined with differential erosion along both fracture sets, has produced rock amphitheaters filled with odd-shaped pinnacles known as *hoodoos*. As an added feature, many of the pinnacles are brightly colored in shades of red, pink, yellow, orange, and white due to small amounts of iron oxide and manganese in the rock. Figure 10.45 is a photograph of some of the brightly colored hoodoos.

Although technically not a fracture pattern, Upheaval Dome is an anomaly that deserves mention. This landform

FIGURE 10.45   A photograph of brightly colored rocks in the Pink Cliffs of Bryce Canyon.

is located on a mesa between the Green and Colorado rivers (Figure 10.46a). A close view of the landform, shown in Figure 10.46b, indicates that the central part is a basin about one mile in diameter that surrounds a central dome. The pattern results from the presence of fractured, easily eroded rock in the core of an anticlinal structure as depicted in Figure 10.46c. Not easily seen in Figure 10.46b is the circular (ring) syncline that completely surrounds the central basin. The two limbs of the syncline form ring cuestas. An inner cuesta overlooks the central basin. An outer cuesta looks outward away from the central basin. The inner cuesta can be seen in Figure 10.46b as a thin sandstone ridge along the lip of the basin. The outer cuesta can also be seen as a sandstone ridge. This unusual landform appears to be the result of a meteorite that slammed into Earth at an uncertain time less than 170 million years ago.

One additional area deserves discussion and that is the Grand Canyon itself. The history of the Grand Canyon is complex and controversial, although fractures and faults have almost certainly played a role. Down-cutting along fractures is evident, not so much along the Colorado River itself, but along its many tributaries. Several of these, including Bright Angel Canyon, form very straight river channels that, through detailed field mapping, have been shown to follow near vertical fault or fracture zones. The Bright Angel fault/fracture zone extends across the rim of the canyon, through Grand Canyon village. These same faults elsewhere form monoclines, the origin of which was outlined in Figure 10.22.

As mentioned earlier, most of the structures on the Colorado Plateau developed between about 75 and 40 million years ago during a strong phase of deformation that also produced many of the structures in the Middle

and Southern Rocky Mountains. But the Colorado Plateau has been far from quiet. Since 40 million years ago, the landscape has been shaped and reshaped by both tectonic and erosional processes. Tectonism less than 10 million years ago has included faulting, particularly normal faulting associated with the Basin and Range, as well as volcanism. There is also evidence for regional uplift potentially on the order of thousands of feet. This very recent uplift is responsible for development and enhancement, if not the creation, of some of the canyons and landforms on the plateau, including those in Arches, Zion, Bryce, and the Grand Canyon. We will discuss the origin of the Grand Canyon in Chapter 18.

THE COASTAL PLAIN

Sedimentary rocks on the Coastal Plain and adjacent continental shelf are, for the most part, hidden beneath unconsolidated beach, river, lake, soil, and shallow marine deposits. Most of the rock layers are less than 100 million years old and are relatively poorly consolidated (soft). The rock layers tilt gently toward the ocean at a slightly higher angle than the slope of the land such that the oldest rocks are close to the Fall Line and the youngest are close to the ocean. The entire rock sequence thickens toward the ocean from zero at the Fall Line to more than 13,000 feet below the Atlantic continental shelf, and to more than 40,000 feet off the Gulf Coast, producing an overall wedge-shaped geometry. Extensive drilling and seismic profiling indicate that there are sedimentary rocks buried deep below the continental shelf that are much older than those exposed at the surface. The oldest sedimentary rocks were deposited about 230 million years ago during initial stages of the rifting event that opened the Atlantic Ocean. Crystalline rocks that

(a)

(b)   **(c)**

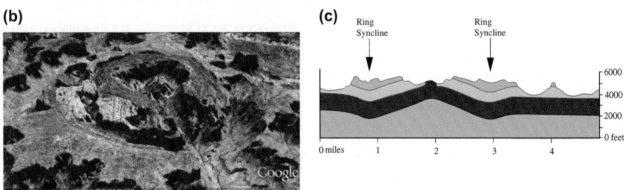

FIGURE 10.46   (a) A Google Earth image looking west across the Colorado and Green rivers. A white line points to Upheaval Dome. Bowknot Bend (Figure 8.1b) is seen at upper right. (b) A close-up view of Upheaval Dome looking southeast. (c) A cross-section of Upheaval Dome. The ring syncline completely encircles the impact structure. Cross-section based on Thornbury (1965).

represent the eroded remnants of the Appalachian/Ouachita Mountain belt underlie these old sedimentary rocks. The presence of Appalachian/Ouachita crystalline rocks at depth indicates that this part of the mountain belt had already eroded and subsided to sea level by 230 million years ago. Figure 5.1 is a cross-section that shows the buried Appalachians overlain by buried sedimentary rock.

The grain size, composition, sorting, and fossils in the exposed rock layers on the Coastal Plain suggest depositional environments that mimic those of the present day. This suggests that the Coastal Plain has been an area of low relief, similar to what it is today, during the entire time the rocks were deposited (more than 100 million years). Three major depositional environments are preserved in these rocks, and their distribution is an indicator of sea-level oscillation. The depositional environments are shoreline, inland continental, and shallow ocean. *Shoreline rocks* contain sediment deposited in estuaries, lagoons, tidal marshes, beaches, and barrier islands. *Inland continental rocks*

contain river, delta, and lake environments with sediment derived from erosion of the Appalachian Mountains. *Shallow ocean environments* include rocks such as limestone and dolostone that were deposited as offshore reefs. The age and distribution of these rocks indicate that sea level was quite high between 100 and 50 million years ago and then began to fall although with many oscillations. A few of the sea-level high stands are shown in Figure 7.1.

The sedimentary rock sequence, because it is weakly consolidated and poorly exposed, does not exert very much landscape control except locally where differential erosion of the tilted layers has produced small, inland-facing cuestas and hills. The north side of Long Island is an example of a cuesta covered with glacial moraine sediment (Figure 9.6). Note that the bedrock cuestas face inland, in a direction opposite that of the beach scarps described in Chapter 9. Most cuestas form low-relief hills and are hardly noticeable. They produce a mild form of bench-and-slope topography. Two areas where cuestas are well developed are the

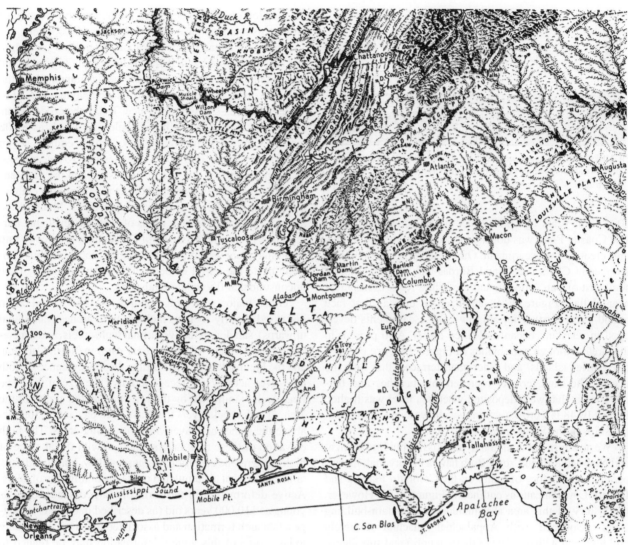

FIGURE 10.47   A landscape map of the Black Belt region.

Black Belt region along the eastern side of the Mississippi embayment and the Texas Coastal Plain. Individual rock units in the Black Belt exert control of the landscape because the layers are relatively thick and they show strong contrasts in erodability. Bedrock control of the landscape is clearly evident on the landscape map of Figure 10.47. Here, the margin of the Coastal Plain swings northward around the southern end of the Appalachian Mountains and across the Appalachian Plateau to form a boundary with the Interior Low Plateaus. The Fall Line, which is distinctive along the Atlantic Coast, no longer forms an escarpment that separates crystalline from sedimentary rock. With sedimentary rock on both sides, the boundary between the Coastal Plain and the Appalachian Plateau becomes indistinct and marked with a line of hills underlain by resistant sandstone referred to as the Fall Line Hills as shown in Figure 10.47 in the vicinity of Tuscaloosa, Alabama. Southwest of the Fall Line Hills a weak limestone known as the Selma Chalk

forms the Black Belt Valley which was once famous for its large mansions and cotton plantations and is still famous for its rich black soil. Beyond the Black Belt Valley lies the distinctive Pontotoc-Ripley cuesta followed by a series of less distinct bedrock-controlled valleys and cuestas. The Pontotoc-Ripley cuesta is an eroded, discontinuous set of hills, 50 to 200 feet above the Black Belt Valley, that are underlain by a resistant green clay (glauconite)-bearing sandstone.

Regarding the Texas Coastal Plain, several escarpments are shown in Figure 9.14 between Dallas and the Rio Grande River. The most obvious is the Balcones escarpment between Del Rio and Waco which separates the Coastal Plain from the Edwards Plateau section of the Great Plains. Unlike many of the bedrock cuestas on the Coastal Plain that face inland away from the ocean, the Balcones escarpment faces the ocean. The escarpment corresponds with a series of normal faults, active between about 25 and

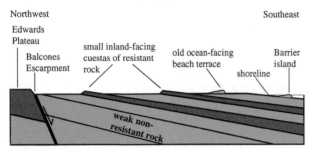

Northwest   Southeast

FIGURE 10.48 A cross-section of the Texas coast from San Antonio southeastward, showing escarpments and terraces. Based on Hunt (1974, p.228).

10 million years ago, known as the Balcones fault zone. The Coastal Plain was down-dropped along the fault zone, thus creating the escarpment. Other escarpments, including the Austin Chalk cuesta and Bordas-Oakville escarpment southeast of San Antonio, and the White Rock escarpment near Dallas (Figure 9.14), are bedrock cuestas that face inland (westward). They are low-lying escarpments developed in resistant, oceanward-tilted layers. Also present in this area, although not obvious in Figure 9.14, are present-day barrier islands, and older sand terraces, that face the ocean. The sand terraces represent old beaches and barrier islands that were deposited when sea level was higher. Figure 10.48 is a cross-section that shows some of the relationships in the San Antonio area.

We now turn our attention to the Mississippi Embayment and New Madrid fault zone. Both have a history that, in part, parallels the history of faults on the Colorado Plateau and Interior Low Plateaus as outlined in Figure 10.22. However, the history of this area also includes a mountain-building event associated with Appalachian-Ouachita mountain building, and with a period of thermal uplift and subsidence associated with a hot spot. Much of our knowledge is based on seismic and drill hole data. Faults at depth below the Mississippi Embayment (including the New Madrid fault zone) can be traced back possibly as far as 800 million years ago when a failed rift zone (an aulocogen) known as the Reelfoot Rift produced a large number of extensional normal faults across the region. The faults became inactive and were subsequently covered with sedimentary layers as depicted in Figure 10.22. At about 300 million years ago, the area was involved in Appalachian-Ouachita mountain-building events that resulted in the reactivation of some of the preexisting faults, and the creation of new faults, resulting in the development of a highland. Like most of the Coastal Plain, the Mississippi Embayment area may have eroded and subsided to a lowland by 230 million years ago when Pangea first began to break apart and the Atlantic Ocean began to open. The area remained a lowland until about 115 to 95 million years ago when it passed over a thermal plume known as the Bermuda hot spot. The addition of heat below the embayment area resulted in thermal isostatic uplift, the

reactivation of older preexisting faults, and the creation of mountains that may have stood as high as 10,000 feet.

The rocks began to cool by about 95 million years ago, and erosion coupled with thermal sinking quickly converted the mountains to a lowland before being inundated by the ocean about 80 million years ago. The embayment area has apparently been sinking ever since. Today the eroded roots of the mountain belt are buried below as much as 8,500 feet of younger sedimentary rock, which equates to an overall subsidence rate of 0.12 inches per 100 years over an 85-million-year time interval. Constant sinking over tens of millions of years has helped direct rivers into the embayment area, specifically the Mississippi River. Much of the area was inundated during times of high sea level and, during those times, the Mississippi Delta emptied directly into the embayment well north of its present location, thus contributing to the burial of bedrock and development of the unconsolidated floodplain landscape.

Other major structural features on the Coastal Plain discussed briefly below include two arches (anticlinal upwarps), salt domes, active normal faults, reefs, and a meteorite impact crater. The Cape Fear Arch in North Carolina, and the Peninsular Arch in Florida, display evidence of both long-lived activity and recent activity. The size of the Cape Fear Arch is indicated by the presence, in the center of the arch, of older rock layers that have been elevated 2,500 feet above the same rock layers in surrounding areas. The high elevation of these rock layers, coupled with the absence of high topographic elevation, suggests a slow, long-lived, sustained uplift that accumulated over tens of millions of years. Active deformation in the form of warped sand ridges as young as 110,000 years old (as described in Chapter 9) suggest that arch formation and associated uplift are continuing today. Some of this young deformation may be related to earthquakes that shake this area from time to time.

The Peninsular Arch is centered on the Ocala dome in north-central Florida (Figure 9.8). This arch is a broad structure with an average regional dip away from its center of about 5 feet per mile. The age of deformed beds suggests that uplift began prior to 40 million years ago and continued until about 5 million years ago, producing some of the highest topography in Florida. The crest of the Peninsular Arch exposes limestone along the entire backbone of Florida from Tallahassee to Orlando, which, in this humid climate, has created a major karst area of sinkholes and springs known as the Lime Sink region (Figures 2.7 and 9.8).

Salt domes are present at depth below the continental shelves of New England and the Carolinas. However, they are most abundant in the Gulf region, both inland on the Coastal Plain and along the coastline, particularly from Mississippi westward to the northern coastline of Texas. As noted in our discussion of the Colorado Plateau, thick layers of salt are unstable at depth. They tend to rise as a column and, in doing so, they fracture, fold, and fault surrounding

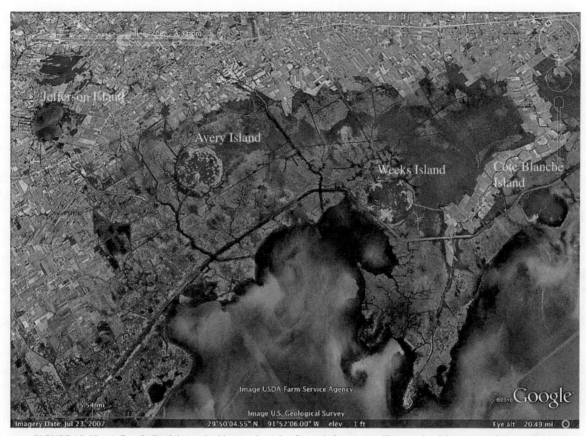

FIGURE 10.49   A Google Earth image looking northward at four salt domes near Florence, Louisiana, west of New Orleans.

and overlying sedimentary layers. The deformation is what creates major oil traps in the region. Many of the domes along the Gulf Coast intersect the surface, creating, in some cases, a circular upland 1 to 4 miles in diameter and a few feet to about 150 feet above surrounding land or, in other cases, a circular lake. Figure 10.49 shows four of the most visible salt domes on the Louisiana coast.

Active normal faults, down-thrown toward the Gulf Coast, are present across southern Texas, Louisiana, and Mississippi. The rocks here are so poorly consolidated that the faulting results primarily in subsidence along the coast. Coupled with rising sea level, this subsidence is bound to cause problems in the near future. One thing the folks in this area do not have to worry about is an earthquake. The poorly consolidated rocks of the Coastal Plain are not strong enough to create an earthquake that could do any damage.

Most of Florida south of Lake Okeechobee is marsh and swampland less than 20 feet above sea level. This is the area of the Big Cypress Swamp and the Everglades. South of the Everglades, the Florida Keys extend for more than 150 miles to Key West (Figure 9.8). The Eastern Keys, from Key Biscayne National Park to the vicinity of Seven Mile Bridge and Big Pine, are built on a dead coral reef that became emergent within the past 100,000 years. With the rise in sea level over the past 18,000 years, a new live coral

reef has formed along the eastern side of the dead reef. The area from Big Pine to Key West is built, not on a reef, but on a limestone shoal (sandbar), most of which is less than 20 feet above sea level.

The Coastal Plain sedimentary sequence was disrupted about 35 million years ago when a meteor, estimated to be between 1.8 and 3.1 miles in diameter, crashed offshore just north of Norfolk near what is today the mouth of Chesapeake Bay. The location is shown in Figure 9.7. Sea level was higher at the time of impact such that the shoreline was in the vicinity of Richmond and the Fall Line. The impact produced a crater more than 53 miles (85 km) in diameter and about 5,000 feet deep, penetrating through sedimentary rock and into the crystalline rock of the eroded Appalachian Mountains. It is known as the Chesapeake Bay Impact Structure. It is the largest impact structure in the United States and the sixth largest in the world. It likely vaporized both itself and the surrounding ocean water on impact. Broken rock material, thrown up as a result of impact, came down in a pile that filled the crater. At about the same time, ocean water swept into the pile of broken rock creating a water-saturated slurry of sand and rubble known today as the *Exmore breccia*. The crater is now buried and out of sight under several thousand feet of sediment that accumulated following impact.

Most of what we know of the crater was gathered from drill core and seismic reflection records. Shocked quartz and melted fragments from drill cores confirm that it is indeed an impact structure. The 35-million-year age of impact is based on radiometric dating of melted fragments. The impact permanently disrupted groundwater flow by truncating fresh water aquifers. Water within the Exmore breccia is 1.5 times saltier than normal seawater, making it useless for drinking or for industry. Today there is greater subsidence over the impact structure than in surrounding areas.

QUESTIONS

1. Beginning with Figure 10.1a, explain why one area might erode as shown in Figure 10.1b and another area as shown in Figure 10.1c. Speculate on how rock type, rock hardness, climate, and surrounding geography might be different in the two areas.
2. Beginning with Figure 10.1d, explain why one area might erode as shown in Figure 10.1e, another as shown in Figure 10.1f, and another area as shown in Figure 10.1g. Speculate on how rock type, rock hardness, climate, and surrounding geography might be different in the three areas.
3. In Figures 10.5 and 10.6, note the degree of dissection in the Missouri Plateau surrounding the Black Hills. Compare the level of dissection with that of the Llano Estacado (Figure 10.8), the Colorado Plateau (Figure 10.11), and the Appalachian Plateau (Figure 10.35). Which of these three areas does the Missouri Plateau most resemble? Explain your reasoning. Why is there more dissection west of the Missouri River than to the east in Figure 10.6?
4. Use Figure 10.4 as a guide to draw boundaries between the High Plains, Colorado Piedmont, Plains Border, and Central Lowlands in Figure 10.7. Pick any two adjacent provinces and describe their physiographic differences.
5. Speculate on the origin of the Colorado Piedmont.
6. If at one time rivers flowed across the Llano Estacado, explain how the Mescalero Escarpment formed.
7. There are several domal mountains on the Colorado Plateau, as indicated in Figures 10.4 and 10.11, including Mt. Ellen, the La Sal Mountains, and the Navaho Mountains. Speculate on their origin.
8. Referring to Figure 10.11, how is the northeast border of the Colorado Plateau in Colorado different from the northwest border in Utah? Describe physiographic differences on either side of the border.
9. Describe changes to the cross-section in Figure 10.14 if the region were to shift to a much more humid climate.
10. Using Figure 10.23 as a guide, describe physiographic differences between the Interior Low Plateaus and the Central Lowlands along the northwestern border near Evansville and compare these differences with the northeastern border of the Interior Low Plateaus with the Appalachian Plateau.
11. Research the Crooked Creek Impact crater and write a one- or two-paragraph report.
12. Using Figure 10.39 as a guide, describe physiographic differences between the Ozark Plateau and the Ouachita Mountains.
13. Research Upheaval Dome and write a one- or two-paragraph report.
14. Briefly explain why the dominant structural style in the United States is nearly horizontal sedimentary rocks.
15. How can mountains form in flat-lying sedimentary layers?
16. Explain how hogbacks form.
17. What is an unconformity? What is the minimum amount of time represented by the unconformity between the oldest sedimentary rocks in the central US and the underlying ancient crystalline rock?
18. What is the maximum age of the landscape of the Great Plains and Wyoming Basin?
19. Name several of the basins that form the Wyoming Basin.
20. What is the significance of the Ogallala formation?
21. Explain bench-and-slope landscape. Draw a labeled example in cross-section.
22. How did monoclines of the Colorado Plateau form? Explain with a cross-section.
23. Compare and contrast the bench-and-slope landscape of the Interior Low Plateaus with that of the Colorado Plateau.
24. The resistant sandstone in Figure 10.29 appears to occupy a greater width on the south side of the Eagle Valley syncline than on the north side. What can you infer from this observation regarding the dip of the sandstone, and why?
25. How do erosional mountains differ from plateaus? How do they differ from the "classic" mountain?
26. Explain the stages of dissection resulting in erosional mountains.
27. Explain why a maturely dissected landscape can be younger in actual years than a less mature landscape.
28. Under what circumstances are erosional mountains most likely to form?
29. Describe the landscape of the Ozark Plateau.
30. What is chert? How do interbeds of chert in the limestone affect the landscape of the Salem and Springfield Plateaus?
31. How does the lithology of the Boston Mountains help explain their existence on the Ozark Plateau?
32. What is the minimum age of the Ozark Plateau? How do we know?
33. Explain the formation of vertical fractures in the Entrada sandstone. Why are they significant to the development of landscape in Arches National Park?

34. Why do arches form? Explain using the following terms: dry climate, vegetation, freeze-thaw, fin, differential weathering.
35. Explain how fractures influence the landscape of Zion National Park.
36. Explain how fractures influence the landscape of Bryce Canyon National Park.
37. Why are the rocks of the coastal plain wedge-shaped?
38. How can sea-level fluctuations (also called eustatic rise and fall or transgression and regression) be discerned from the rocks and fossils of the Coastal Plain?
39. How can a geologist in the field discern between a loosely consolidated bedrock cuesta and a beach scarp?
40. Why do the rock units of the Black Belt region exert notable landscape control?

41. How does the Balcones Escarpment differ from other bedrock cuestas (the Austin Chalk cuesta, Bordas-Oakville Escarpment, White Rock Escarpment)?
42. Define an aulocogen. What famous aulocogen is responsible for the New Madrid fault zone and the faults of the Mississippi Embayment?
43. What may have caused thermal isotatic uplift and high mountains in the Mississippi Embayment 100 Ma?
44. In spite of active normal faults, why shouldn't the residents of Texas, Louisiana, and Mississippi be worried about a devastating earthquake? What should they be concerned about instead?
45. What is the groundwater significance of the Chesapeake Bay Impact Structure?
46. Identify the structure-controlled and erosion-controlled landscape in Figure 10.21. Speculate as to why or how the two landscapes are in such close proximity.

Chapter 11

Crystalline-Cored Mid-Continent Anticlines and Domes

Scattered across the central part of the United States are anticlinal and domal mountains in which erosion has removed the thin, overlying sedimentary sequence, and exposed the underlying crystalline rock in their core. The characteristic landscape is one of massive, rugged highlands. Major domes include the Adirondack Mountains of New York, the Black Hills of South Dakota, the St. Francois Mountains of Missouri, the Wichita and Arbuckle Mountains of southern Oklahoma, the Llano Uplift of West Texas, several domal mountains in the northern Great Plains and Colorado Plateau, and the largest dome structure of all, the Middle-Southern Rocky Mountain region (Figure 8.4 or 10.2). Many are flanked by normal, reverse, or thrust faults that accentuate the domal structure. The crystalline rock in most (but not all) of these areas is part of the ancient North American crystalline shield.

Each area has a slightly different history. Some are ancient uplands hundreds of millions of years old that retain their highland topography precisely because of the presence of hard crystalline rock in the anticlinal center. Some may have evolved as an anticlinal basin similar to the Nashville Basin and, over time, eroded and reincarnated into an anticlinal mountain as depicted in Figure 8.3. A few developed when young igneous intrusions pushed up and domed the sedimentary layers above. The Middle-Southern Rocky Mountains, the Colorado Plateau, and the western part of the Great Plains together form a broad, regional dome that results in some of the highest elevations in the country.

THE ADIRONDACK MOUNTAINS

The Adirondacks form a circular dome 125 miles across with a core of ancient, complexly deformed crystalline shield rock 1.3 to 0.95 billion years old surrounded by younger sedimentary layers (<542 Ma) that dip away from the central dome in all directions forming a series of low-lying cuestas. The Adirondacks are located in upstate New York west of the Champlain Valley (Figure 9.4). The entire area is protected as a state park, the largest in the contiguous United States.

Figure 11.1 is a Google Earth image of the west-central Adirondacks. The figure reveals a complex structural pattern in the rocks that appears to have been etched out by erosion. Many of the curved valleys follow folds in weak rock layers. Straight valleys follow fault or fracture zones. More than 2,000 lakes fill depressions. The landscape that most resembles the central Adirondacks can be found in the glacially scoured region north of the Great Lakes (Figure 6.8c). The topography, therefore, suggests that the entire Adirondack dome underwent erosion beneath a continental glacier. A closer look at Figure 11.1 shows that the eastern half of the dome maintains a mountainous terrain with about a dozen peaks above 5,000 feet and topographic relief on the order of 3,000 feet. Many of the high peaks, including Mt. Marcy, the highest peak at 5,344 feet, are composed of a rare intrusive rock known as anorthosite which consists almost entirely of the mineral plagioclase.

Although glacial erosion features dominate the landscape, glacial depositional features are also present. Most unique among these is a complex system of eskers that trend northeast-southwest from the Saranac Lake area just west of Lake Placid to Stillwater Reservoir and Cranberry Lake (Figure 11.1). An *esker* is a long, sinuous (snake-like) ridge of glacial-derived stream gravel that forms below a stagnant glacier. Figure 11.2 is a Google Earth image of one

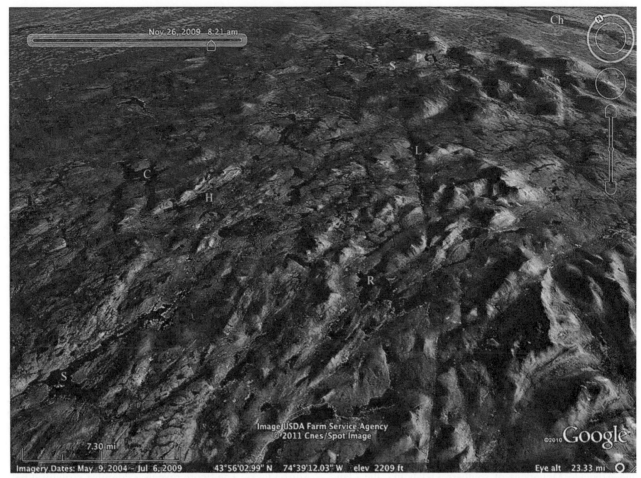

FIGURE 11.1 A Google Earth image looking NNE across the west-central Adirondack Mountains. The small upside-down horseshoe-shaped lake at upper right is Lake Placid (*P*). Mt. Marcy and other high peaks are located just south of Lake Placid. Stillwater Reservoir (*S*) is located at lower left and Cranberry Lake (*C*) at center left. Note the long, nearly continuous lineament located at right center, occupied in part by Long Lake (*L*). The lineament passes through Saranac Lake (*S*), located just to the left (west) of Lake Placid. The large lake at lower-right center, close to the lineament, is Raquette Lake (*R*). The line of eskers trends NE-SW across the image from north of Saranac Lake to Cranberry Lake and Stillwater Reservoir. The esker shown in Figure 11.2 is at Hitchins Pond (*H*) in Sabattis, located just southeast of Cranberry Lake. Lake Champlain (*Ch*) is at extreme upper right.

of these eskers located in Sabattis, New York. The origin of an esker is as follows. As a stream flows below a stagnant ice pack, the surrounding walls of ice prevent the stream from meandering. Sediment falls to the bottom of the stream and builds into a long sinuous pile. Rather than moving off the pile as would be expected, the stream continues to flow over the pile because it is locked in position by the walls of stagnant ice. Once the ice melts, the pile of sediment is left as a long, sinuous ridge. Streams normally produce topographic valleys, but in this case, topography was inverted and a ridge was created. The presence of numerous eskers in the Adirondacks indicates that the glacier that once covered the area did not retreat with the main continental glacier. Instead, it detached from the main glacier, stagnated, and melted in place, thus allowing eskers to develop.

The metamorphic and plutonic rocks that form the Adirondack Mountains are part of the North American crystalline shield. The area was eroded to sea level by about 542 Ma

(i.e., million years ago) and subsequently covered with sedimentary rock. Evidence for this interpretation is found in the small amount of sedimentary rock that is still preserved in down-dropped fault basins within the dome surrounded by crystalline rock. The faults and associated fracture zones were active about 465 million years ago and are responsible for some of the north- to northeast-trending lineaments that were later accentuated by glacier erosion. The lineament that passes through Long Lake in Figure 11.1 is an excellent example. Sedimentary rocks continued to be deposited with intermittent periods of uplift and erosion until at least 300 million years ago, which is the age of the youngest rock that surrounds the dome. The history of the Adirondacks after 300 million years ago is less well known, but some geologists contend that the area has been above sea level since that time.

Insight into the recent history of the Adirondacks has been obtained from apatite fission track and (U-Th)/He dating. The data suggest that the Adirondacks underwent slow

FIGURE 11.2   A Google Earth image looking eastward at the snake-like esker in Hitchins Pond, Sabattis, New York. The esker can be followed in the distance along the south (right) side of the pond.

cooling from about 170 to 130 million years ago, followed by a heating event that lasted until about 105 Ma. This was followed by a more rapid cooling event that lasted until about 95 Ma. The data suggest that the Adirondack landscape was a low-lying flatland or a low plateau that underwent very little uplift or erosion prior to and during the 170- to 130-million-year time period. Crystalline rocks at this time may have already been exposed at the surface or were close to the surface under a thin layer of sedimentary rock. The heating event from 130 to 105 million years ago correlates with the passage close to the Adirondacks of a thermal plume known as the New England (or Great Meteor) hot spot. This is the same thermal plume that would later produce the New England Seamount chain (Chapter 5). Heating associated with the hot spot is believed to be responsible for broad doming and uplift of the Adirondacks, possibly reactivating older faults. These events, in turn, triggered increased rates of erosion, causing dissection that continued until about 95 Ma. The crystalline core likely became widely exposed during this time interval. By 95 Ma, the hot spot was far enough away for the rocks to cool to a normal geothermal gradient. The mountains likely subsided slightly in response to cooling.

The study therefore suggests that the Adirondack Mountains, as we know them today, came into being between 130 and 95 million years ago as a result of broad doming and erosion of a plateau surface. Given this interpretation, the Adirondacks could be considered an erosional mountain underlain by crystalline (rather than sedimentary) rock. The mountains were sculpted, accentuated, and elevated into their present shape by an overall global sea-level drop since 95 Ma and by glaciation during the past 2.4 million years which covered and scoured the entire mountain dome.

THE ST. FRANCOIS MOUNTAINS

The St. Francois Mountains are located south of St. Louis on the Ozark Plateau where crystalline rock crops out in a circular faulted dome surrounded by sedimentary rock of the Salam Plateau. The St. Francois Mountains are shown in Figure 10.39 on a landscape map of the Ozark Plateau and in Figure 11.3 in cross-section.

As is typical of many crystalline-cored domes, the crystalline rocks are part of the ancient nucleus of the United States (the crystalline shield). Most are between 1.55 and

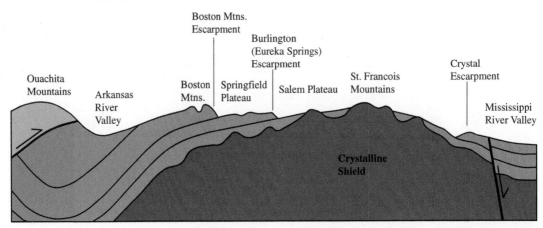

FIGURE 11.3 A schematic cross-section across the Ozark Plateau showing the St. Francois Mountains and the surrounding plateaus. Note the buttress unconformities surrounding crystalline rock in the St. Francois Mountains.

1.3 billion years old, and similar to the Adirondacks, they underlie much younger sedimentary layers of the Salem Plateau across an ancient erosion surface (an unconformity as described in Chapter 10). In addition to granite and a small amount of gabbro, much of what I refer to here as crystalline rock is actually very hard, silicic volcanic rock. I have included this rock in the crystalline category because it is extensively intruded by granitic rock and because it is as hard or harder than the closely associated granitic rock. One interpretation of these volcanic rocks is that, long ago, this area resembled Yellowstone National Park.

An interesting aspect of the St. Francois area is that the erosion surface (the unconformity) between the ancient crystalline rock and the much younger sedimentary rock is not planar. Most erosion surfaces, including those in the Adirondacks, were eroded to nearly a flat plain prior to deposition of overlying rock. What makes the St. Francois unconformity so unusual is that the ancient crystalline landscape may have had as much as 2,000 feet of relief during deposition of the overlying sedimentary layers. The interpretation is that the sedimentary rocks were deposited from a shallow ocean during rising sea level between about 542 and 443 million years ago such that the ancient crystalline land area was slowly drowned. The evidence for this interpretation is the unconformity itself. The type of unconformity present in the St. Francois Mountains and shown in Figure 11.4 is known as a *buttress unconformity* because younger sedimentary layers abut and truncate against the crystalline rock. This type of unconformity is explained in Figure 11.5. The figure suggests that ocean water first covered low areas of crystalline rock while leaving topographically high areas as islands. The ocean would lap up against the islands and deposit a layer of sedimentary rock around their margin perhaps in the form of a beach. As sea level continued to rise, sedimentary layers would progressively cover more and more of the islands. Ultimately, the ocean would drown the islands and a layer of sedimentary rock would cover the entire region.

FIGURE 11.4 A photograph of a buttress unconformity in the St. Francois Mountains where layers of sedimentary rock abut against a knob of older crystalline rock.

Ocean water laps against an ancient hillside and deposits sediment.

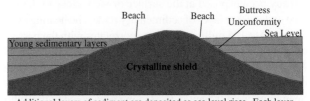

Additional layers of sediment are deposited as sea level rises. Each layer abuts against the ancient crystalline rock creating a buttress unconformity.

FIGURE 11.5 The origin of a buttress unconformity.

The situation just described is similar to what is happening today along the coast of Maine. Here crystalline rock of the Appalachian Mountains trends directly into the ocean. The area is slowly subsiding against rising sea level, resulting in deep ocean inlets and an abundance of coastal

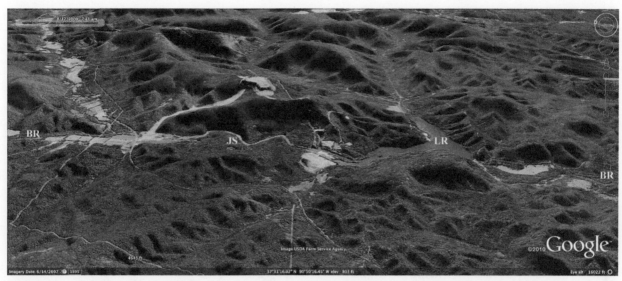

FIGURE 11.6   A Google Earth image looking east at the East Fork Black River (*BR*). The river flows from left to right (southward) across the photo from sedimentary rock on the left, into crystalline rock at Johnson Shut-ins (*JS*) and back into sedimentary rock at Lower Reservoir (*LR*) on the right.

islands. Ocean sediment, deposited at progressively higher intervals along the margins of the slowly drowning crystalline islands and inlets, will produce a buttress unconformity. Another area where we can find a buttress unconformity is near the bottom of the Grand Canyon. Here crystalline rock of the inner gorge locally peaks up through the lowest layers of overlying sedimentary rock.

In the case of the St. Francois Mountains, the buttress unconformity has produced some interesting landscape features. Uplift of the Ozark Dome, coupled with river erosion, has created patches of crystalline rock surrounded by sedimentary rock. The crystalline rock, because of its hardness, forms highlands or knobs. The sedimentary rock is weak and forms flat-floored valleys. Note in Figure 11.3 how crystalline rock sticks up from between layers of sedimentary rock. The East Fork of the Black River flows from left to right across the center of Figure 11.6. On the left side of the image, the river flows through low-lying topography (farmland) underlain by sedimentary rock. In the center of the image, the river enters a gorge of rugged highland topography underlain by crystalline rock (labeled Johnson Shut-ins, JS). On the right side of the image, the river flows back into low-lying topography underlain by sedimentary rock (marked by farmlands and the Lower Reservoir). The interpretation is that the river was in existence and flowing through sedimentary layers prior to exposure of crystalline rock. Over time, the river cut downward into the slowly emerging crystalline rocks while managing to maintain its original course. Because of the strong contrast in erodability, the river had to cut a gorge through the crystalline rock to maintain a downstream gradient. The result is a river that flows from an open valley, into a gorge, and back into an open valley due to contrasts in rock hardness. River gorges that form

in this way are referred to in Missouri as *shut-ins*. The Johnson Shut-ins are a prime example.

The uplift history of the St. Francois Mountains follows that of the Ozark Plateau discussed in Chapter 10. The area has been a highland since at least 265 million years ago and possibly much longer. As erosion has slowly reduced the landscape, more and more of the buried ancient crystalline hills have been exhumed thus creating the shut-ins. This interpretation would imply that some of the rivers are as old as the landscape.

THE WICHITA, ARBUCKLE, AND LLANO STRUCTURAL DOMES

The Arbuckle and Wichita Mountains lie less than 100 miles from each other in the Central Lowlands of southwest Oklahoma just west of the Ouachita Mountains and north of Wichita Falls, Texas (Figure 11.7). Both are old domal mountains complicated by faults. They expose crystalline rock in their core and are surrounded by younger sedimentary layers. The Wichita Mountains are the largest and most conspicuous of the two. The mountains rise 500 to 1,400 feet above the surrounding plain, with Haley Peak, at 2,481 feet, the highest point. The dominant crystalline rock is a red granite, but also present are a variety of other rocks, including gabbro, anorthosite (as in the Adirondacks), and silicic volcanic rocks. A unique aspect to the Wichita Mountains, as shown in Figure 11.8, is the wide exposure of bare granitic rock. The rock is exposed free of overlying soil or vegetation across most of the mountain except the central part which is covered in the natural prairie grasslands of Wichita Mountains Wildlife Refuge. A pair of strong fracture/fault systems oriented approximately west-northwest and north-south within the granitic rock controls much of

FIGURE 11.7 A landscape map of the Wichita, Arbuckle, and Llano areas. Boundaries between the Central Lowlands, Great Plains, and Coastal Plain are shown. The oval-shaped area on the Great Plains west of Austin forms the Llano uplift.

FIGURE 11.8   A Google Earth image looking west at fractured, bare, granite exposures in the Wichita Mountains. The Meers fault is visible as a thin line along the right side of the image. The fault is parallel with the dominant ENE fracture orientation in the Wichita Mountains.

the landscape. The dominant west-northwest-trending system is responsible for the overall trend of the mountain range. The crystalline rock is unusual in the sense that it is only 550 to 600 million years old. Such an age could hardly be considered young, but it is considerably younger than the 1- to 3-billion-year-old crystalline rock that forms the bulk of the North American crystalline shield.

The Wichita Mountains boast one of the very few active faults that reach the Earth's surface east of the Cordillera. Known as the Meers Fault, it forms a 15-mile-long lineament oriented parallel with the dominant west-northwest-trending fracture system. It is located just north of Lawton, Oklahoma, in the northwest part of the range. It can be seen along the right side of Figure 11.8. Trenching along the fault suggests that the last major earthquake occurred between about 1,100 and 1,400 years ago.

The Arbuckle Mountains form a low-lying subdued area that perhaps should be more appropriately referred to as hills or a small dissected plateau. Mountaintops are flat and plateau-like. The highest elevation is 1,412 feet, and there is less than 500 feet of relief between the mountaintop and surrounding lowland. Most of the mountain area retains its sedimentary cover with crystalline rock exposed primarily in the southeastern part. Although dominantly granite, the crystalline rock is weaker (more easily eroded) than rocks in the Wichita Mountains, and about the same age as rock in the St. Francois Mountains (1.3 to 1.4 billion years old). The surrounding sedimentary layers are deformed into an anticline with steeply dipping layers and minor folds and faults. Differential erosion of sedimentary layers has created a series of hogbacks and cuestas along mountain flanks. Abundant limestone at or close to the surface has locally created a karst topography of sinkholes and caves.

The Llano Uplift is located at Llano, Texas, roughly 270 miles south of the Wichita Mountains on the Great Plains, along the eastern margin of the Edwards Plateau. The area is circled in the lower part of Figure 11.7. Although structurally a dome, topographically it is more accurately described as a broad, poorly defined topographic basin with relatively rugged relief, especially in the vicinity of rivers. The structural dome is cored by ancient crystalline rock (metamorphic and granite) between about 1.0 and 1.35 billion years old and surrounded by sedimentary rock. Crystalline rock occupies the lowest parts of the basin. The few crystalline mountains that are present, such as Smoothingiron Mountain and Granite Mountain, are erosional in nature, very small, and mostly less than 500 feet above an expansive basin floor. The pink granite at Enchanted Rock State Park shown in Figure 11.9 is an excellent example. In terms of its relative erosional advancement, the Llano Uplift has progressed beyond that of the Nashville Basin to where crystalline rock is just emerging across a flat plain (Figure 8.3d). Presumably, as additional crystalline rock is unearthed by erosion, the Llano Basin will emerge as a small domal mountain similar in some respects to the St. Francois Mountains.

The Llano Dome, and the Arbuckle and Wichita Mountains, all underwent a similar landscape history. We can be reasonably confident that all three emerged as a highland area sometime between 300 and 265 million years ago because this is when the nearby Ouachita mountain-building phase reached its climax. The rocks were deformed at that time and may have been at their highest elevation. Notice in Figure 11.7 that the southern boundary of the Arbuckle Mountains coincides with the boundary of the Coastal Plain. Along this boundary, young sedimentary rocks of the Coastal Plain

FIGURE 11.9 A Google Earth image looking north across the Llano basin at Enchanted Rock and several other small outcroppings of crystalline rock.

sit directly above ancient crystalline rocks. Detailed mapping suggests that a similar relationship exists in the Llano area. In order for this to happen, the crystalline rock had to have been exposed at the surface when deposition of the overlying sedimentary layers began. Thus, we can infer that, beginning 265 million years ago, the Arbuckles (and by extension, the Wichita and Llano areas) underwent slow erosion punctuated by sea-level change and isostatic adjustment such that, by about 100 million years ago, crystalline rock was exposed at the surface. Following erosion to a lowland, all three areas were drowned by the 100- to 80-million-year-old high sea-level stand, resulting in deposition of young Coastal Plain sedimentary rocks directly above the crystalline rock. The inferred sequence of events is outlined in Figure 11.10. Interestingly, the circa 100-million-year-old sedimentary layers in the Llano area produced buttress unconformities similar to, but far younger than, the buttress unconformities in the St. Francois Mountains.

All three areas emerged from below sea level as a low-lying landscape between 80 and 65 million years ago; therefore, this could be considered the maximum age of the present-day landscape. At that time, crystalline rocks were buried beneath a thin layer of young sedimentary rock. In the Llano Uplift it is possible to approximately date the time that crystalline rock was reexposed at the surface. Clasts (pieces) of crystalline rock are absent in surrounding eroded sediment older than 2 million years but are present in sediment younger than 2 million years. This suggests that crystalline rock was reexhumed to the Earth's surface less than 2 million years ago. Once crystalline rock was reexposed, all three areas likely grew old over time via slow erosion. Given the overall history of these areas, it is possible that the landscape we see today actually resembles the landscape that existed 100 million years ago prior to drowning by a shallow sea (compare Figures 11.10b and d).

THE NORTHWESTERN GREAT PLAINS

North of the Bighorn Mountains, along the fringe of the Great Plains in western Montana, are numerous small domal mountains. From north to south in Figures 10.4 and 10.5, they include the Sweet Grass Hills, Bears Paw, Little Rocky, Highwood, Moccasin, Judith, and the Big Snowy Mountains. Elevations are mostly between 6,000 and 7,000 feet which is 3,000 to 4,000 feet above the surrounding Great Plains. The Big Snowy Mountains rise to nearly 8,700 feet. The mountains consist variably of intrusive, volcanic, and sedimentary rocks. They are part of the Central Montana alkalic province, which implies that the intrusive and volcanic rocks are rich in potassium and poor in quartz and plagioclase relative to a true granite. The rocks are 50 to 70 million years old. All except the Big Snowy Mountains contain at least a small amount of intrusive rock in their anticlinal core. Intrusive rocks are especially well exposed in the Sweet Grass Hills, Moccasin, and Judith Mountains. Volcanic rocks are abundant in the Bears Paw, Little Rocky, and Highwood Mountains.

CRYSTALLINE-CORED DOME MOUNTAINS ON THE COLORADO PLATEAU AND THE COLORADO MINERAL BELT

Domal mountains on or near the Colorado Plateau cored with crystalline granitic rock include the Henry, La Sal, Abajo, Carrizo, Navajo, Ute, La Plata, and Rico Mountains. Some of the mountains, including Mt. Ellen and Mt. Pennell (both in the Henry Mountains), are more than 11,000 feet high. All of the mountains are located in Figure 10.4 and named in Figure 10.11. These mountains, like those on the Great Plains, formed when granitic intrusions punched their way into overlying sedimentary rock, thus creating a dome

(a)

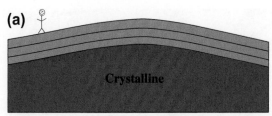

Mild anticlinal uplift between 300 and 265 million years ago. The rocks remain nearly flat-lying.

(b)

Erosion and exposure of crystalline rock by 100 million years ago.

(c)

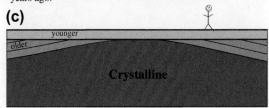

Drowning of the land area by a shallow sea, and deposition of circa 100 to 80 million year old Coastal Plain sedimentary rock unconformably above older crystalline and sedimentary rocks.

(d)

Slight doming and erosion reexposes crystalline rock. Note the angular unconformity between the younger sedimentary rocks and the older, but still nearly flat-lying, sedimentary rocks.

FIGURE 11.10 Sketches a through d that show the sequence of landscape development in the Wichita, Arbuckle, and Llano areas.

structure. Subsequent erosion exposed the granitic intrusive rock. Most of the intrusions are between 20 and 40 million years old, although some, including the Carrizo and Ute intrusions, are circa 70 million years old.

The Colorado Mineral Belt is a northeast-trending, 300-mile long, 15- to 30-mile-wide belt of intrusions and mining districts that extends from the Four Corners region on the Colorado Plateau to Boulder, Colorado. Several intrusions on the Colorado Plateau, including those that form the Carrizo and Ute Mountains, are part of the Colorado Mineral belt as shown in Figures 10.4 and 11.11. The orientation of the mineral belt is unusual because it cuts obliquely across the geologic grain and seems to have no correlation with tectonic or structural elements. Fission-track and argon dating indicate that intrusions associated with this

belt began about 75 million years ago and continued until at least 18 million years ago. The main trend and some of the mineralization were created by intrusions that are more than 43 million years old; however, most of the world-class lead, zinc, silver, gold, and molybdenum mining districts are associated with younger intrusions concentrated within the bulge in the trend near Salida and Leadville (Figure 11.11). The bulge includes the Climax, Henderson, and Red Mountain mines, all of which formed between 40 and 24 million years ago. Mining began in 1858, reached a peak at the turn of the 20th century, and has continued sporadically into the 21st century. Colorado is renowned for its many old abandoned mining camps such as the one shown in Figure 11.12 nestled in the Mosquito Range near Leadville. Note in Figure 10.4 that the bulge in the Colorado Mineral belt occurs at the intersection with the Rio Grande Rift. Given this relationship, it is possible that extensions associated with the rift opened easy pathways for some of the youngest magma to reach the surface. The Rio Grande Rift is discussed in Chapter 15.

THE MIDDLE AND SOUTHERN ROCKY MOUNTAINS

The Middle and Southern Rocky Mountains differ considerably from the Northern Rockies in terms of rock type, structure, and topography. The eastern part of the Northern Rockies is characterized by long linear mountain ridges and intervening valleys in which thick sequences of sedimentary rock (>30,000 feet) have been pushed eastward along major thrust faults. The mountains of the Middle-Southern Rockies are more massive and less continuous. These differences are clearly evident in Figure 10.4.

The structure of the Middle-Southern Rockies, although complicated by reverse, thrust, and normal faults, is primarily one in which crystalline-cored anticlinal mountains alternate with synclinal valleys of sedimentary rock. The thickness of sedimentary rock is less than what is found in the Northern Rockies (mostly <20,000 feet). Most of the crystalline rock is in the age range of 1.4 to 1.8 billion years old and is part of the North American crystalline shield; however, there are younger, 70- to 20-million-year-old, granitic intrusions such as those in the Colorado Mineral Belt. The Middle and Southern Rocky Mountain regions occupy most of Wyoming and Colorado, respectively. The rocks and the structures are similar in these two states, yet the landscape is distinctly different. The Middle (Wyoming) Rockies are dominated largely by wide basins, whereas the Southern (Colorado-New Mexico) Rockies are mountainous. The reason for this difference will become evident at the end of this section.

Figure 10.3 lists the major physiographic features. Major anticlinal mountains are highlighted in Figure 11.11. Figure 11.13 is a cross-section through west-central Colorado that shows the relationship between structure

FIGURE 11.11   A landscape map of the Middle-Southern Rocky Mountains with primary anticlinal mountains highlighted. The heavy lines that trend southwest to northeast enclose the Colorado Mineral Belt.

FIGURE 11.12 A photograph of an abandoned mine camp in the Mosquito Range near Leadville, Colorado.

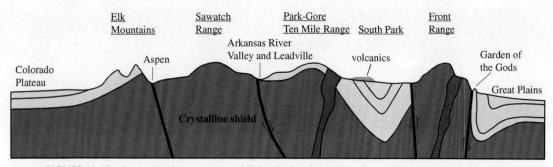

FIGURE 11.13 A cross-section across central Colorado that shows the major anticlinal mountains and valleys.

and landscape. We will survey the Southern Rockies from east to west beginning in the Great Plains. Mountainous topography begins at the western edge of the Great Plains where flat-lying sedimentary layers tilt abruptly toward the vertical as they encounter the anticlinal uplift of the Rocky Mountain Front Range. Tilted sedimentary layers form hogbacks along nearly the entire length of the Front Range and are beautifully displayed at Garden of the Gods in Colorado Springs where erosion of nonresistant sedimentary layers has left resistant layers to form high hogback ridges that today are used for rock climbing. A photograph and schematic cross-section of Garden of the Gods is shown in Figure 11.14.

The Front Range extends from the North Platte River in Wyoming to Royal Gorge on the Arkansas River in Colorado (Figure 11.11). The northern part of the Front Range is known as the Laramie Range. The southern part, south of Denver, is known as the Rampart Range. The Medicine Bow Mountains splay off the Laramie Range across a narrow valley and trend north-northwestward into Wyoming to the edge of the Wyoming Basin. South of the Arkansas River, the Wet Mountains, the Spanish Peaks, and finally the Sangre de Cristo Mountains form the primary range front overlooking the Great Plains. The Spanish Peaks are unlike and separate from the surrounding anticlinal ranges. They consist of two crystalline-cored mountains. West Spanish Peak tops

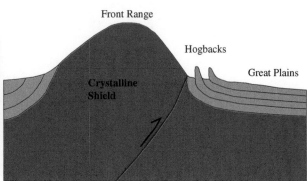

FIGURE 11.14   A photograph and schematic cross-section of Garden of the Gods, Colorado Springs, showing hogbacks of resistant rock layers along the western margin of the Great Plains.

FIGURE 11.15   A photograph looking across the South Park synclinal valley from Wilkerson Pass. Young volcanic rocks form some of the small dark hills.

out at 13,626 feet and East Spanish Peak at 12,683 feet. In the early days of US history, the two peaks formed an easily visible topographic target for settlers crossing the Great Plains. The crystalline rock in this case consists of young intrusions mostly between 27 and 21 million years old. The peaks are well known for the many dikes that radiate in all directions from the central mountain. A dike is a tabular intrusion (like the top of a table) that is often oriented close to vertical. The largest dikes radiate up to 25 miles from the summit area of Spanish Peaks.

West of the Front Range in Colorado, we encounter the wide, flat-floored, mostly treeless synclinal valleys of North, Middle, and South Park (uncolored in Figure 11.11). These are high-altitude valleys; South Park lies between 8,500 and

10,000 feet, North Park at about 8,000 feet. South Park, as seen from Wilkerson Pass in Figure 11.15, is particularly wide and flat with a few volcanic hills. These synclinal valleys separate the Front Range from a semicontinuous line of mountain ridges that includes (from north to south) the Sierra Madre, Park, Gore, Ten Mile, and Mosquito Ranges, and the Arkansas Hills. Farther south, across the Arkansas River at Salida, the Sangre de Cristo Range continues the trend into New Mexico. North of Salida, the south-flowing Arkansas River occupies a narrow, synclinal, and faulted valley that separates the Gore-Ten Mile-Mosquito ranges from the Sawatch Range to the west. The Sawatch Range is the geographic heart of both the Colorado Rockies and the Colorado Mineral Belt. According to the US

FIGURE 11.16 A Google Earth image looking south from south of Denver at the slope-flat-slope topography of the Front Range. The mountains rise abruptly from the Great Plains to a flat that represents the Rocky Mountain erosion surface, and then rise again to Pikes Peak. The Sangre de Cristo Mountains are seen in the distance at upper right.

Geological Survey, there are 58 named peaks in Colorado that rise above 14,000 feet (climbers recognize 53 peaks; they do not recognize nearby peaks separated by less than 300 vertical feet). Fifteen of them, including the highest, Mount Elbert (14,433 feet or 14,440 feet as corrected by the National Geodetic Survey), are in the Sawatch Range. The Sawatch Range ends southward in the wide San Luis Valley which separates the Sangre de Cristo Range from the largely volcanic San Juan Mountains. All of the ranges are anticlinal and cored with crystalline rock, although a few summit areas retain part of their sedimentary cover as depicted in the cross-section (Figure 11.13). The Elk and West Elk mountains lie to the west of the Sawatch Range and are composed largely of sedimentary layers that merge with the Colorado Plateau (not highlighted in Figure 11.11).

Anticlinal mountains in the Southern Rockies lose elevation northward and disappear in the Wyoming Basin only to reappear in the form of the Bighorn, Wind River, Beartooth, and Uinta mountains. The Black Hills, although squarely in the Great Plains, are geologically connected with the Middle Rockies. All except the Uinta Mountains contain ancient crystalline rock in their anticlinal core. The mountains surrounding the Wyoming Basin trend northwesterly except for the Uinta Mountains, which trend almost due west. As shown in Figure 11.11, the mountains appear to radiate from, and curve around, the Colorado Plateau. The anticlinal structures in the Middle and Southern Rockies developed primarily between 75 and 55 million years ago with some deformation continuing to 40 million years ago. This is the same mountain-building phase that formed monoclines and other folds on the Colorado Plateau. The landscape history of the Black Hills and the Bighorn and

Wind River ranges is discussed in more detail later in this chapter. The following discussion explores primarily the landscape history of the Southern (Colorado) Rockies.

A relict erosion surface, referred to here as the Rocky Mountain erosion surface, is present in the Southern Rocky Mountains and, to a lesser extent, in the Middle Rockies. In most areas this surface expresses itself as a tilted, relatively flat, dissected area located either at high elevation or along mountain flanks. It is best seen in the Front Range in the vicinity of Pikes Peak. As a result of this erosion surface, the Front Range has a slope-flat-slope profile. As shown in Figure 11.16, the frontal slope of the Front Range rises abruptly 3,000 feet above the Great Plains to an elevation between 8,000 and 9,000 feet where, across a wide area, topography levels off to form the erosion surface (the flat in the slope-flat-slope profile). The mountain then rises another 5,000 to 6,000 feet to the summit of Pikes Peak. The erosion surface is present in other ranges but is more difficult to recognize because it is tilted, eroded, and occurs at variable elevation.

The presence of this erosion surface at high elevation implies a landscape history that is more complex than simple erosional lowering following circa 75 to 40 Ma mountain building events. One clue to understanding the origin and significance of the erosion surface is to follow the surface northward along the trend of the Front Range to Wyoming. If we do this, the first thing we notice is that, without the presence of a high peak such as Pikes Peak, it is the erosion surface itself that forms the top of the mountain. The second thing to realize is that the erosion surface dips gently northward such that relief along the frontal slope diminishes from 3,000 feet at Denver to zero in the

FIGURE 11.17   A Google Earth image looking south-southwest at the Front Range (*FR*) between Cheyenne and Laramie. The Gangplank (*GP*) forms a stream-dissected ramp that connects the Great Plains with summit area of the Front Range. Note how the Rocky Mountain erosion surface (*RMES*) appears to disappear below unconsolidated sediment of the Great Plains and the Gangplank. I-80 extends up the Gangplank, around the Medicine Bow Mountains (*MB*), and into the Wyoming Basin.

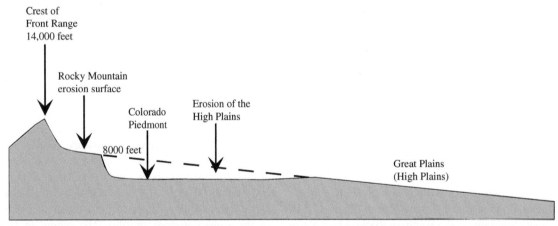

FIGURE 11.18   A profile that suggests that the High Plains were once connected with the Rocky Mountain erosion surface prior to erosion of the Colorado Piedmont. Based on Chronic (1980, p. 20).

area of Cheyenne, Wyoming, just across the Colorado border. These relationships are clearly evident in the Google Earth image of the Northern Front Range shown in Figure 11.17. Here there is no distinction (no break) between the Great Plains (High Plains) land surface and the summit area of the Front Range. The High Plains form a continuous surface (a ramp) to the summit of the Front Range and then into the Wyoming Basin. The ramp is known as the Gangplank. It is the route followed by I-80 and is visible in Figures 10.3, 10.4, 11.11, and 11.17. The connection from the Great Plains, over the Gangplank, to the Wyoming Basin is a primary piece of evidence to suggest that the Wyoming Basin is the westward extension of the Great Plains. Figure 11.18

is a profile across the Colorado Piedmont in the vicinity of Colorado Springs. The figure suggests that, at one time prior to erosion, the High Plains land surface was continuous with the Rocky Mountain erosion surface. Thus, it appears that, prior to uplift and erosion, the Front Range was once a relatively flat part of the Great Plains with only a few relatively small isolated mountain peaks such as Pikes Peak.

Given the existence of the Rocky Mountain erosion surface and its connection with the Great Plains, how do we interpret the landscape history of the Middle-Southern Rocky Mountains? Recall from Chapter 7 that, based on the distribution of marine sedimentary rocks, this entire region was near or below sea level 80 million years ago

(Figure 7.1a). Thus, the landscape we see today must be younger than 80 million years old. Based on the age of deformed rocks versus undeformed rocks, and on the age of sediment eroded from the mountains and deposited into adjacent basins, we can say with strong certainty that the rocks were deformed and uplifted into a mountain range between 75 and 40 million years ago. Perhaps significantly, much of the sediment that was eroded from the mountain was not carried great distances. Instead, the sediment was deposited in adjacent valleys or spread out like an apron on the Great Plains. We know that the Rocky Mountain erosion surface was largely developed by 34 million years ago because this is the age of the oldest volcanic layers deposited on the eroded surface. The question then becomes, was this erosion surface cut into the mountain at its present high elevation, or was the erosion surface cut at low elevation and the entire region (including the Colorado Plateau and eastern Great Plains) subsequently uplifted into a large dome? There is no definitive answer to this question, therefore, we will form a working hypothesis based on available evidence.

The Middle-Southern Rockies (and probably also the surrounding Great Plains and Colorado Plateau) achieved high elevation during the mountain-building phase from 75 to 40 million years ago. This time interval is depicted schematically in Figure 11.19a. The presence of the erosion surface across the Front Range suggests that by about 34 million years ago the mountains were beveled to a nearly flat surface with only a few isolated peaks (such as Pikes Peak) protruding above the erosion surface. Much of the eroded material, rather than being transported away from the mountain, was deposited in adjacent synclinal basins, which had the effect of burying the mountain in its own debris as depicted in Figure 11.19b. It was at this time that the Great Plains extended across the Southern Rockies, Wyoming Basin, and Colorado Plateau with only minor interruption from scattered peaks. This would have been a time when one could travel from Denver to Salt Lake City and see almost no mountains at all! However, we do not know with absolute certainty if we would have been driving across the Great Plains at (present-day) high elevation or at lower elevation.

What we do know is that the Southern Rockies have undergone a great deal of erosion since 34 million years ago, and the effect of this erosion has been to strip the Rockies of their unconsolidated cover. In other words, the erosion has had the effect of exhuming the buried, beveled mountains, thus creating the present-day landscape shown in Figure 11.19c. We can then explain the Wyoming Basin as a remnant of the buried mountain landscape that has not yet undergone complete exhumation.

There is both indirect and direct evidence that broad uplift and doming were associated with post-34 Ma erosion and exhumation. The dome structure is massive. It reaches its crest in west-central Colorado; however, it also affected surrounding areas, including the Wyoming Basin,

Mountain building between 75 and 40 million years ago creates the anticlinal structure and high mountains.

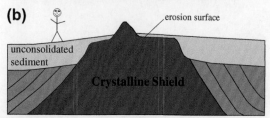

Between about 34 and 10 million years ago the mountain surface was beveled and the mountain was largely buried in its own debris with only a few peaks protruding above the erosion surface. The Rocky Mountain erosion surface is created.

Broad doming across the entire region results in erosion of unconsolidated sediment from around the buried, beveled mountain. The mountain, and the Rocky Mountain erosion surface, are exhumed, thus creating the present day landscape.

FIGURE 11.19 A series of cross-section sketches that show: (a) the formation of the Southern Rocky Mountains between 75 and 40 million years ago; (b) mountain burial and formation of the Rocky Mountain erosion surface; and (c) exhumation of the buried, beveled mountain to form the present-day landscape.

Great Plains, and Colorado Plateau. Indirect evidence for recent broad uplift includes the many hot springs in the Colorado Rockies (as many as 84). The existence of so many hot springs indicates that rocks below the mountain topography are hot and, therefore, isostatically buoyant. Such a possibility is supported by measurements of crustal thickness in the Colorado Rockies which vary from 23 to 35 miles (38 to 56 km). This thickness alone is not enough to support the height of the mountains. Therefore, the height must be partly supported by thermal isostasy. Such a possibility is further suggested by geophysical studies which indicate that the mantle below the crust is also relatively warm and buoyant. Some of the near-surface heat may have been introduced 40 to 20 million years ago during intrusion associated with development of part of the Colorado Mineral Belt. If this is the case, doming could have begun as early as 35 million years ago. Additional evidence for recent doming is the sheer height of the Southern Rocky Mountain region. It is arguably the highest area in North America.

FIGURE 11.20 A photograph looking across the Arkansas River Valley at the flat-topped Mosquito Range and a wedge of unconsolidated sediment from near the summit of Mt. Elbert, Sawatch Range, Colorado.

Although Alaska and Canada have a handful of very high peaks, including Denali (Mt. McKinley), the highest peak in North America at 20,320 feet, there are only 23 peaks total in Alaska and Canada that rise above 14,000 feet. Colorado has between 53 and 58 peaks above 14,000 feet, all within a relatively small area. When the Southern Rockies are viewed from above, as in Figures 1.7 and 10.3, they appear to be a large domal structure surrounded on all sides by flatland of the Great Plains, Colorado Plateau, and Wyoming Basin.

Direct evidence for recent uplift and doming includes the eastward tilt of both the High Plains and the Colorado Piedmont. If we analyze the present-day tilt of young sedimentary layers in these areas, and compare it with the inferred tilt at the time of deposition, we find that there has been as much as 2,230 feet of uplift at the western margin of the Great Plains since about 17.5 million years ago. Approximately 460 feet of this uplift have been attributed to erosion-generated isostatic uplift. The remainder must be tectonic uplift. There are several additional lines of evidence to suggest uplift of the Colorado Rockies on the order of at least 1,640 to 3,280 feet (500 to 1,000 m) within the past 6 to 10 million years, much of it attributed to tectonic uplift (rather than isostatic). The evidence includes high river incision rates of 0.49 to 0.76 inches per 100 years over the past 10 million years, and high apatite fission track and apatite (U-Th)/He derived exhumation rates on the order of 0.59 to 1.18 inches per 100 years over the past 6 to 10 million years coupled with evidence that exhumation rates were lower prior to 10 million

years ago. Traditional leveling surveys employed over the past 80 years suggest that uplift is active and ongoing.

Although the details and timing of uplift are debatable, we can conclude that the Southern Rocky Mountain landscape developed as a result of at least two separate tectonic events. The first ended between 55 and 40 million years ago and is responsible for the structure of the rock (the anticlines, synclines, faults). The mountains that formed at that time may have been above 10,000 feet, but these are not the mountains we see today. These mountains were partly eroded and buried in their own debris by about 34 million years ago. Fossil evidence suggests that elevation remained high throughout this period of erosion, probably within 2,000 to 4,000 feet of its present elevation. A second period of tectonic activity may have begun prior to 17.5 million years ago but accelerated within the past 10 million years and is active today. Tectonic activity during the past 10 million years was not so much a deformational phase as it was a broad domal uplift on the order of 1,640 to 3,280 feet centered in Colorado. The uplift resulted in eastward tilting of the Great Plains, which produced enough of a gradient to allow major rivers to carry sediment away from the Colorado Rockies. In doing so, rivers quickly stripped away the nonresistant unconsolidated sediment from the margins of the buried crystalline mountains thereby uncovering the erosion surface and literally exhuming the same mountains that had existed 40 to 55 million years earlier. The process of mountain exhumation is visible in Figure 11.20, which is a photograph looking eastward across the Arkansas Valley at the Mosquito Range from the eastern flank of Mt. Elbert in the Sawatch Range. Note the very flat summit area of the Mosquito Range in the photograph. This may be the old exhumed erosion surface. Note also the wedge of sediment that climbs up the mountain flank just below the summit area. This is the wedge of sediment that perhaps earlier had buried the Mosquito Range and is now being removed from the flank of the exhuming mountain.

Although the mountains of today are the same as those that existed some 40 to 55 million years ago, there are two major landscape differences. The first is the presence of the Rocky Mountain erosion surface which did not exist in the earlier iteration of the mountain range. The second is the ubiquitous glacial sculpturing that began about 2.4 million years ago. Many of us are quite happy with the existence of a relatively flat erosion surface at high elevation because it allows Colorado to be one of the premier locations in the world for mountain biking. It is quite possible to ride across the mountain range at high elevation without too much worry of overly steep slopes and a quick trip over your handlebars!

If we look for a solution as to why numerous basins occupy the Middle Rocky Mountains, and why the Southern Rockies are mountainous, we can suggest broad domal uplift, primarily within the past 10 million years, as a factor. The Southern Rockies are at the apex of this broad uplift such that most of the sediment that had buried the mountains has since been removed. The Middle Rockies have not been uplifted to the same extent and therefore are still largely buried in their own debris. Other factors may have played a role as well. Perhaps the Middle Rockies were more deeply buried than the Colorado Rockies. Or perhaps the 75 to 40 Ma deformation did not affect the Wyoming Basin to the same extent as in Colorado and there are fewer anticlinal upwarps to be exhumed.

THE WIND RIVER AND BIGHORN MOUNTAIN RANGES

One of the largest and highest mountain ranges in the Middle Rockies is the Wind River Range located at the northwest margin of the Wyoming Basin (Figure 11.11). It is a crystalline-cored anticlinal mountain approximately 125 miles long and 40 miles wide. The crystalline rocks are primarily granitic, about 2.6 billion years old, and part of the North American crystalline shield. The highest point is Gannett Peak, located in the heavily glaciated, remote, north-central part of the range. At 13,804 feet, Gannett Peak is also the highest point in Wyoming. Similar to the Southern Rockies, an erosion surface is present along the southwestern flank of the range that may actually cross the northern part of the range.

Initial uplift of the Wind River Range, and development of the anticline structure, are associated with displacement on a major thrust fault, the Wind River thrust. This fault is mostly buried beneath younger unconsolidated sediment but is located along the southwestern side of the mountain based on seismic reflection data. Rocks in the Wind River Range were displaced about 13 miles southwestward and, in the process, folded into an anticline and elevated into a mountain. The thrust fault was active prior to 49 million years ago based on the age of the oldest undeformed sediment that covers the thrust. Apatite and zircon fission track ages from crystalline rock in the core of the mountain suggest a period of rapid cooling, possibly associated with thrusting and uplift between about 62 and 57 million years ago. Clasts of crystalline rock in the surrounding sediment indicate that the ancient crystalline core of the range was exposed at the surface by about 56 million years ago. Similar to the Southern Rockies, the Wind River erosion surface likely developed soon after thrusting and anticlinal uplift had ended. The area likely existed at low elevation until renewed faulting (on faults other than the Wind River thrust) elevated the mountain beginning about 30 million years ago.

An interesting aspect to the zircon fission track ages is that they are on the order of 600 million years old or more. Recall from Chapter 4 that zircon tracks are erased above about 310°C and retained below about 210°C. The presence of such old zircon tracks indicates that the presently exposed crystalline rocks have not been buried more than 5 to 7 miles (8 to 11 km) since at least 600 million years ago. The area must have been a lowland region of slow erosion and periodically below a shallow sea throughout the past 600 million years.

The Bighorn Mountains form a massive range 200 miles long that stretches across the Wyoming border to Montana (Figure 10.5; only the southern part is shown in Figure 11.11). It consists of sedimentary rock with crystalline shield, as much as 3 billion years old, in its glaciated core. There are two peaks above 13,000 feet. The highest is Cloud Peak (13,167 feet). A thrust fault along the eastern side of the range accentuates the anticlinal structure. Based on the presence of clasts of crystalline rock in the surrounding sedimentary layers, the Bighorn Mountains were elevated, and crystalline rocks were exposed, by 56 million years ago and possibly as early as 65 million years ago. Similar to the Wind River Range, apatite (U-Th)/He ages indicate that the currently exposed crystalline rock has not been deeply buried for at least the past 369 million years.

THE BLACK HILLS

Figures 10.4 and 11.11 locate the Black Hills dome on the Missouri Plateau of the Great Plains just north of the Pine Ridge escarpment between the White and Little Missouri Rivers. The Black Hills are elliptical in shape, approximately 125 miles north to south and 65 miles east to west.

A cross-section of the dome revels an asymmetric structure with the east side dipping more steeply than the west (Figure 11.21). The dome is topographically higher than the Adirondacks, with peaks mostly between 5,000 and 6,500 feet. The highest is Harney Peak, which, at 7,242 feet, rises some 4,000 feet above the surrounding Great Plains. Geologically, the Black Hills are part of the Middle-Southern Rocky Mountains because both share a similar history. Only the central core of the Black Hills is crystalline, consisting mostly of ancient metamorphic and granitic rock that forms part of the North American crystalline shield. Most of the rocks are circa 1.8 billion years old but some are as old as 2.8 billion years. All major mountain peaks, including Harney Peak and Mt. Rushmore (the Presidential Peaks), are composed of this highly resistant rock. These ancient crystalline rocks are host to the famous Homestake gold mine, which was in continuous operation for more than 125 years before closing operations in 2002. The mine remains open for visitor tours.

The crystalline core is overlain by differentially eroded sedimentary layers that form concentric bands along the flanks of the mountain. The lowest sedimentary layer consists of a resistant limestone that forms a wide plateau surface on the gently dipping west side of the mountain, and an inner hogback on the steeper-dipping east side (Figure 11.22). The limestone is overlain by nonresistant red shales that erode to form the Red Valley (also known as the Racetrack). This is a continuous 2-mile-wide valley of red soil that almost completely encircles the dome. Surrounding Red Valley is a highly resistant sandstone that forms a second, outer cuesta/hogback known on its steep eastern side as the Dakota Hogback (or Hogback Ridge). Figure 11.22 is a Google Earth image of the Dakota Hogback at Rapid City, where Rapid Creek cuts a gap through the hogback.

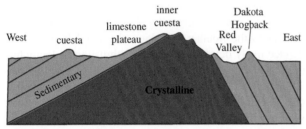

FIGURE 11.21   A cross-section of the Black Hills dome. Note the limestone plateau, Red Valley, and the Dakota Hogback.

FIGURE 11.22   A Google Earth image looking northward at Red Valley, the Dakota Hogback, and Rapid City, South Dakota.

The Black Hills were deformed and uplifted between 75 and 53 million years ago and have remained a positive area overlooking the Great Plains ever since. There is little evidence of an erosion surface similar to what is found in the Colorado Front Range. Rock units younger than 40 million years are tilted, indicating that the Black Hills have continued to rise. Leveling surveys suggest uplift may be continuing today.

An interesting feature in the Black Hills is Wind Cave, which is the fifth-longest cave system in the world. Located in limestone layers in the southeastern part of the Black Hills, Wind Cave has nearly 140 miles of passageways and may be one of the oldest cave systems in the world, having begun to form as early as 320 million years ago. Jewel Cave, located northwest of Wind Cave, is ranked as the second-longest cave system in the world with 160 miles of mapped passageways. Much of Jewel Cave remains unexplored, so who knows, it could eventually rival Mammoth Cave as the largest cave system in the world once all passageways have been mapped.

A few small crystalline-cored dome structures are present at the north end of the Black Hills, including Bear Lodge Mountain and Devils Tower (Figure 11.12). Bear Lodge Mountain and a few nearby smaller domes such as Citadel Rock and Crow Peak contain a small amount of ancient crystalline rock. However, most of the crystalline rock in these areas consists of young (70 to 50 Ma) intrusions that have punched up into the surrounding sedimentary rock to create small dome structures. These young intrusions are probably related to the larger and more numerous intrusions of the same age that dot the northwestern Great Plains of Montana. As shown in Figure 11.23, Devils Tower is a vertical rock edifice that rises nearly 1,300 feet above the Great Plains. Much of the tower is composed of near-vertical five- or six-sided columns of a somewhat rare igneous rock known as phonolite. The columns formed as a result of magma cooling and shrinking in place. Devils Tower is a popular location for rock climbers except in June, when it becomes a ceremonial destination for American Indians. The origin of Devils Tower has been debated. Some have suggested that it represents the throat or feeder tube of an old, eroded volcano (a volcanic neck). However, the absence of volcanic material in the area suggests instead that the tower is an eroded remnant of a shallow intrusion around which sedimentary rocks have been eroded. The Missouri Buttes form a similar landscape only a few miles to the northwest (Figure 11.23).

WATER GAPS IN THE ROCKY MOUNTAINS

A *water gap* forms where a river flows directly through a mountain or ridge rather than around it. The flow of Rapid Creek through the Dakota Hogback at Rapid City is an example (Figure 11.22). There are several water gaps in the Rocky Mountains, particularly in the Middle Rockies. Major water gaps include Sheep Canyon, where the Bighorn River cuts across the north end of the Bighorn Mountains; Wind River Canyon at the east end of the Owl Creek Mountains; Devil's Gate at the east end of the Granite Range; Lodore Canyon (Split Mountain Canyon) at the east end of the Uinta Range; and several in the Front Range, including Royal Gorge on the Arkansas River near Canon City (Figure 11.11). Water gaps are also present on the Colorado Plateau. For example, the San Juan River cuts directly across the Raplee Anticline, and the Colorado River cuts across the Kaibab Uplift to form the Grand Canyon (Figure 10.11). Water gaps are also common in the Appalachian Mountains.

The existence of a water gap is perhaps a bit surprising because rivers normally go with the flow, so to speak. In other words, a river will seek the easiest route to the sea,

FIGURE 11.23   A Google Earth image looking west at Devils Tower (foreground) and the Missouri Buttes. The snow-covered Bighorn Mountains can be seen in the distance behind Devils Tower.

and in this case that should have been around the mountain. The explanation for most of the rivers in the Rocky Mountain region is that the rivers are older than the present-day land-scape. Recall that the mountains in this area have been exhumed from below unconsolidated sediment. It has been suggested that the river course was established within the overlying unconsolidated material prior to exhumation. The river simply maintained its course as it cut downward into the slowly exhuming anticlinal mountain. Note that in this explanation, the river cuts downward into a buried but preex-isting mountain structure. There are other mechanisms that can produce water gaps, and at least one different mechanism may be responsible for some of the water gaps in the Rocky Mountains. I could discuss other mechanisms at this time but would rather present it as a question at the end of the chapter for you to research. We will revisit this problem when we discuss the Appalachian fold and thrust belt (Chapter 12).

QUESTIONS

1. If you look closely at the Long Lake lineament in Figure 11.1 you will notice that topography is signifi-cantly higher to the immediate east of the lineament than to the west. This relationship appears to be consis-tent along the entire width of the Adirondacks. Suggest several hypotheses that might account for this land-scape pattern.

2. In Google Earth, fly to the following locations in New York and see if you can locate eskers. The loca-tions are Five Ponds, Massawepie, Spectacle Pond, Onchiota, and Mountain Pond. How far can you rea-sonably follow eskers in each area?

3. In Google Earth, fly to Beaver River, New York. What lake are you on? Save an image centered on the lake but include the area around the lake (about 19 by 14 miles). Print the image and sketch the linear and curved land-scape trends. How many linear trends are there? What is their relationship with the orientation of the lake? What type of rock might characterize the curved land-scape trends? Hypothesize on the origin of the linear trends. Do they pre-date or post-date the curved trends?

4. Use tracing paper to trace the crystalline rock and the sedimentary layers in Figure 11.4. Label the buttress unconformity.

5. What is the nature of the treeless area at the top of the mountain across the river from the Johnson Shut-ins in Figure 11.6? Why does the treeless area extend down the mountain to the river? Research this by flying to Johnson Shut-ins State Park, Missouri, in Google Earth and by searching on "Johnson Shut-ins flood."

6. Use a yellow pencil to color all lowland areas (farm-land and lakes) in Figure 11.6. Assuming these areas are underlain by sedimentary rock and the highlands by crystalline rock, are there any locations, other than

Johnson Shut-ins, where a river or stream appears to flow from sedimentary rock into crystalline rock and back into sedimentary rock?

7. Speculate as to why the Meers fault in Figure 11.8 is parallel with the dominant fracture orientation in the Wichita Mountains. Is it possible for the Meers fault to be an ancient reactivated fault? Explain.

8. Note in Figure 11.9 that the pink granite is visible in a series of hills that form a circular (oval) shape. Note also that the ground is darker to the right of the granite outcroppings than to the left. Suggest reasons to explain the outcrop pattern and the change in ground color.

9. Describe the landscape history of the Arbuckle Mountains.

10. What is the primary piece of evidence to suggest that the present-day Colorado Front Range was, at one time, mostly buried beneath young, unconsolidated sediment?

11. What type of rock makes up most of the Middle and Southern Rocky Mountains? What other types of rocks are present in the mountains?

12. Summarize some of the evidence that suggests that the Southern Rockies have been elevated within the past 10 million years.

13. In Google Earth, fly to Pinedale, Wyoming. North and east of Pinedale are a series of four lakes at the foot of the Wind River Range. Examine these lakes and explain their origin.

14. In Google Earth, fly to Pinedale, Wyoming. Locate the erosion surface along the western flank of the Wind River Range. How far does the erosion surface extend along the western part of the range? What is the rela-tionship between the erosion surface and the adjacent basin? Is there evidence in the northern part of the range to suggest that the erosion surface crosses to the eastern side of the range?

15. A water gap can form if a river first establishes its course within overlying unconsolidated material and then maintains its course as it cuts downward into a pre-existing but buried mountain or ridge. Can you think of other possibilities that could produce a water gap?

16. Research the term *wind gap*. What is a wind gap? Where do they occur? How do they form?

17. Given an average river incision rate of 0.5 inches per 100 years, what is the total incision in feet and in meters over a 10-million-year period?

18. Draw the boundaries of the Colorado Plateau on Figure 11.11. Locate and highlight crystalline-cored domes on the Colorado Plateau. Use Figures 10.4 and 10.11 as a guide. Research the history of one of the domes.

19. Within the Middle Rocky Mountains, describe one method used to determine how long ago crystalline shield rock first became exposed at the surface.

20. What are the Missouri Buttes? Where are they located? How did they form?

Foreland Fold-and-Thrust Belts

From the perspective of the geologist, a thrust fault is one of the most common and most obvious structures within a mountain system. Thrust faults can be present within sedimentary, crystalline, or volcanic rock along the entire length of a mountain system and across its entire width. Thrust faults permeate both the Appalachian and Cordilleran mountain systems. The term *foreland* refers to sedimentary rocks at the front of a mountain system where the front refers to the area closest to the continental interior. A *foreland fold-and-thrust belt*, therefore, is found at the front of the mountain system where thick piles of sedimentary rock (>30,000 feet) have been transported toward the continental interior along major thrust faults. We will have more to say about the geology of foreland fold-and-thrust belts in Part III of this book. In this chapter we concentrate on the landscape that is produced as a result of thrust faulting of sedimentary layers.

The most conspicuous foreland fold-and-thrust belt in the United States is the Appalachian Valley and Ridge. Recall from Figure 2.10 that this is the only sedimentary province in the Appalachian physiographic region. The Valley and Ridge shows the characteristic landscape of a foreland fold-and-thrust belt which is one of long linear mountains (or ridges) separated by linear valleys. In addition to the Valley and Ridge, this combination of rock type, rock structure, and resulting landscape is found in the Ouachita Mountains, the Marathon Basin region of West Texas, northern Montana-Idaho, and in an area known as the Idaho-Wyoming thrust belt west of the Wyoming Basin. These areas are highlighted in Figures 8.4 and 10.2a.

The geometry of a foreland thrust fault consists of a flat that is nearly horizontal, a ramp that dips at angles between about 20 and 45 degrees, and an upper nearly horizontal flat at the top of the ramp. The process of thrust faulting is illustrated in Figure 12.1. Rock layers that begin movement along the lower flat are bent parallel with the ramp and then bent back to horizontal as they encounter the upper flat. Notice that rocks on the lower plate underneath the fault

remain undeformed and horizontal. Notice also that ramping has the effect of stacking older sedimentary layers above younger ones. The stacking, in turn, results in burial and overall shortening of the land surface. It is quite possible for foreland thrust faults to shorten an area that was originally 400 miles wide to one that is less than 200 miles wide.

Faulting and stacking begin at depth and, with increasing displacement, the fault grows and climbs by stair-stepping (ramp-flat-ramp) up the layers until it reaches the surface, usually along a ramp. The geometry is illustrated in Figure 12.2. It is not unusual for rock to be transported 20 miles or more along a single thrust fault that extends for more than 100 miles. Major thrust faults can extend the length of a mountain belt and can carry rock hundreds of miles. Note in Figure 12.2 that stair-stepping produces an anticline above ramps and a syncline along flats. Ramps can be continuous over great distances, thus producing a characteristic landscape of long linear (anticlinal) mountain ridges and intervening narrow valleys. Additionally, the thrust fault itself is a zone of weakened, broken rock that will often erode to form a linear valley wherever it intersects the surface.

THE CORDILLERAN (SEVIER) FOLD-AND-THRUST BELT

There are significant differences between the Cordilleran fold-and-thrust belt in Idaho, Montana, Wyoming, and Utah, and the Appalachian (Valley and Ridge) and Ouachita-Marathon fold-and-thrust belts. Primary among these are (1) the age of the Cordillera, which is considerably younger than the Valley and Ridge-Ouachita-Marathon belts, and (2) the extent to which each area has been modified via younger tectonic processes. Tectonic modification is extensive in the Cordillera but nearly absent in other regions. The result of tectonic modification in the Cordillera is the near-complete destruction of the typical linear ridge-and-valley landscape.

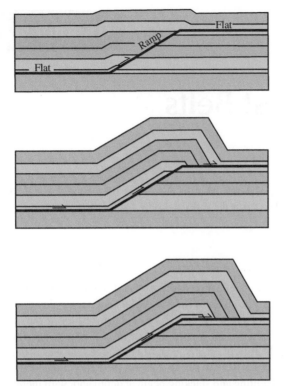

FIGURE 12.1 A schematic series of cross-sections that shows the process of thrust faulting and development of an anticline above the ramp.

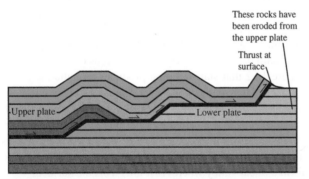

FIGURE 12.2 A schematic cross-section showing how a thrust fault might stair-step to the surface.

Cordilleran foreland thrust faulting was active between about 115 and 55 million years ago during what is known as the Sevier mountain-building event and, because of this, the thrust belt is often referred to as the *Sevier fold-and-thrust belt*. Geologically speaking, the US Cordilleran foreland fold-and-thrust belt extends the length of the Cordillera from Canada to California. The frontal (easternmost) trace of the thrust system through the US is shown in Figure 12.3. As shown in this figure, the thrust belt forms the Rocky Mountain front with the Great Plains through northern Montana. It then crosses the Snake River Plain and continues within the eastern part of the Basin and Range around the west and southwest sides

of the Colorado Plateau to southern California. Thrust faults are present in rocks along this line and to the west of this line such that, by 55 million years ago, a linear belt of foreland fold-and-thrust mountains was in existence from Canada southward to California. Nearly all of this mountain belt was subsequently destroyed and reincarnated via younger Basin and Range and Snake River Plain normal faulting and volcanism. The linear fold-and-thrust belt landscape survives primarily in two areas: northern Montana and Idaho (referred to here as the Northern Rocky Mountain fold-and-thrust belt) and the Idaho-Wyoming fold-and-thrust belt west of the Wyoming Basin at the mutual borders of Utah, Idaho, and Wyoming. Both areas are highlighted in Figure 12.4, but even these areas have undergone erosion and some tectonic modification.

A zone of normal faults is responsible for the separation of the Northern Rocky Mountain thrust belt into an eastern section and a western section (Figure 10.4). Major normal faults, such as those in the Flathead River Valley and along the base of the Mission Range, have north-northwest trends similar in orientation with the trend of thrust faults. This similarity preserves some of the original linear landscape although down-dropped normal-fault basins, such as the basin occupied by Flathead Lake, are typically wider than fold-and-thrust fault valleys (Figure 12.4).

High topography is located in the area east of the zone of normal faults and this is where foreland fold-and-thrust landscape is best developed. The difference in topography between the eastern section and the western section is obvious in Figure 12.5 which is a Google Earth image looking south across the Northern Rocky Mountain Ranges from the Canadian border. Coeur d'Alene Lake is at right. Flathead Lake is at left-center. Note the beautifully developed fold-and-thrust landscape along the Rocky Mountain front east (left) of Flathead Lake. This is an area of closely spaced thrust ramps known as the *Montana Disturbed Belt* (MDB) where elevation of a few peaks exceeds 9,000 feet. Erosion has preferentially attacked and partly removed elevated rocks along the crest of anticlines thereby producing a characteristic asymmetric shape to the mountain with steep slopes or cliffs on the eastern side due to partial erosion of the anticline, and relatively gentle dip-slopes on the western side parallel with the dip of rock layers above the ramp. Thrust faults occupy narrow intervening valleys. A Google Earth image and schematic cross-section of the structure are shown in Figure 12.6. The sedimentary rocks in this area are less than 542 million years old, well layered, and much thicker than their counterparts on the Great Plains (>30,000 feet). South of the Montana Disturbed Belt the fold-and-thrust landscape becomes heavily disrupted and reincarnated by younger volcanism, igneous intrusion, normal faulting, and anticlinal uplifts (Figure 10.4).

A linear fold-and-thrust landscape is present but not well developed in the larger, more massive, and heavily glaciated Lewis and Livingston Ranges located along the Rocky

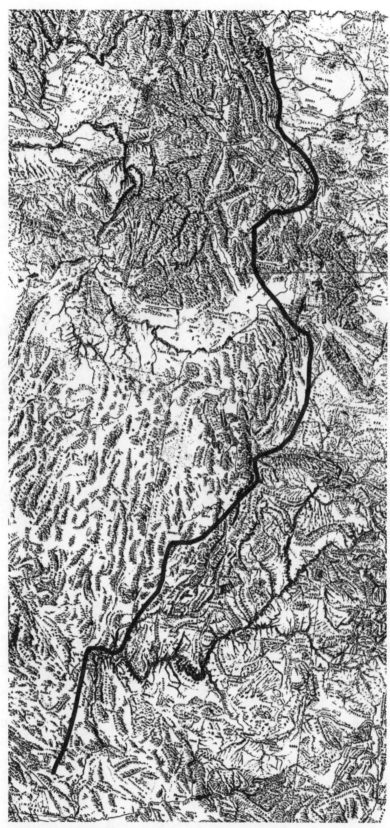

FIGURE 12.3 A landscape map that shows the frontal (easternmost) trace of the Cordilleran foreland fold-and-thrust belt. Thrust faults are present in rocks along this line and to the west of this line; however, much of the landscape has undergone tectonic reincarnation via younger Basin and Range and Snake River Plain normal faulting and volcanism.

FIGURE 12.4 A landscape map of the Northern Rocky Mountains with the foreland fold-and-thrust belt structural province highlighted.

FIGURE 12.5 A Google Earth image looking south across the Northern Rocky Mountains. Coeur d'Alene Lake is at right. Flathead Lake is just left of center. The line across the bottom of image is the boundary with Canada. GNP = Glacier National Park, MDB = Montana Disturbed Belt.

Mountain front due north of the Montana Disturbed Belt. This entire area constitutes Glacier National Park (GNP, Figure 12.5) and includes some of the highest elevations in the US Northern Rockies with six peaks over 10,000 feet. The structure across the area is a broad syncline underlain by the folded Lewis thrust, which crops out along the eastern front of the range in several areas, including Chief Mountain. A schematic cross-section and a Google Earth image showing the Lewis thrust and the synclinal structure are presented in Figure 12.7. Note the correspondence of high elevation with structurally high areas along the eastern and western flanks of the syncline.

The Lewis thrust is one of the largest in the Cordillera. It extends northward into Canada for 250 miles and has displaced rock a maximum of about 60 miles. One of the most significant aspects of the Lewis thrust is that it carries sedimentary layers that are significantly older than any sedimentary rock previously discussed. These rocks are 1,470 to 1,370 million years old and are known as the Belt Supergroup (Purcell Supergroup in Canada). The rocks are as old or older than some of the rocks that form the North American crystalline shield. They are, in fact, some of the oldest sedimentary layers in the Cordillera. The absence of a strong linear trend in Glacier National Park is due to the presence of a single large, synclinally folded thrust sheet below the entire region rather than a series of thin thrust sheets as is seen in the Montana Disturbed Belt. The Belt Supergroup is thick-bedded and more resistant relative to the younger sedimentary rock layers

that form the Montana Disturbed Belt, and this contributes to the lack of landscape linearity.

Returning to Figure 12.5, note the wide valley occupied by Flathead Lake and its extension northward to the Canadian border. Note also that the valley continues south of Missoula. These areas are part of a continuous valley known as the *Rocky Mountain trench* that extends for hundreds of miles through Canada. In this usage, the word trench refers to a long, continuous valley. It is not a subduction zone. A second long valley, the Purcell trench, extends northward from Coeur d'Alene, Idaho, into Canada, where it merges with the Rocky Mountain trench. The extent of the Rocky Mountain and Purcell trenches is evident in a Google Earth image of western Canada shown in Figure 12.8. Note in this figure that neither valley is particularly well developed in the US. Both valleys are underlain by a combination of thrust faults and younger normal faults, and both were widened and deepened by glaciers.

The western segment of the Northern Rocky Mountain fold-and-thrust belt west of Flathead Lake is deformed into a broad, composite anticline known as the *Purcell anticlinorium*. This anticline forms the structural counterpart to the syncline that underlies Glacier National Park. The region is mountainous, relatively dry, and forested (except where clear-cut), but elevations are considerably lower than those in Glacier National Park. Most of the peaks are between 6,000 and 7,000 feet. As shown in Figures 12.4 and 12.5, the region does not display a particularly strong

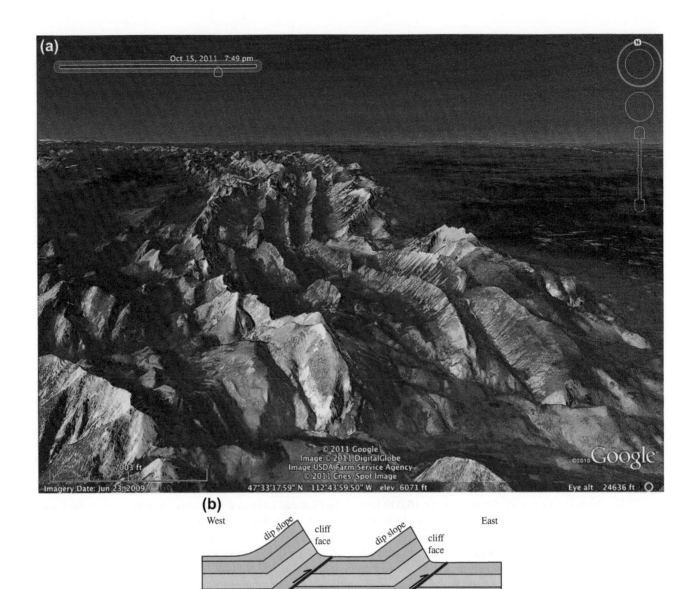

FIGURE 12.6 (a) A Google Earth image looking north along the Montana Disturbed Belt. (b) A schematic cross-section that shows the general structure of the Montana Disturbed Belt.

linear landscape. There are several reasons for this. First, the area is underlain by the thick-bedded, resistant Belt Supergroup; second, thrust faults are widely spaced relative to those in the Montana Disturbed Belt, and third, the fold-and-thrust landscape is strongly modified by younger normal faults, some of which cut at an angle to the thrust faults. One large man-made feature worth mentioning is Lake Koocanusa. The lake is long, narrow, and visible at the bottom center of Figure 12.5. It extends for 93 miles along the Kootenai River well into British Columbia behind Libby Dam which was completed in 1974.

A close look at Figure 12.5 reveals an odd feature that appears to be a scar across the land. It extends southeastward (lower right to upper left) across the image from

Coeur d'Alene Lake through Missoula and beyond toward the Great Plains. The scar corresponds with a zone of concentrated right-lateral strike-slip faults known as the Lewis and Clark Line. It is drawn on Figure 10.4 and is clearly evident in Figure 12.4 from Coeur d'Alene to Helena. It is believed to have originated hundreds of millions of years ago within the underlying North American crystalline shield and has periodically reactivated ever since. Some of the faults are active today.

The Idaho-Wyoming fold-and-thrust belt, in spite of its location adjacent to the Basin and Range and Snake River Plain, has not been tectonically modified to any great extent by younger normal faults or volcanism. Although strongly eroded and partly buried in its own debris, it still retains

FIGURE 12.7   (a) A Google Earth image looking north-northwest along the Lewis (right) and Livingston ranges in Glacier National Park. The heavily glaciated peaks consist of ancient Belt Supergroup sedimentary rocks. The synclinal structure is evident in the layering of rocks at the center of the image. Flathead Valley is at left. The line in the distance is the Canadian border. (b) A schematic cross-section across Glacier National Park showing the Lewis thrust and the synclinal structure. Based on Kiver and Harris (1999, p. 553).

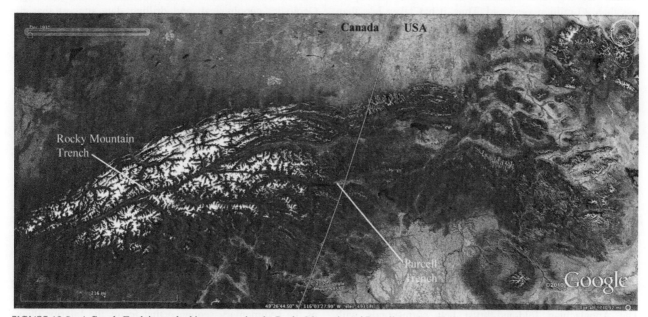

FIGURE 12.8   A Google Earth image looking eastward at the Rocky Mountain and Purcell trenches. Flathead Lake and Coeur d'Alene Lake are visible at center of image. Yellowstone Lake is at upper right.

FIGURE 12.9 A Google Earth image looking north across the Idaho-Wyoming fold-and-thrust belt from southwestern Wyoming. The Wyoming Basin (*WB*) is at right. The Teton Mountains (*T*) are at center-top of image. The border with Utah (not shown) crosses the upper-left corner.

its linear character which is well displayed in Figure 12.9. The area is a small surviving island of fold-and-thrust landscape surrounded on all sides by younger tectonics.

It is amazing to see how relatively quickly the landscape south of the Idaho-Wyoming fold-and-thrust belt was reincarnated. Basin and Range and Snake River Plain normal faulting and volcanism are, for the most part, less than 20 million years old. Thus, it has taken less than 20 million years of active tectonics to totally obliterate a preexisting landscape that probably consisted of tall mountains, much like the present-day Montana Disturbed Belt. An interesting aspect to the effect of reincarnation is seen in the area of active normal faulting located south of Butte, Montana. A quick glace at Figure 12.4 shows that this is an area where the trend of linear thrust-fault mountains becomes completely disoriented. The effect starts to become apparent south of Missoula as shown in Figure 12.5. Because of the strong influence of several generations of normal faulting, this area is known as the Rocky Mountain Basin and Range and is shown as part of the normal fault-dominated structural province in Figures 8.4 and 10.2. This area is discussed in Chapter 15.

THE APPALACHIAN FOLD-AND-THRUST BELT

The Appalachian fold-and-thrust belt corresponds with the physiographic Valley and Ridge province. It extends the length of the US Appalachians from Lake Champlain,

Vermont, to Tuscaloosa, Alabama. Similar rocks, structure, and landscape are present in the Ouachita Mountains west of the Mississippi River and in the Marathon region of southwest Texas. These areas collectively form the Appalachian-Ouachita foreland fold-and-thrust belt. In contrast to the Cordilleran fold-and-thrust belt, this is an ancient belt, having formed between about 335 and 265 million years ago. Geologic, geophysical, and drilling data suggest that the three now separate regions formed a continuous fold-and-thrust belt at the end of Appalachian mountain-building events 265 million years ago. Intervening areas, and the entire Marathon region, were eroded and covered with younger rock primarily during high sea-level stands over the past 265 million years. The Marathon region has since been uncovered. Much of the Valley and Ridge and Ouachitas have existed as a highland since their initial formation. In all of the presently exposed areas, there have been changes in elevation and relief due to sea level and isostatic variation, but there has not been any major tectonic modification. The structure we see today is nearly identical with the structure that existed some 265 million years ago. The primary change to the landscape has been erosional lowering. Today these regions form an erosion-controlled linear landscape of resistant sandstone ridges and nonresistant limestone and shale valleys. Elevation rarely exceeds 3,000 feet. All of the rocks are sedimentary, less than 542 million years old, and typically more than twice as thick as the sedimentary sequence that forms the Interior Plains and Plateaus region. Here we see a clear distinction between an ancient Appalachian-Ouachita fold-and-thrust system that

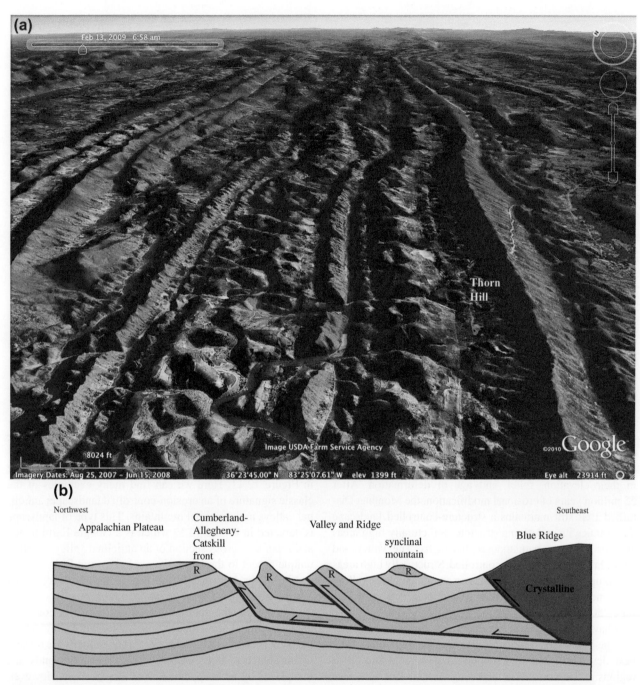

FIGURE 12.10   (a) A Google Earth image looking northeastward across the Tennessee fold-and-thrust belt (Valley and Ridge) from a short distance south of the Virginia border. The Cumberland front is shown in the upper-left corner. (b) A schematic cross-section of an erosion-controlled fold-and-thrust belt in which hard rock forms topographic highs. *R* signifies resistant rock.

has grown old over time, and a relatively young Cordilleran fold-and-thrust system that was tectonically reincarnated into a landscape that, with a few exceptions, is unrecognizable relative to its previous incarnation.

The structure and landscape of the Valley and Ridge vary from south to north. The southern part from Alabama to Virginia is a classic fold-and-thrust belt in which rocks have been pushed west-northwestward well over 50 miles along as many as 10 major thrust faults. Thrust faults are long and semicontinuous with typical flat-ramp-flat geometry. Topographic high areas do not necessarily correlate with anticlines but instead correlate with areas of hard rock. Because the thrust faults dip east, most of the sedimentary layers also dip east producing long, asymmetric ridges of hard sandstone with steep slopes on their west side, and more gentle slopes on their east side as illustrated in the schematic cross-section and a Google Earth image shown in Figure 12.10.

FIGURE 12.11   A Google Earth image looking northeastward across the Pennsylvania Valley and Ridge from west of Harrisburg. The major river is the Susquehanna. Note how the Susquehanna cuts directly across resistent ridges.

The Tennessee-Virginia Valley and Ridge mirrors that of the Montana Disturbed Belt. Both contain thick layers of sedimentary rock that have been transported many miles toward the continental interior. The landscapes differ mainly in their elevation, relief, and age of deformation. After 55 million years of erosional modification, the Montana Disturbed Belt still maintains a structure-controlled landscape that mimics its underlying structure. Structurally high areas are, for the most part, topographic highs. The Valley and Ridge landscape is erosion-controlled. Structurally high areas are composed of hard rock and, as a result, anticlinal valleys and synclinal mountains are more common. Perhaps at one time 250 million years ago, the Valley and Ridge landscape may have been similar in grandeur to the Northern Rockies.

The Tennessee-Virginia Valley and Ridge is one of the best developed and best-studied foreland fold-and-thrust regions in the world. It is surprising that this structure does not extend northward into Pennsylvania where we have to stretch the definition of a foreland fold-and-thrust belt just to include it. There are plenty of folds, and in fact, it is the manner in which the folds are eroded that makes the Pennsylvania Valley and Ridge rather geologically unique. The problem is that there are very few thrust faults. It is possible, and even probable, that thrust faults exist at depth, but certainly they are not present at the surface. A thrust fault can break and displace rocks tens of miles. Folds, on the other hand, squeeze rocks, which usually implies far less total movement. On this basis, we can generalize and say that the tectonic collision that created the fold-and-thrust belt

was more intense in the south, where thrust faults are present, than in Pennsylvania. Perhaps this is our good fortune because erosion of the folds has created a truly amazing zig-zag pattern of resistant bedrock ridges as is clearly evident in Figure 12.11. Here we can find many examples of the classic signature of an erosion-controlled landscape: anticlinal valleys and synclinal mountains. This type of landscape is depicted in Figure 12.12 which is a Google Earth image and schematic cross-section of an anticlinal valley. Another unique aspect to the Pennsylvania Valley and Ridge is that nearly all the known US reserves of hard anthracite coal occur within these folded layers.

Northward, the fold-and-thrust belt narrows considerably as it enters the Hudson and Champlain valleys along the New York-New England border. Thrust faults are present in this area; however, the classic valley-and-ridge landscape is replaced by what appears to be a single valley. However, a close look at Figure 9.4 reveals a drainage divide located in the valley at the center-right margin of the Adirondack Mountains (see also Figure 1.9). The single valley is actually two valleys. Lake Champlain drains northward into the Richelieu - St. Lawrence River system, and the Hudson River flows southward to the Atlantic Ocean. Again, referring to Figure 9.4, notice that the Hudson River flows across its own valley just north of New York City on its way to the Atlantic Ocean. Rather than ending at the Atlantic Ocean, the valley continues southward into Pennsylvania toward Harrisburg. If we then took the trouble to look at Figures 9.7 and 10.23, we could follow this valley along the western

FIGURE 12.12 (a) A Google Earth image looking westward along the Nittany Valley, Pennsylvania. Rock layers dip away from the center of the valley creating a classic anticlinal valley. (b) A schematic cross-section showing anticlinal valleys and synclinal mountains. *R* signifies resistant rock.

side of the Blue Ridge all the way to Alabama. It is really a composite of several river valleys that includes the Hudson Valley, Shenandoah Valley in Virginia, and the Tennessee Valley. Collectively, it is known as the *Great (Appalachian) Valley*. It is, in a sense, the eastern counterpart to the Rocky Mountain trench. It is underlain by thrust faults along most of its length, and it forms the boundary between sedimentary rocks of the Valley and Ridge and crystalline rocks of the Blue Ridge and New England provinces. Limestone in the Great Valley has produced extensive areas of karst topography particularly from Allentown to the vicinity of Roanoke, including Crystal Cave, Indian Echo Cave, and the Baker, Shenandoah, Massanutten, and Dixie caverns. In most areas of karst topography, such as Florida and the Interior Low Plateaus, the limestone layers are approximately

flat-lying. Karst topography in the Valley and Ridge, however, is unusual because it is developed in dipping layers.

THE OUACHITA AND MARATHON FOLD-AND-THRUST BELTS

The Ouachita fold-and-thrust belt is something of a composite of the Tennessee-Virginia and Pennsylvania regions. There are major thrust faults that produce long, linear ridges, and there are folds, some of which produce a zigzag pattern while others show a dome structure. All three patterns are clearly evident in the shaded relief image and landscape maps shown in Figure 12.13. On the basis of its landscape, the Ouachita fold-and-thrust belt can be divided into four areas: the Arkansas River valley in the north,

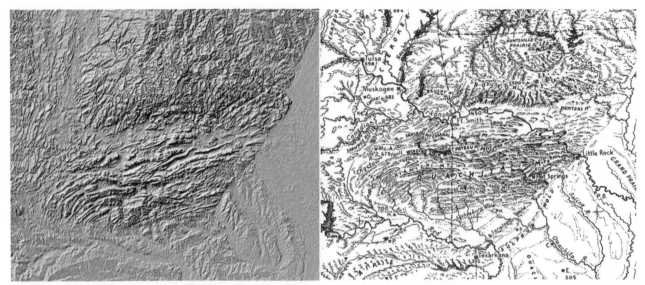

FIGURE 12.13   Shaded relief and landscape maps of the Ouachita region.

FIGURE 12.14   A Google Earth image looking eastward across the Arkansas Ouachita Mountains showing the main divisions. The Arkansas River is at upper left. Note the many arms of Lake Greeson on the Athens Plateau at far right. The boundary of the Ouachitas with the Coastal Plain is at the southern end of Lake Greeson. The line at lower left is the Oklahoma state line.

followed southward by a belt of linear mountains known in Arkansas as the Fourche Mountains, followed by the Central and Western Ouachita Mountains where fold structures are well developed, and finally, the Athens Plateau in the south. Characteristic landscape in each of the four areas is visible in Figure 12.14.

The Arkansas River valley actually lies to the north of the main deformational front of the fold-and-thrust belt between the Boston Mountains of the Ozark Plateau and the Ouachita Mountains proper. It is also geologically between these two areas. This region contains nearly flat-lying layers similar to those in the Ozarks, as well as thrust faults and folds similar to those in the Ouachitas. It appears to have been an upland area that was eroded and dissected by the Arkansas River and its tributaries. There are isolated erosional (flat-topped) mountains and plateaus (mesas) composed of flat-lying rock as well as more complex erosional

leftovers of folded and faulted mountains. These features are visible in Figure 12.13. Perhaps surprisingly, it is the Arkansas River Valley that boasts the highest point in Arkansas: Mount Magazine (2,753 feet).

The Ouachita Mountains are one of the few east-west oriented mountain ranges in the United States; for this reason there are visible contrasts in vegetation across ridges. South-facing slopes are warm, sunny, and dry. They boast large areas of pine forests. North-facing slopes are wet and cool with plenty of hardwood forests. As can be inferred from Figure 12.14, the northern part of the Fourche Mountains forms the main thrust belt and a strong linear landscape. Here, the ridges consist of hard sandstone, and the valleys of interlayered shale and sandstone. In contrast to the Valley and Ridge, limestone is uncommon. The Arkansas River cuts through these ridges in the vicinity of Little Rock. Southward we enter the main Central and Western

Ouachita Mountain region which contains thrust faults but which also boasts an amazing assortment of fold ridges similar to the Pennsylvania Valley and Ridge. This area is well known for a type of chert (microcrystalline quartz) referred to as *novaculite*. In prehistoric times, this rock was used for arrowheads and spear points and as a sharpening stone. It could also be used to spark a flame when struck with steel. Novaculite is a very hard, very resistant rock. It occurs in layers and is often found along ridge crests.

Still farther south we drop down in elevation to below 1,000 feet onto the Athens Plateau. The landscape here is somewhat flat, but the rocks are just as folded and faulted as those to the north. Resistant layers create low-lying, tightly spaced ridges across the region. Several rivers flow southward directly across the low-lying ridges which creates rapids and waterfalls and, therefore, really good whitewater kayaking and rafting conditions. Several of the rivers have been dammed, thereby creating lakes such as Greeson Lake, which can be seen at the right edge of Figure 12.14 with many arms that extend up valleys between resistant ridges. The low-lying topography and cross-cutting rivers suggest that this part of the Ouachita Mountains was at one time covered by younger, flat-lying Coastal Plain sedimentary layers. This landscape, therefore, is much younger than the topographically higher areas to the north and has recently emerged from below the younger, flat-lying layers.

Now we turn our attention southwestward to the Marathon region which is shown in Figures 8.4, 10.2, and at the bottom of Figures 10.4 and 10.8 between the Glass and Santiago Mountains west of the Edwards Plateau. The Marathon region is a pivotal area from both a geological and a landscape point of view. It is part of the Appalachian-Ouachita mountain belt, yet it is located at the edge of the Cordilleran belt. It is a structural dome and a physiographic basin, and therefore is similar to the Llano Uplift and the Nashville Basin. Unlike the Llano Uplift, the central part of the basin contains no crystalline rock. Nearly flat-lying rocks surround the Marathon Basin, but unlike the Nashville Basin they are not present within the basin itself. Instead, the central part of the basin, the core of the dome structure, exposes thrust-faulted and folded sedimentary layers and a linear landscape that is essentially identical to what is found in the Ouachita Mountains. The rocks are of the same type, age, and structure, including the presence of novaculite. Drilling confirms the connection with the Ouachitas. Clearly, the Appalachian Mountain system at one time extended continuously southward and westward around the Gulf region where it must have interacted with rocks that later became part of the Cordilleran Mountain system.

The influence of Cordilleran tectonics is also present. The northwestern margin of the Marathon basin is flanked by 35-million-year-old volcanic rocks of the Trans-Pecos volcanic field (the Davis Mountains). The southwestern margin is flanked by a normal fault associated with extension in the Basin and Range. Across the border in Mexico there may

even be remnants of the Cordilleran fold-and-thrust belt. Figure 12.15a is a Google Earth image of the northeastern and eastern edges of the Marathon Basin. One can clearly see the upturned edges of folded layers in the bottom and middle-right of the image. Look closely at the folded layers at right-center and how these layers disappear beneath the nearly flat-lying layers that form the entire upper-left portion of the image. Note especially how the landscape changes its morphology as a result of the change in rock structure. The nearly flat-lying layers in Figure 12.15a are part of a younger Great Plains sedimentary sequence. The folded layers are much older but still younger than 542 million years. Figure 12.15b is a close-up view in which one of the upturned edges of older rock is cut off and covered by younger, nearly flat-lying layers. Can you find this location in Figure 12.15a? The only way to explain this geometry is to conclude that the older layers were deposited, folded, and eroded prior to deposition of the younger, nearly flat-lying layers. Recall from Chapter 10 that geologists refer to an erosion surface within a sequence of rock as an *unconformity*. In Chapter 10 we described a buttress unconformity. The type of unconformity shown here is an *angular unconformity* because it separates two layers of rock that are oriented at an angle with each other. An angular unconformity between younger and older nearly flat-lying sedimentary rocks is also shown in the bottom two panels of Figure 11.10.

The angular unconformity so beautifully displayed in Figure 12.15b shows the exact same relationship that we would see at the eastern and western margins of the Ouachita fold-and-thrust belt as well as at the southwestern margin of the Appalachian belt in Alabama. In all three areas the older faulted and folded sedimentary layers disappear beneath younger, nearly flat-lying layers of the Coastal Plain or Great Plains. Unfortunately, in these areas, the unconformable contact is not so beautifully exposed.

By now you should already know the significance of the unconformable contacts discussed above. It means that Appalachian mountain building and subsequent erosion to sea level of this part of the mountain belt had already occurred before the young, nearly flat-lying layers were deposited. In the case of the Marathon Basin, the nearly flat-lying rocks are domed up slightly such that the underlying rocks are now reexposed. It is a situation that mirrors the history of the Llano Uplift shown in Figure 11.10 although there are some important differences. Both areas were deformed between about 335 and 265 million years ago. Deformation in the Llano area, however, was very mild. A thin sequence of (older) sedimentary layers was broadly arched into an anticline and then eroded such that crystalline rocks were exposed in the core of the anticline as depicted in the top two panels of Figure 11.10. These older sedimentary layers, however, are tilted only a few degrees. They remain nearly flat

FIGURE 12.15   (a) A Google Earth image looking southeastward across the eastern side of the Marathon Basin. Note the many folded resistant ridges in the Marathon fold-and-thrust belt at lower right. Younger, nearly flat-lying rocks cover the upper left of the image. Note how the fold ridges disappear below the nearly flat-lying rocks and the abrupt change in landscape. (b) A Google Earth image looking northeastward at a close-up of the angular unconformity visible in (a).

lying. On this basis, we would have to conclude that the Llano area was not involved in foreland fold-and-thrust belt mountain-building processes. The Marathon region, on the other hand, involves a much thicker sequence of sedimentary rocks that were strongly deformed. They are definitely part of the Appalachian-Ouachita foreland fold-and-thrust belt.

WATER GAPS IN THE VALLEY AND RIDGE AND OUACHITA MOUNTAINS

The Valley and Ridge and Ouachita areas are well known for rivers that cut directly through ridges to form water gaps. Perhaps the most famous is the Delaware Water Gap on the Delaware River along the Pennsylvania-New Jersey state line. A close look at Figure 12.11 provides additional examples, including several along the Susquehanna River.

There are several methods by which a water gap can form. We mentioned one such method in Chapter 11 with respect to the Middle Rocky Mountains by which a river establishes a channel in flat-lying layers and maintains that channel as it cuts downward into underlying folded rock layers. This same method has been suggested for some of the water gaps in the Valley and Ridge, but such a possibility is difficult to prove because there are no flat-lying rock layers in close proximity to the water gaps. This method does seem to be reasonable to explain water gaps on the Athens Plateau and possibly in other areas within the Ouachitas.

Another possibility is that a river first establishes itself in flat-lying rock and then maintains its channel as the rock is thrust-faulted and folded. This is a possible method, but it seems unlikely because the fold-and-thrust belt is hundreds of millions of years old. It would require that the river not change course appreciably during all that time.

A third possibility that has been suggested for water gaps on the Susquehanna River is for the river to establish itself in a thick, folded layer of weak shale. The idea is that the shale layer was thick enough to cover the entire area and, in spite of being folded, exerted very little influence on the course of the Susquehanna River which was able to cut a path across some of the folded layers on its way to the sea. The river was strong enough to maintain its channel as it stripped away the shale and cut across similarly folded resistant layers underneath, thus producing a water gap.

A final favored possibility for some of the water gaps, including the Delaware Water Gap, is simply the exploitation of a fractured, weak rock along the crest of a resistant ridge. The idea is that water will channel itself into areas on the ridge that are strongly fractured and therefore weaker than other points on the ridge. The weak part of the ridge simply wears down, allowing the river to cross the ridge and capture another river on the other side. The sequence is depicted schematically in Figure 12.16a. The presence of fractured rock within many of the water gaps lends credence to this possibility.

Also present, particularly in the Valley and Ridge, are wind gaps. A *wind gap* is a low-lying mountain pass through which a river at one time passed but is no longer present. A wind gap can form if a river cuts partly across a ridge prior to abandoning its channel for lower ground as depicted in Figure 12.16b. This scenario seems to have occurred in the Valley and Ridge, but one must be careful because a wind gap can also form via stream piracy as shown in Figure 12.16c.

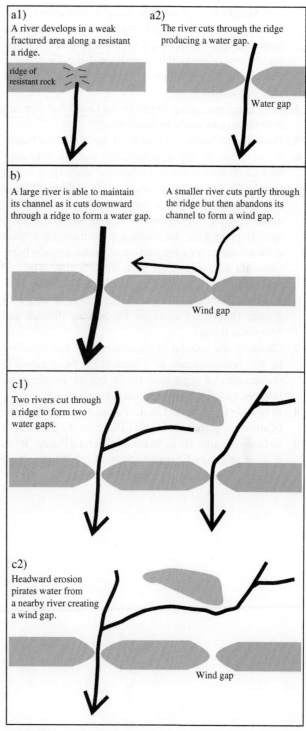

FIGURE 12.16 Schematic maps showing (a) formation of a water gap by headward erosion of a river through a fractured, weakened section of a ridge; (b) formation of a wind gap where a smaller river is defeated in its attempt to maintain its channel through an emerging ridge; and (c) formation of a wind gap by headward erosion and river piracy.

QUESTIONS

1. Use Figure 10.4 to trace the area of normal faulting in the Northern Rockies onto Figures 12.4 and 12.5. Trace the Lewis and Clark Line on both figures.

2. What is the name and origin of the major valley in the upper left of Figure 12.7a?

3. What geologic (surface) evidence might you look for in order to conclude that the Valley and Ridge, Ouachita,

and Marathon regions were once a continuous fold-and-thrust belt? As a helpful hint, let's assume that you know the age of the rocks in all three areas, including the age of rocks that have been thrust-faulted and folded as well as the age of rocks that were deposited after deformation.

4. Why are north-facing slopes in the Ouachitas wet and cool relative to south-facing slopes?

5. Describe the landscape history of the Marathon Basin.

6. Some areas of the Appalachian-Ouachita fold-and-thrust belt have remained emergent since their initial formation. Other areas, such as the Athens Plateau and the Marathon Basin, were covered by a shallow ocean at some time between 135 and 80 million years ago. Explain how the landscape in these two areas differs from other emergent areas in the Appalachian-Ouachita fold-and-thrust belt. Specifically, why are emergent areas in the Appalachian-Ouachita fold-and-thrust belt generally at higher elevation and with greater relief than areas on the Athens Plateau and Marathon Basin?

7. Calculate the amount of shortening that has occurred in the bottom panel of Figure 12.1 by measuring the amount of separation (in inches or centimeters) between two formerly adjacent layers.

8. In Google Earth, fly to Oriole, Pennsylvania. What type of structure surrounds Oriole? Describe the landscape.

9. In Google Earth, fly to Shamokin, Pennsylvania. What type of structure surrounds Shamokin? What human activity is occurring at this location?

10. In Google Earth, fly to Strasburg, Virginia. What is the name of the river that flows along the southern side of Strasburg? Suggest hypotheses to explain why the river meanders along the side of the mountain. What type of structure forms the mountain south of Strasburg? Note that another river on the south side of the mountain also meanders. What is the name of this river?

11. In Google Earth, fly to Powells Crossroads, Tennessee. What type of structure surrounds Powells Crossroads? What physiographic province are we in?

12. In Google Earth, fly to East Signal Peak, Oklahoma. What type of structure surrounds East Signal Peak? What physiographic province are we in?

13. Using a copy of Figure 12.15a, trace the boundary between the fold-and-thrust landscape and landscape underlain by nearly flat-lying sedimentary rocks.

14. What is novaculite and where does it occur?

15. What is the Belt Supergroup? Where does it occur? What is the age of the Belt Supergroup?

16. Major valleys are present in the Northern Rocky Mountain fold-and-thrust belt and in the Valley and Ridge. What is the name of these valleys? What is their origin?

17. Can you pick out the angular unconformity in Figure 11.10?

18. If the Llano area was not involved in foreland fold-and-thrust faulting and the Marathon region was involved, draw the inferred front (the northernmost extent) of the foreland fold-and-thrust belt from the Valley and Ridge to the Ouachitas and then to the Marathon region.

Crystalline Deformation Belts

Crystalline rock is different from sedimentary and volcanic rock primarily because layering and differential erosion of layers play a less prominent role in landscape development. Layering in rocks creates a certain amount of predictability in both the geology and the landscape of an area that may not be present in crystalline rock. Instead, the landscape may be monotonous, somewhat random, massive, or shaped by structural influences. Batholiths and other areas of intrusive rock, in particular, tend to be nonlayered and unpredictable in their landscape characteristics. Metamorphic rocks are somewhat more predictable, primarily because most are characterized by a certain degree of layering and structural form.

The distribution of crystalline rock across the United States is shown in Figure 2.9. It is clear from this figure that crystalline rocks occur in a variety of structural settings and in all but a handful of physiographic provinces. We can very broadly subdivide crystalline rock in the US into four categories; (1) rocks greater than 1 billion years old that form the ancient crystalline shield of North America, (2) batholiths, mostly between 55 and 220 million years old, (3) rocks associated with Cordilleran mountain building, metamorphosed and deformed less than 400 million years ago, and (4) rocks associated with Appalachian mountain building, metamorphosed and deformed mostly between 450 and 265 million years ago.

Rocks that form the North American crystalline shield were introduced in Chapter 10 (Figure 10.2). These rocks underlie the entire Interior Plains and Plateaus region, including the Superior Upland, and are present below at least part of the Appalachian, Cordilleran, and Coastal Plain physiographic regions. They form the crystalline core of nearly all mid-continent anticlines and domes including most of the Middle and Southern Rocky Mountains. They crop out at the bottom of the Grand Canyon, in the southern and southwestern Basin and Range, and in the Teton Mountains. In the Appalachians they form part of the western Blue Ridge Mountains and the Green Mountains of western Vermont.

Batholiths are large areas of intrusive (usually granitic) rock typically associated with subduction zones or hot spots. There are three batholiths of note within the Cordillera and one in the Appalachians. Those in the Cordillera are the Sierra Nevada of eastern California, the Peninsular batholith of southern California, and the Idaho batholith. All three are between 55 and 220 million years old and are associated with subduction of oceanic lithosphere beneath western North America. Within the Appalachian Mountains, the White Mountain batholith of central New Hampshire is between 100 and 200 million years old and is associated with rifting of North America following Appalachian mountain building and with the passing of the Great Meteor (New England) hot spot.

Areas that were metamorphosed during Cordilleran and Appalachian mountain building include both native North American rocks and rock material that was accreted (added) to North America during mountain building. Native metamorphic rocks in the Appalachians are restricted to the western Blue Ridge and Green Mountains. In the Cordillera, these rocks are scattered across primarily the eastern part of the mountain belt in Montana, Idaho, and Utah. Accreted crystalline terranes make up most of the Appalachians including the eastern Blue Ridge, the Piedmont Plateau, and New England east of the Green Mountains. In the Cordillera, crystalline accreted terranes are present in California, Oregon, and Washington.

Crystalline rocks in many parts of the US are best discussed in the context of their structural setting. For example, we have already described the crystalline-cored anticlines and domes of the Middle and Southern Rocky Mountains (Chapter 11). In later chapters we will describe several additional prominent areas of crystalline rock in context with their structural setting including the Sierra Nevada and Klamath Mountains. This chapter is concerned primarily with the largest continuous area of crystalline rock in the country, the Appalachian Mountains, with a large area of crystalline rock in the Pacific Northwest, and with the Superior Upland. These areas are shown in Figure 8.4.

In Chapter 12 we described the structure of sedimentary rocks within the Appalachian Valley and Ridge as a foreland fold-and-thrust belt. Crystalline rock in the Blue Ridge, Piedmont Plateau, and New England Highlands (both native and accreted) form the rear of the foreland fold-and-thrust belt where thrust faults were able to cut deep enough into the crust to carry crystalline rock toward the surface to be exposed via erosion. This entire region can be referred to as a *crystalline deformation belt*. These areas are characterized by deformation that is far more complex than what is seen in the adjacent foreland fold-and-thrust belt. In addition to thrust faults, crystalline deformation belts are often subjected to multiple generations of folds, metamorphism, and intrusion, any of which may occur prior to, during, or following thrust faulting. Strike-slip faults are common, and in some cases normal faults are present. These complications play a prominent role in the history of compressional mountain building and are addressed in Part III of this book. In this chapter we will concentrate on the character and recent evolution of the landscape.

Within the Cordillera we will look in particular at the landscape of the Northern Rocky Mountain-Cascade crystalline belt, which consists of the Idaho batholith and a belt of crystalline rock that extends from northernmost Idaho westward to the North Cascades. In terms of compressional mountain building, this area has an origin similar to that of the crystalline Appalachians in that both areas developed at the rear of a foreland fold-and-thrust belt and both areas include native and accreted rock. The Northern Rocky Mountain-Cascade crystalline belt, however, is markedly different from the Appalachian crystalline belt. In the Appalachians, the structure produced during compressional mountain building, although relatively old, has not been substantially altered by later tectonic events. The structures are still well preserved. This is not the situation in the Cordillera. Although the crystalline deformation belt is considerably younger than the Appalachian belt, the structures produced during compressional mountain building have largely been overprinted by even younger tectonic features that include an early period of Basin and Range normal faulting, volcanism, extensive intrusion, and later periods of deposition. It is the structure of these young tectonic features, and not the structure produced during earlier periods of mountain building, that controls much of the present-day landscape.

We end this chapter with a description of the largest area of crystalline rock in the Interior Plains and Plateaus region, the Superior Upland. The rocks of the Superior Upland are part of the crystalline shield. They form part of an ancient crystalline deformation belt, one that is well over 1 billion years old. The glacial landscape of the Superior Upland was described in Chapter 6. In this chapter we will describe some of its crystalline rock characteristics.

THE CRYSTALLINE APPALACHIANS

Formation of the US Appalachian Mountains culminated between 300 and 265 million years ago with collision of North America, Africa, Europe, and South America to form the supercontinent Pangea. During final collision, the Appalachians may have reached elevations of 20,000 feet or more. Since that time, the landscape has undergone erosion without substantial tectonic modification. As a result, much of what was once a high mountainous region has been reduced to a low-lying landscape. The Blue Ridge Mountains, the Piedmont Plateau, and the New England Highlands form the crystalline core of the once mighty Appalachians. East of the Piedmont, the erosional stump of more than half of the Appalachian mountain belt is now buried beneath sedimentary rock of the Coastal Plain and continental shelf (Figure 5.1). Only the Blue Ridge and New England Highlands retain their mountainous character, and these are low, rounded, eroded mountains without the jagged peaks that characterize many of the younger mountain ranges of the Cordillera. The rocks are dominantly metamorphic including an abundance of metamorphosed granitic rock (granitic gneiss), most of it quite old (>265 Ma). The structure is complex with an assortment of thrust faults, strike-slip faults, and folds that are evident both at the scale of an outcrop and at the scale of the entire mountain range. Major faults produce zones of weakness such that the physiographic grain (trend) of the mountain belt parallels the strike of faults.

A poorly understood aspect of the Appalachians is that it is still a mountainous region some 265 million years after mountain-building events ended. The presence of mountainous topography after all this time suggests that overall erosional lowering has been interrupted by punctuated thermal isostatic changes, sea-level changes, and climate changes, all three of which can cause elevation (and relief) gain or loss. This implies that the Appalachians were likely lower at times in the past. One possible time period was between about 135 and 50 million years ago when sea level was higher than it is today (Figure 7.1). We will have more to say about the ups and downs of the mountain range later in this chapter. For now we can conclude that the erosional landscape we see today is controlled by a combination of rock hardness, structure, and thermal isostatic warping, punctuated with sea-level and climate changes.

The Blue Ridge, which includes the Great Smoky Mountains, is widest and highest in the south with at least 34 peaks above 6,000 feet, including Mt. Mitchell in North Carolina (6,684 feet, the highest in the Appalachians), Clingmans Dome in Tennessee (6,643 feet), and Mt. Guyot on the North Carolina-Tennessee border (6,621 feet). It is a massive mountainous region as much as 80 miles across. Figure 13.1 is a view looking across the northwestern side of the Great Smoky Mountains at more than 5,500 feet of relief between Pigeon Forge and the high peaks. On the

FIGURE 13.1   A Google Earth image looking southeastward at the northwestern front of the Great Smoky Mountains from above Pigeon Forge, Tennessee (*PF*). The city of Gatlinburg (*G*) and several high peaks are named.

FIGURE 13.2   A Google Earth image looking northwestward across the southern-central Appalachians. *BR* = northern end of the Blue Ridge; *BRE* = Blue Ridge Escarpment; *FL* = Fall Line; *G* = Great Smoky Mountains; *GV* = Great Valley; *PF* = Pigeon Forge; *R* = Roanoke. The following cities, from south to north, trace the Fall Line: *A* = Augusta, *C* = Columbia, *Rh* = Raleigh, *Rd* = Richmond, *W* = Washington, DC, *P* = Philadelphia.

opposite (southeastern) side of the mountain range, the Blue Ridge Escarpment (or Blue Ridge Front) forms a near-continuous ridge with more than 1,000 feet of relief that overlooks the Piedmont Plateau. This feature is visible in the Google Earth image of the Blue Ridge shown in Figure 13.2. A discussion of the origin of the Blue Ridge Escarpment constitutes a separate section later in this chapter. Resistant granitic gneiss forms much of the western Blue Ridge highland, but rocks, in general, are poorly exposed

in the heavy forest and deep soil. Much of the rock is more than 1 billion years old and constitutes a remobilized (remetamorphosed and deformed) part of the ancient North American crystalline shield.

The Blue Ridge narrows to less than 14 miles across in the vicinity of Roanoke, Virginia, where only a few peaks top out above 4,000 feet. Northward, the Blue Ridge is nothing more than a westward-sloping upland surface separated from low rolling hills of the Piedmont Plateau by a subdued

Blue Ridge highland areas are shaded. This figure shows the Shenandoah River prior to lengthening along the west side of the Blue Ridge and prior to the capture of Goose Creek tributaries. North is to the top of page.

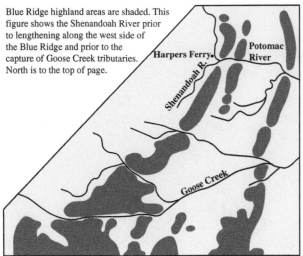

This figure shows the modern Shenandoah River and development of the Snickers, Ashby, and Manassas wind gaps.

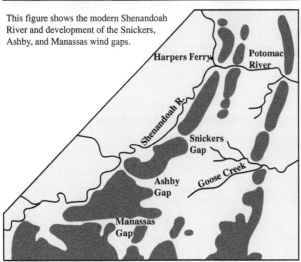

FIGURE 13.3 A sketch map that shows the development of wind gaps by progressive piracy of the Shenandoah River. Modified from Thornbury (1965).

Blue Ridge escarpment and several outlying mountain ridges. The Roanoke, James, and Potomac rivers cross this part of the Blue Ridge on their way to the Atlantic Ocean forming water gaps. Also present are wind gaps such as Manassas Gap that formed via stream piracy by the Shenandoah River as depicted in Figure 13.3. The Blue Ridge continues to lose elevation northward and ends as a distinctive physiographic province just south of Harrisburg in southern Pennsylvania. A landscape map that outlines the Blue Ridge in its entirety is shown in Figure 13.4. The total length of the Blue Ridge is about 600 miles.

The Piedmont Plateau, perhaps because it is closer to the ocean, has been almost completely eroded to a low-lying, relatively flat plateau surface. Elevations in its eastern part are mostly less than 500 feet with little relief (<50 feet) beyond the immediate confines of river channels. Higher elevations and rolling hills are present on the western side,

especially in the south where a few small isolated mountains of particularly resistant rock protrude above the surroundings. These erosional remnants include Stone Mountain in Georgia, Kings Mountain, South Mountain, Moores Knob, and Brushy Mountain in North Carolina, and Big Cobbler in Virginia (Figure 13.4). A mountain or rocky mass, such as Stone Mountain, that has resisted erosion and stands isolated, surrounded by flatland, is referred to as a *monadnock*. A Google Earth image of two monadnocks near Kings Mountain is shown in Figure 13.5. With the exception of stream channels, road cuts, and a few summit areas, bedrock is poorly exposed in the Piedmont due to millions of years of weathering in a dominantly humid climate that has locally produced soil horizons up to 300 feet thick.

Geologically, the New England Highlands are an extension of the Blue Ridge and Piedmont provinces. However, unlike the southern Appalachians, the New England Highlands were completely covered by glaciers during the past 2.4 million years. The result is a hilly landscape in the south with elevations increasing northward toward Vermont and New Hampshire where glacial erosion has exposed considerable bedrock. There are three mountainous areas: the Taconic Mountains along the New York state line from northern Connecticut to southern Vermont; the Berkshire and Green Mountains in western Connecticut, Massachusetts, and Vermont; and the White Mountains in north-central New Hampshire. Mt. Katahdin (5,268 feet), the highest point in Maine by more than 1,000 feet, is isolated enough to be considered a monadnock separate from other ranges. All four areas are shown in Figure 13.6 which is a Google Earth image of northern New England. A landscape map of New England is presented in Figure 9.4.

The Taconic Mountains are situated along the eastern margin of the Hudson Valley, which is part of the Great (Appalachian) Valley. The Taconics are a low-lying mountain range composed of sedimentary and metamorphic rock with maximum elevations less than 3,000 feet in the south and below 4,000 feet in the north. Major peaks include Mt. Equinox (3,850 feet) and Dorset Mountain (3,772 feet). The eastern margin of the range north of Bennington is marked by a distinctive, flat-floored valley known as the Valley of Vermont, which separates the Taconic Range from the Green Mountains. This valley, from Rutland to Shaftsbury, is visible in Figure 13.7. Southward, the Taconics merge with the Berkshire Mountains across narrow valleys. Dorset Mountain is home to the largest underground marble quarry in the world, where some of the finest marble is produced. Since around 1902 it has been the marble of choice for our national cemeteries and for many of our national buildings in Washington, DC.

The Berkshire (or Hoosac) Mountains of western Connecticut and Massachusetts, and the Green Mountains of western Vermont, form the backbone of New England and some of the best skiing in the northeast. In terms of rock

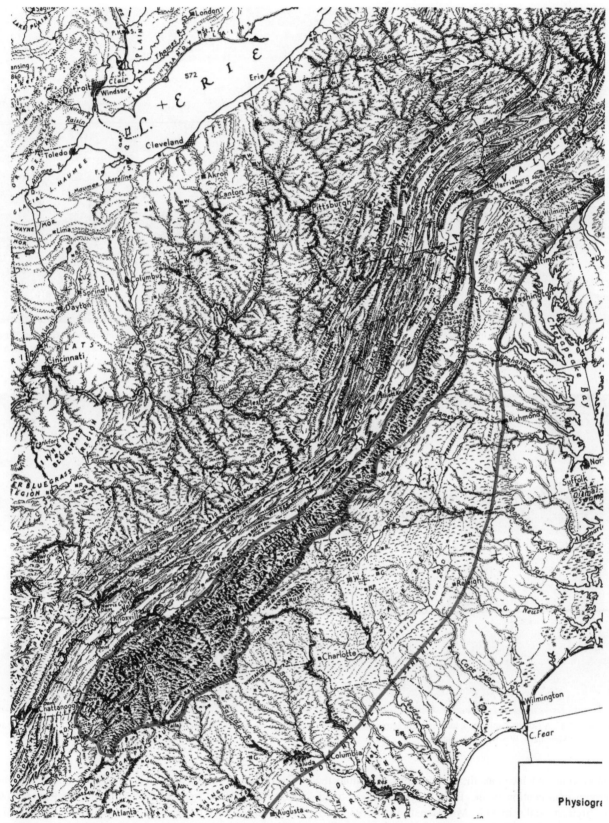

FIGURE 13.4 A landscape map that shows the Fall Line and total extent of the Blue Ridge. Landscape maps of the New England and southern Appalachians are shown in Figures 9.4 and 10.47, respectively.

FIGURE 13.5   A Google Earth image looking northwestward at two monadnocks on the Piedmont Plateau near Kings Mountain, North Carolina. King's Pinnacle (1,625') is at left and Crowder's Mountain (1,706') at right. Both stand 700 to 900 feet above the plateau. The Blue Ridge Escarpment can be seen in the distance at the skyline.

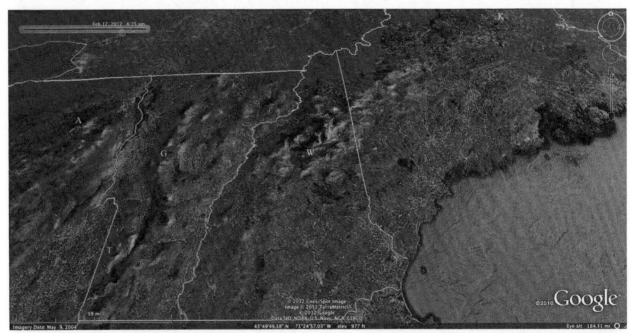

FIGURE 13.6   A Google Earth image looking north across northern New England. *A* = Adirondack Mountains, *G* = Green Mountains, *K* = Mt. Katahdin, *T* = Taconic Mountains, *W* = White Mountains.

composition and location, these mountains constitute the northern continuation of the Blue Ridge mountain trend. The highest peaks include Mt. Mansfield (4,393 feet), Camels Hump (4,088 feet), and Killington Peak (4,235 feet), all in the Green Mountains of Vermont and all more than 2,000 feet lower than the highest peaks in the Smoky Mountains.

The mountain range is less than 35 miles wide and thus recreates the narrow trend of the northern Blue Ridge. However, rather than forming an escarpment, as in the Blue Ridge, the first-order structure of the mountain range is that of an anticline. Similar with the Blue Ridge, the southern part of the Green Mountains exposes ancient (>1 billion

FIGURE 13.7   A Google Earth image looking north at western Vermont and the Champlain Valley. The ski areas Stratton (*S*), Bromley (*B*), and Killington (*K*) are in the Green Mountains. The Valley of Vermont extends from Shaftsbury (*S*) to Rutland (*R*) and separates the Green from the Taconic Mountains. Dorset Mountain (*D*) is in the Taconic Mountains. Lake Champlain, the Adirondack Mountains, and the Canadian border can be seen at upper left.

years old) North American crystalline shield in its core. Two narrow ridges of ancient crystalline rock continue south of New England into New York and Pennsylvania. These are the Reading Prong, which extends from the Connecticut-New York border to Reading, Pennsylvania, and the Manhattan Prong, which forms Manhattan Island. A water gap is present where the Reading Prong crosses the Hudson River.

The Connecticut River Valley extends along the Vermont-New Hampshire border and serves to separate the Green Mountains from a massive granitic batholith that forms the White Mountains of north-central New Hampshire (Figure 13.6). It is these mountains that boast the highest peaks in New England. There are seven peaks that exceed 5,000 feet, including Mt. Adams (5,774 feet), Mt. Lafayette (5,249 feet), and the only peak above 6,000 feet, Mt. Washington (6,288 feet). An additional 41 peaks rise above 4,000 feet. Most have rounded summit areas typical of the Appalachians. Although elevations are not impressive by Cordilleran standards, the weather at the top of these peaks can be very impressive. By that I mean cold, windy, and nasty. The highest peaks are at least 1,000 feet above tree line. The summit of Mt. Washington boasts an observatory that records some of the windiest weather anywhere. On April 12, 1934, the wind at the observatory was clocked at 231 miles per hour. This is the highest surface wind speed ever recorded! The granitic batholith that forms these mountains is considerably younger than the ancient rock that forms the Berkshire and Green Mountains. The first and most significant phase of magmatism

occurred between 200 and 165 million years ago during the breakup of Pangea. A second period of magmatism occurred between about 130 and 100 million years ago when the Great Meteor (New England) hot spot passed below New Hampshire. Farther northeast, in upper Maine, the landscape becomes subdued to almost flat, with many lakes interspersed between small isolated mountain peaks of resistant rock (monadnocks). This is an area of glacial scouring and erosion followed by glacial deposition. We have already noted the highest monadnock, Mt. Katahdin, which marks the end of the Appalachian Trail. The mountain is underlain by a large, resistant, circa 400-million-year-old granite intrusion.

There are three notable exceptions to the dominance of crystalline rock across the Appalachian mountain belt. The first are the narrow Triassic Lowland valleys composed of sedimentary rock, basalt, and gabbro located in Southern New England and the Piedmont Plateau. These are normal fault valleys and are described in Chapter 15. The second area is in northern Maine north of Mt. Katahdin where the grade of metamorphism is so low that the rocks retain their sedimentary character. The third area is in the Blue Ridge where there are flat-floored basins known as *coves* underlain by sedimentary rock. Coves were key locations for understanding the geology of the Appalachians in the early days of exploration. The appearance of sedimentary rock on the floor of these coves has an origin related to thrust faulting. During Appalachian mountain building, major thrust faults carried crystalline rock over sedimentary rock of the Valley and Ridge. The entire package was then folded and subsequently eroded such that crystalline rock (and the underlying thrust fault) was removed from the crest of an anticline, thereby exposing sedimentary rock beneath the thrust sheet. The formation of a cove is shown in Figure 13.8 with a series of three cross-sections and a map. A structure such as this is referred to in geological parlance as a *window* in the sense that it is a window through which rocks below the thrust sheet can be seen.

EROSIONAL HISTORY OF THE APPALACHIAN MOUNTAINS

The erosional history of the Appalachian Mountains during the 265 million years following mountain building is not entirely clear. One way to gain a better understanding of this history is to look at the age and thickness of sediment that has been eroded from the mountain. The idea is to correlate periods of high sediment accumulation with periods of regional uplift. One has to be careful here because changes in climate or sea level could also cause changes in sediment accumulation rates. In the case of the Appalachians, we have a good enough understanding of climate and sea-level history to take these changes into consideration. When we do this, we find that isostatic uplift and subsidence are major influences on landscape history. Isostatic adjustments result from many factors, one of which is the temperature of rock below the mountain. An increase in temperature, perhaps due to an upwelling of hot material from deep within the Earth or from a granitic intrusion, will lower the density of rock and result in isostatic uplift in the form of a broad, anticlinal bulge. We have already discussed numerous examples of this phenomenon, such as the Adirondacks, the Mississippi Embayment, and the Middle-Southern Rocky Mountains. The location of warm rock beneath the mountain can migrate over time, implying that the crest of an anticlinal bulge, and therefore the location of high topography in the mountain belt, can also migrate.

Large basins off the eastern US coastline, such as the Baltimore Canyon Trough and the Carolina Trough, trap sediment eroded from the Appalachian Mountains. Both troughs extend parallel to the coastline along the shelf-slope break. The Baltimore Canyon Trough extends from New Jersey southward to Maryland. The Carolina Trough parallels the North and South Carolina coastlines. Fossils are used to date the sediment which is as much as 180 million years old. These troughs, particularly the Baltimore Canyon Trough, record erosion primarily from New England and the Central Appalachians. Sediment thicknesses suggest that more than 4.3 miles (7 km) of crust has been removed from the Appalachian Mountains in the past 180 million years (equivalent to 0.15 inches per 100 years). The data indicate high rates of sediment accumulation at about 155, 130, 85, and 15 million years ago (Ma). These

Crystalline rocks of the Blue Ridge are thrust over sedimentary rock of the Valley and Ridge.

The thrust fault is folded.

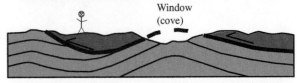

The thrust fault is eroded across an anticline exposing sedimentary rock in a window (cove) below the thrust.

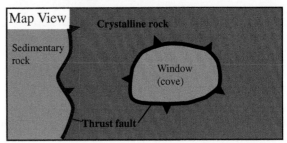

Map view of a window in the Blue Ridge Mountains. Teeth are on the upper plate of the thrust fault.

FIGURE 13.8   Cross-section sketches and a map that shows formation of coves in the Blue Ridge.

time intervals are understood to correlate with periods of relatively rapid uplift/erosion (that is, exhumation). The data, however, must be interpreted with the understanding that exhumation did not occur uniformly along the length of the Appalachians during the past 180 million years but rather was more likely concentrated in one area or another at different times. All four sedimentation pulses likely correlate with local or regional areas of thermal isostatic uplift due to a rising mantle heat source or due to the rise of a deep-seated crustal intrusion. The circa 15 Ma sediment accumulation pulse continues today at somewhat lower rates, implying that present-day elevations in the Appalachians are a recent (<15 My old) phenomenon.

Another clue to the landscape history are the sedimentary rocks of the Coastal Plain that overlie the eroded stump of the Appalachians. The age of these rocks indicates how long ago this part of the Appalachians was eroded to sea level. Additionally, they provide clues to the way sea level has changed over time. The oldest oceanic rocks on the Coastal Plain suggest that this part of the Appalachians was already at or below sea level by 130 million years ago. Sedimentary rocks on the Coastal Plain generally become younger toward the ocean, suggesting an overall retreat of the sea (a relative lowering of sea level) since 85 Ma.

Given this information, we can suggest a scenario for the erosional history of the Appalachian landscape. Following the culmination of mountain building some 265 million years ago, part of the mountain belt may have already been reduced to a lowland by 230 million years ago, prior to the rifting event that produced the Triassic Lowlands and the Atlantic Ocean. Actual separation of North America from Africa and Europe occurred between 230 and 150 million years ago in a part of the mountain belt that is currently below the Coastal Plain (east of the present-day Appalachians). The rifting event likely produced elevation gain in the Appalachians through thermal isostatic uplift, although elevations were nowhere near the height attained during the earlier continental collision. The first pulse of sedimentation at 155 million years ago may correlate, in part, with elevation gain during the late stages of the rifting event, or with the heat associated with intrusion of the White Mountain batholith. It is typical for a region to undergo thermal cooling and slow subsidence following a rifting event. This phenomenon, coupled with erosion, would have again worn the Appalachians down to a lowland probably by around 140 million years ago. The pulse of sedimentation at 130 million years ago was relatively small and thus, probably does not signal a significant gain in elevation. A high sea level stand had begun by about 110 million years ago and lasted until 34 million years ago. It was during this time interval that elevations were likely lower than today and probably the lowest in the history of the Appalachians. Much of the Coastal Plain and perhaps part of the Piedmont Plateau was below sea level. Because of high sea level, the 85-million-year spike in erosion/sedimentation rates probably correlates with thermal isostatic-induced uplift but not with significant elevation gain. Any uplift generated at this time would instead have been neutralized by increased rates of erosion and sedimentation. Given these inferences, we can surmise that, as early as 80 million years ago, the Appalachian landscape was reduced to a peneplain (a low-lying, flat, erosion surface), with isolated highlands in the vicinity of the Blue Ridge and New England Highlands. Part of the peneplain still exists in dissected form, although it is difficult to recognize.

Elevations likely remained subdued and below present-day elevation for the next 60 to 70 million years as sea level retreated, thereby slowly exposing the Coastal Plain and marginally increasing overall elevation. The final relatively strong pulse of sedimentation began about 15 million years ago and has continued to the present day. This pulse is correlated with regional bulging of the Appalachians, possibly due, again, to thermal isostatic changes brought on by the upward flow of warm rock from the mantle.

Thus, the present-day topography and elevation of the Appalachian Mountains could be the result of uplift that is only 15 million years old. If this is the case, the Appalachian landscape we see today, although shaped by 265 million years of erosion, may actually be relatively young at least in terms of its elevation. In a sense we can say that the Appalachians are an old mountain range that has found the fountain of youth and has reverted back to a younger stage. The Roanoke, James, and Potomac rivers, among others, form water gaps through the Blue Ridge. It is possible that these rivers were already in existence long before 15 million years ago and were able to maintain their channel by down-cutting through buried structures during regional uplift to produce the water gaps. The more recent glaciation would have, at times, lowered sea level substantially enough to cause river incision and therefore increase overall relief, thus contributing to the present-day landscape.

One interesting aspect of this analysis is that it is likely that the southern Blue Ridge, and perhaps other parts of the Appalachians, have remained above sea level since mountain-building events began more than 450 million years ago. Thus, in spite of its mountainous character, the Blue Ridge could form some of the oldest continually exposed landscape in the United States.

THE FALL LINE

A generic *fall line* is one that marks a distinct drop in land elevation. The *Fall Line* in the Appalachians is a semicontinuous break in slope, with relief between 100 and 300 feet, that results directly from erosional differences between hard crystalline rock of the Piedmont Plateau and soft sedimentary rock of the Coastal Plain. The Fall Line is visible in Figure 13.2 and in landscape maps shown in Figures 13.4, 9.4, and 10.47. It is the easternmost of three major escarpments in the eastern US, the others being the Cumberland/Allegheny/Catskill Front and the Blue Ridge Escarpment.

Crystalline rock of the Piedmont Plateau resists erosion and therefore sits higher than weak sedimentary rock of the Coastal Plain. Most of the rivers in the eastern US flow eastward across the Fall Line to the Atlantic Ocean. The abrupt change in rock type from crystalline to sedimentary rock at the Fall Line disrupts the river's ability to maintain a smooth gradient to the sea. Rivers are forced to cut downward into the crystalline rock in order to meet the lower topography of the Coastal Plain. In doing so, rivers form gorges, rapids, and waterfalls. Once a river reaches the lower gradient of the Coastal Plain, it will slow down and deposit material. In contrast to strong crystalline rock, the weak sedimentary rock of the Coastal Plain allows rivers to meander, resulting in a wide, navigable waterway that extends all the way to the Atlantic Ocean. These relationships are depicted in Figure 13.9.

In the early days of colonization, most freight was moved by ship. The ships could navigate upriver as far as the Fall Line but no further because they could not navigate through rapids. The rapids and waterfalls also provided

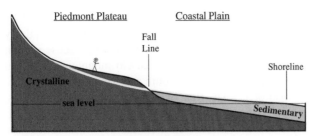

FIGURE 13.9 A cross-section sketch across the Fall Line. A stream seeks a smooth gradient to the sea along the yellow line. Hard crystalline rock of the Piedmont resists erosion and, therefore, sits higher than the weak sedimentary rock of the Coastal Plain. This disrupts the river's ability to maintain a smooth gradient. The river is forced to cut downward (toward the yellow line) to meet the lower topography of the Coastal Plain. In doing so, the river forms gorges, rapids, and waterfalls at the Fall Line.

water-based power for growing industries. The result was the development of a string of major cities along the Fall Line including Philadelphia, Baltimore, Washington, Richmond, Raleigh, and Macon (Figures 13.2 and 13.4). A few of the rivers, including the Roanoke, Neuse, Cape Fear, and Peedee, have cut downward into crystalline rock at the Fall Line producing rapids for up to 20 miles into what would normally be regarded as part of the Coastal Plain. In a sense,

this stripping of sedimentary layers is expanding the Piedmont eastward at the expense of the Coastal Plain. The Fall Line is most conspicuous between Washington, DC and Philadelphia where the Delaware Bay, and an arm of the Chesapeake Bay, extend inland all the way to the Fall Line. A mini version of a Fall Line is present in the St. Francois Mountains of Missouri as was previously discussed in Chapter 11.

THE BLUE RIDGE ESCARPMENT

The Blue Ridge Escarpment is a major east-facing scarp that separates the Blue Ridge Mountains from low-lying hills of the Piedmont Plateau. It extends for more than 300 miles from northern Georgia to southern Maryland and averages between 1,000 and 2,000 feet in relief. It is clearly visible south of Roanoke on landscape maps (Figure 13.4) and in Google Earth images such as Figure 13.10. Along its southern part it separates rivers that drain into the Gulf of Mexico from rivers that drain eastward directly into the Atlantic Ocean. Rivers farther north flow eastward across the escarpment on their way to the Atlantic Ocean forming water gaps as previously mentioned.

FIGURE 13.10 A Google Earth image looking west along the Blue Ridge Escarpment from just south of Roanoke, Virginia. *F* = Ferrum, *W* = Woolwine. The view crosses into North Carolina southwest of Woolwine.

The origin of the Blue Ridge Escarpment is a bit of a mystery. Crystalline rocks are present on both sides, so it is not held up by especially hard rock. In fact, part of the escarpment is held up by easily eroded schist and gneiss (the Alligator Back Formation). The escarpment does not appear to be the result of recent faulting (such as Basin and Range type normal faulting) as there is no evidence for a fault near its base and no evidence that nearby mapped faults have been active during at least the past 50 million years.

Apatite (U-Th)/He and apatite fission track-dating techniques have been applied to rocks on both sides of the escarpment in the North Carolina-Virginia area in order to look for differences or trends in exhumation rates that may shed some light on the origin of the escarpment. Given an estimated geothermal gradient of 25°C per kilometer of depth, the apatite fission track clock starts at about 3.2 km (1.95 miles) depth and the apatite (U-Th)/He clock starts between 1.6 and 2 km (1.0 to 1.2 miles) depth. An apatite fission track and (U-Th)/He date, therefore, indicate how long a rock has been within 1 or 2 miles of the surface.

Calculated apatite fission track ages range from 152 to 111 Ma, and apatite helium ages from 204 to 68 Ma. These rather old ages suggest that the rate of exhumation in this part of the Appalachians has been quite slow for at least the past 150 million years. For example, 1 mile of exhumation in 100 million years in the Blue Ridge is equivalent to about 0.06 inches per 100 years. However, given the fact that some of the apatite helium ages are older than some of the fission track ages from the same area, any conclusion regarding absolute numbers and rates must be viewed with caution.

Data from this study, however, do show some significant trends that can be used to infer a possible origin for the escarpment. The ages of both analyses are older in the Blue Ridge than in the Piedmont. Apatite fission track ages range from about 152 to 129 Ma in the Blue Ridge and from 127 to 111 Ma in the Piedmont. Apatite helium ages show a stronger separation with ages between 204 and 122 Ma in the Blue Ridge and between 106 and 68 Ma in the Piedmont.

On the basis of these trends, it has been suggested that the escarpment formed originally between about 230 and 190 million years ago during normal faulting associated with development of the Triassic Lowlands. The normal faults created an escarpment much like the escarpment that forms along normal faults in the Basin and Range. This escarpment originally was far to the east of where it is today and, over time, has retreated to its present position via erosion. This explains the relatively younger ages in the Piedmont. As the escarpment erodes and retreats, rocks directly below the escarpment are brought closer to the surface faster than rocks that remain below the Blue Ridge (Figure 13.11).

Thus, one explanation for the escarpment is that it formed initially on the up-thrown block of a normal fault that, over time, has retreated westward such that the width of the Piedmont has grown at the expense of the Blue Ridge. What is unusual is that the escarpment has persisted as a topographic feature even though it has undergone erosional retreat for upward of 200 million years. One suggestion for the persistence of the escarpment is that it is due in part to microclimate. The idea is that moisture becomes trapped on the Piedmont below the escarpment. This entrapment results in concentrated precipitation along the escarpment face which, in turn, focuses preferential erosion on the escarpment face. The escarpment, over time, will continue to retreat westward at the expense of the Blue Ridge.

THE NORTHERN ROCKY MOUNTAINS AND NORTH CASCADES

Earlier we noted that the Blue Ridge and Piedmont crystalline provinces form the rear of the Valley and Ridge foreland fold-and-thrust belt where thrust faults were able to cut deep enough into the crust to carry crystalline rock toward the surface. It is obvious from Figure 2.9 that the crystalline deformational belt extends the entire length of the Appalachians along the eastern margin of the foreland fold-and-thrust belt. The situation is

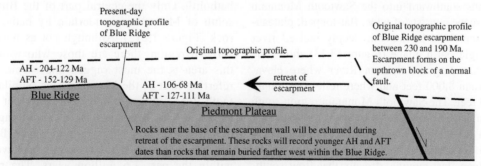

FIGURE 13.11 A topographic profile across the Blue Ridge Escarpment in the North Carolina-Virginia area with apatite fission track (*AFT*) and (U-Th)/He (*AH*) ages. As the escarpment erodes and retreats, rocks directly below the escarpment are brought closer to the surface faster than rocks that remain below the Blue Ridge. This explains the relatively young ages on the Piedmont. Based on Spotila et al. (2004).

different in the Cordillera where exposure of crystalline rock is not as widespread or continuous. One possible reason for this is that Cordilleran mountain-building events are younger than those in the Appalachians and the rocks are not yet eroded to a deep enough level to expose crystalline rock. Given the level of erosion, it is likely that crystalline rock was never exposed as a continuous belt along the length of the Cordillera in the same manner it is today in the Appalachians. However, a far more important factor in Cordilleran landscape has been the nearly complete disruption, burial, and reincarnation of the fold-and-thrust landscape that resulted from massive volcanism, widespread Basin and Range type normal faulting, and subsequent deposition. The part of the US Cordillera least affected by these later tectonic events is the far northern region of Idaho and Washington. Here the transition from a sedimentary fold-and-thrust belt to the crystalline deformation belt, although tectonically altered, is at least well exposed.

Areas of crystalline rock in the Pacific Northwest are shown on a landscape map in Figure 13.12. The figure shows the main areas of the Idaho batholith as well as surrounding areas of crystalline rock. It should be noted that the diagram is generalized and that there are areas shown as crystalline where sedimentary rocks are present or where the intensity of metamorphism is so low that the rocks appear to be sedimentary.

The Idaho batholith is one of the largest areas of massive granitic rock in the US and it forms a large part of the Idaho Northern Rocky Mountains. The rocks vary widely in age from about 180 to 45 million years old with most in the 100- to 65-million-year range. As shown in Figure 13.12, the three largest exposures occur, from south to north, in the Salmon River-Sawtooth Mountains, the Clearwater-Bitterroot Mountains extending into Montana, and the Selkirk Mountains extending into northern Washington. The lack of layering in these rocks produces a somewhat monotonous, random landscape shaped in some areas by younger normal faults.

The largest continuous part of the Idaho batholith is known as the Atlanta lobe, and it covers most of the Salmon River Mountains southward into the Sawtooth Mountain area. The landscape consists of wide, flat-topped, plateau-like surfaces separated by narrow, deeply incised river valleys. A Google Earth image (Figure 13.13) shows the landscape surrounding the Salmon River where elevations are less than 8,000 feet and where relief commonly exceeds 3,000 feet. The Sawtooth Mountain region south of the Salmon River shows a landscape similar to that of the Salmon River area except that it is more strongly dissected into erosional mountains, perhaps due to the presence of steep, fast-moving streams that flow southward into the topographically low Snake River Plain.

The topography is similar to, but less evolved, than the erosional mountains of the Appalachian Plateau. One interpretation is that the area was a low-lying, nearly flat erosion surface that formed as long ago as 50 million years. The present-day landscape developed less than 5 million years ago when the area was broadly uplifted and incised. The landscape appears to be an example of erosional mountains developed in massive crystalline rock.

Part of the Atlanta lobe is shaped by active normal faults associated with the Rocky Mountain Basin and Range. One such area is the Sawtooth Range proper where peaks such as Thompson Peak (10,751 feet), Decker Peak (10,704'), and Mt. Cramer (10,716') rise sharply along the western flank of Stanley Valley (Figure 13.12). These mountains are distinctly different from the Salmon River Mountain region in that they are higher in elevation and glaciated (eroded) into sharp, serrated ridges from which the mountains receive their name. The Sawtooth Range is small (approximately 30 miles long, 15 miles wide), very high (at least 33 peaks above 10,000 feet), and with a steep, precipitous eastern slope that leads down across a normal fault to the Stanley Valley. The fault shows signs of activity within the past 15,000 years. The strong rock and steep cliffs of the Sawtooth Range provide a superb playground for mountaineering and backcountry rock climbing. Figure 13.14 shows the view from the summit of Decker Peak.

A belt of metamorphic rock separates the Atlanta lobe of the Idaho batholith from the Bitterroot section and associated rocks in Montana. These rocks consist primarily of remetamorphosed sections of the North American crystalline shield as well as younger rocks metamorphosed during Cordilleran mountain-building events.

The Bitterroot Range extends for nearly 450 miles to form the sinuous border between Idaho and Montana. The southern part, known as the Beaverhead Mountains, forms the Continental Divide. This area boasts many peaks above 10,000 feet including Scott Peak (11,393'), the highest in the Bitterroot Range. The Beaverhead Mountain area, however, is not underlain by the Idaho batholith. Only the central part of the Bitterroot Range south of Missoula is underlain by batholithic granitic rock (Figure 13.12). Although not as high as the Beaverhead region, there are those who would argue that this area is the most rugged and scenic in the range, referring to it as the Montana Alps. Similar to the Sawtooth Range, this section of the Bitterroot Range is part of the Rocky Mountain Basin and Range. The steep eastern flank drops down to the wide Bitterroot Valley across a normal fault thought to have been active within the past 1.6 million years. The ruggedness of this part of the range is likely a result of the resistant,

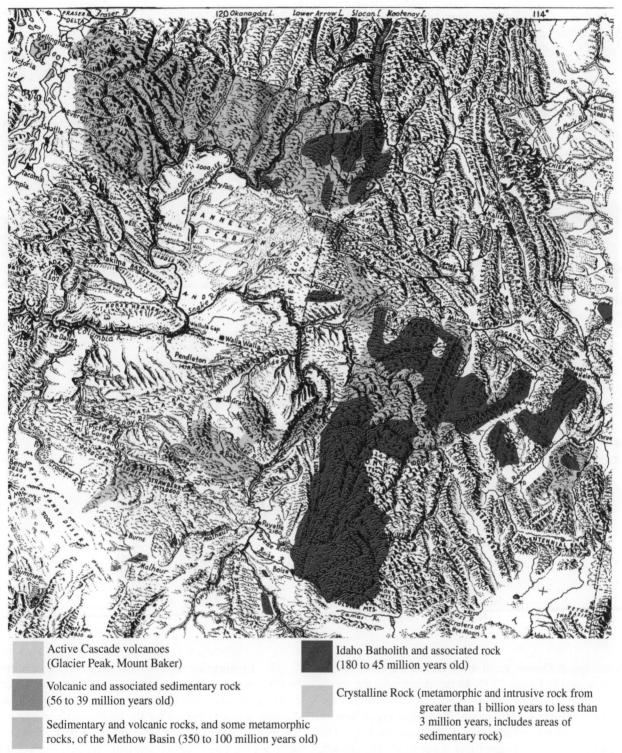

Active Cascade volcanoes (Glacier Peak, Mount Baker)	Idaho Batholith and associated rock (180 to 45 million years old)
Volcanic and associated sedimentary rock (56 to 39 million years old)	Crystalline Rock (metamorphic and intrusive rock from greater than 1 billion years to less than 3 million years, includes areas of sedimentary rock)
Sedimentary and volcanic rocks, and some metamorphic rocks, of the Methow Basin (350 to 100 million years old)	

FIGURE 13.12 A landscape map that shows areas of crystalline rock in the Pacific Northwest.

massive nature of the underlying granitic rock. This area is also unique in that it is part of the Selway-Bitterroot Wilderness, which lies just north of the Frank Church-River of No Return Wilderness. Together these two wilderness areas cover nearly 5,800 square miles, larger than the state of Connecticut.

The belt of crystalline rocks that extends across northern Washington is quite varied, both geologically and in its

FIGURE 13.13 A Google Earth image looking west at a plateau-like surface within the Salmon Mountains that is deeply incised by the Salmon River.

FIGURE 13.14 A photograph that shows the view from the summit of Decker Peak.

landscape. A Google Earth image across the entire region is presented in Figure 13.15. In this figure note the strong topographic difference in landscape east and west of the Okanogan River (*o*). Elevation and overall ruggedness of terrain are clearly greater west of the Okanogan River. The area east of the Okanogan River is one of broad, rounded, relatively low-lying mountains. There are at least 10 peaks above 9,000 feet in the North Cascades, many peaks above 8,000 feet in the Okanogan Range, one peak (Mt. Bonaparte) above 7,000 feet in the South Okanogan Highlands, five peaks above 7,000 feet in the Kettle River Range, and many peaks above 7,000 feet but none above 8,000 feet in the US Selkirk Mountains.

The topographic distribution is evidence that the North Cascades have undergone more recent or more rapid uplift relative to areas east of the Okanogan River. Another possible reason for the elevation distribution is the relative hardness and erodability of rock. However, such a possibility seems unlikely because rock hardness is not significantly different across the region. For example, the fault-bound Methow Basin is occupied primarily by sedimentary and volcanic rocks (rather than crystalline), yet it forms part of the high mountain region west of the Okanogan River (Figure 13.12). Clearly, the Methow Basin is a structural basin and not a topographic basin. Its presence at high elevation seems to favor recent

North Cascade   Okanogan   Okanogan   Kettle River
Mountains   Range   Highlands   Range   Selkirk Mountains

FIGURE 13.15 A Google Earth image looking north across northern Washington and Idaho. Major mountain ranges are named. State boundaries and the international boundary are shown. Abbreviations are *c* = Lake Chelan, *cr* = Columbia River, *E* = Everett, *o* = Okanogan River, *r* = Ross Lake.

uplift as a controlling factor on elevation rather than rock hardness. Climate can influence elevation, and, in this respect, the Cascade Mountains receive far more precipitation than areas farther east, resulting in an abundance of alpine glaciers. Glaciers in the Cascades have excavated valleys and carved around mountain peaks contributing immensely to the scenic beauty of the range. One possibility is that isostatic uplift resulting from the excavation of valleys has lifted uneroded mountain peaks to slightly higher elevations. Another possible factor with bearing on the elevation distribution is the continental glaciation that preceded alpine glaciation. Most of the region was covered by the southern edge of the Cordilleran Ice Sheet (Figure 6.7). Erosion could have flattened high spots, and deposition could have filled in low spots, particularly east of the Okanogan River where subsequent alpine glaciation was not a major landscape contributor.

There is additional evidence for recent uplift of the North Cascades. Circa 16 million year old Columbia River Basalt is absent along the crest of the North Cascades, suggesting that topography was high enough by about 16 million years ago to block or deflect lava as it flowed westward. In addition to an absence of Columbia River basalt, there is also a general absence of Cascade volcanic material in the North Cascades beyond the confines of two active volcanoes, which also implies relatively rapid erosion and, therefore, the existence of both elevation and relief. Thus, it is possible that the North Cascades were already in existence as a highland by 16 million years ago. Additionally, there are river sediments and other lines of evidence to suggest that the North Cascades have been an exposed landmass of uncertain elevation for at least the past 70 million years.

(U-Th)/He and fission-track ages suggest rather rapid exhumation rates in the North Cascades of 1.96 to 3.93 inches per 100 years between about 12 and 4 million years ago. If we correlate rapid exhumation with rapid uplift, this time period may correlate with an important phase of uplift of these mountains. Uplift, although probably still active, has slowed over the past 4 million years. The past 2.4 million years have been a time of glaciation and periodic sea-level changes of 400 feet or more. Even with slow uplift rates, the combination of high elevation, northern latitude, high precipitation, strong rock, and periodic sea-level drop has resulted in intense glacial and river erosion, focused primarily in valleys. Rather than producing typical U-shaped glacial valleys, the combination of uplift and focused erosion has produced deep glacial troughs that accentuate some of the most rugged and vertical landscape in the US. Here we can find incredibly jagged mountain peaks 8,000 to 9,500 feet high separated by deep narrow valleys with 5,000 to 6,000 feet of steep relief. There are currently as many as 700 glaciers in the North Cascades, more than anywhere in the lower 48 states. Few roads traverse this great land which is a Mecca for alpine mountaineers. Figure 13.16 is a detailed landscape map of the area with geology and the only three roads that cross the terrain. Some of the high peaks include Mt. Stuart (9,415 feet), Bonanza Peak (9,511 feet), and Mt. Shuksan (9,131 feet). Figure 13.17 is a photograph from above Washington Pass on the North Cascades Highway (Rt. 20) that captures some of the rugged beauty.

The North Cascades extend from Snoqualmie Pass northward into Canada and include two major volcanoes, Glacier Peak (10,541 feet) and Mt. Baker (10,781 feet), both of which are part of the Cascadia volcanic arc system (Chapter 16). There are three major structural areas, a western domain west of the Straight Creek fault occupied by thrust slices composed of sedimentary rock, moderately metamorphosed rock, and intrusive rock; a central

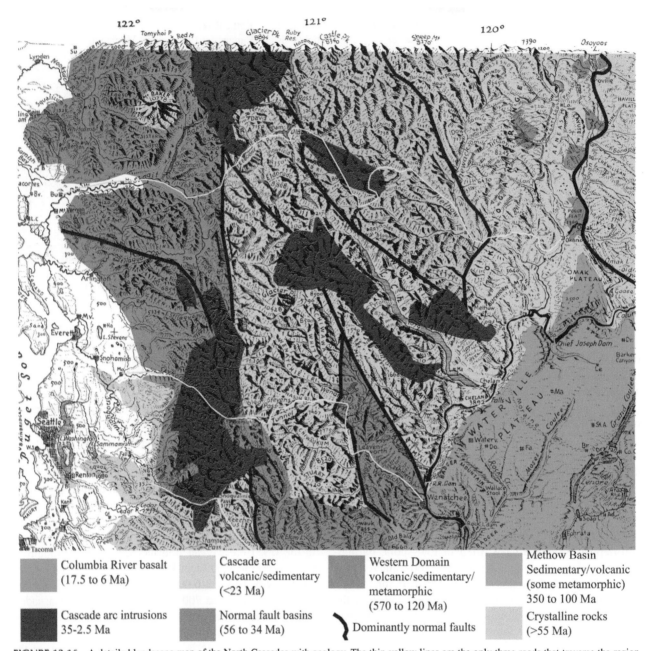

| | Columbia River basalt (17.5 to 6 Ma) | | Cascade arc volcanic/sedimentary (<23 Ma) | | Western Domain volcanic/sedimentary/ metamorphic (570 to 120 Ma) | | Methow Basin Sedimentary/volcanic (some metamorphic) 350 to 100 Ma |
| | Cascade arc intrusions 35-2.5 Ma | | Normal fault basins (56 to 34 Ma) | ⌐ | Dominantly normal faults | | Crystalline rocks (>55 Ma) |

FIGURE 13.16 A detailed landscape map of the North Cascades with geology. The thin yellow lines are the only three roads that traverse the region. From north to south, they are Route 20 (North Cascades Highway), Route 2 over Stevens Pass, and Interstate 90.

crystalline core of hard metamorphic and intrusive rock; and the fault-bound Methow structural basin, occupied by sedimentary, volcanic, intrusive and metamorphic rocks. The Straight Creek fault is a right-lateral strike-slip fault, active primarily prior to about 35 million years ago. The Methow Basin is bound by the Ross Lake and Pasayten faults, both of which are inactive. In spite of the presence of these faults, and in spite of the variation of rock type across each fault, a glance at Figures 13.15 and 13.16 reveals a rather random landscape with no obvious structural control (see also Figure 3.7). A major reason for the rather random landscape

is the ubiquitous presence of intrusive rocks, many of which are associated with the active Cascade volcanic arc. Young Cascade arc intrusions are shown in Figure 13.16 as cutting across all three structural domains, partly obliterating the Straight Creek and Ross Lake faults. These rocks represent the uplifted and eroded batholitic roots of old, now vanished Cascade volcanoes and are as young as 2.5 million years. The presence of these rocks at the surface is another indicator of recent uplift.

A unique feature of the North Cascades is the presence of two long, linear lakes, both of which represent

FIGURE 13.17 A photograph looking west from near Silver Star Mountain. The location is above Washington Pass and the North Cascades Highway at the southern end of the small pluton shown in Figure 13.16 that intrudes rocks in the Methow Basin northwest of Gardner Peak.

old river valleys that were gouged out and straightened during glaciation. Lake Chelan is 55 miles long, about 1 mile wide, and almost 1,500 feet deep. The lake surface is at an elevation of 1,098 feet, which implies that the bottom of the lake is 400 feet below sea level. This was a natural lake originally dammed by glacial moraine deposits. Today, a manmade dam controls lake level. Ross Lake is a dammed reservoir near the border with Canada. It is 23 miles long, 1.5 miles wide, and 540 feet deep. The northern ends of both lakes are remote and best reached by foot, small plane, or boat. There are no roads that connect the northern end of either lake to the North Cascades Highway (Rt. 20). Both lakes are visible in Figures 13.15 and 13.16.

The area east of the Okanogan River shows a general north-south linearity to the landscape as shown in Figure 13.15. Like so much of the Northern Rocky Mountains, the landscape of this area is influenced by normal faults. However, unlike faults that border the Sawtooth and Bitterroot Mountains, these faults are not active. They were active primarily between about 56 and 39 million years ago, and because they represent the final tectonic deformation, they exert landscape control. These faults trend generally north-south and border the South Okanogan Highlands, the Kettle River Range, and the Selkirk Range. A variety of crystalline rock is exposed, including metamorphosed sedimentary rock, young intrusions associated with the normal faults, and older intrusions associated with the Idaho Batholith.

Previously, we noted the rather low-lying mountain topography of northeastern Washington compared to the North Cascades and concluded that the region likely has not undergone the same degree or recentness of uplift as the Cascades. In the Cascades, it was this uplift, and subsequent

erosion, that exposed crystalline rock and created the rather random landscape topography. In other words, the crystalline rocks were exposed via erosional exhumation. Such a simple explanation seems less satisfactory for northeastern Washington. Instead, the evidence suggests that large displacement along normal faults is largely responsible for surface exposure of crystalline rock. The rocks in all three mountain areas (South Okanogan, Kettle River, and Selkirk) are exposed in the lower plate of a series of normal fault structures known as *crystalline* (or *metamorphic*) *core complexes*. There are three core complexes, one in each mountain area, each bordered by normal faults. A discussion of the origin of a core complex is best left for the chapter on normal faults (Chapter 15), but one could obtain a visual understanding of how crystalline rock is exposed by viewing Figure 15.11. On the basis of Figure 15.11 alone, we can infer that a core complex forms as a result of large displacement along nearly horizontal normal faults, and subsequent doming of underlying crystalline rock. The South Okanogan, Kettle River, and Selkirk mountain core complexes are not the only ones present. Crystalline core complexes are present in the Bitterroot Range south of Missoula and in the Raft River-Anaconda Range west of Anaconda, Montana. A core complex is also present in the Pioneer Mountains near Hyndman Peak, just north of Craters of the Moon, Idaho (Figure 13.12). Rather than erosional exhumation, the crystalline rock in these examples is exposed largely as a result of tectonic exhumation.

Normal faults in northern Washington are also responsible for the development of down-dropped basins filled with sedimentary and volcanic rock of approximately the same age as the faults. The largest of these basins follows the Sanpoil River as shown in Figure 13.12. Several smaller basins are shown in Figure 13.16.

Mountains east of the Okanogan River likely formed between 56 and 34 million years ago during development of the crystalline core complexes, which themselves may have reincarnated an earlier sedimentary fold-and-thrust landscape. Since that time, the mountains appear to have grown old via erosion without tectonic modification. Note in Figure 15.11 that crystalline core complexes produce a dome structure similar to mid-continent domes such as the Adirondack Mountains. The major difference is that the mid-continent domes developed via broad, anticlinal warping of ancient, cold, crystalline rock. Development of a crystalline core complex typically involves normal faulting, metamorphism, and intrusion. The presence of gentle, broad dome structures in the mountains east of the Okanogan River has most likely contributed to the overall low-lying mountain topography relative to the North Cascades. Figure 13.18 is a schematic sketch of crustal blocks cut by vertical faults on the left, and an intact dome structure on the right. These two areas may act differently under conditions of broad, isostatic uplift. The presence of vertical faults may allow crustal blocks to move up or down independent of each other, potentially creating isolated areas of high peaks such as what we see in the Cascades. An intact dome structure, on the other hand, would distribute uplift across the entire area, thus creating low-lying, subdued mountains such as what is seen east of the Okanogan River.

Finally, we can look briefly at the largely crystalline mountains of Idaho and Oregon in the area west of the Idaho batholith Atlanta lobe. These mountains exist as islands mostly surrounded by volcanic rock that includes the Columbia River Basalt. They are the Seven Devils Mountains of Idaho; the Wallowa, Elkhorn, Greenhorn, Strawberry, Aldrich, and Ochoco Mountains of Oregon; and the Blue Mountains, which extend north from the Elkhorn Mountains into Washington state east of Walla Walla (Figure 13.12). All except the Ochoco Mountains were glaciated. Peaks in the Strawberry, Elkhorn, Wallowa, and Seven Devils Mountains rise above 9,000 feet. The Wallowa Mountains boast at least 31 peaks above 9,000 feet and are known as the Oregon Alps. The rocks in these mountains are part of various accreted terranes but include intrusive rocks and young volcanic and sedimentary rocks that were deposited following accretion. The older accreted terrane rocks feature sedimentary and volcanic rocks that have undergone various degrees of metamorphism and intrusion. Some of the rock is weakly metamorphosed and still retains part of its sedimentary or volcanic character. Geologically the rocks can be correlated with rocks in the Klamath Mountains, Rocky Mountains, and North Cascades. Major intrusions associated with the Idaho batholith are present in the Wallowa and Elkhorn Mountains. The uplift history of this area is poorly constrained. Faulted, folded, and tilted basalt flows along the margins of some of the mountains suggest that at least some of the uplift

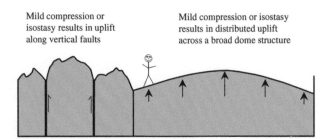

FIGURE 13.18   A cross-section sketch that shows how the style of uplift might vary dependent on the presence or absence of preexisting vertical zones of weakness.

has occurred within the past 12 to 10 million years. Faults active during the past 1.6 million years are especially common surrounding the higher Wallowa and Elkhorn Ranges as well as the Blue Mountains. These faults imply active uplift.

THE SUPERIOR UPLAND CRYSTALLINE PROVINCE

The Superior Upland crystalline province extends across northern Wisconsin, eastern Minnesota, and Upper Michigan surrounding Lake Superior. The region is low-lying, mostly less than 2,000 feet in elevation, with less than 500 feet of relief. The province is defined primarily by the presence of underlying ancient crystalline rock of the North American shield. These ancient rocks extend southward from Canada into the US in the form of two gentle upwarps. One arm extends southward to form the Adirondack Mountains. A second arm forms the Superior Upland. The surrounding sedimentary layers in both areas dip gently away from the upwarp, creating a series of small cuestas that wholly or partly surround the crystalline rock.

As shown in Figure 13.19, the Superior Upland province is divided into three areas with a few small outliers to the south. The western area, which includes most of northern Minnesota, consists largely of rocks that are 2.5 billion years old and older. One of the oldest rock formations in the world, the 3.6-billion-year-old Morton Gneiss, crops out in the Minnesota River Valley west of Minneapolis. Younger, weakly metamorphosed sedimentary rocks are also present, including the 1.8- to 1.9-billion-year-old banded iron formations of the Mesabi and Vermilion Ranges, the largest iron deposits in the United States.

Of special interest is the central area surrounding the southwestern part of Lake Superior near Duluth where relatively young rocks, the 1.1-billion-year-old Keweenawan basaltic volcanics and the intrusive Duluth gabbro, crop out in association with sedimentary rocks and normal faults. This is one of the few areas in the central US where ancient, relatively unaltered volcanic and sedimentary rocks are found at the surface. The volcanism produced some of the purest copper deposits in the world. The rocks represent a failed

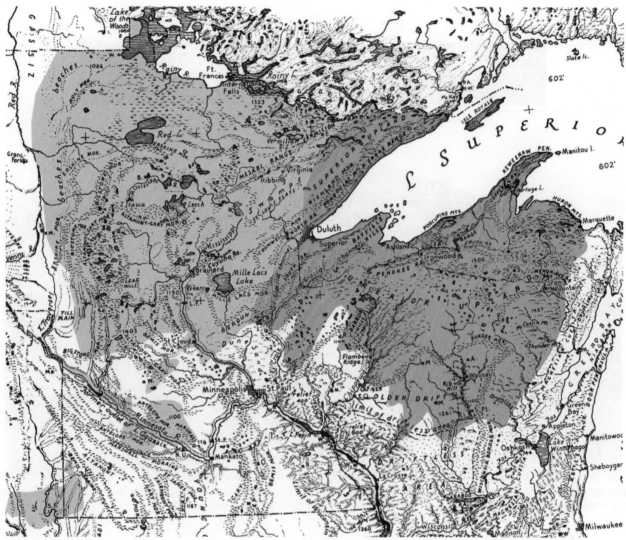

FIGURE 13.19 A landscape map that shows the location and distribution of ancient crystalline and volcanic rock in the Superior Upland. The area to the west of Lake Superior and in the Minnesota River Valley consists mostly of 2.5-billion-year-old (and older) crystalline rock. The central area surrounding the southern part of Lake Superior consists of the 1.1-billion-year-old Keweenawan volcanic rocks and Duluth gabbro. These rocks form the trough of a broad syncline as depicted in Figure 13.20. The eastern area consists mostly of 1.8- to 1.9-billion-year-old crystalline rock.

attempt by the North American continent to separate (rift) into two pieces. The attempted separation produced a narrow version of Basin and Range normal fault topography, with sedimentary, volcanic, and intrusive rock filling the down-faulted troughs. Geophysical data and drilling indicate that the Keweenawan Rift (also known as the Mid-Continent Rift) extends, at depth, south of Lake Superior along the Minnesota-Wisconsin border and then southwestward across central Iowa to eastern Nebraska and northeast Kansas. It is not exposed at the surface beyond the Lake Superior region.

The eastern area, in northern Wisconsin and Michigan south of Lake Superior, consists of crystalline rocks that were strongly deformed between about 1.88 and 1.83 billion years ago during a major mountain-building event referred to as the *Penokean event*, after the Penokee Range near Ironwood,

Michigan. The rocks include metamorphosed basaltic and silicic rocks, metamorphosed sedimentary rocks, abundant granitic intrusions including anorthosite (as in the Adirondack Mountains), and a few areas of ancient (>2.5 billion year) granitic and metamorphosed volcanic rock.

The ancient deformation found in the crystalline rock of the Superior province is not the primary structure that produces the present-day landscape. Rather, it is a much younger and much weaker structure, a broad syncline, that controls today's landscape. Following the failed Keweenawan Rift attempt, all of the rocks were folded into a broad syncline oriented northeast-southwest through the center of Lake Superior. The syncline is asymmetrical, with dips greater than 35 degrees on the southeast side and about 15 degrees on the northwest side. As shown in Figure 13.20,

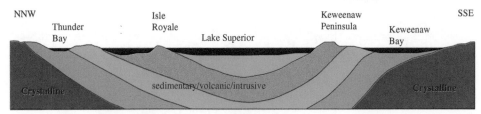

FIGURE 13.20   A cross-section sketch of the Keweenawan syncline.

FIGURE 13.21   A Google Earth image looking eastward along the Keweenaw Peninsula. Southeast-facing cuestas are developed on the southeastern side of the Keweenawan syncline. The large cuesta near the center is underlain by the Duluth gabbro.

erosion of the synclinal structure has produced a series of outward-facing cuestas (cuestas with steep slopes that face away from Lake Superior) within particularly resistant rock of the Keweenawan volcanics, some of which includes the Duluth gabbro. It is these tilted resistant rocks that form small mountains on either side of Lake Superior (Figure 13.19). Northwest-facing cuestas along the northwest shore of Lake Superior form the Fond du Lac and North Shore Mountains (also known as the Sawtooth Mountains), which includes the highest point in Minnesota (Eagle Mountain, 2,301 feet). Northwest-facing cuestas are also present on Isle Royale. Southeast-facing cuestas are present in the Porcupine Mountains and along the southeastern shore of Michigan's Keweenaw Peninsula as shown in Figure 13.21.

Perhaps the most notable aspect of this region is its recent glacial history which was discussed in Chapter 6. The northern part near Canada was scraped, scoured, and flattened during the glaciation. Today it is characterized by numerous lakes, marshes, and slow-moving streams. The Great Lakes themselves are the result of deep glacial erosion. Each lake

was once a river valley that was gouged, widened, and deepened during the glaciation. Lake Superior lies at an elevation of 602 feet above sea level but reaches depths greater than 700 feet below sea level. North of Lake Superior, the Mesabi and Vermilion Ranges are erosional mountains underlain by resistant rock that rise a few hundred feet in relief to form part of a triple divide between rivers that flow northward to the Hudson Bay, southward to the Mississippi River, and eastward to Lake Superior and the St. Lawrence River. Most of the province is characterized by a glacial depositional landscape with an abundance of glacial moraines.

QUESTIONS

1. Name and briefly describe the different types of crystalline rocks and their location in the United States.
2. Name the state that belongs to each of the following cities located in Figures 13.1 and 13.2: Gatlinburg,

Pigeon Forge, Roanoke, Augusta, Columbia, Raleigh, Richmond, Philadelphia.

3. In Google Earth, fly to Woolwine, Virginia. What is the relief across the Blue Ridge Escarpment at this location? How far south can you follow the escarpment? What is the relief across the escarpment at Lowgap, North Carolina? How does the escarpment change between Lowgap and Millers Creek, North Carolina? How has the escarpment changed between Millers Creek, North Carolina, and Dahlonega, Georgia? What happens to the Blue Ridge Mountains and the escarpment between Woolwine and Syria, Virginia?

4. What is the relief between the Blue Ridge and the Piedmont at the northern end of the Blue Ridge near Dillsburg, Pennsylvania?

5. What is the relief between Mt. Katahdin and Upper Togue Pond at the foot of the mountain?

6. What are some of the differences, in terms of rock type and relief, between the Fall Line and the Blue Ridge Escarpment?

7. Explain differences in the rocks and topography between the Green Mountains of Vermont and the White Mountains of New Hampshire.

8. In Google Earth, fly to Doubtful Lake, Washington (in North Cascades National Park). Name the various glacial landforms in the area.

9. What is the origin of the many islands along the coast of Maine?

10. Referring to Figure 13.12, what rock type or feature forms the southern limit of the Northern Rocky Mountain-Cascade crystalline belt?

11. What is the origin of Lake Chelan and Ross Lake?

12. Note the orientation of the Mesabi and Vermilion ranges in Figure 13.19. Speculate as to why these ranges survived flattening during glaciation.

13. What is the Fall Line? Where is it located? What is its significance with respect to American history?

14. What is a monadnock? What famous monadnock exists in Maine? In Georgia?

15. Speculate on the origin of the line of lakes in Figure 13.6 that stretches south and east of the White Mountains across the New Hampshire-Maine border.

16. What is the evidence for recent uplift in the North Cascades?

17. What is the Methow Basin? Describe its topography.

18. What type of structure forms the landscape of the Kettle River Range?

19. Trace the cuestas in Figure 13.21. How many are present? What is their origin?

20. Research the Mid-Continent Rift. What is its extent? How was it traced beneath the surface? How did it form and how long ago?

21. Explain why there is seemingly no landscape difference or landscape change between areas of the US underlain by crystalline rock in Figure 13.19 and areas (not highlighted) underlain by sedimentary rock.

22. It is typical for a region to undergo thermal cooling and slow subsidence following a rifting event. Why?

Young Volcanic Rocks of the Cordillera

Volcanic rocks have unique attributes as well as attributes of both sedimentary and crystalline rock. Their unique attribute is that they form a distinctive set of landforms that includes major volcanoes such as those in the Cascade Mountain range and Hawaii, lava plateaus and plains such as on the Columbia River Plateau and Snake River Plain, and other smaller volcanic features such as cones, lava fields, domes, and calderas. Volcanic rocks, particularly basaltic lava flows, are similar to sedimentary rocks in the sense that they are commonly layered and originally nearly flat-lying. However, they do not produce well-developed bench-and-slope landscape primarily because there are no strong contrasts in erodability between the layers and because the layers are less continuous and less planar in comparison to sedimentary rock. Volcanic rocks typically are more resistant than sedimentary rocks. Where volcanic rocks are interlayered with sedimentary rocks, it is the volcanic layers that form resistant caprock, although there are exceptions. Volcanic rocks are also unique in the sense that they can reincarnate a landscape in a matter of weeks or even days in some instances.

The distribution of major landscape-forming volcanic rocks is shown in Figure 14.1. To a certain extent, recent or active volcanic rocks influence landscape in all provinces of the Cordillera. One reason for this is that the volcanic rocks are so young. Nearly all are less than 60 million years old, and many are less than 20 million years old. With the exception of those in the Basin and Range, most are flat-lying or only weakly deformed. Nonexplosive basalt is the most common volcanic type, but explosive silicic rocks are widespread and significant. The Columbia River Plateau in particular forms a distinctive volcanic landscape that consists dominantly of nearly flat-lying basaltic lava flows. This vast area of basalt is continuous with basaltic and silicic volcanic rock of the Snake River Plain whose volcanism is still active within Yellowstone

National Park. Both provinces bury and reincarnate what was likely a mountainous terrane similar to the Rocky Mountain region to the north. Much of this chapter is devoted to discussion of these two provinces. The Southern-Central Cascade Mountains form a third area of dominantly volcanic rock, but this province is more properly considered with respect to the larger Cascadia volcanic arc system. It will be discussed in Chapter 16. A fourth province where volcanic rocks are widespread is the Basin and Range. In this region volcanic rocks are cut by normal faults that exert a greater control on the landscape. We will discuss the significance of some of the volcanic rocks in the Basin and Range, but the structure and evolution of the area is reserved for Chapter 15. Selected volcanic areas in other provinces of the Cordillera will also be discussed.

Landscape-forming volcanic rocks in the Appalachian Mountain system are largely restricted to the Triassic Lowlands of the Piedmont Plateau and New England Highlands where circa 210-million-year-old basaltic and gabbroic rock is interlayered with sedimentary rock. These areas developed via normal faulting and are discussed in Chapter 15.

Older volcanic rocks in both the Cordilleran and Appalachian mountain belts are commonly intermingled with sedimentary or plutonic rocks and are strongly deformed or metamorphosed. They form many of the accreted terranes in both mountain belts, but by themselves they are not major landscape formers. They are discussed in Part III of this book in the context of mountain building.

The largest area of volcanic rocks outside the Cordilleran and Appalachian Mountains are the 1.1-billion-year-old Keweenawan volcanics in the Superior Upland. These rocks, along with a smaller domal area of circa 1.4-billion-year-old volcanic rocks in the St. Francois Mountains of Missouri, are discussed in the context of their surrounding crystalline rocks in Chapters 13 and 11, respectively. Outside the contiguous

United States, the Hawaiian Islands and the Aleutian Islands (along the southern coastline of Alaska) are composed primarily of young landscape-forming volcanic rock.

MAGMA TYPES AND LAVA DOMES

Previously we distinguished between volcanic rocks of basaltic and silicic (andesite-rhyolite) composition. In this section we distinguish between andesite and rhyolite and introduce another volcanic rock known as *ignimbrite*. Basaltic magma is fluid and can flow quietly for miles across a surface without too much fanfare. This type of magma is derived from partial melting in the mantle below the crust. *Rhyolite*, on the other hand, is derived primarily from partial melting of continental crust. It is the volcanic equivalent of granite and tends to erupt explosively. *Andesite* is less fluid and more explosive than basalt but generally more fluid and less explosive than rhyolite. It is intermediate between basalt and rhyolite. Andesite is also significant because it is found primarily in volcanoes above subduction zones. Keep in mind that for more than 300 million years, oceanic lithosphere off the US West Coast has subducted beneath the Cordilleran continental margin. Thus, the age and distribution of andesite across the Cordillera will, to a first approximation, tell us when and where subduction was occurring.

Rhyolite does not flow very easily, so when it reaches the surface it can pile up at the vent and form a mound which we refer to as a *lava dome*. Once formed, a lava dome can freeze in place or grow higher, at which point it will either begin to flow as lava or simply tip over and blow up. If it blows up, it could produce a hot plume of suspended rock particles and gas known as a *pyroclastic flow*. A pyroclastic flow is not lava. It is a high-velocity volcanic cloud that hugs the ground following a violent explosion. You do not want to be in the way of one of these because it will kill you in short order. The resulting pyroclastic rock is composed of pumice, ash, and many rock fragments. Large deposits are known as *ignimbrites*. A large pyroclastic flow can enter a river channel where the volcanic material mixes with sediment and flows for miles down-river. The resulting deposit is referred to as a *volcanic breccia*. The location where the lava dome blows up may sink as a result of loss of magma. The largest eruptions produce a circular hole in the ground known as a *caldera* that is usually surrounded by curved normal faults along which the center has collapsed.

FIGURE 14.1 A landscape map showing areas of landscape-forming volcanic rocks. Many of these areas include intermingled sedimentary, intrusive, and metamorphic rocks.

THE COLUMBIA RIVER PLATEAU

The Columbia River Plateau is covered almost entirely by the Columbia River flood basalts. The term *flood basalt* refers to an outpouring of lava that is fluid enough to flow several tens or possibly several hundreds of miles, and large enough to essentially bury (flood) the preexisting landscape. These rocks, together with volcanic rocks of the Snake River Plain and Cascade volcanic arc system cover more than half of Washington and Oregon and about one third of Idaho. A Google Earth image of this vast area is shown in Figure 14.2.

The extent of Columbia River flood basalts and their continuation into the Snake River Plain is shown in a landscape map in Figure 14.3. The lava flows have been extensively studied and radiometrically dated. Extrusion of lava began about 17.5 million years ago and ended 6 million years ago, initially covering more than 200,000 square miles to an average thickness of almost 2 miles. About 85% of the flows were extruded between 16.8 and 15.6 million years ago, averaging one flow per 10,000 years. Nearly all of the lava was extruded from a series of long fissures located southeast of Walla Walla, Washington, near the Oregon-Idaho border. The flows were nonexplosive and very fluid. They filled topographic low spots, buried pre-existing Northern Rocky Mountain and Cascade landscape, and redirected major rivers during this 1.2-million-year interval that reincarnated a once mountainous terrain. If we assume that an average thickness of 9,000 feet of basalt was achieved across the plateau in 1.2 million years, the average rate of accumulation was about 9.0 inches of lava per 100 years. The end result was development of a relatively flat plateau surface mostly 1,000 to 3,000 feet above sea level.

Primarily between 16.8 and 14.5 million years ago, basaltic lava flowed westward through the Columbia River channel all the way to the Pacific Ocean. In doing so, the lava repeatedly choked, dammed, and displaced the river northward. Today the remains of these flows have created more than 70 waterfalls, including Multnomah Falls, which, at 620 feet, is one of the tallest in the US.

Although no longer a location of volcanic activity, the relatively young age of the basalt flows, their resistance to erosion, and the arid (rain shadow) climate have allowed the plateau surface to remain relatively undissected except in the vicinity of major rivers such as the Columbia and Snake rivers, which have cut canyons 1,000 to 2,000 feet deep into the volcanic landscape. One exceptionally deep canyon is Hell's Canyon along the Snake River at

FIGURE 14.2 A Google Earth image looking vertically down across the Columbia Plateau and Snake River Plain. State boundaries and the international boundary with Canada are shown. North is to the top of the image. The dark area on the Snake River Plain to the right of the *SR* symbol are the 2,000-plus-year-old lava flows of Craters of the Moon National Monument. Snow-capped volcanic peaks are visible in the Cascade Mountains. *BM* = Blue Mountain anticlinal area; *c* = Columbia River; *CM* = Cascade Mountains; *h* = Hell's Canyon; *NB* = northern Columbia basin; *SB* = southern Columbia basin; *sn* = Snake River; *SR* = Snake River Plain; *Y* = Yellowstone.

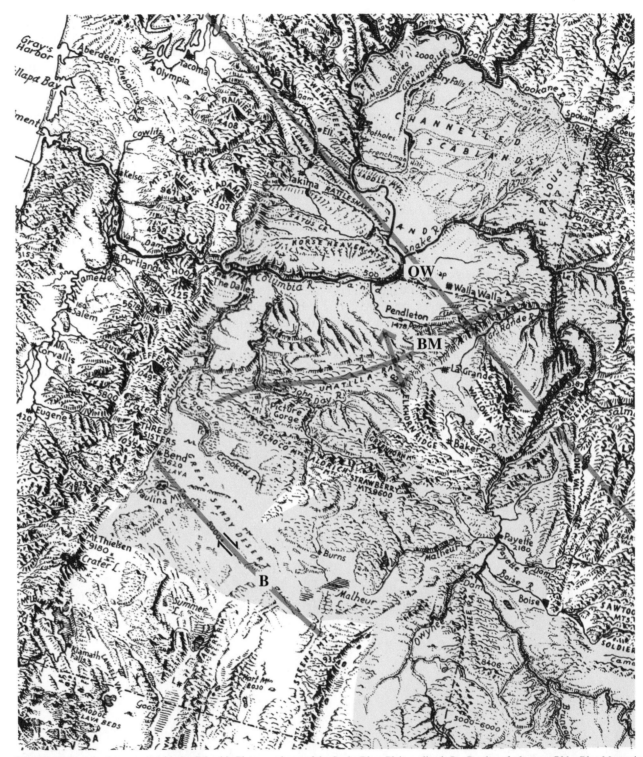

FIGURE 14.3   A landscape map with the Columbia Plateau and part of the Snake River Plain outlined. *B* = Brothers fault zone; *BM* = Blue Mountain anticline; *OW* = Olympia-Willowa lineament.

the edge of the plateau between Oregon and Idaho east-southeast of Walla Walla in the area west of Seven Devils Mountain (Figures 14.2 and 14.3). This canyon is deeper than the Grand Canyon. A Google Earth image of Hells Canyon in Figure 14.4 shows significant landscape

differences between the nearly flat-lying Columbia River basalt on the west (right) and the steep, rugged, and glaciated Seven Devils Mountain peaks, which consist of circa 250-million-year-old deformed volcanic rocks (accreted terranes) and minor younger intrusions. The highest peaks

FIGURE 14.4 A Google Earth image looking southward along the Snake River at Hell's Canyon on the Idaho-Oregon border. The plateau area at right (west) consists of layered Columbia River basalt. The dissected high peaks along the upper-left skyline are part of the Seven Devils Mountains volcanic accreted terrane.

are He Devil and his bride, She Devil, both of which are approximately 9,400 feet in elevation and can be seen at the upper left skyline of the image.

Figures 14.2 and 14.3 show a series of mountain ranges that cross the south-central part of the Columbia Plateau including the Wallowa, Elkhorn, Greenhorn, and Blue Mountains in eastern Oregon, and the Strawberry, Aldrich, and Ochoco Mountains in central Oregon. Many of the peaks exist as islands of sedimentary and crystalline rock surrounded by a sea of Columbia River basalt. Just north of these mountains, in the Umatilla Range, the dominantly basaltic rocks form a broad, anticlinal structure known as the Blue Mountain anticline, which must be younger than the circa 15-million-year-old rocks that it deforms. The Blue Mountain anticline and the nonvolcanic mountains to the south separate the Columbia River Plateau into two subplateaus, one to the north and the other to the south (Figures 14.2 and 14.3). Both are down-warped into basins that result, at least in part, from isostatic adjustment under the weight of the volcanic rocks.

The northern basin includes a unique feature known as the *channeled scablands*. This is an area of bare basalt eroded by gigantic flood waters that poured through the region between about 15,500 and 12,500 years ago when ice dams holding back large glacial lakes suddenly failed. The scablands are clearly visible in the Google Earth image of the northern basin shown in Figure 14.5. One of

the largest of the glacial lakes was Lake Missoula, which formed partly in the Rocky Mountain Trench at the present-day town of Missoula, Montana. The evidence for this lake is seen in the surrounding mountains where wave-cut terraces abound. The lake formed when the Cordilleran Ice Sheet dammed the Clark Fork River. The lake grew to the size of Lake Michigan, 250 miles long and 1,000 to 2,000 feet deep. Apparently there were many renditions of the lake. As the lake grew, the ice dam would periodically rupture, releasing a wall of water estimated to be several hundred feet high that poured through the area in a sudden torrent at 30 to 50 miles per hour removing soil and creating the scabland landscape. There may have been as many as 30 Lake Missoula floods over the 3,000-year period. This amounts to one major flood every 100 years.

As the mass of floodwater moved across the area, it would coalesce into preexisting or previously abandoned river channels and erode them into much larger channels which, today, are known as *coulees*. Two of the largest are Grand Coulee—50 miles long, 1,000 feet deep, and between 0.5 and 4 miles wide—and the smaller Moses Coulee. Both are located in the northwest part of the plateau (Figure 14.5). A Google Earth image of Moses Coulee is shown in Figure 14.6 where you can see the deep, wide canyon floor with tilted layers of basalt in the canyon walls on both sides. In addition to coulees, the scablands

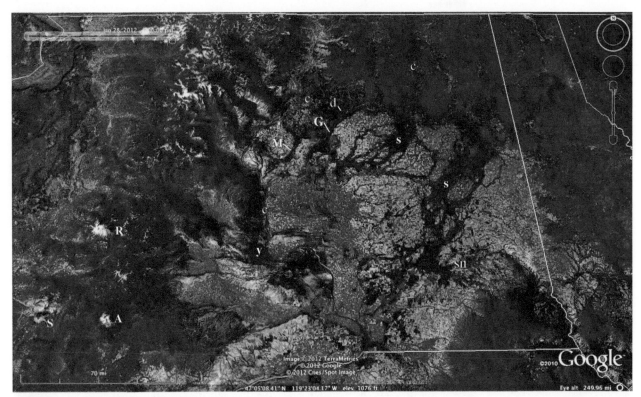

FIGURE 14.5 A Google Earth image looking north across eastern Washington. State boundaries and the international border are shown. *A* = Mount Adams; *c* = Columbia River; *d* = Dry Falls, *y* = Yakima fold belt; *G* = Grand Coulee; *M* = Moses Coulee; *R* = Mount Rainier; *s* = scablands; *S* = Mount St. Helens; *sn* = Snake River.

FIGURE 14.6 A Google Earth image looking south-southwestward at Moses Coulee and surrounding Columbia River basalt near Palisades, Washington. From right to left, the three volcanoes in the distance are Mount Rainier, Mount Adams, and Mount Hood.

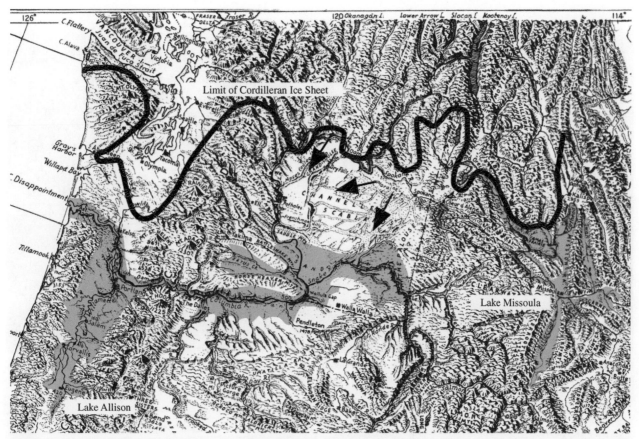

FIGURE 14.7 A landscape map that shows glacial Lake Missoula and associated floodwater lakes. Arrows point in the direction of floodwater flow across the scablands. The thick line is the southern limit of the Cordilleran Ice Sheet.

contain huge dry waterfalls with plunge pools, gravel bars, and giant streambed ripples up to 50 feet high. The raging water plucked large blocks of basalt, eventually creating Dry Falls on Grand Coulee, which is reported to be one of the largest waterfalls known to have existed. During a flood, water would have dropped 400 feet across a sheer cliff nearly 3.5 miles across. Floodwaters would drain into and overflow the Columbia River channel forming a series of downstream lakes that include Lake Allison, which flooded the northern Willamette Valley to a depth of more than 400 feet. Figure 14.7 is a landscape map that shows glacial Lake Missoula and some of the floodwater lakes.

We now turn our attention to what appear to be actively developing landforms near Yakima just north and west of the Columbia River. The area is located with a *y* in Figure 14.5 and is known as the *Yakima fold belt*. These are anticlinal ridges that have undergone very little erosion. They rise to as much as 2,000 feet above the surrounding lowland and are separated by synclinal valleys. A close-up view of the area is presented in Figure 14.8 where several of the ridges are labeled. The smallest ridge, Frenchman Hills (barely visible in the figure), rises just 500 feet above its surroundings and has a fault along its northern side that has been active in the past 1.6 million years. The north side of Saddle Mountain

is also faulted with activity within the past 130,000 years. The absence of erosion and the presence of recently active faults suggest that the folds are active and that these ridges are young and in the process of forming.

In addition to folds, there are also lineaments that cross the Columbia Plateau. A *lineament* is any visible scar or alignment of features that can be followed across the Earth's surface. A lineament could represent a single fault or fracture, or it could represent an alignment of folds, faults, fractures, or volcanic features. A lineament could have no known explanation or could conceivably be a figment (an optical illusion) of the camera angle. There are two major lineaments on the Columbia Plateau, both of which are labeled in Figure 14.3. Although neither is especially obvious on the landscape map, both are associated with zones of active normal and right-lateral strike-slip faults. Erwin Raisz, the person who drew the landscape maps used in this book, first identified the Olympic-Wallowa lineament. Note in Figure 14.3 that this lineament cuts across the Yakima fold belt and the Blue Mountain anticline.

The Brothers fault zone extends across the Great Sandy Desert in the southern Columbia basin. It is seismically active and perhaps a bit unusual because, in addition to active faults, it has more than 100 silicic, explosive, volcanic

FIGURE 14.8   A Google Earth image looking northwest at a series of recent and active anticlinal ridges in the Yakima fold belt near Yakima, Washington. Mt. Rainier is at upper left. Note that the Columbia River cuts directly through Saddle Mountain.

FIGURE 14.9   A Google Earth image looking eastward across the Newberry Caldera. East Lake is at left, Paulina Lake at right. Note the large volcanic cone and rhyolitic flow between the lakes. Note also the additional flows behind East Lake. The Big Obsidian flow is at right behind Paulina Lake. Several volcanic cones can be seen in the distance.

centers associated with the fault zone. The volcanic rocks become progressively younger from the southeast, where they are more than 10 million years old, to the northwest, where they are less than 2,000 years old. The youngest volcanic features include the Newberry Caldera located in the Paulina Mountains (Figure 14.3). The Newberry Caldera is a low-lying volcanic pile of basalt (a shield volcano) with a central depression (a caldera) five miles across that contains very young, explosive, volcanic cones and lava flows composed of pumice and obsidian. A particularly impressive feature is the Big Obsidian flow which erupted only about 1,300 years ago. The flow covers 1.7 square miles, consists of 90 percent pumice and 10 percent obsidian, and shows beautiful flow ripples which are visible in the Google Earth image shown as Figure 14.9. Many additional young and very well-preserved volcanic landforms surround the Newberry Caldera including buttes, small explosion calderas lava domes, and lava caves. This is an excellent location to view and study a variety of volcanic features within a small area. Additionally, there are spectacular views of Columbia River basalt at nearby Cove Palisades, where the Crooked, Metolius, and Deschutes rivers meet, and there are views of

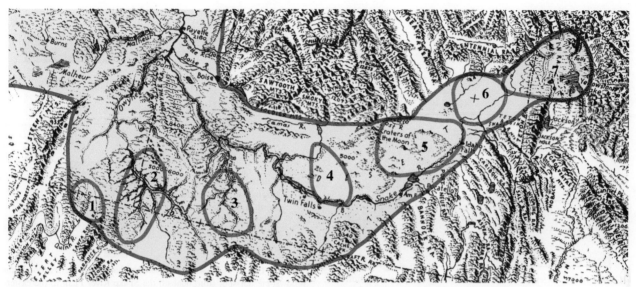

FIGURE 14.10 A landscape map with the Snake River Plain outlined. The ages of major caldera volcanic eruptions are as follows: (1) 16.5 to 15 Ma; (2) 15 to 13 Ma; (3) 12 to 10.5 Ma; (4) 10.5 to 8.6 Ma; (5) 10 to 7 Ma; (6) 6.5 to 4.3 Ma; (7) 2.2 to 0.6 Ma.

Cascade volcanic tuff at Smith Rock near Bend. Although currently dormant, the Newberry volcanic area is an active volcanic landscape that could awaken at any time.

The origin of the two lineament features is not well understood. It is possible that they are associated with old, buried structures that are reactivating. Alternatively, some of the faults may be associated with the northward migration of Basin and Range-type extension into the region as discussed in Chapter 16, or with deformation associated with rotation of the Sierran plate as discussed in Chapter 17.

In addition to interesting volcanic landforms, a particularly interesting sequence of sedimentary rocks known as the John Day fossil beds is located in central Oregon in the John Day River valley, just north of the Ochoco and Aldrich Mountains (Figure 14.3). Actually, there are three locations of fossil beds all located in the John Day River valley; the main location is near the town of Dayville, and the other two are near Mitchell and Fossil. The rocks consist primarily of a succession of volcanic ash beds, thick soil horizons, and lake and stream sediment that was deposited mostly between 36 and 18 million years ago, with some beds as old as 54 million years and others as young as 6 million years. The beds are unique in that they contain a great variety of plant and mammal fossils including dogs, rodents, and horses.

THE SNAKE RIVER PLAIN

The Snake River Plain forms the southeastern extension of the Columbia Plateau without a distinct boundary. It is a flat landscape consisting almost entirely of undeformed basaltic lava flows dotted with several small volcanic cones and a series of explosive rhyolite calderas. Basaltic and silicic volcanism on the Snake River Plain began about 16.5 million years ago during initial outpouring of the Columbia River

flood basalts. It remains active near its eastern terminus at Yellowstone National Park. Basin and Range normal faulting extends into the area, especially in the west, but most of the faults are covered by younger lava flows. The area is outlined on a landscape map in Figure 14.10, which also shows the succession of rhyolitic calderas.

Basaltic lava flows, in general, are unique in that they flow into lowland areas normally occupied by rivers. Because the lava crystallizes in a valley, it will invert topography and displace a preexisting river valley to the margin of the flow as shown in Figure 14.11. An excellent example is the displacement of the Snake River toward the southern margin of the Snake River Plain in the area east of Twin Falls due to a series of lava flows that include the 15,000- to 2,000-year-old basalt flows at Craters of the Moon (Figure 14.10). Here, the Snake River cascades through a series of waterfalls that formed where the river flows over the edges of lava flows.

An interesting aspect to the Snake River rhyolite calderas is that they become progressively younger toward the east. This same feature is seen in the Hawaiian Island chain and in the New England Seamount chain. In all three cases, the position of active volcanics relates to movement of a tectonic plate over a stationary hot spot (Figure 5.9). In the case of Hawaii, the moving tectonic plate is the Pacific (oceanic) plate. In the case of the Snake River Plain, the moving tectonic plate is the North American (continental) plate. Those who live in Hawaii know that volcanic eruptions associated with the Hawaiian hot spot are not especially explosive. Basaltic magma generated in the mantle can easily penetrate the thin, dense, oceanic lithosphere and reach the surface virtually unchanged. Thus, the Hawaiian Islands are composed dominantly of quiet, fluid basalt lava flows. Basaltic magma generated below continental lithosphere has a much harder time reaching the surface due in part to

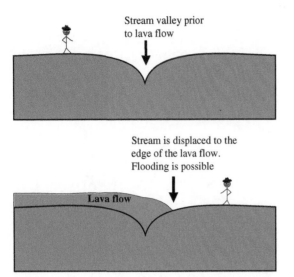

FIGURE 14.11 A cross-section sketch of topographic inversion and displacement of a preexisting river valley to the margin of a volcanic lava flow.

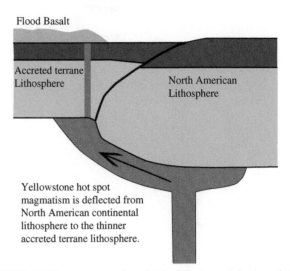

FIGURE 14.12 A cross-section sketch that shows deflection of the Yellowstone hot spot from beneath the Snake River Plain to the eastern edge of the Columbia River Plateau 16.5 million years ago to produce the Columbia River flood basalts. Based on Winter (2010).

the thickness of the lithosphere and to buoyancy contrasts between the magma and the low-density continental crust. Under these conditions it is possible for some of the basaltic magma to pond and cool at depth. Cooling releases heat that can melt continental crust to create silicic (rhyolitic) magma and explosive volcanism at the surface. Figure 14.10 shows the locus of major explosive volcanism over a period of 16.5 million years across the Snake River Plain. Since its initial outpouring, active volcanism has migrated eastward at a rate of 8.5 feet per 100 years to its present position below Yellowstone National Park. Each caldera in the chain essentially represents an ancient Yellowstone National Park. Given enough time, active volcanism associated with the Yellowstone hot spot will migrate across southeastern Montana, into North Dakota, and eventually into Canada.

As noted earlier, the 16.5-million-year-age of Snake River volcanism corresponds with the time of initial massive outpouring of lava on the Columbia Plateau. This suggests that the Yellowstone hot spot is not only responsible for Snake River Plain volcanism, it may also be responsible for at least the initial outpouring of Columbia River volcanism, even though the outpouring was not in line with the track of the hot spot. In this interpretation the magma was deflected northward from the Snake River tract when it first neared the surface due to differences in the structure and composition of the deep crust between the two areas. Deep crust below the Snake River Plain forms part of the ancient crystalline shield of North America, which is very strong, rigid, and thick. The deep crust below the Columbia Plateau, on the other hand, consists of accreted terranes that were added to the edge of the North American crystalline shield. These rocks form crust that is thinner, weaker, younger, and probably with through-going fractures and faults that allow easy access for magma to reach the surface. The situation is depicted in Figure 14.12.

The active Yellowstone caldera forms a nearly circular welt that rises 3,000 feet above the Snake River Plain. Cooling magma less than 7 miles below the surface provides thermal isostatic buoyancy that elevates the region to between 7,000 and 8,000 feet above sea level and also provides heat and energy for the famous geysers and hot springs. As shown in Figure 14.13, the center of the caldera is a depression occupied by Yellowstone Lake, one of the highest lakes in the country. Magma movement at depth causes the caldera to act as if it is alive. The ground bulges as magma moves underneath, and subsides when magma moves away. It is almost as though the Earth is breathing.

The present-day Yellowstone hot spot is located in a somewhat unusual area flanked by several different mountain types that include normal fault mountains associated with the Rocky Mountain Basin and Range, anticlinal mountains of the Middle Rockies, and foreland fold-and-thrust belt landscape of the Northern and Middle Rocky Mountains (Figure 14.13). Individual mountains at the southern flank of Yellowstone include the Teton Range, which is a crystalline normal fault block mountain, the Idaho-Wyoming foreland fold and thrust belt, and the anticlinal Wind River Range. The north flank boasts the crystalline, sedimentary, and volcanic Gallatin and Madison Ranges, both of which are normal fault-block mountains, and the anticlinal Beartooth Mountains. The largely volcanic Absaroka Range lies along the eastern flank of Yellowstone.

Rhyolitic magma has produced three major explosions in the Yellowstone area at 2.059, 1.285, and 0.639 million years ago. These eruptions were gigantic compared with the 1980 eruption of Mount St. Helens. Most of the brightly colored stone in the Grand Canyon of the Yellowstone consists of interlayered rhyolite and tuff generated during volcanic eruptions and subsequently

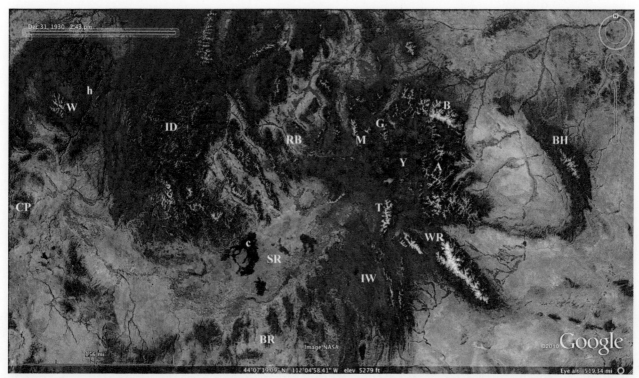

FIGURE 14.13   A Google Earth image looking northward across the Yellowstone area. *A* = Absaroka Mountains; *B* = Beartooth Mountains; *BR* = Basin and Range; *BH* = Bighorn Mountains; *c* = Craters of the Moon; *CP* = Columbia Plateau (southern basin); *G* = Gallatin Range; *h* = Hell's Canyon and the Seven Devils Mountains; *ID* = Idaho Batholith; *IW* = Idaho-Wyoming Fold and Thrust belt; M-Madison Range; RB-Rocky Mountain Basin and Range; *SR* = Snake River Plain; *T* = Teton Mountains; *W* = Wallowa Mountains; *WR* = Wind River Range; *Y* = Yellowstone.

altered by the ubiquitous hot waters of the region. If you look closely, however, you might also see a few interlayers of basalt. The most recent major activity occurred 150,000 and 75,000 years ago when a series of rhyolite lava domes and lava flows developed in the area directly east of Old Faithful.

The eastern part of the Snake River Plain contains what is probably the largest aquifer west of the continental divide. It is the sole source of drinking water for most of the population across the region. Water for the aquifer originates from precipitation and snowmelt in the mountains north and east of the Snake River Plain. Once in the aquifer, groundwater moves from northeast to southwest through rubble zones along contacts between buried basalt flows. There are two large natural discharge areas (springs); the first is near American Falls Reservoir, and the second (larger) one is at Thousand Springs. Collectively, these springs greatly increase the volume of water in the Snake River.

CORDILLERAN VOLCANIC AREAS BETWEEN 60 AND 20 MILLION YEARS OLD

Basaltic, andesitic, and rhyolitic volcanic rocks occur in abundance in the Basin and Range, Colorado Plateau, and Rocky Mountains. In this section, and in the section that follows, we will discuss some of the larger and more significant volcanic areas of the Cordillera, excluding those of the

Columbia River Plateau, Snake River Plain, and Cascade volcanic arc. The volcanic areas are located and named in Figure 14.14. Rocks older than 20 million years are shown in yellow.

In many areas the volcanic rocks occur in association with crystalline domes composed of shallow intrusive rock. These intrusions, as you might expect, are roughly the same age as the volcanic rocks and likely represent magma that didn't quite reach the surface. They are exposed today through erosion. Excellent examples of intermingled volcanic and intrusive rocks occur in the isolated mountains of the northwestern Great Plains (compare Figures 10.4 and 14.14). Additional examples occur on the Colorado Plateau and in the Southern Rocky Mountains.

Some of the oldest volcanic rocks occur discontinuously from northeast Washington across Idaho and western Montana to the Yellowstone area of Wyoming (Figure 14.14). These rocks are part of the Challis magmatic arc complex, named for volcanic rocks in central Idaho. The rocks erupted between 55 and 40 million years ago and are dominantly andesite and rhyolite-tuff. The Absaroka volcanic field is the largest volcanic center. Rocks of the Absaroka field occupy the eastern and northern edge of Yellowstone National Park, extending from the eastern shores of Yellowstone Lake eastward and northward into the heart of the Absaroka and Gallatin ranges. These volcanic rocks obviously are far older than the hot-spot-generated volcanic rocks that form the Yellowstone calderas and much of the national park. The

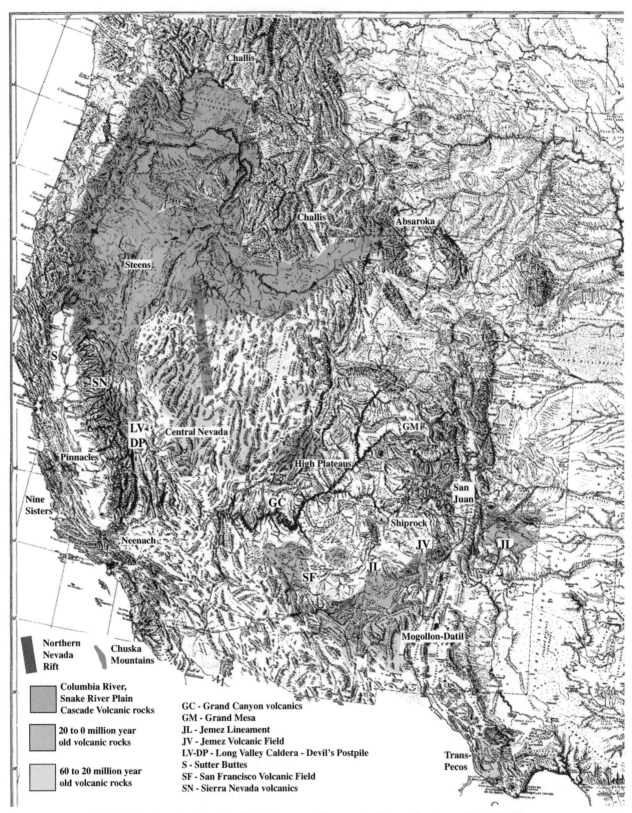

FIGURE 14.14   A landscape map that shows some of the more significant volcanic areas in the Cordillera.

presence of andesite suggests that the Absaroka volcanic field is related to subduction. From a landscape point of view, the Absaroka Mountains are impressive. They constitute one of the largest portions of roadless areas in the contiguous United States. Much of the area is a high volcanic plateau dissected into erosional mountains. Several mountain peaks top 11,000 feet including what appear to be old, extinct, explosive volcanoes not unlike the volcanoes in the Cascade Range. Volcanism in the Absaroka region began as early as 53 million years ago with massive eruptions occurring between 49.5 and 47.5 million years ago. The volcanic rocks encase some spectacular fossils, including an area within Yellowstone National Park known as Specimen Ridge. Here, volcanic ash and flows have fossilized as many as 27 separate forests. Figure 14.15 is a photograph of a tree 12 feet tall that was first smothered and buried by volcanic ash, then petrified and turned to chert by silica-rich fluids.

Relatively old volcanic rocks are present in the central Cordillera although seemingly none are as old as those in the Challis magmatic belt. Argon dates indicate most are between 43 and 24 million years. Some of the rocks are intermediate (andesite) in composition, not unlike those of the Challis belt and therefore probably related to subduction. However, the most significant volcanic rocks of this time frame are the more than 100 separate rhyolitic calderas that also are most likely associated with subduction. Explosive volcanism of this type is present in the Basin and Range of central Nevada, the Mogollon-Datil region of southwestern New Mexico-southeastern Arizona, the San Juan Mountains of southwestern Colorado, and the Trans-Pecos (Davis Mountains) region of west Texas. All four of these areas are labeled in Figure 14.14. This was a massive volcanic episode referred to in the literature as the *ignimbrite flare-up* (or mid-Tertiary ignimbrite flare-up). This particular flare-up was especially intense between about 38 and 27 million years ago with more than a dozen giant explosions perhaps on a scale equal to or greater than the more recent explosions at Yellowstone. It is likely that by 24 million years ago large areas of the west, including most of Nevada, Colorado, Arizona, and New Mexico, were covered in volcanic ash hundreds of feet thick. Interestingly, also present between about 34 and 27 million years ago was a giant sand dune field located in the Four Corners area of the Colorado Plateau known as the Chuska erg. At its largest, the dune field may have stretched from the Colorado River area along the Utah-Arizona border to the Mogollon-Datil volcanic field and the San Juan field. Much of the evidence for the Chuska erg has since been eroded. The largest remnant is in the Chuska Mountains along the Arizona-New Mexico border just south of the Four Corners where deposits reach thicknesses of about 1,750 feet. The Chuska Mountains are located in Figure 14.14.

The ignimbrite flare-up in Colorado is associated with intrusions of the Colorado Mineral Belt. The largest intrusion

FIGURE 14.15 A photograph of a 12-foot-high tree now fossilized to silica (chert) on Specimen Ridge, Yellowstone National Park.

forms part of Mt. Princeton, a 14,000-foot peak in the Sawatch Range. Also associated with the ignimbrite flare-up, and tucked away in the San Juan Mountains near Creede, Colorado, is one of the most unique areas in the United States. The Wheeler geologic area is a spectacular region of odd-shaped pinnacles, domes, and arches made of tuff (volcanic ash) and carved by water. The ash was deposited 27.8 million years ago following the explosive eruption of the nearby La Garita Caldera. This area is a challenge just to reach. One would have to walk more than 7 miles along a footpath, or drive more than 14 miles along a nearly impassible four-wheel-drive road, just for a well-deserved glimpse of the area.

Many of the volcanic calderas in the Central Nevada field are no longer recognizable in the landscape due to reincarnation via younger Basin and Range normal faulting. The calderas are recognized instead by virtue of their rock deposits. The fact that many of the calderas have been reincarnated implies that the ignimbrite flare-up occurred prior to development of normal fault Basin and Range landscape in Nevada. The Cordilleran landscape (moonscape?) 24 million years ago must have looked very different from

FIGURE 14.16   A Google Earth image looking eastward at Shiprock volcanic neck and radiating dikes on the Colorado Plateau, New Mexico, near the Four Corners area.

today. Chapter 15 explores the timing of Basin and Range normal faulting in Nevada and what the landscape may have looked like during the ignimbrite flare-up prior to Basin and Range reincarnation.

There are a few other areas of note where volcanic rocks older than 20 million years are present. Shiprock is an isolated edifice in the Four Corners area of New Mexico that rises some 1,700 feet from the desert floor of the Colorado Plateau. The rocks are not strictly volcanic but are the erosional remnants of a volcano that likely became extinct between 27 and 25 million years ago. The main edifice represents the central feeding pipe (the neck) of a now vanished volcano. The neck is composed of broken pieces of volcanic rock in the form of volcanic breccia. A series of six vertical dikes radiate like giant curved walls from the central edifice. The dikes are composed of an unusual rock known as *minette*, consisting of mica, carbonate, and pyroxene. The main edifice and several dikes are shown in Figure 14.16. The volcanic neck and dikes likely developed some 2,500 feet below the surface and are exposed due to erosional stripping of overlying nearly flat-lying sedimentary layers.

A few small volcanic fields are also present across the California transpressional structural province (Chapter 17). Pinnacles National Monument in the Gabilan Mountains east of Salinas Valley in the Coast Ranges represents part of a circa 23-million-year-old rhyolitic volcanic field. The area is famous for its massive monoliths, spires, and sheer walled cliffs exploited by rock climbers. This small volcanic area is located west of the San Andreas fault. The other half of the volcanic field is located near Neenach, which is east of the San Andreas fault in the Mojave Desert, about 200 miles to the south. The field was split in half by displacement along the San Andreas fault.

A related set of extinct volcanic peaks in the Coast Ranges are known as the Nine Sisters or the Morros. As shown in Figure 14.17, these are small volcanic peaks located between Morro Bay and San Luis Obispo. Like the Pinnacles, the Nine Sisters are rhyolitic in composition and between 20 and 28 million years old. The volcanic cones are probably associated with subduction of oceanic lithosphere beneath the North American plate. The linear orientation of the Sisters suggests that magma

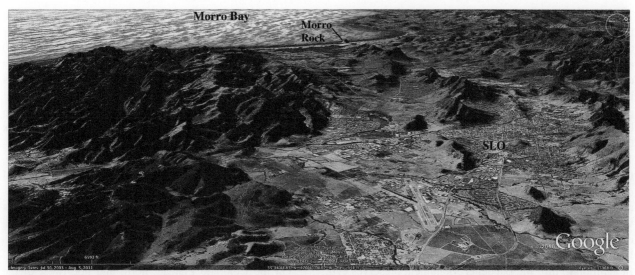

FIGURE 14.17   A Google Earth image looking north-northwest at the Nine Sisters. The volcanoes extend for 12 miles through the center of the image, from San Luis Obispo (*SLO*) to Morro Rock. The two high peaks overlooking San Luis Obispo are Cerro San Luis Obispo (1,292 feet) and Bishop Peak (1,559 feet), which is the highest in the chain. The barely visible peak at the coastline is Morro Rock (576 feet), home to the peregrine falcon. The shape of Morro Rock is the result of quarrying, which continued until 1969. The San Luis Range lies along the shoreline to the left and the Santa Lucia Range to the right.

may have found an easy path to the surface along a fracture or fault zone.

CORDILLERAN VOLCANIC AREAS YOUNGER THAN 20 MILLION YEARS

Volcanic rocks younger than 20 million years old include previously discussed Columbia River and Snake River Plain volcanism as well as rocks associated with the Cascadia arc system, to be discussed in Chapter 16. Additional volcanic rocks of this age are widely distributed across the Cordillera, including large areas of the Basin and Range, Colorado Plateau, and Southern Rocky Mountains. Their distribution is shown in Figure 14.14. We will survey these rocks beginning in and around the Colorado Plateau where some of the volcanism is very young or active.

The San Francisco volcanic field (*SF* in Figure 14.14), located at the southern margin of the Colorado Plateau near Flagstaff, Arizona, consists of more than 600 volcanoes, cinder cones, and lava flows that range in composition from basalt to rhyolite. This field includes Arizona's highest mountain, the 1.0 to 0.4 million-year old Humphreys Peak (12,633'), as well as Arizona's youngest volcano, Sunset Crater, only 950 to 850 years old. The San Francisco field is thought to have developed above an ancient fault or fracture zone along which magma was able to reach the surface.

Another area of young volcanism occurs east of the San Francisco field in eastern Arizona and northern New Mexico. The Jemez lineament (*JL* in Figure 14.14) is a chain of volcanic centers that extend from the Arizona-New Mexico border area east of Phoenix, across Mt. Taylor, to the Raton volcanic field near the New Mexico-Colorado-Oklahoma border. The lineament does not follow an obvious geologic trend but rather cuts obliquely across several structural provinces beginning in the Basin and Range, extending along the southeast margin of the Colorado Plateau, and then crossing the Rio Grande Rift zone and Southern Rocky Mountains, ending on the Great Plains. The Rio Grande Rift is an area of normal faults connected with but separate from the Basin and Range. It is discussed in Chapter 15. There are at least 74 volcanic vents along the Jemez lineament composed mostly of basaltic lava but also with significant explosive silicic volcanism. Volcanism began sometime prior to 13 million years ago and remains active. Much of the volcanism has occurred in the past 4 million years. There is no age progression along the lineament; therefore, it is not considered the product of a hot spot. Seismic evidence has revealed discontinuities in the lower crust that suggest that magma reached the surface along an ancient, circa 1.65 to 1.68 billion-year-old fault zone.

There is one area in particular along the Jemez lineament where explosive rhyolite volcanism has occurred at a scale on par with the Long Valley and Yellowstone eruptions, both of which are discussed later in this section. This area is known as the Jemez volcanic field (*JV* in Figure 14.14), and it lies just west of Santa Fe, along the western edge of the Rio Grande Rift zone. It is the largest volcanic field along the Jemez lineament, approximately 50 miles across. Much of the Jemez field consists of a single large volcano with a central crater 7 to 12 miles across, known as the Valles Caldera (or Sierra de los Valles). An older, smaller, and partly destroyed crater, known as the Toledo Caldera, is preserved at the northeast and southeast margins of the Valles Caldera.

The Jemez field has shown nearly continuous volcanism for the past 13 million years with more than half of the volcanic rocks accumulating between 10 and 7 million years ago. Figure 14.18 shows the volcanic pile with the Valles and Toledo Calderas labeled. The different colored lines in the figure represent faults that have been active in the past 1.6 million years. The yellow faults near Los Alamos have been active during the past 130,000 years. The Valles Caldera formed via two explosive eruptions approximately 1.45 and 1.12 million years ago. Both explosions produced thick deposits of pumice and ash known today as the Bandelier Tuff. The second explosion was the largest and is responsible for producing the final shape of the caldera as well as the greatest thickness of Bandelier Tuff. Excellent exposures of Bandelier Tuff can be seen at nearby Bandelier National Monument, about 6 miles east of Los Alamos. In addition to tuff, the Bandelier Monument preserves evidence of human presence that dates back as far as 11,000 years. The area was home to the ancient Pueblo people from 1150 to 1600. Ceremonial structures and cliff dwellings are preserved in the soft tuff layers.

Originally it was thought that the smaller Toledo Caldera formed during the 1.45-million-year explosion. However, more recent analysis based on the distribution, thickness, and flow direction of associated volcanic deposits suggests instead that volcanic debris of this age was ejected from the Valles Caldera. It is now understood that the Toledo Caldera formed a short time prior to 1.45 million years ago.

Note the several round, dome-shaped protrusions within the central depression of the Valles Caldera in Figure 14.18.

These are lava domes of obsidian, rhyolite, and ash that pierced the surface and then froze beginning sometime after the 1.12-million-year explosion but prior to about 530,000 years ago. Formation of the lava domes was followed by almost 500,000 years of dormancy that ended 57,000 years ago with the creation of additional rhyolite lava domes at the southern edge of the caldera. Although there has been no subsequent volcanic activity, seismic imaging suggests the presence of a magma body within 7 miles of the surface, implying that the region remains volcanically active. A drilling project to look into the feasibility of geothermal energy reached temperatures of 340°C at a depth of 2 miles (3.2 km). Although these are very hot temperatures, it was determined that the reservoir was too small to support a commercial project. The Valles Caldera became a government-owned national preserve in 2000.

On the Colorado Plateau, north and northwest of the Jemez lineament, basalt flows form caprock on several mesas and plateaus including the 10-million-year-old basalt that caps Grand Mesa east of Grand Junction (*GM* in Figure 14.14), and several plateaus in the High Plateaus region including the Aquarius Plateau. One pretty amazing site is in the western part of the Grand Canyon where basalt flows less than 750,000 years old are seen frozen in time tumbling down the canyon walls of the Colorado River. A Google Earth image of this site is shown in Figure 14.19.

From the Colorado Plateau, we jump to the northern part of the Basin and Range where a conspicuous volcanic area of rhyolite and basalt is known as the Northern Nevada Rift. This area, located in Figure 14.14 as a thick, straight line,

FIGURE 14.18 A Google Earth image looking north at the Valles Caldera. Colored lines represent faults that have been active during the past 1.6 million years. Los Alamos (*LA*) sits near the eastern rim of the caldera. Note the concentric ring of lava domes within the central depression.

extends for about 300 miles from the Nevada-Oregon border through central Nevada. It consists of an alignment of volcanic flows, dikes, normal faults, and gold-silver deposits mostly between 16.5 and 15 million years old. The rift is understood to be associated with initiation of the Yellowstone hot spot. The interpretation is that the hot spot created a bulge that stretched and cracked the crust, creating a linear fracture along which magma could reach the surface. The volcanism was short-lived and had ended by 15 million years ago presumably when this part of the North American plate had migrated sufficiently far from the hot spot.

West of the Northern Nevada Rift, we can find a great variety of volcanic rocks over a large area of southern Oregon, eastern California, and western Nevada, including the northern and central parts of the Sierra Nevada. This region is labeled *Steens* (for Steens Mountain) and *SN* (for Sierra Nevada) in Figure 14.14. It includes basalt, rhyolite, and andesite of all ages, from 20 million years old to the present, as well as some volcanic rocks that are older than 20 million years. The Steens Mountain area lies just south of the Columbia River Plateau between Cascade arc volcanic rocks and silicic rocks associated with the Snake River Plain-Yellowstone hot-spot track. The oldest rocks consist of rhyolite lava domes and andesite between about 24 and 17 million years old that may be associated with the Cascade arc. A large volume of basalt between 16.6 and 15.3 million years old, likely associated with the Yellowstone hot spot, overlies the andesite-rhyolite sequence and forms most of Steens Mountain. The basalt is partly intermingled with, and partly overlain by, andesite and ignimbrite deposits as young as 15 million years old. Minor volcanism has continued intermittently almost to the present day. Considering the large amount of basalt, this area

could be considered part of the Columbia River Plateau physiographic province. It is not physiographically part of the Columbia Plateau because of extensive Basin and Range normal faulting that began less than 17 million years ago. These faults, rather than the volcanism, exert primary control on the landscape. Steens Mountain is the largest normal fault-block mountain in the northern Basin and Range.

Volcanic rocks extend southward from Steens Mountain along the California-Nevada border into the Sierra Nevada where they are intermingled with older sedimentary, granitic, and metamorphic rocks. Within the Sierra Nevada, the Lovejoy Basalt consists of a series of flows that erupted from a vent at Thompson Peak in the Diamond Mountains west of Honey Lake, and flowed westward into the Sacramento Valley near Lake Oreville. This is the largest single volcanic unit in California. It is 15.4 million years old and is understood to be associated with the Columbia River flood basalts and the Yellowstone hot spot. In addition to the Lovejoy Basalt, there are numerous additional locations in the Sierra Nevada where andesite and rhyolite form small volcanic domes, ignimbrites, and volcanic breccias. These rocks have filled ancient stream beds, thus displacing stream water to the side of the channel, as shown in Figure 14.11. The rocks are between 14 and 6 million years old and are interpreted to represent the final stages of subduction-related volcanism when the Juan de Fuca plate extended farther south than its present position. Volcanism ended when the San Andreas Fault grew longer and replaced the subduction zone with a strike-slip system.

Additionally, there are many volcanic fields along the western edge of the Basin and Range directly east of the Sierra Nevada. Included are the 100,000-year-old basalt

FIGURE 14.19   A Google Earth image looking northward at basalt flows of the Unikaret volcanic field in the western Grand Canyon. A few of the young basalt flows are seen spilling over the rim near Vulcan's Throne.

flows that form five- and six-sided polygonal columns at Devil's Postpile near Mammoth Lakes, and the Long Valley Caldera, which is a large and very significant non-basaltic volcanic feature also in the Mammoth Lakes area. The Long Valley Caldera is visible in Figure 14.20 as a 9-by-19-mile elliptical depression filled with silicic volcanic rocks that formed via violent explosion about 758,900 years ago. The eruption is understood to have lasted about 90 hours based on the distribution of ash. The caldera remains active with smaller volcanic explosions occurring roughly every 200,000 years. The caldera is second only to Yellowstone in terms of its capability to create a large volcanic explosion. Ash and associated rocks from a series of volcanic eruptions between 220,000 and 50,000 years ago cover much of the western caldera floor. These same eruptions built Mammoth Mountain, a popular ski destination on the rim of the caldera. The most recent volcanic activity has been along the Inyo-Mono Craters volcanic chain, which extends northward for 30 miles beginning just west of the Long Valley Caldera and ending at Mono Lake. This chain has produced numerous small ash explosions and silicic lava flows, including obsidian (volcanic glass) within the past 5,000 years. Part of the chain was built between 600 and 250 years ago. The volcanic chain is visible as small hills (labeled *IM*) in Figure 14.20.

Although the last major eruption occurred 250 years ago on Paoha Island in Mono Lake, the Long Valley Caldera itself has shown signs of life as recently as 1989. Following several strong earthquakes in the area, geologists noticed that the central part of the caldera was rising. This was taken to indicate that magma beneath the ground was moving—a potential precursor to an eruption. Several months later, areas of dead and dying trees were detected in and around Mammoth Mountain. It was determined that the trees were suffocated by exceptionally high concentrations of CO_2 emitted into the soil by cooling underground magma. The caldera floor has risen more than 31 inches since 1978, suggesting that magma is present at a depth of as little as 1.8 miles. The US Geological Survey, other government agencies, and universities continue to monitor this potentially explosive situation.

Finally, we look westward at an isolated volcanic exposure in the Central Valley of California. The Sutter (Marysville) Buttes (*S* in Figure 14.14) are probably the most famous volcanic landmark in the Central Valley. The buttes form a series of small andesite and rhyolite lava domes near Yuba City. The volcanoes were active between 2.5 and 1.5 million years ago as part of the Cascade subduction system when the Juan de Fuca plate extended farther south than its present position. The buttes formed via subduction just like

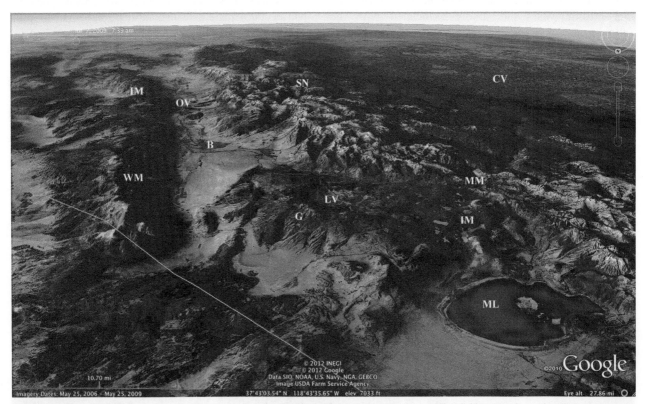

FIGURE 14.20 A Google Earth image looking south along the Nevada-California border. The Long Valley Caldera (*LV*) forms the elliptical depression rimmed by rhyolite and obsidian of Glass Mountain (*G*) and Mammoth Mountain (*MM*). The Inyo and Mono Craters (*IM*) form the series of lava domes that lead toward Mono Lake (*ML*). The White Mountains (*WM*) host the Bristlecone Pine forest (the oldest trees on Earth). Also visible is the town of Bishop (*B*), the Inyo Mountains (*IM*), the Sierra Nevada (*SN*), Owens Valley (*OV*), and the Central Valley (*CV*). The white line is the Nevada-California border.

other volcanoes in the Cascade chain. However, since that time, the southern boundary of the subduction zone (the Mendocino triple junction) has migrated northward, causing subduction below the buttes to cease. Today the buttes stand as an isolated relict landscape that can be seen for miles rising from the flat valley floor. There are other areas of volcanic rocks that will not be discussed in detail such as the 19 to 13 million-year old rocks in the vicinity of Los Angeles and in the Channel Islands, the 33 to 43 million-year old rocks in the South Park region of Colorado, and additional volcanic rocks in the Basin and Range, to name a few (Figure 14.1).

QUESTIONS

1. The Appendix contains an uncolored version of Figure 14.1. Color this map and describe the distribution of volcanic rocks across the US.
2. What are the characteristics of the following volcanic types: basaltic lava, andesite, rhyolite, ignimbrite?
3. What is a lava dome?
4. Using Figure 14.10, where 1 inch equals 72 miles, or using Google Earth, calculate the velocity and direction of North American plate movement between 14 and 9 million years ago and between 9 and 5 million years ago based on the Yellowstone hot spot track.
5. Speculate as to why the Columbia River flows around the margin of the Columbia Plateau rather than through the center.
6. Based on Figure 14.3 alone, how might one infer the flow direction of floodwaters across the Channeled Scablands?
7. In Google Earth, fly to Dry Falls, Washington. Describe what you see. What are the dimensions of the falls and its height as measured in Google Earth?
8. Trace all of the anticlinal ridges visible in Figure 14.8. Given that these are actively forming folds, what can you infer about the present-day direction of maximum stress in this part of the Earth's crust?
9. What type of faults might you expect at the base of these folds?
10. In Figure 14.8, what is unique about Saddle Mountain?
11. Suggest possible hypotheses to explain why the Columbia River cuts through Saddle Mountain but is deflected eastward around ridges north of Rattlesnake Mountain, as shown in Figures 14.3 and 14.8.
12. In Google Earth, fly to Pinnacles National Monument in California and then to Neenach. Use the Google Earth ruler to measure the distance between the two. Given that the volcanic rocks in both areas are 23 million years old and that they were once adjacent to each other, calculate the rate of displacement in feet per 100 years, miles per million years, and kilometers per million years, along the San Andreas Fault.
13. Use Figure 18.1 as a guide to trace the Hurricane Fault across Figure 14.17. How is the fault expressed in the landscape?
14. What is odd about the age progression of volcanic rocks in the Brothers Fault zone relative to the Yellowstone hot-spot track? Can you explain the age progression?
15. What is different (structurally) about the Central Nevada volcanic field relative to other fields?
16. What is the age (or ages) of volcanic rocks at Yellowstone National Park?
17. Describe the San Francisco volcanic field.
18. Research Specimen Ridge in the Absaroka volcanic field. What is special about this area?
19. Why hasn't the Valles Caldera's geothermal potential been utilized?
20. Research any of the following and write a few paragraphs on it: Mammoth Mountain, Long Valley Caldera, Devil's Postpile, Inyo-Mono Craters, Mogollon-Datil volcanic field, San Juan volcanic field.

Normal Fault-Dominated Landscapes

Active and recently active normal faults form perhaps the most conspicuous and widespread landscape in the Cordillera. Not only are they ubiquitous across the Basin and Range physiographic province, they have infiltrated nearly every province in the Cordillera. Normal faults cut the Grand Canyon region of the Colorado Plateau, they cut into the heart of the Colorado Rockies, and they cut lava flows and volcanic features on the Snake River Plain, Columbia Plateau, and Cascade Range. They are present in crystalline rock of the North Cascades, and they permeate nearly the entire Northern Rocky Mountains. Normal faults are partly or wholly responsible for some of the highest mountains in the country, including the Sawatch Range in Colorado, the Sierra Nevada of California, the Wasatch Range in Utah, and the Teton Range in Wyoming.

The physiographic province that most exemplifies normal fault landscape is the Basin and Range. In this chapter we will discuss the Basin and Range physiographic province and some aspects of its infiltration into surrounding provinces. We will discuss an active arm of the Basin and Range known as the Rio Grande Rift that extends into the Southern Rocky Mountains. We will also look in more detail at two active normal fault mountains, the Teton Range and the Wasatch Range. There are no active normal faults east of the Cordillera, but we will discuss an area of ancient normal faults in the Appalachian Mountains known as the Triassic Lowlands where topography is primarily erosion-controlled rather than structure-controlled. Another area of ancient normal faults, the 1.1-billion-year-old Keweenawan rift zone in the Superior Upland, was previously discussed in conjunction with crystalline rock in Chapter 13.

Rocks, as we all know, are rigid and brittle near the surface. If tectonic forces act to pull them apart, they will break into blocks separated by steeply dipping faults as shown in Figure 15.1. Blocks that have the shape of a downward-tapered wedge will slide downward along fault planes under the force of gravity. The adjacent block will remain elevated, thus producing the most distinctive set of landforms associated with normal faults, a basin and a range. In geology, we refer to the down-dropped basin as a *graben*. The range is known as a *horst*. We will discover, however, that normal faults are not always steeply dipping and that isostasy and volcanism play important roles in landscape development.

Returning to Figure 15.1, did you notice that extension and normal faulting results in an increase in surface area? Compare the width of the cross-section in Figure 15.1a with 15.1d. This is an important consequence of normal faulting.

THE BASIN AND RANGE

One of the most striking aspects of the physiography of North America is the wide desert area of the Basin and Range province centered in the state of Nevada, with its alternating mountain ranges and flat-floored valleys. This is an active structure-controlled landscape that has been under extensional stress for at least 40 million years, although much of the deformation has occurred only within the past 17 million years. As shown in Figure 1.5, the Basin and Range extends from the Colombia Plateau southward into Mexico, and from the Sierra Nevada eastward to the Rocky Mountains and Great Plains. It is easily the largest physiographic province in the Cordillera. The characteristic basin-and-range signature is clearly evident on the shaded relief map of the Cordillera shown in Figure 15.2. Note especially how the province wraps around the southern margin of the Colorado Plateau. Sedimentary and volcanic

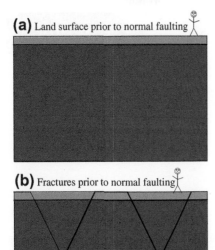

(a) Land surface prior to normal faulting

(b) Fractures prior to normal faulting

(c) Extension creates additional land surface

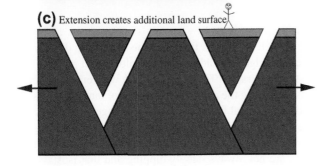

(d) Blocks drop down under force of gravity to fill void created by extension. Horst and graben (range and basin) topography is created

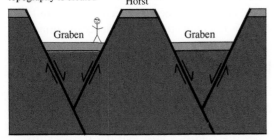

FIGURE 15.1 A series of schematic cross-sections (a through d) that show the mechanics of near-surface normal faulting.

FIGURE 15.2 A digital shaded-relief image of the Cordillera. From Thelin and Pike (1991) and downloaded from the National Center for Earth-Surface Dynamics (www.nced.umn.edu).

rocks are widespread but have secondary landscape influence to the much stronger influence of normal faults. The province is too large to show on a single landscape map and instead is shown in a series of three maps (Figures 15.3, 15.4, and 15.5).

An analysis of Figures 15.2 through 15.5 reveals several characteristics of Basin and Range landscape. First, in Figure 15.2, note the strong north-northeasterly trend of mountain ranges in Utah and Nevada as well as south of the Colorado Plateau. Note how this trend is broken at the eastern margin of the Basin and Range, near California. The area where the northerly trend is broken is known as the *eastern California–Walker Lane belt*. The eastern edge of the belt is drawn in Figures 15.3 and 15.4. As shown in Figure 8.4, this line forms the boundary with the California transpressional structural province, which is discussed in Chapter 17.

In all the figures, note that mountains in the southern part of the Basin and Range from the Mojave Desert in southern California eastward to New Mexico are generally smaller and less obvious than those in Utah and Nevada. This feature is especially obvious in Figure 15.2 where the ranges in Arizona look almost as though they have been buried. We will discuss this aspect of Basin and Range landscape in more detail later in this section.

Finally, note in Figure 15.3 that the Northern Nevada Rift zone does not have obvious topographic expression and actually appears to cut across the grain of present-day ranges. The rift is defined not so much by its topography but by a strong magnetic signature that suggests the presence of a deep, linear, graben-like structure filled with (magnetic) silicic and basaltic volcanic rocks. Some of these rocks crop out at the surface and have been dated between 18.6 and 13.6 million years old. The rift zone has bearing on the landscape history of the Basin and Range and will be considered

FIGURE 15.3 A landscape map of the Nevada-Utah Basin and Range showing physiographic boundaries. The Albian-Raft River-Grouse Creek, Ruby, and Wheeler Peak (Snake Range) crystalline core complexes are highlighted, as are the Tobin, Cortez, Ruby, Wasatch, and White Mountains, and Mojave Desert.

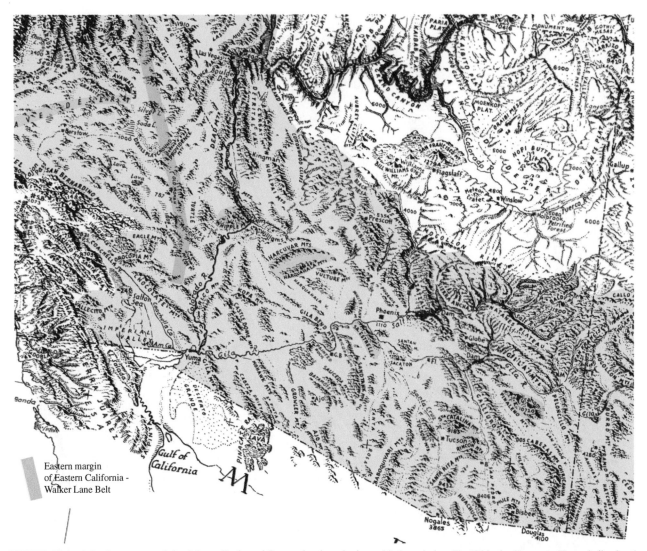

FIGURE 15.4 A landscape map of the Arizona Basin and Range showing physiographic boundaries. The Whipple, Harcuvar, Harquahalla, South Mountains-White Tank, Catalina-Rincon, and Santa Teresa-Pinaleno crystalline core complexes are highlighted, as are the Sierra Estrella and part of the Mojave Desert.

later in this section. For now, we begin our discussion with a look at the large-scale consequences of extension and its influence on landscape.

Figure 15.6 is a schematic cross-section that shows a series of normal faults prior to significant erosion. Ranges are bordered on one or both sides by normal faults that dip steeply near the surface. If two adjacent faults dip in opposite directions, horst-and-graben geometry is formed. If surface faults dip in the same direction, half-grabens are formed, producing asymmetric mountains. Although it is obvious that extension allows the basin to drop down under the force of gravity, perhaps less obvious is the simultaneous uplift of the mountain block. This occurs due to flexural rebound and isostatic unloading when part of the crust breaks, allowing one part to act

independently of the other. The process is depicted in Figure 15.7 and is similar to what was discussed regarding Figure 13.18.

The large-scale process of extension is more complex than formation of simple horst-and-graben or half-graben geometry. Although steep at the surface, most faults curve and flatten at depth into a single, horizontal fault known as a *detachment* (or *décollement*) into which all overlying faults merge. A horizontal detachment surface is shown in Figure 15.6. We noted a similar geometry with foreland fold-and-thrust belts (Figure 12.1). A curved fault that is steep near the surface and horizontal at depth is known as a *listric fault*. The crustal plate above the detachment surface is pulled horizontally in one direction while the crustal plate below the detachment is pulled horizontally in the opposite direction. This results in crustal delamination (the splitting apart of layers) as the lower crust is pulled out from below the upper crust. The upper crustal plate above the detachment stretches and breaks, producing normal faults. The lower crustal plate below the detachment remains largely intact.

Horizontal delamination, coupled with listric geometry, creates two additional characteristics of Basin and Range landscape. The first involves the fault blocks themselves, which tend to rotate (tilt backward) as they slide down the fault surface. Note that the originally horizontal sedimentary layers in the half-grabens shown in Figure 15.6 have rotated toward the vertical. They will continue to rotate as they slide down the listric fault surface. Note also that the shape of the mountain block is asymmetric with a gentle west-facing slope and a steep east-facing slope. The second characteristic involves development of a rollover anticline on the upper plate close to a listric normal fault surface. As rocks on the upper plate are pulled horizontally away from the curved fault, they tend to fold, fault, and collapse into the void that was created between the fault surface and retreating upper plate. As shown in Figure 15.8, this action produces a slight bulge in the landscape known as a *rollover anticline* located a short distance from the fault, as well as a low area at the base of the fault where the rocks have collapsed. It is not uncommon for lakes to form above the collapsed area adjacent to the mountain front. Notice that the process of delamination resembles a ship leaving harbor. As the ship (the upper plate) moves away, it creates a void that, in the case of a normal fault, is filled by the collapse of the bow (front) of the ship to create the rollover anticline.

The western edge of the physiographic Basin and Range is located at the base of the Sierra Nevada (within the California transpressional structural province). As suggested in Figure 15.6, this is the surface location where the upper crustal plate (the Basin and Range) has detached from the lower plate (the Sierra Nevada). Flexural uplift of the lower (Sierra Nevada) crustal plate, cupled with faulting and collapse of rocks in the upper (Basin and Range) plate, has tilted the Sierra Nevada mountain block westward to produce the greatest relief anywhere in the contiguous United States. The eastern front of the Sierra Nevada towers more than 10,000 feet above the adjacent Owens Valley as is evident in the Google Earth image shown in Figure 15.9.

The eastern edge of the physiographic Basin and Range is located where normal faults become less numerous and basin-and-range landscape becomes less obvious. In northern Utah, this location is at the base of the Wasatch Range east of Salt Lake City (Figure 15.3). Farther south, this location coincides with the western edge of the Colorado Plateau, which has managed to remain largely intact as a weakly deformed region in spite of its position on the upper plate of the detachment fault (Figures 15.4 and 15.6). With the margins of the physiographic Basin and Range defined,

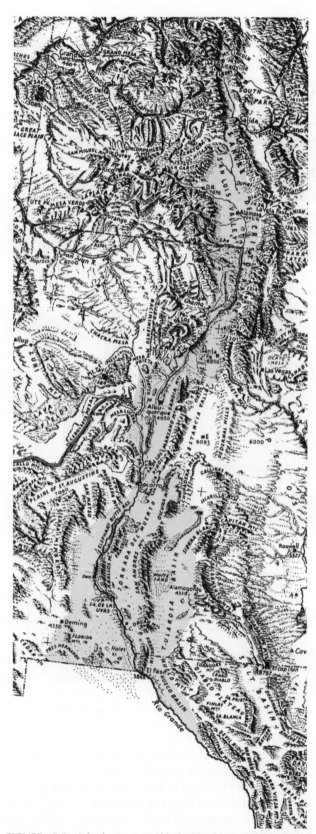

FIGURE 15.5 A landscape map with the Rio Grande Rift highlighted. The edge of the physiographic Basin and Range is shown with a thick line.

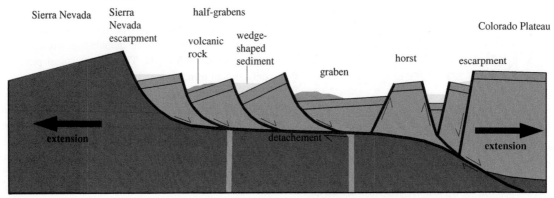

FIGURE 15.6 A schematic cross-section of Basin and Range landscape. Faults on the left side of the cross-section show listric geometry, steep near the surface, curving to near horizontal at depth.

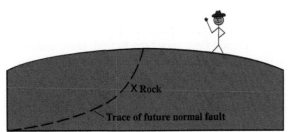

Broad flexure of the crust occurs prior to normal faulting. A rock in the lower plate will undergo tectonic exhumation.

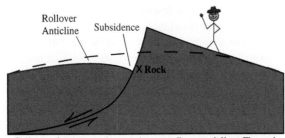

Collapse of the upper plate produces a rollover anticline. Flexural and isostatic rebound of the lower plate produces uplift and a tilted fault block. The rock in the lower plate is exhumed to the surface. The dashed line is the original land surface prior to faulting.

FIGURE 15.7 Flexural rebound and isostatic unloading along a normal fault. The area to the left of the fault is the upper plate.

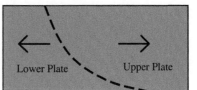

Before extension. The future normal fault is dashed

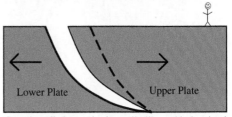

Crust is pulled apart horizontally like a ship leaving harbor. A gap is created between the upper plate and lower plate. A future secondary fault is dashed.

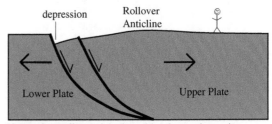

Upper plate collapses into the void to form a depression against the lower plate. A rollover anticline forms on the upper plate.

FIGURE 15.8 Development of a rollover anticline.

we are now in a position to look at some of the characteristics of the interior region.

There are more than 150 fault-block mountains in the Basin and Range, with several peaks above 13,000 feet, including White Mountain Peak (14,246') and Wheeler Peak (13,063'). Total vertical displacement (up and down movement combined) along some of the faults is between 26,000 and 33,000 feet. Averaged over 40 million years, this implies between 0.78 and 0.99 inches of vertical movement every 100 years. However, none of the faults have been active over the entire 40-million-year-period. Faults that have been active for 17 million years or less have displacement rates on the order of 2 to 4 inches per 100 years.

Valleys are constantly filled with sediment derived from erosion of uplifted mountain blocks. Some valleys contain more than 20,000 feet of sediment. Subsiding valleys that do not accumulate sediment fast enough will sink below sea level. An example is Death Valley, which, at 282 feet below sea level, is the lowest point in the Western Hemisphere. Death Valley was occupied by Lake Manly about 15,000 years ago during the end of the glaciation. Today

FIGURE 15.9   A Google Earth image looking southward at Owens Valley. The steep Sierra Nevada Escarpment is at right. The Inyo Mountains are at left. The 1872 Owens Valley earthquake rupture (*OVF*) forms a linear scar that extends south of the town of Independence (*IN*).

the valley is bone-dry with summer temperatures well above 100°F. Consequently there is little erosion of surrounding mountain blocks except during thunderstorms.

In Figure 15.1 we pointed out that extension and normal faulting result in an increase in surface area. The amount of increased surface area is not trivial. Total east-west extension in the state of Nevada is on the order of 155 miles. If all of this extension occurred in the past 40 million years, then this amounts to an average of just over 2 feet of increased land width every 100 years. This, again, is an underestimate because large parts of Nevada were not active until sometime after 17 million years ago. Actual measured rates of extension vary from 19.7 inches to 6.6 feet per 100 years. Figure 15.10 shows the presumed location of the Pacific coastline prior to Basin and Range extension. Most of the extension has occurred across the state of Nevada, which has essentially doubled in width.

Extension and delamination in the Basin and Range have thinned the crust to between 14 and 22 miles thick in most areas. This is thinner than the average crustal thickness of 25 to 31 miles. As a result, the warm mantle asthenosphere has risen close to the surface, which, in turn, has resulted in a geothermal gradient (the increase in temperature with depth in the Earth) that is approximately three times normal. This makes the Basin and Range a prime location for development of geothermal energy. The warm interior also allows the entire region to rise isostatically. Basin floors in many parts of the Basin and Range are between 3,000 and 6,500 feet in elevation.

Associated with a warm interior is lots of volcanism. Volcanic rocks less than 45 million years old are ubiquitous across the Basin and Range where they are intermingled with generally older sedimentary rocks. Many are less than 20 million years old as described in Chapter 14. Crystalline rocks are also present. The most significant crystalline landscapes are known as *crystalline* (or *metamorphic*) *core complexes* because they expose a dome-shaped core of crystalline rock at their center, surrounded by a zone of normal faults and younger sedimentary rocks. We have already noted the existence of core complexes in the Northern Rocky Mountains (Chapter 13). A core complex is the product of extreme lithospheric extension, delamination, and isostatic adjustment. Essentially what happens is that the sedimentary layers on the upper plate are stretched, thinned, and rotated during the delamination process. As the crust becomes thinner, it weighs less and therefore rises isostatically to the point where the lower crystalline crustal plate reaches the surface in the form of a dome. The process is depicted in Figure 15.11. Note that, in this case, crystalline rocks have reached the Earth's surface as a result of tectonic processes (normal faulting) rather than by erosion. This is a form of tectonic exhumation rather than erosional exhumation. Three core complexes are highlighted and listed in Figure 15.3 and six in Figure 15.4.

Before we discuss the timing of normal faulting in the Cordillera, let us return to the White Mountains in eastern California where we can find the oldest living trees (and perhaps the oldest living things) on Earth. The Ancient Bristlecone Pine Forest includes trees that approach 5,000 years old. Although bristlecone pines grow throughout the arid West and regularly reach ages of 1,000 to 2,000 years, they

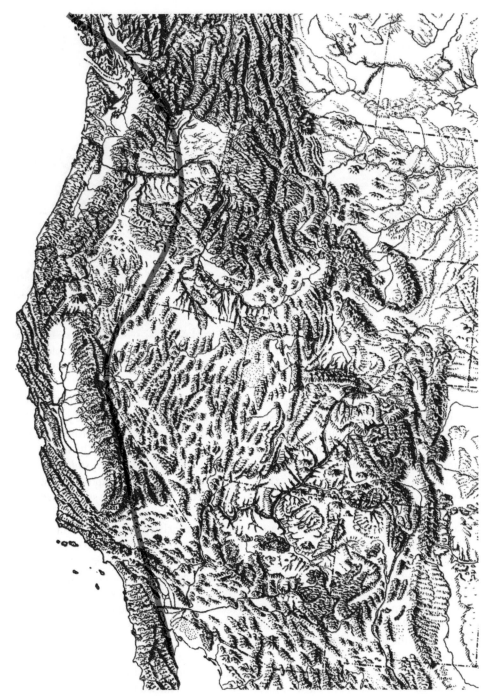

FIGURE 15.10 A landscape map showing the inferred location of the West Coast prior to Basin and Range extension. Based on Moores and Twiss (1995, p. 91).

are particularly old in this part of California because they grow in a harsh, dry climate at elevations between 10,100 and 11,200 feet on a dolomite soil that lacks important nutrients. Virtually nothing else can grow under these conditions, so the trees have no competition. Bristlecone pines are particularly resilient trees that seemingly refuse to die. If a tree blows over, a branch will burrow into the ground to create a root system. If part of a tree dies, another part survives to

the point where it resembles living driftwood. An example is shown in Figure 15.12.

The timing of faulting virtually anywhere can best be determined by comparing the depositional age of rocks deformed by a particular fault with those that are undeformed. It should be obvious to you by now that a particular normal fault must have been active after deposition of the deformed rocks, but before deposition of undeformed

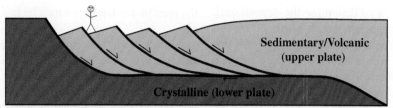

During the delamination process, steeply-dipping faults at the surface bend at depth into listric geometry and merge with a single detachment fault that separates the upper and lower plates.

With additional extension, the blocks rotate and spread apart. This has the effect of thinning the crust which begins to rise isostatically into a dome structure.

Fault blocks either slide off the dome or are eroded thus exposing crystalline rock of the lower plate below the detachment fault. The crystalline rock is surrounded by the normal fault detachment surface.

FIGURE 15.11 A schematic cross-section that shows the origin of a crystalline (metamorphic) core complex. Based on Van der Pluijm and Marshak (2004, p. 392–393).

FIGURE 15.12 A bristlecone pine tree in the Patriarch Grove, White Mountains, California.

rocks. A second method is to determine the depositional age of rocks derived from erosion of fault-generated mountains. The age of these rocks (or sediment) indicates the time when the mountain was in existence. A third method is to dig a trench across the fault trace so that cross-cutting relationships within unconsolidated sediment are clearly visible. This method allows geologists to date very recent fault activity, within the past several hundred years to several million years.

When field analysis is performed across the Cordillera, it becomes apparent that normal faulting has been a dynamic process that has migrated, expanded, and reactivated over time. There are areas that were active 10, 20, and even 50 million years ago but are no longer active. Conversely, there are areas that were active 40 to 50 million years ago, then became inactive, and then later became active again.

The earliest known Basin and Range-type normal faulting in the US began as early as 56 million years ago in the South Okanogan, Kettle River, and Selkirk Mountain region of the Northern Rocky Mountains (Chapter 13). Also active at this time, although initiating somewhat later, is another area in the Northern Rockies located northwest of Yellowstone National Park. This is the area referred to briefly in Chapters 12 and 13 as the Rocky Mountain Basin and Range. A quick gander at this region in Figure 12.4 shows a distinctive montage of seemingly random mountain blocks that strongly disrupts the better-established linear fold-and-thrust landscape to the north. This phase of normal faulting had ended by about 34 million years ago. The landscape in the Rocky Mountain Basin and Range suggests that we have an example of an area that was in the process of transitioning (reincarnating) from a fold-and-thrust landscape to a normal-fault landscape prior to the cessation of normal-fault activity. At present, there is a new phase of normal-fault activity in the Rocky Mountain Basin and Range that began less than 10 million years ago and that may eventually finish the job of reincarnation. In any case, it is pretty amazing to realize that while normal faulting was occurring in both the Rocky Mountain Basin and Range and in the far northwestern Northern Rockies circa 40 to 56 million years ago, a foreland fold-and-thrust belt was completing development in the eastern Northern Rockies and Middle Rockies, and crystalline-cored anticlines were continuing to develop in the Middle and Southern Rockies.

Extensional normal faulting began in the Basin and Range physiographic province between 40 and 35 million years ago at about the time that normal faulting ended farther north. The process of extension in the Basin and Range, however, appears to be more complex than simply beginning 35 or 40 million years ago and continuing to the present. Although some extension undoubtedly occurred prior to 17 million years ago, there is a body of field evidence from Nevada and adjacent areas that suggests major extension did not begin in earnest until about 17 million years ago and that

the present-day landscape may be less than 10 million years old.

Recall from our earlier discussion of the Northern Nevada Rift zone that volcanism, and therefore normal faulting, began sometime around 17 million years ago and that the rift cuts obliquely across present-day mountain ranges. One interpretation according to field evidence is that normal faulting across the Basin and Range began about 17 million years ago, and it was during this period that Basin and Range topography first came into existence across large parts of Nevada and surrounding areas. These mountains, however, are not the mountains we see today. The mountains and valleys that were produced during this phase of normal faulting were oriented roughly parallel with the trend of the Northern Nevada Rift zone. The interpretation is that these mountains were destroyed and reincarnated with a different (north-northeasterly) orientation beginning less than 10 million years ago along faults, many of which are still active. Thus, the normal-fault ranges we see today are a second-generation tectonic (structural) reincarnation of an earlier set of normal-fault ranges. Evidence for this interpretation is seen in the oblique, cross-cutting trend of the earlier Northern Nevada Rift basin relative to present-day ranges, and the presence of earlier-generated depositional basins in uplifted parts of present-day mountain ranges as depicted in Figure 15.13. Geologists are not sure if the preceding scenario can be applied to the entire Basin and Range, but what we can say is that much of the province does appear to have developed no earlier than about 17 million years ago.

Active normal faults, and Basin and Range-type landscape, are currently expanding into surrounding physiographic provinces. Even the Colorado Plateau now boasts several active and recently active normal faults that are beginning to chop away and lower the western margin of the plateau, particularly in the vicinity of the Grand Canyon. Normal faults are strong landscape-formers. Therefore, there is no doubt that the Cordilleran landscape will look

Early (<17 Ma) faulting creates basin and range topography.

Later (<10 Ma) faulting reincarnates the Basin and Range topography along a more northerly orientaion. Part of the older basin is uplifted and part of the older mountain is down-dropped.

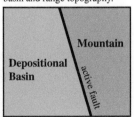
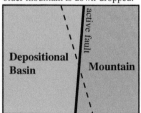

FIGURE 15.13   Sketch maps that show how earlier (<17 Ma) Basin and Range topography in northern Nevada was reincarnated along younger (<10 Ma) north- to northeast-trending normal faults, thereby uplifting earlier-formed depositional basins. The old, inactive fault is dashed in the right-side panel.

significantly different several million years from now as the process of normal faulting continues to develop and expand.

Let us now look to see where active normal faulting and reincarnation encroach on surrounding provinces. The US Geological Survey, in conjunction with state geological surveys, has surveyed all known active faults in the country and placed them on a Website with a Google Earth link. Fault displacement tends to be episodic; therefore, the definition of active is relative. The US Geological Survey provides five maps, each based on the age of last known movement. Three of the maps are shown in Figure 15.14. Most, but certainly not all, of the faults on each map are normal faults. The Website (given in the figure caption) provides additional detailed information on each individual fault. Figure 15.14a shows fault activity during the past 15,000 years. Faults are concentrated in the western Nevada-eastern California region extending to southern California with arms that extend into eastern and west-central Oregon. Those in west-central Oregon extend into the Cascade volcanic arc as far north as Mt. Hood. These regions boast normal and strike-slip faults. Normal faults are once again active in the Rocky Mountain Basin and Range of Idaho and Montana, possibly reactivating old fault scars that were active 45 to 35 million years ago. Normal faults in this area extend westward into Oregon and southward through Yellowstone, the Teton Range, and central Utah to the western margin of the Colorado Plateau. There is also a line of normal faults in central Colorado that extends discontinuously southward through New Mexico into Texas. These faults are part of the Rio Grande Rift system discussed below. And let us not forget the Meers Fault in Oklahoma, which is the easternmost known active fault trace in the US. Note also the active fault on the High Plains of Colorado. The Meers Fault is strike-slip. The Cheraw Fault on the High Plains is a normal fault. Active faults along the California coast are dominantly strike-slip and associated with the San Andreas Fault system. These faults, and those in the California-Nevada border region (the eastern California-Walker Lane belt), are part of the California transpressional system (Chapter 17). Thrust faults are present offshore in Oregon and Washington associated with subduction of the Juan de Fuca Plate. Finally, there are a few faults in the vicinity of Portland and Seattle.

Fault activity over the past 130,000 years mostly fills in these same trends with the addition of faults in eastern Nevada (Figure 15.14b). Fault activity over the past 1.6 million years covers most of the Nevada Basin and Range, with additional faults on the Colorado Plateau in the Grand Canyon region, in northeastern Oregon, and in the eastern part of the Northern Rocky Mountains, including the Flathead Lake area (Figure 15.14c). Note the absence of active faults in the Basin and Range of southern Arizona, southeastern California, and part of the Mojave Desert south of the Sierra Nevada.

Now let us turn our attention to landscape evolution and how the look of a mountain block changes depending on the recency of fault activity. First, we must realize that Basin and Range topography is evolving in a dominantly arid climate where there are few perennial rivers. Erosion occurs not so much by wind but during infrequent thunderstorms, and the progress is slow. Fault scarps that formed 50 to 100 years ago remain well exposed today. The Pleasant Valley earthquake scarp along the western side of the southern Tobin Range (located in Figure 15.3) is one of several in the Basin and Range that are visible in Google Earth imagery. As shown in Figure 15.15, the scarp is present discontinuously along the base of the entire mountain. The fault scarp formed during a series of three large earthquakes within seven hours on October 2, 1915. In a more humid climate these same fault scarps would be eroded and hidden in thick vegetation.

Much of the Basin and Range lies within the Great Basin such that any precipitation that does fall will have no outlet to the sea (Figure 1.9). Thunderstorm-spawned streams will wash loose sediment through mountain canyons as far as the valley floor at the base of the mountain where it will accumulate. Excess water collects in a low part of the basin where it evaporates, forming a saltpan or dry lakebed known as a *playa*. The net effect of erosion is the transfer of sediment from the interior mountain to the adjacent basin. Thus the mountain is eroded and simultaneously buried in its own debris. The process is similar to what has been proposed for the crystalline-cored domes of the Middle and Southern Rocky Mountains (Figure 11.19). This self-burial is an important characteristic of the Basin and Range.

The extent to which a mountain is buried in its own debris is dependent on the rate of mountain uplift relative to the rate of erosion and burial. The rate of uplift, in turn, depends on the level of activity along the range-front fault. The term *range-front fault* refers to a fault at the base of a mountain located roughly at the boundary between the mountain-front and the adjacent valley. It is the fault that separates the basin from the range. Recall from Chapter 3 that a steeply dipping plane (a normal fault in this instance) will produce a relatively straight line (or scar) across the Earth's surface (Figure 3.14). Also keep in mind that, because of the dry climate, the rate of uplift across an active fault will very likely outpace the rate of erosion. Thus, if we have constant displacement along a steeply dipping, active, range-front normal fault, the net effect of that displacement will be to maintain a relatively straight and abrupt transition from valley floor to mountain front. This type of transition is seen in the southern Tobin Range (Figure 15.15) and in the southern Sierra Nevada (Figure 15.9).

A sharp transition between mountain front and valley floor is not maintained once the fault becomes inactive. Instead, the front of the mountain is eroded back while at the same time the valley is filled with sediment. The

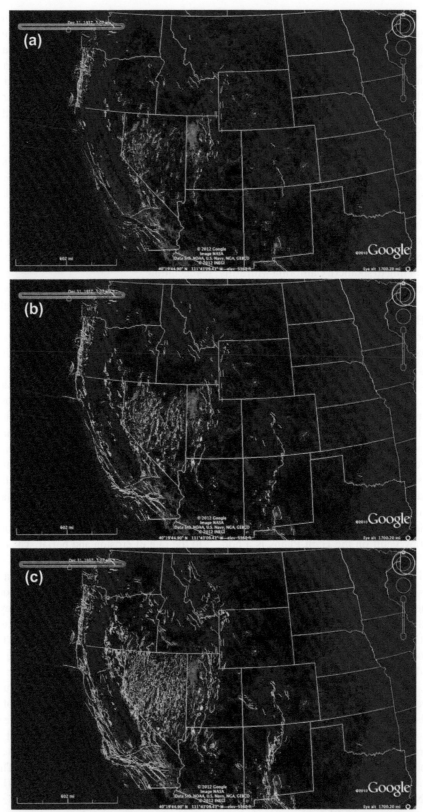

FIGURE 15.14   Google Earth images showing faults in which the age of last known movement is (a) less than 15,000 years ago, (b) less than 130,000 years ago, and (c) less than 1.6 million years ago. Data obtained from the US Geological Survey, the Arizona, California, Colorado, Idaho, and Utah Geological Surveys, the Montana and Nevada Bureau of Mines and Geology, the New Mexico Bureau of Mines and Mineral Resources, and the Texas Bureau of Economic Geology, 2006. Quaternary fault and fold database for the United States, accessed January 18, 2012, from the USGS Website: http://earthquake.usgs.gov/regional/qfaults/.

FIGURE 15.15 A Google Earth image looking southeast at a well-preserved fault scarp along the southern Tobin Range. The fault scarp formed during a series of three large earthquakes on October 2, 1915.

mountain and the range-front fault are progressively buried in their own erosional debris, creating an increasingly sinuous (curvy) mountain front.

The progressive burial of a mountain is shown schematically in both map view and cross-section in Figure 15.16. A mountain block with an active range-front fault on both sides is shown in Figure 15.16a. Fault scarps on both sides of the mountain are exposed because the average rate of fault displacement is greater than the rate of erosion. The high dip angle of the fault produces a sharp valley-to-mountain transition.

Once the normal fault becomes inactive, stream erosion will have enough time to erode and bury the fault surface, producing a sinuous, embayed mountain front as depicted in Figure 15.16b. As erosion lowers the mountain, and the mountain front retreats, a gently tilted erosion surface known as a *pediment surface* is cut into the mountainside at about the same elevation as the valley floor. Canyon streams sweep sediment out onto the pediment surface in the form of an alluvial fan. The term *alluvium* (or *alluvial*) refers to any sediment deposited by a stream. The term *fan* refers to the shape of the deposit, which develops as stream water spreads out into the basin. As the mountain continues to be eroded and buried, alluvial fans grow larger and coalesce to form a continuous alluvial apron known as a *bajada* on the pediment surface. Burial is complete when a bajada on one side of a mountain begins to merge with a bajada on the other side. At this point, only isolated peaks protrude above the alluvium, as shown in Figure 15.16c. Notice in Figure 15.16c that sediment forms a steers-head geometry in cross-section—thick on the down-thrown (valley) side of the fault and thin on the up-thrown (mountain) side where

the pediment surface is developed. Notice also that the mountain shape still exists in cross-section although it is buried beneath sediment. Finally, notice that even though the structure is different, the erosional process, and the resulting slope-flat-slope mountain profile, is identical to what happened in the Middle and Southern Rocky Mountains some 30 to 40 million years earlier.

Different mountains in the Basin and Range are at different stages of erosion and self-burial depending on the recentness and magnitude of activity along range-front faults relative to the rate of erosion and burial. By comparing active mountain fronts with inactive ones at various stages of burial, we can see, at the scale of an individual mountain, how a normal fault landscape slowly grows old.

Let us look first at an area of active faulting. Figure 15.17 is a Google Earth image of northern Nevada from the Tobin Range in the east to the Ruby Mountains in the west. The barely visible thin colorful lines are faults that have been active within the past 1.6 million years. The two white lines enclose the Northern Nevada Rift basin that has been uplifted in some of the younger mountain ranges such as the Cortez Range (as depicted in Figure 15.13). The location of the 1915 Pleasant Valley scarp is also shown. This area is highlighted in Figure 15.3, in which Interstate 80 follows the Humboldt River. The first thing to notice in Figure 15.17 is that active faults occur mostly along the western side of the ranges and that the ranges are tilted eastward. This type of geometry implies that faults dip west in a series of half-grabens, each with the form depicted in Figure 15.7 (and in the opposite direction of that shown in Figure 15.6). The Cortez Range (shown in more detail in Figure 15.18) exemplifies this type of tilted mountain-block landscape. In Figure 15.18 we can

Cross-Section View

Map View

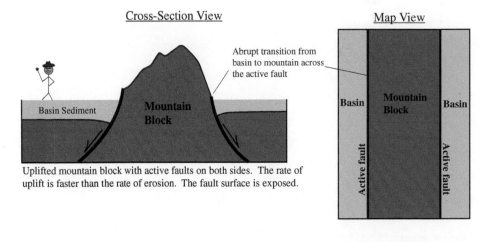

Uplifted mountain block with active faults on both sides. The rate of uplift is faster than the rate of erosion. The fault surface is exposed.

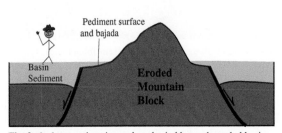

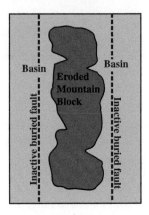

The faults become inactive and are buried beneath eroded basin sediment. The mountain front retreats producing a pediment surface on both sides of the mountain covered with a thin veneer of alluvial sediment.

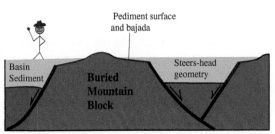

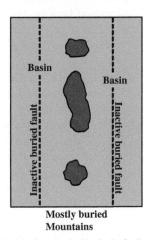

The mountain is buried in its own debris with only isolated peaks protruding above the bajada. Sediment forms a steers-head geometry between two buried mountain blocks.

Mostly buried
Mountains

FIGURE 15.16   A sequence of schematic maps and cross-sections that show erosion and burial of mountain blocks in the Basin and Range.

compare a sharp, active range-front on the west side of the mountain, with a sinuous, eroded eastern side where no active faults exist. Note also that the highest peaks are along the extreme western side of the range, directly above the active fault. One can use the Cortez Range as a template to interpret other tilted mountain ranges.

Examples of eroded and mostly buried range fronts can be found in the southern Basin and Range region where

active normal faults are absent. Figure 15.19 is a Google Earth image of Sierra Estrella located just south of Phoenix, Arizona (Figure 15.4). The mountain has a highly irregular range front on both sides, suggesting a long period of erosion without fault activity. If fault activity remains dormant, we would expect this mountain to be progressively lowered and buried in its own debris. This final stage of mountain evolution is depicted in Figure 15.20, which shows mostly

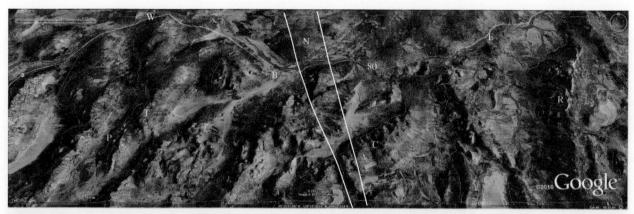

FIGURE 15.17   A Google Earth image looking northward across north-central Nevada at the Tobin Range (*T*), Cortez Range (*C*), Ruby Mountains (*R*), and the location of the 1915 Pleasant Valley scarp (*P*). The white lines outline the Northern Nevada Rift basin (*N*). Note that the rift basin is uplifted in the Cortez Mountains. Also shown as are the towns of Winnemucca (*W*), Battle Mountain (*B*), and Elko (*E*). The very faint, colorful lines (with the exception of I-80) are active faults from USGS Website: http://earthquake.usgs.gov/regional/qfaults/.

FIGURE 15.18   A Google Earth image looking northeastward at the Cortez Mountains, north-central Nevada. The mountain block is tilted eastward. An active normal fault on the west side of the range maintains a sharp transition from valley to mountain front. There is no fault along the western side, producing an irregular range front. The white area in the valley at left is a dry lakebed (playa). The range-front fault was active 2,500 to 3,000 years ago.

buried mountains in the Mojave Desert west of Barstow, California.

Although it is possible to predict the recency of active faulting along range fronts, it is important to realize that the actual time required for a complete cycle will vary depending on rate of fault activity versus the rate of climate-driven weathering and erosion. It is possible for one landform to be numerically younger than another but still be farther along in the erosion cycle. Additionally, the cycle can be interrupted at any time by renewed fault activity.

If Basin and Range landscape developed sometime after 17 million years ago and was then reincarnated less than

10 million years ago, what did the landscape look like 20 or 40 million years ago? In Chapter 12 we noted that the Cordilleran fold-and-thrust belt once extended the length of the US Cordillera from Montana to Arizona. We also noted that the Idaho-Wyoming thrust belt stands as an island of relict landscape surrounded by areas of younger deformation, volcanism, and deposition. Thus, prior to Basin and Range landscape reincarnation, the entire eastern half of the Basin and Range was characterized by a landscape that resembled the Montana Disturbed Belt or the Idaho-Wyoming Thrust Belt. Evidence for this earlier, now vanished, landscape is seen in individual fault blocks (horsts) that contain complexly thrust-faulted and

FIGURE 15.19 A Google Earth image looking northeast at eroded mountain fronts along Sierra Estrella near Phoenix, Arizona.

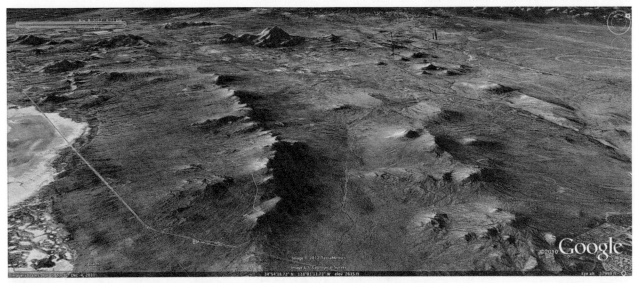

FIGURE 15.20 A Google Earth image looking west at what appear to be buried mountains in the Mojave Desert, California. A large playa is at left; part of Edwards Air Force Base is at the lower-right corner.

folded sedimentary layers typical of a fold-and-thrust belt. A schematic cross-section that shows the internal structure of a fault block is presented in Figure 15.21. We might also assume that 20 million years ago, just prior to extensive Basin and Range extension, much of the region was blanketed in a layer of volcanic ash and ignimbrite expelled from the many calderas that were active in Nevada and surrounding states.

But what did landscape look like in the western Basin and Range prior to the reincarnation that began 17 million years ago? Here again, the calderas and their ignimbrite deposits show some critical relationships. The most significant relationship is the fact that pyroclastic material ejected from calderas in central Nevada can be followed, along river valleys, across the eastern escarpment that now separates the Sierra Nevada from the Basin and Range, all the way to the western slope of the Sierra Nevada. What this means is that, at the time of the ignimbrite flare-up 43 to 24 million years ago, the great eastern escarpment of the Sierra Nevada did not exist. Central Nevada was topographically higher than

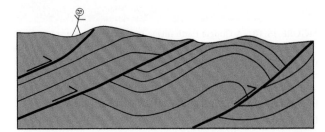

FIGURE 15.21 A schematic cross-section that shows the internal fold-and-thrust belt structure of individual normal fault (horst) blocks in the Basin and Range.

the Sierra Nevada, and, as argued previously, much of the normal fault topography that characterizes the present-day western Basin and Range did not yet exist.

Certain clay minerals found within stream channel volcanic breccias are capable of incorporating water in their chemical structure. This water can be extracted and used for oxygen and hydrogen isotope analysis, which in turn gives

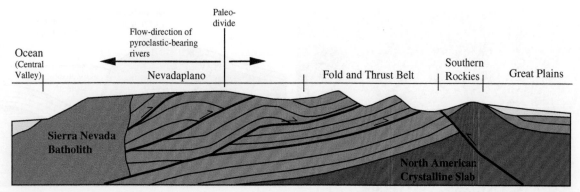

FIGURE 15.22   A cross-section that shows the Cordilleran landscape as it may have looked approximately 25 million years ago. The Nevadaplano and fold-and-thrust belt are covered in volcanic ash. The great eastern escarpment of the Sierra Nevada does not exist. The Rocky Mountain pediment surface covers part of the Southern Rockies. The Nevadaplano has reincarnated an older fold-and-thrust belt. Compare with Figure 17.19.

information on the relative change in elevation across an area. When these studies are combined with field and paleobotanical studies, they indicate that both the Sierra Nevada and the western Basin and Range were at high elevation in what was probably a warm, dry, rain shadow climate. The evidence suggests that the entire region was part of a high-elevation, low-relief plateau that was in existence from well before 40 million years ago to at least 17 million years ago. The exact elevation of this plateau cannot be obtained, but estimates are in the range of 7,000 feet to more than 12,000 feet. The understanding is that the Sierra Nevada formed the western edge of the plateau, which dropped down to the ocean at the present-day location of the Central Valley. Geologists refer to this now vanished plateau landscape as the *Nevadaplano*. Figure 15.22 is a cross-section that depicts the landscape of the Basin and Range at circa 25 million years. Remember, at this time the state of Nevada would have been about half as wide as it is today and the plateau would have been blanketed with ignimbrite deposits.

RIO GRANDE RIFT

The Rio Grande Rift zone consists of a series of interconnected grabens (basins) extending from the Arkansas River Valley at the eastern flank of the Sawatch Range in the heart of the Colorado Rocky Mountains southward into Mexico (Figure 15.5). It is considered to be a separate, independent part of Basin and Range extension. In addition to the Arkansas River Valley, other down-dropped basins include the San Luis Valley between the Sangre de Cristo and San Juan Mountains in Colorado, and basins farther south now occupied by the cities of Santa Fe, Albuquerque, Socorro, and El Paso. The rift is unusual in that it extends along the virtual crest of the Southern Rocky Mountains. The height of the Sawatch Range, which boasts some of the highest peaks in Colorado, is partly a product of normal faulting associated with the Rio Grande Rift.

Across central and southern New Mexico, the Rio Grande Rift zone separates two relatively stable areas of nearly flat-lying rock, the Colorado Plateau and the Great Plains. On the basis of field analysis, the rift zone began

fault activity about 30 million years ago and various parts have remained active ever since. In a few places the rift zone follows and reactivates older faults. The most recent phase of extension began between 16 and 10 million years ago and it is this phase that is responsible for most of the landscape.

There are areas of widespread volcanism surrounding the Rio Grande Rift zone such as the 40- to 20-million-year-old volcanic rocks of the San Juan Mountains in southwest Colorado, and silicic volcanism less than 13 million years old along the Jemez Lineament which cuts across the Rio Grande Rift; however, a direct link between these volcanic rocks and rifting is debatable. Younger volcanism, including numerous basalt flows from 5 million years to a few thousand years old, are scattered along the rift zone and are probably more directly associated with rifting. Typical of most rift zones, the continental crust is relatively thin, particularly in the southern part of the rift zone near the Mexican border. Heat flow (the geothermal gradient) is high. Seismic and gravity studies indicate that the crust is less than 19 miles thick beneath El Paso, Texas, but thickens progressively northward to less than 25 miles thick at the New Mexico-Colorado border. This is in contrast to the crust below the adjacent Colorado Plateau which is at least 27 miles thick, and below the Great Plains which is more than 31 miles thick. The relationships suggest that the rift zone began, or at least shows greatest extension, in its southern part, and has propagated northward. Given enough time, the rift will likely continue to propagate northward into the Middle Rocky Mountains.

THE TETON RANGE

The Teton Range is an actively uplifting tilted normal-fault mountain block approximately 40 miles long and 10 to 15 miles wide, composed of ancient (>2,500 Ma) North American crystalline shield. It is located just south of Yellowstone National Park. A landscape map and cross-section are presented in Figure 15.23. As shown in both the cross-section, and in a Google Earth image shown in

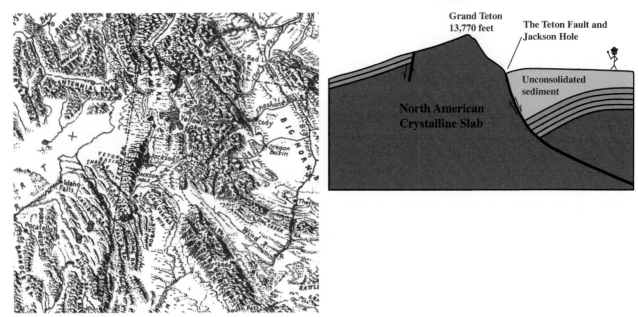

FIGURE 15.23 A landscape map and cross-section of the Teton Range. The Teton Fault is listric and forms a rollover anticline and depression (Jackson Hole) at the base of the escarpment. The location of the Teton Range is just beyond the upper-right corner of Figure 15.3.

FIGURE 15.24 A Google Earth image looking north along the Teton Range. The fault block tilts westward, creating a high escarpment along the eastern flank. The high peak is Grand Teton; *J* = Jenny Lake.

Figure 15.24, the mountain block tilts gently westward such that the highest peaks are concentrated near the eastern edge. This steep eastern escarpment is one of the most magnificent sights in the world. From the photograph in Figure 15.25, one could imagine that a wall of mountain has risen from the flatlands. There is nearly 7,000 feet of relief between Jenny Lake (6,795') at the base of the escarpment and Grand Teton (13,770') across a horizontal distance of less than 3.5 miles. Major peaks in addition to Grand Teton include South Teton (12,514'), Middle Teton

FIGURE 15.25 A photograph looking westward at the Cathedral Group and the steep eastern front of the Teton Range. The peaks from left to right are Mt. Teewinot, Grand Teton, and Mt. Owen.

(12,804'), Mt. Owen (12,928'), Mt. Teewinot (12,325'), Mt. Moran (12,605'), and Thor Peak (12,016'). The range is terminated to the south by landscape associated with the Idaho-Wyoming fold-and-thrust belt, and more gradually terminated to the north where crystalline rocks plunge gently below young volcanic rocks of Yellowstone National Park.

Although it is located geographically within the Middle Rocky Mountain province, the Teton Range owes much of its elevation and relief to extension associated with normal faulting. The Teton Fault is an active normal fault that extends along the base of the steep eastern escarpment. The fault has listric geometry such that a low area (Jackson Hole) has formed at the base of the mountain producing Jackson, Leigh, and Jenny Lakes (Figure 15.23). In terms of its activity and geology, the Teton Range could be considered part of the Rocky Mountain Basin and Range, even though it is separated from this region by the Yellowstone volcanic field.

Initiation of normal faulting and total displacement (total cumulative upward and downward motion) associated with the Teton Fault are not known with certainty. It is generally agreed that displacement is on the order of 29,500 feet based on the vertical separation of crystalline rock on either side of the fault; however, an unknown amount of this displacement may have occurred more than 35 million years ago, during the earlier period of Rocky Mountain Basin and Range fault activity. There are two competing theories. Some geologists suggest that normal faulting and initial uplift of the present-day Teton Range began between 13 and 5 million years ago, with total cumulative displacement between 19,680 and 29,520 feet. This is equal to between 1.8 and 7.1 inches of displacement every 100 years. Other geologists suggest normal faulting was active some 40 million years ago, and again beginning between 3 and 2 million years ago. Total cumulative displacement on the most recent phase of activity is estimated to be between 8,200 and 11,480 feet (equal to between 3.3 and 6.9 inches of displacement every 100 years). In either case, displacement occurs in spurts of 6 feet or more during a single earthquake. Exposures of the fault surface, and offset glacial and landslide deposits along the entire length of the fault, suggest that there has been about 100 feet of displacement during the past 15,000 years (equal to 8 inches of displacement per 100 years). Geological evidence suggests that 8 to 10 earthquakes have occurred during that time span, which is equivalent to one every 1,500 to 2,000 years, with an average displacement of 10 to 12 feet per event.

There have been no earthquakes along the northern part of the fault for at least the past 1,500 years, and no earthquakes in the southern part for more than 5,000 years. Overall, it appears that earthquakes were more common before 8,000 years ago. An interesting hypothesis suggests that isostatic adjustment following the melting of glaciers may have accelerated fault displacement. A large ice cap, more than a half-mile thick, covered nearby Yellowstone National Park 22,000 years ago. At the same time, alpine glaciers were present in the valleys of the Teton Range. It is suggested that initial removal of the weight of the glaciers, beginning about 16,000 years ago and ending 14,000 years ago, caused isostatic uplift of the crust. Rather than broad uplift of the entire region, the hypothesis suggests that movement was concentrated along the Teton normal fault, resulting in an acceleration of earthquakes during the time the crust was isostatically rebounding.

THE WASATCH MOUNTAINS

The Wasatch Mountains extend for about 230 miles from southern Idaho to central Utah directly east of several major cities that include Logan, Salt Lake City, Provo, and Nephi, at the northeastern edge of the Basin and Range (Figure 15.3). The mountains are a source of water and recreation for more than 80% of Utah's population. Similar to the Teton Range, the Wasatch Range is an asymmetric, tilted, normal fault-block mountain, with a steep western face that locally towers more than 7,000 feet above the populated valleys to the west. In contrast to the Teton Range, the rocks are dominantly sedimentary, but areas of young intrusive rock and ancient metamorphic (crystalline) rock are also present. The highest peaks, including Mt. Nebo

(11,928') and Mt. Timpanogos (11,749'), occur along the western face. The base of the western face is traditionally considered the boundary between the Basin and Range to the west and the Rocky Mountains to the east. It is also the site of the Wasatch Fault, one of the longest and most active normal faults in the country with 16 major earthquakes in the past 5,600 years (one earthquake every 350 yrs.). Paleoseismic field studies suggest that accumulated vertical displacement (total uplift and subsidence across the fault) as a result of these earthquakes has averaged between 3.9 and 6.7 in/100 yrs during the 5,600-year period.

The east side of the mountain is rugged with deeply incised glacial valleys that belie the gentle eastward tilt of the range. Figure 15.26 shows that the northward trend of the Wasatch Range between Provo and Ogden is perpendicular to the trend of the adjacent Uinta Mountain anticline. Flexure of the Uinta anticline has produced an especially high, broad, and rugged region within the adjacent Wasatch Range where some of the best skiing in the world can be found. The anticlinal flexure, however, does not cross the Wasatch fault.

The Wasatch fault zone is one of the most intensely studied normal fault systems in the world. The fault dips steeply westward at the surface (65° to 80°), but the dip angle likely decreases with depth. Several investigations have constrained rates of fault displacement and rates of exhumation and denudation on time scales from several thousand to several million years, particularly along the central part of the fault trace. Here, we need to make a point about the surface trace of faults that has not yet been stressed. Like nearly every major fault in the world, including the San Andreas Fault, the Wasatch Fault zone is segmented in the sense that the surface trace of the fault is not continuous. In some cases, there are gaps between fault segments where

FIGURE 15.26   A Google Earth image looking east at the western front of the Wasatch Range, Utah. The Uinta Mountain anticline forms the snow-covered bulge at top-center. The trend of the anticline is perpendicular to the orientation of the Wasatch front. *GSL* = Great Salt Lake, *P* = Provo, *SLC* = Salt Lake City, *UL* = Utah Lake; *d* = area where isotopic studies were conducted. The vague thin lines are active surface-fault traces from the USGS Website: http://earthquake.usgs.gov/regional/qfaults/.

no surface trace of a fault can be found. In other cases, two segments overlap but are separated horizontally by up to several miles. The segmented nature of the fault trace is evident in Figure 15.26. Segments along a particular fault are often grouped and named based on their earthquake recurrence interval over short time scales of thousands of years. Some segments are more active than others. The Wasatch system is composed of 10 segments, each named after a city or mountain in the vicinity. Four of the central segments, from Weber to Nephi, were active between 1,250 and 600 years ago but have no historic earthquakes. The three southern segments have been active during the past 15,000 years. The three northern segments have not been active in the past 15,000 years.

Apatite (U-Th)/He ages from the base of the mountain block directly east of the five central fault segments (from Brigham City to Nephi) were used to determine the exhumation history during the past few million years. The average apatite (U-Th)/He age along all five fault segments is 5.3 million years, with one anomalous 1.6-million-year date from the southern end of the Salt Lake City segment. These dates, because they are on the up-thrown block close to the fault surface, are interpreted as dating tectonic exhumation of rocks (not erosional exhumation). The relationships are shown in a cross-section of the range in Figure 15.27 (review Figure 15.7 to see an example of tectonic exhumation at the base of an up-thrown normal fault block). If we apply depth estimates to the apatite (U-Th)/He data, the best estimate on the rate of exhumation is <0.79 to 1.57 in/100 yrs over the 5.3-million-year period, with a maximum rate of about 4.0 in/100 yrs for the rapidly exhumed southern end of the Salt Lake City segment (that is, the sample with the 1.6 Ma date).

The Salt Lake City segment forms the highest and most rugged part of the range, and it coincides with the westward extension of the Uinta Mountain anticline. It makes sense,

therefore, that this is also the most rapidly exhuming part of the range. Because of its anomalous character, this part of the range has been investigated in great detail. Apatite fission track, apatite (U-Th)/He, and zircon fission track ages have been determined along a west-to-east traverse across the range. All three methods show an increase in age from the Wasatch Fault eastward across the mountain. Apatite fission track ages increase from 3.4 to 39.6 Ma, zircon fission track ages increase from 9.3 to 37 Ma, and apatite He ages increase from 1.6 to 23 Ma. As shown in Figure 15.27, this relationship is consistent with the observation that the fault block is tilted eastward, resulting in greater exhumation (and higher elevations) along the western front of the range. The increase in ages from west to east is not smooth. There is a jump in age data across the Silver Fork-Superior fault zone which lies within the mountain block east of the Wasatch fault. The age jump across the fault disappears in data that is less than 10 million years old. This suggests that the Silver Fork-Superior Fault was active prior to about 10 Ma, that it is no longer active, and that the locus of fault activity has jumped eastward from the Silver Fork-Superior Fault to the Wasatch Fault beginning between 10 and 12 Ma. In other words, the Wasatch Fault began activity 10 to 12 million years ago, implying that uplift of the present-day Wasatch Mountains did not begin until 10 or 12 million years ago. The timing is reasonably consistent with a less than 10-million-year age for most of the Basin and Range landscape.

A fluid inclusion study of hydrothermally altered rocks was also conducted at the base of the Salt Lake City segment where the 1.6 Ma apatite (U-Th)/He age was determined. This type of study is a direct measure of the amount of pressure the rock was under when the fluid inclusion was trapped. The study determined that the base of the western face of the mountain had been exhumed from a depth of 36,080 feet (11 km). This equates to an

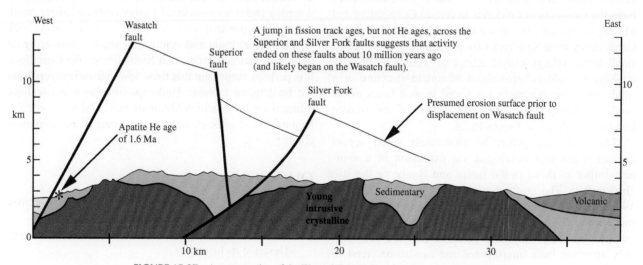

FIGURE 15.27   A cross-section of the Wasatch Range. Based on Perry and Bruhn (1987).

exhumation rate over the 10- to 12-million-year period of 4.33 to 3.61 inches per 100 yrs. These rates would be lower if some of the exhumation occurred prior to 12 Ma. Nevertheless, the rate agrees rather nicely with the exhumation estimate of 4 inches per 100 yrs deduced from the 1.6 Ma apatite (U-Th)/He age obtained in the same area. Figure 15.27 shows the amount of rock estimated to have been eroded from the southern Salt Lake City segment. Note the decrease in both total uplift and erosion toward the east.

The Wasatch Fault offers a natural laboratory to study the geometry and uplift/exhumation history along a well-exposed, active normal fault. We are reasonably sure that the Wasatch Fault began activity between about 10 and 12 million years ago, and we can guess that most of the 7,000-plus feet of relief between mountaintop and valley floor were generated during that time. This equates to development of relief at a rate of 0.70 to 0.84 inches per 100 yrs over the 10- to 12-million-year period. We know that the Wasatch Fault is active, but we do not know if the mountain range has reached a steady state. Therefore, we do not know if the mountain will continue to gain relief or if it has reached its maximum height.

TRIASSIC LOWLANDS OF THE APPALACHIAN MOUNTAINS

Nestled within crystalline rock of the Piedmont Plateau and New England Highlands are narrow, isolated lowland valleys underlain by red sandstone, basaltic lava flows, and shallow gabbroic intrusions (dikes and sills). These rocks, particularly the sandstone, are not nearly as resistant as the surrounding crystalline rock and thus, they tend to form discrete lowlands in which the red color of the sandstone contrasts sharply with the surrounding lighter-colored crystalline rock. Lowland valleys cored with these rocks extend from Massachusetts to North Carolina. They form the Connecticut Lowlands in central Connecticut and Massachusetts, and the Newark Basin which parallels the Great Valley from New York City to Virginia. Farther south, small linear valleys extend across the Piedmont Plateau, crossing the older Appalachian Mountain structure at a small angle. Similar rocks are found in well holes as far south as Florida. The surface distribution of the Triassic Lowlands is shown in Figure 15.28.

The rocks are gently to moderately tilted across normal faults that developed via extension in a manner similar to those in the Basin and Range or the Rio Grande Rift. The faults, however, are between 230 and 190 million years old and have not been active since that time. Consequently, they produce an erosional topography. Basaltic rocks, because they are considerably stronger than interlayered red sandstone, tend to form high hills and cliffs within the lowland valleys. Figure 15.29 is a schematic cross-section that shows the relationships. Figure 15.30 is a Google Earth image looking eastward across the Connecticut Valley from Long Island Sound northward well into Massachusetts. The strong crystalline walls of the valley are clearly evident in the figure. Note the long, nearly continuous ridge that extends for more than 75 miles from just north of New Haven, Connecticut, to well north of Springfield, Massachusetts. This is a tilted ridge of basalt known as the Metacomet Ridge. Other tilted basalt cliffs include the Palisades along the western bank of the Hudson River in New Jersey, and the nearby Watchung Mountains also in New Jersey. Basalt flows in the Watchung Mountains have a total thickness of approximately 1,850 feet.

The Connecticut Valley is also host to about 2,000 dinosaur tracks that range from 10 to 16 inches in length and are spaced 3.5 to 4.5 feet apart. The tracks are about 200 million years old and are named Eubrontes. They are believed to have been left by the carnivorous dinosaur *Dilophosaurus* although no fossil remains are present at the site, located at Dinosaur State Park in Rocky Hill, just south of Hartford. This is one of the largest dinosaur track sites in North America.

In Chapter 8 we noted that the Appalachian-Ouachita Mountains formed some 265 million years ago as a result of continental collision between North America, Europe, Africa, and South America, and that this collision produced the supercontinent Pangea (Figure 8.5). It was the rifting of Pangea and the separation of North America from Africa and Europe beginning about 230 million years ago that eventually produced the Atlantic Ocean. The Triassic Lowlands are a remnant of this ancient rifting event. The Lowlands represent narrow down-dropped basins (grabens) flanked by normal faults, much like the Rio Grande Rift. The Triassic Lowland rifts developed too far inland to go to completion. Instead, the eventual split between North America and Africa occurred farther eastward along much larger rift basins that are now buried beneath younger rock of the Coastal Plain and continental shelf. Considering all the recent tectonic events that have affected the Cordillera, it is perhaps surprising that these lowland valleys represent the final major tectonic landscape-forming event to have affected the Appalachian Mountain belt. And this particular tectonic event affected only a small part of the mountain chain.

QUESTIONS

1. Draw a simple cross-section that shows horst-graben and half-graben geologic structures.
2. Where is the Rocky Mountain Basin and Range? Describe its history.

FIGURE 15.28 A landscape map of the eastern US showing the location of Triassic Lowland valleys.

3. How long ago was the earliest known Basin and Range normal faulting, and where did it occur?

4. What is the Northern Nevada Rift zone, and why does its history suggest that present-day Basin and Range landscape developed less than 17 Ma?

5. Locate and describe several areas in Figure 15.3, 15.4, or 15.5 where there is a sharp boundary between the Basin and Range province and an adjacent physiographic province. Also locate several areas where the boundary is diffuse and poorly defined. Speculate as to why the boundary is sharp in one area and diffuse in other areas.

6. Different mountains are at different stages of erosion and self-burial, depending on the recentness of activity along range-front faults. What other factors control the rate of self-burial?

7. In Google Earth, fly to Mt. Moses, Nevada. Comment on the stage of erosion and self-burial of the range front.

8. Use Figure 15.6 as a guide and draw a cross-section across the area shown in Figure 15.17, from the Tobin Range to the Ruby Range.

9. In Google Earth, fly to Mt. Tobin, Nevada. The 1915 Pleasant Valley Fault scarp is located along the range front 8 miles southeast of the peak. Compare the morphology of the eastern faulted side of Pleasant Valley with the western side. Is there evidence for an active fault along the western side of the valley?

10. In Google Earth, fly to to the Toiyabe Range, Nevada. The Big Smokey Valley lies just south of the peak. Compare the morphology of the eastern side of the valley with the western side in terms of recentness of faulting.

11. Speculate on the parallelism of the Central Nevada Rift and the Walker Lane belt evident in Figure 15.3. Are the two in any way related?

12. Describe the landscape expression of the Rio Grande Rift in Figure 15.5. How is the landscape distinctive from the surrounding area?

13. In Google Earth, fly to Dixie Valley, Nevada. This area is just southwest of the Tobin Range. Compare the morphology of the mountain fronts on both sides of the valley. On which side would you expect to find more recent faulting, and why? Zoom down to the mountain front, where you might expect more recent faulting, and look for the 1954 Dixie Valley Fault scarp. Describe the morphology of this scarp.

14. In Google Earth, fly to Borah Peak, Idaho, in the Rocky Mountain Basin and Range. Examine the morphology of mountain fronts on both sides of the valley that lies to the west of the peak. On which side would

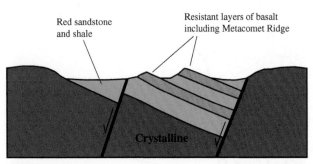

FIGURE 15.29   A schematic cross-section of the Triassic Lowlands. Based on Hunt (1974).

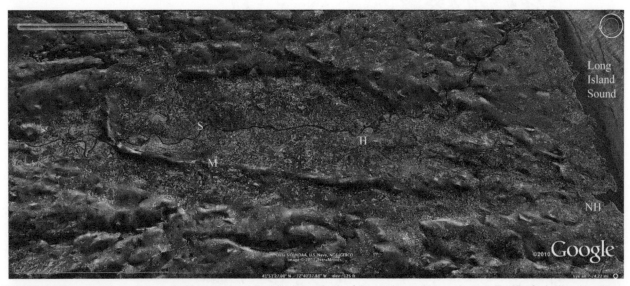

FIGURE 15.30   A Google Earth image looking east across the Triassic Lowlands of Connecticut and Massachusetts. *H* = Hartford, *NH* = New Haven, *S* = Springfield. The crystalline margins of the lowland basin are evident. A tilted layer of resistant basalt extends for more than 75 miles through the valley to form Metacomet Ridge (M). Several state attractions, including Sleeping Giant and Hanging Hills north of New Haven and Mt. Tom north of Springfield, are present along the ridge. The large lake at upper left is the Quabbin Reservoir. The smaller lake at lower center is the Barkhamsted Reservoir. Both reservoirs are situated within crystalline rock. The Connecticut River flows through the lowlands, then turns east into crystalline rock.

you expect to find more recent faulting, and why? This is the site of the 1983 Borah Peak earthquake. Zoom down to the mountain front, where you might expect to see evidence of recent faulting, and look for the Lost River Fault scarp. This one is more difficult to find than the Dixie Valley scarp. Describe the morphology of the scarp. *Note:* Go to the USGS Website given in Figure 15.14 to more accurately locate these scarps. Now fly to Hebgan Lake, Gallatin National Forest, Montana, site of the 1959 Hebgan Lake earthquake rupture. It is very difficult to locate this fault scarp without help from the USGS Website. Don't confuse the earthquake scarp with the road that runs along the lake. Look for a highly visible scar just below the summit of Kirkwood

Ridge. Explain the location of the scar so high on the mountain.

15. Explain why the Rocky Mountain erosion surface discussed in Chapter 11 is or is not a pediment surface.
16. What is the Nevadaplano?
17. Describe the earthquake history of the Teton Range.
18. What evidence suggests that activity on the Wasatch Fault did not begin until about 10 or 12 million years ago?
19. What is the Metacomet Ridge? Where is it located? How did it form?
20. From a tectonic plate movement point of view, explain why it is not too surprising that development of the Triassic Lowlands was the last major tectonic event to affect the Appalachian Mountains.

Cascadia Volcanic Arc System

The North Cascades, the Central-Southern Cascades, the Washington-Oregon Coast Range and Valleys, the Olympic Mountains, and the Klamath Mountains together constitute the Cascadia volcanic arc system where the Juan de Fuca plate is actively subducting beneath the North American plate. The structure is complex and includes folds as well as thrust, normal, and strike-slip faults, many of which are active. The overall topography depicted in Figure 16.1 is one of coastal mountains, an inland valley, and a tall, volcanic, mountain range. This is a classic subduction-related tectonic (structure-controlled) landscape in which accretion has caused uplift along the coast, subduction has caused volcanism in the Cascade Mountains, and the intervening Puget Sound/Willamette Valley forms the forearc basin. The relationship between topography and tectonics is shown in Figure 16.2 which is a cross-section across central Oregon. If you are unfamiliar with the tectonics of subduction zones, now would be a good time to review Chapter 5.

A narrow, linear trench less than 100 miles off the Northern California-Oregon-Washington coastline marks the location where the Juan de Fuca plate is subducting beneath the North American plate as shown in Figure 16.3. The subduction zone is known as the *Cascadia Trench*. However, it does not have the topographic expression of a deep linear valley. There are two factors that explain the absence of a topographic trench. The subducting oceanic lithosphere is between 26 and 2 million years old and therefore is young, thin, warm, and isostatically buoyant. Seismic modeling suggests that the plate is subducting at an initial angle of 10 to 15 degrees. Apparently it does not sink very easily and therefore, does not create a large indentation (that is, a trench). Second, the Willapa, Columbia, Umpqua, Rogue, and other rivers contribute copious amounts of sediment to the coastline, which fills the shallow trench.

It was once thought that the absence of a trench, coupled with the absence of destructive historic earthquakes, indicated that the Juan de Fuca plate was too warm and ductile to generate a great earthquake as it slipped below North America. Instead of snapping and fracturing, it was thought that the subducting plate was deforming like warm wax. This assumption has since been proven false. Key evidence for large destructive earthquakes was the discovery of suddenly drowned marshes and tree stands. Analysis of recently deposited sediment suggests as many as 12 powerful subduction-related earthquakes in the past 7,700 years, or about one every 642 years. The last major earthquake occurred more than 300 years ago on January 26, 1700, but smaller earthquakes, some large enough to be felt, occur on average every five years or so. An earthquake of magnitude 6.8 occurred about 11 miles below Olympia, Washington on February 28, 2001. It was powerful enough to destroy buildings and cause hundreds of injuries. Even this strong earthquake is rather small compared to what is now expected based on new research.

The explosive volcanic mountains of the Cascade Range present another potential threat to surrounding communities. The calamity that a large volcanic eruption would cause is obvious. Less obvious is the potential damage of small eruptions, landslides, and avalanches. The problem with these is that they could result in down-valley mudflows capable of destroying towns and cities that surround Puget Sound. Mt. Rainier is probably the most dangerous volcano in this respect because of the relatively large population in its lowland drainages. Landslides and avalanches off the slopes of Mt. Rainier have produced at least seven large mudflows in the past 5,600 years, including the Osceola Mudflow 5,600 years ago and the Electron Mudflow only 560 years ago. The Osceola Mudflow was by far the largest, extending all the way to the Seattle suburbs.

Landscape Evolution in the United States. http://dx.doi.org/10.1016/B978-0-12-397799-1.00016-6

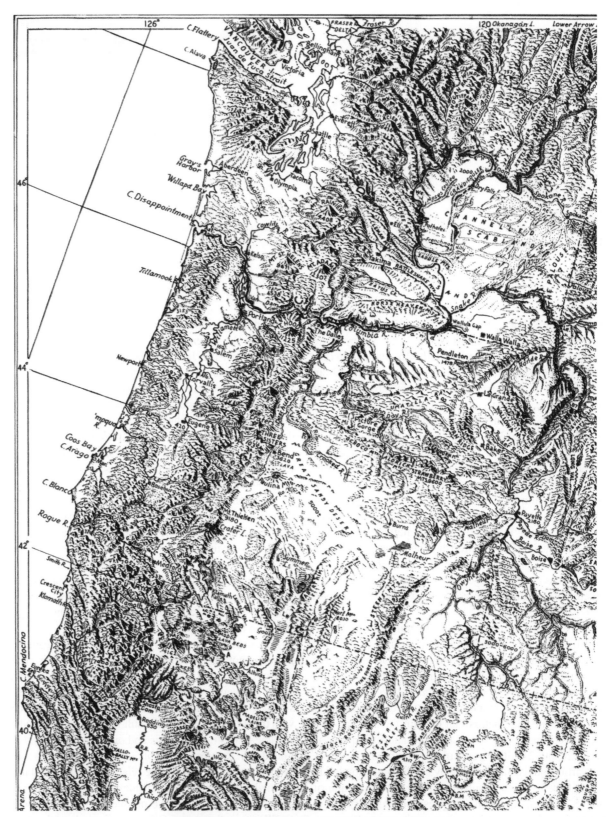

FIGURE 16.1   A landscape map of the Cascadia Arc System.

The volcanic arc is a direct consequence of subduction of the Juan de Fuca plate. It is not surprising, therefore, that the northern terminus of the volcanic arc at Mt. Meager in British Columbia, and the southern terminus at Mount Lassen in northern California, correspond with the northern and southern termini of the Juan de Fuca plate. However, it is important

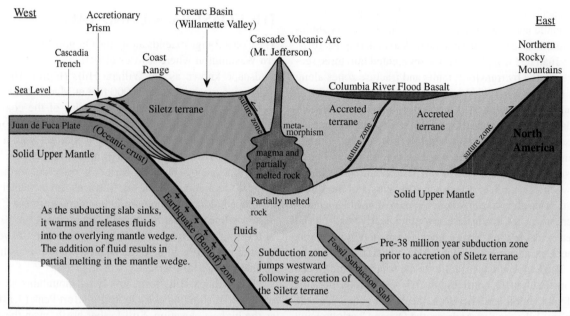

FIGURE 16.2 A cross-Section of the Cascadia Subduction Complex at Mt. Jefferson, Central Oregon.

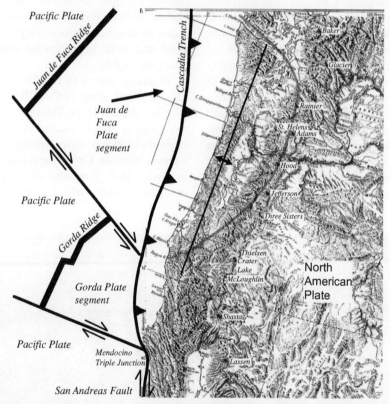

FIGURE 16.3 A map of the Juan de Fuca plate with major Cascade volcanoes labeled. The arrow shows the direction of convergence at the trench. The thick curved line that encloses Eugene and Seattle shows the inferred shape of the embayed coastline prior to accretion of the Siletz terrane as well as the approximate area currently underlain by the Siletz terrane (based on Parsons et al., 2005). Others have shown the Siletz terrane extending farther east below part of the Columbia River Plateau. The heavy line with double arrow along the Coast Range follows the crest of an anticlinal flexure.

to understand that direct melting of the subducting Juan de Fuca plate is not the cause of most of the volcanism in the Cascade Range. Instead, it is the solid upper mantle directly above the subducting plate that is melting. Although shallow at the trench, the angle of the subducting plate steepens to between 60 and 80 degrees within about 200 miles inland where it reaches a depth of 62 miles. Beyond this depth, the subducting plate becomes hot enough to undergo metamorphic reactions.

These reactions release fluids (H_2O and CO_2) into the solid upper mantle above the subducting plate that disturb equilibrium conditions, causing the mantle rock to melt (Figure 16.2).

The Juan de Fuca plate itself is segmented into three sections separated by transform faults and fracture zones along which the spreading ridge is offset. The Gorda plate forms the southern segment. It is currently subducting beneath the Klamath Mountains (Figure 16.3). The main Juan de Fuca segment is subducting beneath most of Oregon and Washington, and the Explorer segment (not shown in Figure 16.3) subducts beneath Vancouver. The rate of convergence is between 13.1 and 14.8 feet per 100 years and the direction is east-northeastward (about 69° east of north) such that northern areas in Canada are converging nearly head-on and southern areas in the US are converging at an oblique angle.

The Cascadia trench ends along transform strike-slip faults. The San Andreas Fault lies to the south and the Queen Charlotte Fault to the north (Figure 5.5). Both faults mark the end of subduction and, therefore, the end of active Cascade volcanism. Like the San Andreas, the Queen Charlotte Fault allows the Pacific plate to move northward relative to the North American plate. The Queen Charlotte Fault itself ends at the southern coastline of Alaska where the Pacific plate is subducted beneath Alaska to produce volcanoes and earthquakes along the Alaskan coast and the Aleutian Islands.

In this chapter we describe landscape that is directly attributable to subduction of the Juan de Fuca plate. Specifically, we will discuss the Oregon-Washington coastal area, the Olympic Mountains, the central-southern Cascades, and the Klamath Mountains. The North Cascades boast two volcanic cones, Mt. Baker and Glacier Peak. However, the landscape and geology of the North Cascades as a whole is best described as a continuation of crystalline Rocky Mountain geology and landscape (Chapter 13).

THE COAST RANGE AND VALLEYS

The Coast Range extends along the Oregon coast into southern Washington where it loses elevation to form a subdued landscape known as the Willapa Hills (Figure 16.1). The mountains reach just over 4,000 feet at Mary's Peak (4,097 feet) west of Corvallis, within 20 miles of the coast. The Oregon Coast Range shows many characteristics indicative of active uplift. The coastline is straight with a series of slopes and benches that step inland, not as bench-and-slope topography, but as a series of wave-cut terraces up to 1,600 feet above sea level. Terraces are especially well developed along the southern Oregon and northern California coastlines. Present-day uplift rates based on 50 years of leveling are between 0.3 and 3.0 inches per 100 years in most areas, although at Cape Blanco the rate is as high as 35 inches per 100 years.

The Willapa Hills form low-lying mountains along the southern Washington coast, with Boistfort Peak (3,110 feet) as the highest elevation. Still farther north, even these hills disappear into an area of low relief traversed by rivers draining the Olympic Mountains. Inland bays such as Grey's Harbor, Willapa Bay, and the wide Columbia River outlet create an irregular coastline very different from the Oregon coast and more suggestive of subsidence and drowning rather than uplift.

A unique aspect of the Oregon coast is the presence of sand dunes north of Coos Bay that extend for at least 50 miles (Figure 16.1). The sand is brought to the coast by rivers and by coastal erosion. As the sand moves along the coast via longshore drift (Figure 9.5), it becomes trapped in a recess between Cape Arago and Heceta Head as shown in Figure 16.4. The sand piles up to create sand dunes that move onshore with the prevailing wind. The shifting

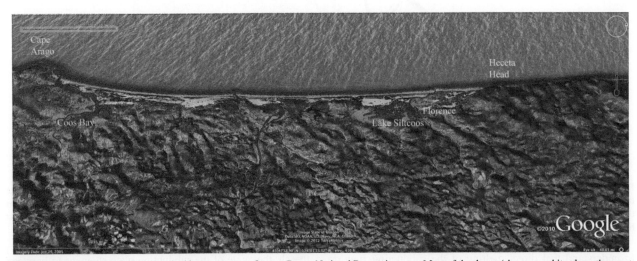

FIGURE 16.4 A Google Earth image looking west across Oregon Dunes National Recreation area. Most of the dunes (shown as white along the coastline) occur where sand is trapped in a recess between headlands at Cape Arago and Heceta Head. The dunes block inland streams, creating Lake Siltcoos (labeled) and Lake Tahkenitch, located just south of Lake Siltcoos. The cliffs at Heceta Head are composed of Siletz terrane basalt.

dunes have blocked small streams from reaching the coast, thereby creating a chain of freshwater coastal lakes that include Lake Siltcoos and Lake Tahkenitch. Smaller areas of coastal dunes are present farther south near Bandon and to the north on both sides of the Columbia River estuary.

Most of the rocks that form the Coast Range in both Oregon and Washington are part of a single piece of accreted material known by several names that include the Coast Range terrane, the Peripheral rocks, the Crescent terrane (in the Olympic Mountains), and the Siletz terrane (in Oregon). For simplicity, it will be referred to here as the *Siletz terrane*. The rocks are more than 65,000 feet thick in the east, thinning to less than 10,000 feet toward the west. They consist of basalt and overlying clastic sedimentary rock deposited between 64 and 38 million years ago as part of a series of volcanic islands and plateaus (seamounts) located on the Juan de Fuca plate offshore from the Washington-Oregon coast. This terrane was transported to the edge of North America on the Juan de Fuca plate and accreted by about 38 million years ago. Following accretion, the subduction zone jumped westward about 50 miles to its present location as depicted in Figure 16.2. You may wish to review Figure 5.7, which shows the process of accretion. Today, the Siletz terrane is well exposed along the length of the Coast Range where it forms the oldest rock. It is buried beneath forearc basin sediment in the Puget Sound-Willamette Valley, and beneath volcanic rock in the Cascade Mountains farther east (Figure 16.2). This was the most recent accretion event to the western US and it changed the shape of the coastline from an embayed outline prior to accretion, to the present-day straight coastline. The inferred extent of the Siletz terrane, and the shape of the coastline prior to accretion, are shown in Figure 16.3.

Following the westward jump in subduction, a new accretionary prism developed via off-scraping and underplating along the western edge of the newly enlarged North American plate, and by 36 million years ago, the modern-day Cascade volcanic arc was taking shape. The associated accretionary prism (known as the *Cascadia Subduction Complex*) includes all the rocks above the subducting Juan de Fuca plate and below the Siletz terrane as shown in Figure 16.5a. The rocks reach a maximum thickness of more than 100,000 feet in the east and taper westward toward the offshore trench. Geophysical studies suggest that nearly the entire 100,000 feet of accreted material in the accretionary prism consists of sedimentary rock scraped from the top of the subducting Juan de Fuca plate. Virtually none of the underlying oceanic basaltic crust has been accreted.

Accreted material added to the underside of the Siletz terrane is one of at least two forces that drive coastal uplift and create some of the previously mentioned wave-cut terraces. The mechanism, known as *underplating*, is outlined schematically in Figure 5.8. Note in Figure 5.8 that wedging of material in the subduction zone results in rotation and landward tilting of overlying rock. This, in turn, causes uplift near the coastline and simultaneous subsidence of inland valley regions. It is safe to conclude, therefore, that areas of active uplift along the coast, such as the southern Oregon Coast Range, are likely associated with recent or active underplating in the subduction zone. Areas that are subsiding or where uplift is occurring at a slow rate, such as the central Washington coastline, may be associated with a part of the subduction zone where underplating has not occurred recently or is not actively occurring. Uplift in the Oregon Coast Range has created a broad anticline that roughly follows the topographic crest of the range (Figure 16.3).

A second factor that causes uplift and subsidence in the Coast Range is the subduction of seamounts carried to the subduction zone on the Juan de Fuca plate. The presence of a seamount (an extinct volcanic area) creates a bulge within an otherwise relatively thin down-going plate. These particular seamounts are smaller than the accreted Siletz seamount terrane and therefore are not buoyant enough to accrete. Instead they are subducted with the down-going plate. As the seamount bulge subducts below the Coast Range, it causes uplift. An interesting aspect to this is that, as the seamount sinks further into the mantle beyond the Coast Range, the Coast Range will subside with its passage. The process brings to mind a snake swallowing a large object such that it creates a bulge as it passes through the snake.

Rocks of the Siletz terrane are the oldest and most widespread in the Coast Range. However, they are not the only rocks in the Coast Range. Also present are large bodies of 30 million-year-old gabbro (an intrusive rock compositionally equivalent with basalt) that may be related to early phases of Cascade volcanism. A gabbro intrusion 700 feet thick caps Mary's Peak. Saddle Mountain (3,283 feet), located northwest of Portland, is slightly different from other high peaks in the Coast Range in that it is composed of young (about 15 My old), highly fractured, Columbia River plateau flood basalt (Chapter 14). These lavas flowed westward more than 200 miles from eastern Oregon and Washington, via the Columbia River channel, to the coastline where they froze and fractured upon entering the ocean. The youngest basaltic rocks in the Coast Range and Valleys are a series of volcanic cinder cones and vents in the Willamette Valley of Oregon and southern Washington known as the Boring Lava Field. At least 74 vents have been named. Most are less than 3 million years old. Three of the cones, including the 1.2-million-year-old Rocky Butte, lie within Portland city limits. Figure 16.6 shows all three cones within Portland City limits as well as a part of the volcanic field southeast of Portland. These young volcanoes are somewhat unusual considering their location in the forearc basin. They are likely related to the Cascade volcanic arc.

(a) Coast Range

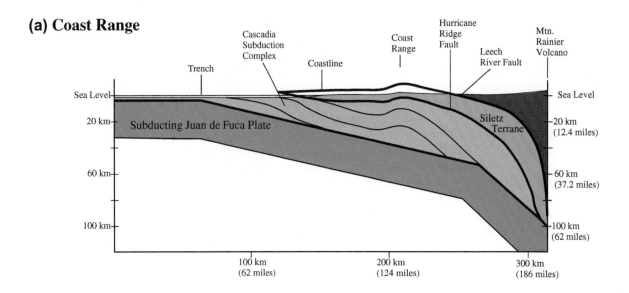

(b) Olympic Mountains

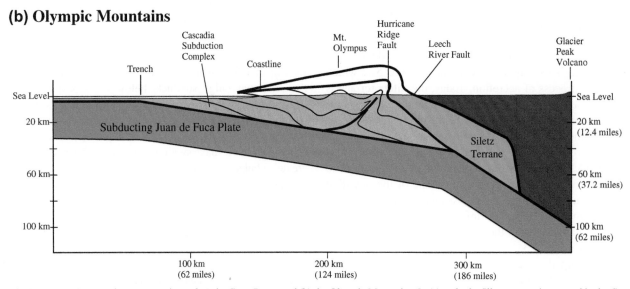

FIGURE 16.5 Comparative cross-sections of (a) the Coast Range and (b) the Olympic Mountains. In (a), only the Siletz terrane is exposed in the Coast Range where the depth to the top of the Juan de Fuca plate is about 25 miles based on the location of earthquakes. In (b), the Cascadia subduction complex is exposed at the surface in the Olympic Mountains due to greater deformation and uplift. Depth to the top of the Juan de Fuca plate is only about 18 miles below Mt. Olympus, which suggests that the Juan de Fuca plate is arched upward below Mt. Olympus. The distance from Mt. Olympus to Glacier Peak is 40 miles longer than from the Coast Range to Mt. Rainier. This suggests that the angle of subduction is steeper below the Coast Range than below Mt. Olympus. Based on Brandon and Calderwood (1990).

THE OLYMPIC MOUNTAINS

The two highest areas along the coast are the Klamath Mountains and the Olympic Mountains. Both are young, active mountain ranges that have undergone recent uplift of more than 7,000 feet. The Olympic Mountains form the Olympic Peninsula in the northwest corner of Washington where Mt. Olympus (7,965 feet) and other high peaks tower over the coastline. A landscape map with simplified geology is shown in Figure 16.7. Known for its temperate rain forests, the western side of the range receives as much as 150 inches of precipitation per year. Although not especially

high, the Olympic Mountains are far enough north to be home to more than 200 glaciers, some of which are evident in the Google Earth image shown in Figure 16.8.

Notice in Figure 16.2 that rocks of the accretionary prism underlie the Siletz terrane. Although the Oregon-Washington Coast Range is arched into an anticline, the amount of uplift and erosion associated with development of the anticline has not been sufficient to expose the accretionary prism. This is not the case in the Olympic Mountains where the Siletz terrane (known here as the *Crescent terrane*) forms a crescent or horseshoe shape that encloses exposed rocks of the accretionary prism. Figure 16.7 shows

FIGURE 16.6 A Google Earth image looking southeastward across the Boring Lava field at Portland, Oregon. Rocky Butte, Kelly Butte, and Mt. Tabor are all within Portland city limits.

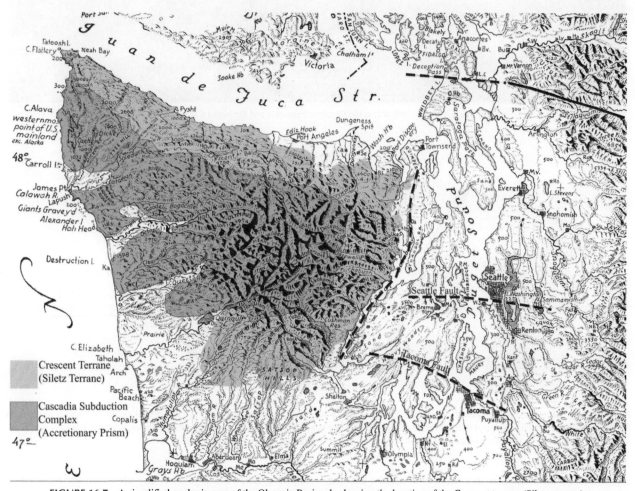

FIGURE 16.7 A simplified geologic map of the Olympic Peninsula showing the location of the Crescent terrane (Siletz terrane).

the distribution of both the Siletz terrane and accretionary prism across the Olympic Peninsula. Figure 16.5 compares a cross-section of the Olympic Peninsula with that of the Coast Range.

It is clear from their size and elevation, and from the presence of deep-seated accretionary prism rocks, that the Olympic Mountains have undergone a greater amount of uplift and erosion than the Coast Range. An interesting

FIGURE 16.8   A Google Earth image looking north across the Olympic Peninsula and the North Cascades. *B* = Mt. Baker, *G* = Glacier Peak.

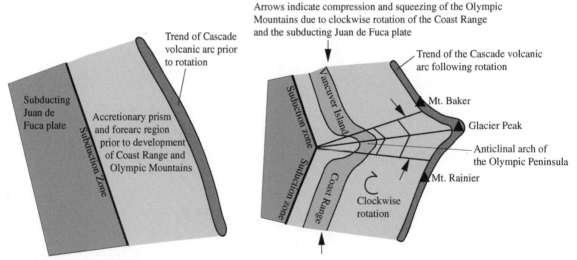

FIGURE 16.9   Sketches that show how the Basin and Range extension beginning 16–18 million years ago has caused clockwise rotation of the Coast Range, compression, and development of an anticlinal arch in the Juan de Fuca plate below the Olympic Mountains. The arch creates a shallow subduction angle such that Glacier Peak is farther inland than other volcanoes. Based on Brandon and Calderwood (1990).

aspect to this conclusion is that there is evidence that the Olympic Mountains have been in a steady state for the past 14 million years. Recall from Chapter 4 that a steady-state mountain is one where the rate of uplift is approximately equal to the rate of erosion such that, in this case, even though the mountain has undergone uplift, it has remained

approximately at its present day elevation for the past 14 million years.

One theory for the origin of the Olympic Mountains has to do with changes in the curvature of the subduction zone. Prior to Basin and Range extension, the subduction zone between the US and Canada was relatively straight. Figure 16.9 suggests that beginning about 16 to 18 million years ago, extension in the Basin and Range province forced a clockwise northward rotation of both the southern part of the Coast Range and the subduction zone. The rotation caught the Olympic area in a sort of vice grip between the rotating Coast Range and non-rotating Vancouver Island. The result was compression in the Olympic area, which caused the Juan de Fuca plate to arch upward, thus creating the Olympic Mountains.

For this interpretation to be valid, the Juan de Fuca plate below the Olympic Mountains should be at a shallow depth (that is, it should be arched upward) relative to areas in the north and south. Two lines of evidence support this assertion. The first is the present-day depth of subduction-related earthquakes (the Benioff zone), which is shallow below the Olympic Mountains relative to areas in the north and south. The second is the location of subduction-related volcanoes. Note in Figures 16.3, 16.5, and 16.8 that Glacier Peak is further inland than either Mt. Baker to the north or Mt. Rainier to the south. A subducting slab typically must reach depths greater than 62 miles to become hot enough to initiate melting and volcanism. The location of Glacier Peak farther inland, coupled with the presence of a relatively shallow Benioff zone, are both consistent with a more shallow angle of subduction below Mt. Olympus. The Oregon Coast Range continues to rotate clockwise at a rate of 1.5° per million years, perhaps maintaining the steady-state Olympic Mountains.

The uplift-erosion history of the Olympic Mountains has been intensely studied using a number of different methods that include field data, apatite and zircon fission track ages, apatite (U-Th)/He ages, and river incision rates. These data provide exhumation rates averaged over time scales that range from 14 million years to about 100,000 years. In addition, there are leveling and tide-gauge studies that provide present-day uplift rates. A significant conclusion is that exhumation rates are roughly the same regardless of whether rates are averaged over a 14-million-year period or over 100,000 years. This suggests that uplift and erosion are balanced and that the mountain has been in a steady state such that, on average, it has not grown taller nor decreased in elevation for the past 14 million years. We noted earlier that one of the dominant driving forces for uplift is under-plating of subducted material against the base of the Cascadia subduction complex. A steady-state mountain, in this case, not only implies that uplift is balanced by erosion, but also that the amount of material added to the underside of the Olympic Mountains via accretion and underplating over a given time interval is roughly equal to the amount of material eroded from the mountain mass.

If we assume that exhumation is tied to uplift, then the rate of both uplift and exhumation should be highest near the central topographic high and should decrease toward the coastline. In other words, exhumation rates should mimic the present-day topography. Modeling of fission track and

(U-Th)/He ages coupled with depositional ages supports this contention. Calculated erosional exhumation rates over the past 14 million years based on fission track and (U-Th)/He ages mimic present-day topography in the sense that the highest rates are in the central, highest, part of the peninsula, and the lowest rates are close to the shoreline. Additionally, the presence of 5 to 12 million-year-old shallow-marine sedimentary deposits along the shoreline indicates that coastal areas have not been elevated significantly above sea level during the past 5 to 12 million years.

Zircon and apatite fission-track modeling, along with geological field data, suggest that initial emergence of the Olympic Peninsula from below sea level occurred approximately 18 million years ago and that the mountain had reached a steady state by 14 Ma. This implies an uplift rate of 2.4 to 3.2 in/100 yrs over the initial 3 to 4 million-year period in order for the mountain to reach its present elevation by 14 million years ago. Zircon and apatite fission track data from the central elevated part of the mountain suggest an average exhumation rate of 3.2 inches per 100 yrs between 14 and 7 million years ago, and a rate of 2.6 inches per 100 yrs from 7 Ma to the present. Apatite (U-Th)/He ages averaged over the past 2 to 3 million years are consistent with the fission-track data. These data suggest that exhumation rates have been relatively constant and steady over the entire 18 My history of mountain building. The rather rapid exhumation rates in the central part of the mountain decrease sharply toward the shoreline to rates that are considerably less than 1 inch per 100 yrs. Integration of all exhumation rates across the entire peninsula suggests that erosion (exhumation) has been occurring at an overall rate of about 1 inch per 100 yrs over the past 7 million years.

Recent and present-day rates of erosion can be compared with the previously stated long-term exhumation rates to see if they support the idea that the Olympic Mountains are in a topographic steady state. Ideally, recent and present-day erosion rates should match long-term exhumation rates. Incision rates on the Clearwater River, which drains the western side of the Olympic Mountains, have been estimated over a 140,000-year period by dating river terraces as outlined in Figure 4.3. The results suggest incision rates of 3.5 inches per 100 yrs in the central mountain area, decreasing to less than 0.39 inches per 100 yrs at the coast—rates that are consistent with long-term exhumation rates. Present-day denudation rates have been calculated for the Hoh River system, which drains the west side of the mountain just north of the Clearwater River, and for the Elwha River system, which drains the north side of the mountain. In both cases, the estimates were calculated by dividing the amount of sediment leaving the drainage basin by the area of the drainage basin as outlined in Figure 4.2. The calculated denudation rates are 1.26 inches per 100 yrs for the Hoh River, and 0.71 inches per 100 yrs for the Elwha River drainage basins. These rates are consistent with an overall exhumation rate of 1 inch per 100 yrs calculated for the entire peninsula over the past 7 million years. Based on this evidence, we can conclude that recent and present-day rates of erosion are consistent with the idea that the Olympic Mountains are in a steady state.

Leveling and tide-gauge data over the past 60 years suggest that the entire Olympic Peninsula is currently uplifting at a rate greater than 3.15 inches per 100 yrs, with rates between 7.87 and 9.45 inches per 100 yrs in the vicinity of Mt. Olympus. The highest rates, however, are along the northwestern coastline, which is rising at a maximum rate greater than 12.6 inches per 100 yrs. The western side of Puget Sound, by contrast, is rising at rates of less than 3.15 inches per 100 yrs, and the eastern side of Puget Sound appears to be subsiding by an equal amount. The leveling and tide-gauge data are higher and in conflict with long-term exhumation rates, especially along the coastline, which shows minimal net long-term exhumation. The explanation is that the rather high uplift rates are transient in nature and not indicative of long-term rates. They are thought to result from elastic flexure of the subducting plate (the storing of energy) prior to a large earthquake. Recall earlier that one line of evidence for a great, prehistoric earthquake is a sudden drop in elevation, particularly along the shoreline, where one can find evidence for the sudden drowning of marshes and low-lying areas. The idea is that the subducting plate is initially locked such that as stress builds, the plate flexes upward. Stress is relieved when a large earthquake occurs, at which time the plate then rebounds to its original shape, thereby causing the ground to rapidly subside. The idea is shown schematically in Figure 16.10. The effect of flexure, earthquake, and rebound over the long term, is little or no net uplift along the coast. Here we see an example in which present-day uplift rates are not necessarily indicative of long-term rates. It points out the perils of extrapolating present-day rates back in time without cor-

roborating evidence and, at the same time, suggests that stress is accumulating for a potentially large earthquake.

From this discussion we can conclude that increased compression in the Olympic area has created less room to accommodate the total volume of accreted rock above an arched and more gently dipping subducting plate. This situation has resulted in earlier and more rapid uplift of the Olympic Mountains relative to surrounding areas. Field geology and exhumation data imply that the first significant land area to emerge from below sea level along the Washington-Oregon coast was the central part of the Olympic Peninsula beginning about 18 million years ago. Uplift progressed slowly northward and southward as the Olympic Mountains grew higher. It is estimated, based on sediment ages, that coastal areas of the Olympic Peninsula first emerged from below sea level about 12 million years ago and that the Oregon Coast Range first began to emerge between 10 and 5 million years ago. If we assume an average exhumation rate between 3.2 and 2.6 inches per 100 yrs, the total thickness of rock removed from the top of the Olympic Peninsula since 18 million years ago is about 7.5 miles. This is considerably more than the estimate for the Coast Range, which is about 1.9 miles. The Coast Range has not yet reached a steady state. However, if tectonic and climatic conditions do not change appreciably in the next several million years, it is possible that the Coast Range will follow a similar but slower uplift history as that followed by the Olympic Peninsula, and that the Coast Range will eventually reach a steady state. The slower uplift rate, however, implies that steady-state topography will likely be lower than the elevation gained in the Olympic Mountains.

THE KLAMATH MOUNTAINS

The Klamath Mountains form a 150 by 100 mile welt along the Oregon-California coast that separates the Oregon Coast Range and the Willamette Valley from the California Coast Range and the Central Valley. The mountains form part of the Cascade-Klamath-Sierra Nevada rain shadow. They are wet on their western side and relatively dry on the east. The eastern side of the mountain range ends in a valley that separates the Klamath Mountains from Mt. Shasta of the Cascade Range (Figure 16.1). The Klamaths are a geologically complex region composed of thrust-faulted and folded sedimentary and metamorphic rock intruded by plutonic igneous rocks, partly covered by younger sedimentary rock, and partly covered by volcanic rock associated with the Cascade volcanoes. Most of the rock consists of accreted terrane oceanic material including rocks that formed in an island volcanic arc environment. Rocks similar to those in the Klamaths are present at the northern end of the Sierra Nevada and in the Blue Mountain region of Oregon.

The topography of the Klamath Mountains is one of a tectonically young, uplifted, heavily dissected plateau

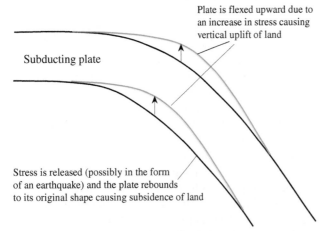

FIGURE 16.10 A sketch that shows upward flexure of a subducting slab followed by earthquake-induced pressure release and subsidence.

surface with isolated, locally glaciated peaks that approach 9,000 feet. The mountains are traversed by swift-flowing, deeply incised rivers including the Klamath, Rogue, Trinity, and Smith, that have cut canyons 1,500 to 2,500 feet deep, especially in the western Klamaths. Figure 16.11 is a Google Earth image looking southeastward across the range. Note the incised river valleys, the flat-topped mountains, and the wave-cut terraces and escarpments along the coast.

The uplifted plateau surface is known as the *Klamath peneplain*. On the basis of fieldwork, it is understood to have existed as a flat erosion surface at or below sea level as late as 5 million years ago. There have been between 6,000 and 8,000 feet of uplift since that time, with most of the uplift occurring in the past 2 million years at an average rate between 1.9 and 4.8 inches per 100 yrs. Present-day uplift rates in some areas of the Klamaths are more than 1 foot per 100 yrs. Rapid uplift and high rainfall on the western side have combined to produce erosional mountains similar to the Appalachian Plateau although the Klamaths are considerably younger, topographically higher, glaciated in part, and composed of a variety of structurally complex rock types rather than simple nearly flat-lying sedimentary rocks. As noted earlier, such active uplift may be associated with underplating or with the passage of a seamount bulge below the Klamath Mountains. Note in Figure 16.3 that the Klamaths lie directly above a small subducting subplate (the Gorda plate) that may be acting independent of the larger Juan de Fuca plate. It is possible that the Gorda plate is subducting at a more shallow angle thereby causing some of the uplift. Another potential factor driving uplift may be tectonic activity and compression associated with the Walker Lane Belt as discussed in Chapter 17.

THE CASCADE MOUNTAINS

The North Cascade Mountains and the Central-Southern Cascade Mountains both contain high volcanoes. However, the rock type, elevation, relief, and structure of the

FIGURE 16.11 A Google Earth image looking SSE across the southern Klamath Mountains, California. At least two wave-cut terraces are visible (labeled *T1* and *T2*); a lower one at the elevation of Crescent City 20 to 50 feet above sea level and a higher one 200 to 500 feet above sea level, in sharp contact with the lower one. The higher terrace extends to the ocean at extreme right to form an escarpment that overlooks the ocean. Crescent City is just off the image in the lower-right corner. The Smith River is at lower left, the Klamath River at upper right. Note the flat-top mountains in the vicinity of the Smith River. The Trinity Alps form the highest glaciated part of the Klamaths.

surrounding landscape in the two areas are very different. Physiographic differences are clearly evident in Figure 16.8, which shows the transition just north of Mt. Rainier. The North Cascades are a rugged, glaciated, high relief, snow-capped mountainous terrain with jagged peaks of crystalline rock that reach 9,500 feet in elevation. The Central-Southern Cascade landscape is a heavily dissected, forested highland of volcanic rock that, with the exception of the volcanic cones themselves, is less than 6,000 feet in elevation, lower than the Southern Appalachian Blue Ridge Mountains.

Differences between the North Cascades and the Central-Southern Cascades are related to the composition of accreted material in the deep crust below the surface of the two terrains, and with the level of erosion. About 75 million years ago, prior to accretion of the Siletz terrane, the northwest coastline of the US had the shape of a large embayment as shown in Figure 16.3. This embayment area would eventually be filled by the accreted Siletz terrane and then covered by sedimentary and volcanic rocks. Thus, the deep crust within the embayment area consists of the same weak, thin oceanic crust of the Siletz terrane that is presently exposed along the Oregon coast.

The deep crust in the North Cascades is considerably different. The crust consists of accreted terranes, but these terranes are older, thicker, strongly metamorphosed, and intruded. They are composed of granitic crystalline rock. It is this deep crystalline crust that has been exhumed to the surface to form the rugged North Cascade landscape. Granitic crystalline rock is present in the deep crust of the Southern Cascades south of the embayment area; however, the level of erosion is not yet deep enough to expose these rocks. Only the younger, overlying volcanic and sedimentary rocks are exposed at the surface in the Southern Cascades. Either the North Cascades were subject to more rapid or more prolonged exhumation, or they were never buried as deeply as the Southern Cascades.

The Southern Cascades at least as far north as the Three Sisters volcanoes in central Oregon have been strongly affected by Basin and Range normal faulting which has thinned the deep crust and allowed primarily basaltic magma to reach the surface. East-west extension is occurring at a rate of about 3.9 inches per 100 yrs and is currently propagating northward into the thinner crust of the Siletz terrane thus allowing basaltic magma to reach the surface as far north as Mt. Adams in southern Washington. It is these basaltic rocks, and not the more silicic rocks of the large volcanic cones, that cover large parts of the Central-Southern Cascades. The greatest concentration of volcanic vents occurs in the vicinity of the Three Sisters. This is where the Brothers fault zone (Chapter 14) intersects the extensional basins of the High Cascades, thus creating many avenues by which magma generated at depth can reach the surface.

The North Cascades have not yet been affected by this phase of extensional normal faulting. Here the crust is up to 31 miles (50 kilometers) thick, much thicker than the crust

within the embayment area. As a result, volcanic material, particularly basaltic volcanic material, is far less common. Note in Figure 16.3 that the two volcanoes in the North Cascades (Mt. Baker and Glacier Peak) are separated somewhat from the main volcanic chain to the south. In contrast to the Central-Southern Cascades, volcanism in the North Cascades is restricted almost entirely to these two volcanic cones which stand as sentinels above the jagged crystalline landscape. We discussed the crystalline North Cascade landscape in Chapter 13. Here we discuss the volcanic landscape of the Central-Southern Cascades.

THE CENTRAL-SOUTHERN CASCADE MOUNTAINS

Although volcanoes have existed in the Cascades for more than 36 million years, most of the present-day volcanoes are less than 800,000 years old. Mount St. Helens, one of the youngest active cones, is less than 40,000 years old. The volcanoes are known as *composite* (or *strato*) volcanoes because of their tendency to produce both silicic lava flows and explosive ash. These are dangerous volcanoes that tend to build tall, beautiful, shapely cones and then, within a few 100,000 or million years, either blow themselves up or go extinct and erode. Field evidence suggests that the volcanoes can build rather quickly. This is obvious with respect to Mount St. Helens, which built a beautiful cone nearly 10,000 feet high within 40,000 years (an average rate of 25 feet/100 yrs). These studies also suggest that a single volcano can undergo a series of closely spaced volcanic spurts that can last more than 100,000 years, and then lie dormant for an equally long time, only to awaken again from its long slumber. During periods of dormancy, or of limited volcanism, the volcano can erode at extremely high rates (feet per 100 yrs). Rapid erosion results from a variety of factors including the abundance of fractured and weathered rock, rock that has been weakened and altered by hot fluids, the high, steep slopes, high precipitation, and the presence of glaciers. Magma movement at depth, coupled with seismicity, ground movement, and minor eruptions, has the effect of destabilizing slopes, causing landslides. The most feared type of erosion involves mass movement in the form of mudflows (known as *lahars*) and debris avalanches. These can remove huge sections of a mountain and, in a matter of minutes, create a mudflow that can travel more than 50 miles, burying any town in its path. A large earthquake-generated avalanche on the north side of Mount St. Helens occurred seconds before the 1980 volcanic explosion. The sudden release of pressure by the avalanche caused gases in the magma just below the surface to explode as though somebody just unlocked a pressure cooker. Field studies suggest that a similar debris avalanche and explosion may have occurred on Mount St. Helens 2,800 years ago.

As Cascade volcanoes formed and died over the past 36 million years, the locus of volcanic activity has migrated eastward over time. Exactly why this has occurred is not known with certainty. Melting typically does not occur until the subducting slab reaches a depth of 62 miles. In the case of the Juan de Fuca subducting slab, if the angle of subduction were to become more shallow over time, the locus of magmatism would migrate eastward. Regardless of the cause, the eastward migration of volcanism has produced an active eastern part of the Cascade Range and an older western part. The inactive Western Cascades represent the uplifted locus of eroded, extinct, and blown-apart volcanoes. The landscape tilts gently westward such that elevation gradually increases from less than 2,000 feet at the edge of the Willamette Valley to about 5,800 feet at the base of the active volcanic cones along the crest of the Eastern Cascades. Although not especially high, the area is rugged due to deep stream dissection. It is the gentle tilt of the mountain range, and not the high volcanic cones, that create one of the most pronounced rain shadows in the country. Annual precipitation averages more than 100 inches on the western slope but is typically less than 16 inches on the eastern slope.

All the active and recently active volcanic mountains reside in the Eastern (High) Cascades, which sit along the crest of the mountain range. The largest of the volcanoes are listed from south to north in Table 16.1 and are highlighted in Figure 16.3. In addition, there are numerous smaller volcanic vents, buttes, cinder cones, and lava fields, some only a few hundred years old. Many of the volcanoes sit in large, normal fault basins (grabens) that are, on average, tens of miles wide, 2,000 feet deep, and filled with piles of basaltic volcanic rock. The basins formed as a result of clockwise rotation of the Coast Range, incursion of Basin and Range extension into the area, and isostatic sinking due to the weight of the accumulating basaltic pile. A beautifully exposed basaltic flow, only about 1,500 years old, is shown in Figure 6.12. The flow sits on McKenzie Pass at the crest of the Cascade Mountains northwest of Bend, between Three Sisters and Mt. Jefferson.

As noted earlier, older basaltic rocks, similar to those at McKenzie Pass, form the massive substratum to the younger, smaller, explosive, silicic volcanic cones. The basaltic rocks take the shape of a shield volcano (a broad, low-lying volcanic dome) similar to the Hawaiian volcanoes except, in the case of the Cascades, the shield volcanoes are capped with a centralized high-elevation silicic volcanic cone (a strato volcano). It is the silicic strato volcano, and not the underlying shield volcano, that most people visualize when they think of a volcano. The Cascade silicic cones are exceptionally high and tower as much as one mile above the surrounding basaltic landscape. Mount Jefferson, shown in Figure 16.13, is a beautiful example.

TABLE 16.1 South-to-North List of Major Cascade Volcanoes

	Active in Past 2,000 Years	Active in Past 200 Years	Elevation (Feet)	Notes
California				
Mt. Lassen	xxx	xxx	10,457	
Mt. Shasta	xxx	xxx	14,162	
Medicine Lake	xxx		7,795	
Oregon				
Mt. McLaughlin			9,495	
Crater Lake			6,176	
Mt. Thielsen			9,182	
Mt. Bachelor			9,065	
Broken Top			9,175	
South Sister	xxx		10,358	
Middle Sister			10,047	
North Sister			10,085	
Mt. Washington			7,794	
Three-Fingered Jack			7,844	
Mt. Jefferson			10,497	
Mt. Hood	xxx	xxx	11,239	
Washington				
Mt. Adams	xxx		12,276	
Mt. St. Helens	xxx	xxx	8,363	9,677 before eruption
Mt. Rainier	xxx	xxx	14,411	
Glacier Peak	xxx	xxx	10,541	North Cascades
Mt. Baker	xxx	xxx	10,781	North Cascades
Canada				
Mt. Gariboldi			8,786	
Mt. Cayley			7,799	
Mt. Meager			8,793	

FIGURE 16.12   A photograph of the edge of a 1,500-year-old basalt flow on McKenzie Pass, Oregon.

FIGURE 16.13   A photograph of Mt. Jefferson, central Oregon. Note how the volcano towers over the surrounding landscape.

Medicine Lake volcano and the Modoc lava beds, located about 30 miles northeast of Mount Shasta, are examples of a Cascade basaltic pile (a shield volcano) without a centralized high-elevation silicic cone (Figure 16.1).

The size of Medicine Lake volcano is enormous. It is 22 miles east to west and 30 miles north to south with a basaltic pile more than 3,000 feet thick. The central part of the volcano is a large depression (a caldera) 4 to 8 miles wide that includes

Medicine Lake itself, along with several large obsidian flows. The flanks of the volcano are littered with cinder cones, fissure vents, and associated lava flows.

Although the high volcanic cones are certainly the most conspicuous landforms in the Cascades, it is the massive underlying shield volcanoes and associated basaltic rock that form most of the volcanic rock in the Central-Southern Cascades. Based on the age of the basaltic rocks, we can estimate that normal faulting beneath the High Cascade range, and the filling of the basins with basaltic lava, began less than 10 million years ago.

The tilting of the western Cascades and development of a rain shadow apparently also occurred in the last 10 million years. Studies of fossil plant life along the east side of the Cascade Range and on the Columbia Plateau suggest that plants growing in these areas had changed from summer-wet to summer-dry species by about 8 million years ago, thus dating the initiation of the rain shadow. Another method of studying the formation of a rain shadow is to look at progressive changes in oxygen isotope ratios. Studies of soil samples from the Columbia Plateau suggest a gradual decrease in precipitation between 16 and 4 million years ago. Presumably, this is the time span over which the Cascade rain shadow developed, which is consistent with fossil plant studies. A third line of evidence has to do with the Columbia River basalts, which extend almost to the crest of the Western Cascades and are tilted westward. The basalt flows are mostly between 16.8 and 15.6 million years old, and the presence of these rocks so far up the flank of the Cascade Range suggests that the mountain could not have existed at this time (that is, the basalts could not have flowed uphill). Recall from Chapter 13 that the North Cascades may have already been a highland with elevation and relief by 16 million years ago and that it likely underwent an additional period of uplift between 12 and 4 million years ago. This later period of uplift apparently also affected the Southern-Central Cascade Range.

The Columbia River has probably been in existence since the beginning of Cascade volcanism some 35 to 40 million years ago. Where it crosses the Cascade Mountains between the Dalles and Portland, it cuts a spectacular 75-mile-long gorge through the volcanic pile (Figure 16.1). It is likely that the gorge was cut within the past 10 million years during uplift and tilting of the Western Cascades. Thus, we can conclude that the Cascade Mountains form a relatively young tectonic landscape less than 10 million years old with conspicuous volcanic cones less than 800,000 years old.

Of the 23 major volcanoes listed in Table 16.1, 10 have erupted in the past 2,000 years and seven in the past 200 years. Mt. Rainier is the highest and possibly the oldest. Radiometric dating of lava flows indicates that the present-day volcanic cone began to form more than 500,000 years ago. The presence of older silicic rock along the flanks of the volcano suggests that earlier volcanic cones had existed at the same location for the past 2.9 million years. The present-day volcano is active and may have been higher 75,000 years ago (15,500 to 16,000 feet) prior to eruptions that lowered its summit. Mount St. Helens is the youngest and most active volcano in the chain. This now famous volcano erupted catastrophically on May 18, 1980, losing 1,314 feet of summit elevation in only a few minutes. Crater Lake occupies the location of ancient Mt. Mazama. Following a massive explosive eruption some 7,700 years ago, Mt. Mazama collapsed into a 6 mile-wide caldera that filled with water to form beautiful Crater Lake. This is the deepest freshwater lake in the US (1,996 feet deep). Several of the other volcanoes and volcanic fields, including Broken Top, Mt. Thielsen, Three-Fingered Jack, North Sister, and Mt. Gilbert (Goat Rocks), are strongly eroded, suggesting they have not had major activity in 50,000 years or more.

QUESTIONS

1. Google "mudflow potential on Mt. Rainier, Washington," or go to http://vulcan.wr.usgs.gov/Volcanoes/Rainier/Lahars/Historical/description_osceola.html and describe recent activity and the potential effects of a mudflow emanating from Mt. Rainier.

2. Google "Cascades Volcano Observatory" or go directly to http://vulcan.wr.usgs.gov/ and describe recent activity on any of the following volcanoes: Mt. Lassen, Mt. Shasta, Mt. Hood, Mt. St. Helens, Mt. Rainier, Glacier Peak, Mt. Baker. What do all of these volcanoes have in common?

3. Trace the boundary of the Cascadia volcanic arc system in Figure 16.1.

4. Describe the Siletz volcanic complex. Speculate on its origin.

5. Speculate on the origin of the Juan de Fuca Straight north of the Olympic Peninsula.

6. Should the town of Corvallis be worried about a volcanic eruption from Mt. Jefferson?

7. What evidence is there, or evidence you would look for, to determine whether the Klamath Mountains were a lowland 5 million years ago? What evidence is there for recent rapid uplift?

8. Research the Boring lava field—its age, origin, distribution, composition.

9. Why are there no volcanoes in the Klamath Mountains?

10. Speculate as to why Oregon has a coastal mountain range versus the coastal area south of the Olympic Mountains, which is a lowland.

11. What is the origin of the Columbia River Gorge?

12. How does one distinguish between an active volcanic cone and one that has been dormant for a long time?

13. Name the highest volcano in Washington, Oregon, and California.

14. Why is there no trench to mark the site of subduction of the Juan de Fuca plate?

15. Why is basalt so abundant in the Cascade Range?

16. Research the 1980 Mt. St. Helens volcanic eruption and write a one- or two-page report.

17. Research the eruption that produced Crater Lake and its aftermath and write a one- or two-page report.

18. Research the Seattle or Tacoma fault (Figure 16.7) and write a one- or two-page report. Are they active? What is the potential for a major earthquake?

19. Describe the uplift history of the Olympic Mountains.

20. What is the evidence for the timing of uplift of the Cascade Range?

21. How might the substrate of Glacier Peak and Mt. Baker (in the North Cascades) differ from the basaltic substrate of volcanoes to the south?

22. Why are there so many volcanic landforms in the vicinity of Three Sisters?

23. If volcanism has occurred at the site of the Cascade Mountains for the past 36 million years, why are none of the volcanoes more than 1 million years old?

24. Why was the Siletz terrane accreted and not subducted?

25. How could one prove that volcanism has occurred in the Cascade Range for the past 36 million years?

26. How does the deep crust of the North Cascades differ from the Central Cascades and from the Southern Cascades?

27. Provide an explanation as to why short-term measured uplift rates may not be indicative of long-term rates.

Chapter 17

California Transpressional System

Chapter Outline

The California Borderland, the Sierra Nevada, and the western margin of the Basin and Range together constitute the *California transpressional system*. A landscape map of the area is shown in Figure 17.1 coupled with a shaded relief image shown in Figure 17.2. This structural province is fundamentally different from all other provinces in that it is the only one that straddles a plate tectonic boundary. The Pacific plate is sliding northward relative to the North American plate along the San Andreas Fault. Displacement occurs in lurches during earthquakes in some areas, and slowly and constantly without earthquakes in other areas. The current rate of displacement is about 15.8 feet per 100 yrs. What this means is 15.8 feet of displacement must be accounted for along faults in California every 100 years. The San Andreas Fault system, including all active faults in the Coast Ranges, accounts for about 11.2 feet. Faults east of the Sierra Nevada account for about 3.0 feet, and the remaining 1.6 feet probably occur off the California coast. Displacement of this magnitude creates one of the most geologically diverse and active regions in North America. Landscapes are decidedly structure-controlled rather than erosion-controlled, this in spite of the misconception that California is sliding into the sea. In addition to a major tectonic plate boundary, there is also a subplate known as the *Sierran plate* that moves independent of both the North American and Pacific plates, there are areas of very rapid uplift and subsidence, there are relict landscapes, and there are examples of nearly every type of structure and rock mentioned in this book. The rocks are primarily sedimentary and crystalline (both plutonic

and metamorphic), but volcanic rocks are an important part of the landscape, particularly within and east of the Sierra Nevada. Several volcanic areas were briefly discussed in Chapter 14 including the Long Valley Caldera and the Sutter Buttes.

Recall from Chapter 3 that a shear stress produces strike-slip faults; a compressional stress produces folds, reverse faults, and thrust faults; and an extensional stress produces normal faults. Much of the landscape across California has been affected by either a combination of shear stress and compression, or a combination of shear stress and extension. Geologists use the term *transpression* for areas where shear is combined with compression. Such a combination produces both mountains and valleys. Conversely, the term *transtension* is used for areas where shear is combined with extension to produce basins.

The California Borderland province is really a grouping of several smaller landscape areas, each with unique characteristics but all part of the same active tectonic system. It includes the California Coast Ranges, which extend from the Klamath Mountains in northern California to the vicinity of Point Sal at Santa Maria; the Transverse Ranges, which form a series of east-west mountains between Point Sal and Los Angeles; the Peninsular Ranges and Valleys south of Los Angeles; and the Central Valley of California. The Sierra Nevada is included in the transpressional system as is the entire western margin of the Basin and Range as far east as westernmost Nevada, including the Mojave Desert, Death Valley, and Lake Tahoe. All of these areas have been shaped, at least in part, by transpressional or transtensional strike-slip tectonics. The eastern margin

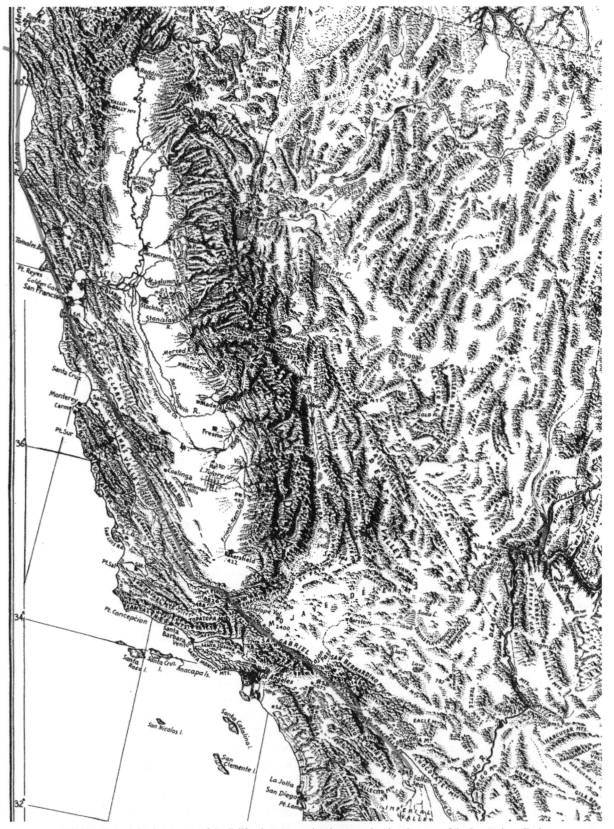

FIGURE 17.1   A landscape map of the California transpressional system showing the trace of the San Andreas Fault.

THE SAN ANDREAS FAULT SYSTEM

The San Andreas Fault is one of the most intensely studied faults on the planet. It is a continental transform that separates the Pacific plate from the North American plate along its entire trace. The trace of the fault through California is shown in Figure 17.1. Its northern point begins at the Mendocino triple junction near Cape Mendocino. This is the location where the North American, Pacific, and Juan de Fuca plates all meet at a point or at least within a small area. Here the San Andreas merges with the Mendocino fault zone and extends out to sea as a transform fault that separates the Juan de Fuca plate from the Pacific plate (Figure 16.3). From Cape Mendocino the San Andreas Fault extends southward just off the coastline before reaching land again at Point Arena. From Point Arena it forms a straight lineament through Tomales Bay into San Francisco and then through the Santa Cruz Mountains to an area east of Monterey Bay where the fault makes a slight bend to the east. It then maintains a straight line diagonally through the Coast Ranges to the southern end of the Central Valley between Bakersfield and Santa Barbara. From here the San Andreas Fault makes a second larger and more abrupt "Big Bend" to the east across part of the Transverse Ranges east of Mt. Pinos. It then turns southward along the northeastern side of the San Gabriel Mountains. East of Los Angeles the San Andreas splits into two primary branches for a distance of about 60 miles within and along the southwestern edge of the San Bernardino Mountains before rejoining into a single fault just north of the Salton Sea. After about 600 miles, the San Andreas Fault ends along the eastern shoreline of the Salton Sea near the town of Bertram (Figure 17.1).

As with the Wasatch Fault in Utah, and with any large fault, the surface trace of the San Andreas is not continuous. Instead, the surface trace is segmented such that individual fault segments may act independently of one another. One or several segments may break during an earthquake. Each segment is separated from another by several tens of feet to several miles. Nearly all of these segments have one thing in common: The dip of the San Andreas Fault is vertical or nearly vertical along nearly its entire length, at least to the depth of its deepest earthquake which is less than 12 miles. A zone of weakened, crushed rock that, in some areas can be up to a mile wide, marks the fault zone itself. This zone is easily eroded and, in combination with the vertical dip of the fault, produces a strong topographic scar (lineament) that is easily seen in the shaded relief map of Figure 17.2.

Figure 17.3 and Table 17.1 identify a few of the many faults in the transpressional system, all of which have been active in the past 1.6 million years. The distribution of several major fault zones is partly controlled by the two bends in the San Andreas. The Calaveras (10), Greenville (11),

FIGURE 17.2 A shaded relief map of the California transpressional system.

of the transpressional system is defined by a change in the orientation of Basin and Range mountain blocks from a north-northeast trend in central Nevada to a north-northwest trend in western Nevada. This change is best seen in western Nevada in the vicinity of Pyramid and Walker Lakes (Figures 17.1 and 17.2).

Strike-slip faults overprint older, now relict landscapes that include a subduction complex physiographically similar to the Cascadia volcanic arc system, and a high plateau landscape (the Nevadaplano) that existed in western Nevada-eastern California prior to widespread Basin and Range extension and prior to the present-day Sierra Nevada landscape (Figure 15.22). In many areas, the process of reincarnation is in its early stages such that older structures still impart at least some control on the landscape. In this chapter we will concentrate on the San Andreas Fault, the Transverse Ranges, the Coast Ranges, and the Sierra Nevada.

284

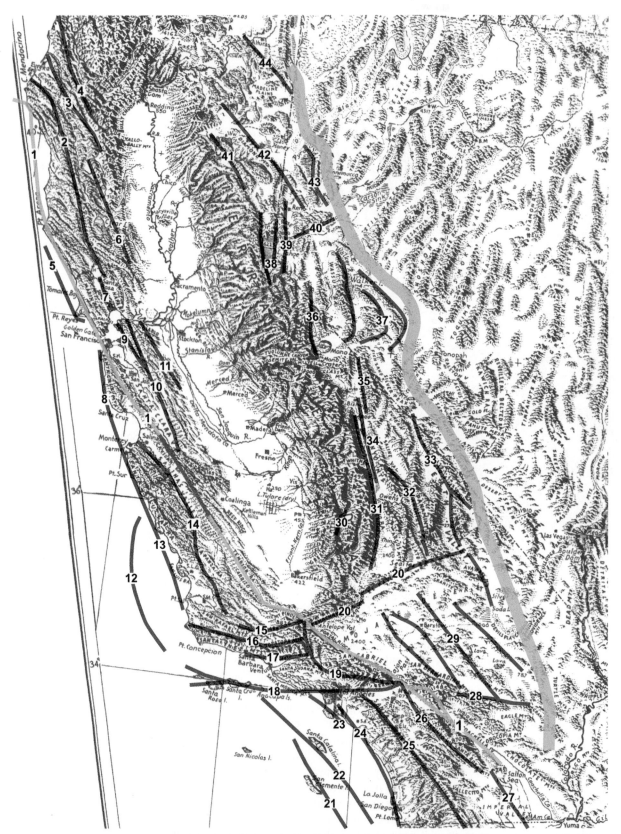

FIGURE 17.3 A landscape map showing major active faults in the California transpressional system. The approximate eastern boundary of the Walker Lane Belt is shown with a thick line. The creeping section of the San Andreas Fault is shown with a slightly darker line. Numbers correspond with fault names shown in Table 17.1. The map is rotated counterclockwise.

TABLE 17.1 Major Faults Shown in Figure 17.2

1. San Andreas	23. Palos Verdes
2. Maacama	24. Rose Canyon
3. Eaton Roughs	25. Elsinore
4. Grogan	26. San Jacinto
5. Point Reyes	27. Imperial Valley
6. Bartlett Springs	28. Pinto Mountain
7. Rogers Creek	29. Eastern California system (several faults)
8. San Gregorio	30. Kern Canyon Fault
9. Hayward	31. Sierra Nevada Frontal Fault System
10. Calaveras	32. Panamint Valley
11. Greenville	33. Fish Lake Valley-Furnace Creek-Death Valley
12. Santa Lucia Bank	34. Owens Valley
13. Hosgri	35. White Mountains
14. Reliz–Rinconada–East Huasna	36. Mono Lake
15. Big Pine	37. area with numerous faults
16. Santa Ynez	38. West and East Tahoe
17. Mission Ridge–San Cayetano–Del Valle	39. Genoa
18. Santa Cruz Island–Santa Rosa Island–Malibu Coast–Raymond	40. Carson Lineament
19. San Gabriel	41. Butt Creek
20. Garlock	42. Honey Lake
21. San Clemente	43. Pyramid Lake
22. Santa Catalina	44. Likely

Hayward (9), Rogers Creek (7), Bartlett Springs (6), Maacama (2), Eaton Roughs (3), and Grogan (4) faults all extend northwestward from the vicinity of the small bend near Monterey Bay. The Hayward Fault trends just east of Oakland and was the site of a strong earthquake in 1868. Together, the Hayward and San Andreas faults frame San Francisco Bay.

The fault system near the Big Bend is more complex. The Garlock Fault (20), a major left-lateral strike-slip fault that separates the Sierra Nevada from the Mojave Desert, intersects the San Andreas at a high angle. Several other faults, including the Big Pine (15), Santa Ynez (16), and Mission Ridge–San Cayetano–Del Valle (17) faults also intersect the San Andreas at a high angle. These faults, along with the San Gabriel (19), Santa Cruz Island–Santa Rosa Island–Malibu Coast–Raymond (18), and Pinto Mountain faults (28), form the Transverse Ranges.

Numerous other faults are present at the Big Bend that are semiparallel with the San Andreas. The most important is the San Gregorio-Hosgri Fault (8, 13), which skirts the coastline from San Francisco to Point Sal. This fault, together with the Reliz–Rinconada–East Huasna Fault (14), frames the Santa Lucia Range south of Monterey. Additional faults south of the Transverse Ranges include the San Clemente (21), Santa Catalina Ridge (22), Palos Verdes (23), and Elsinore (25) faults. The San Jacinto Fault (26) splays off the San Andreas at San Bernardino, and the two faults frame the Jacinto and Santa Rosa Mountains (part of the Peninsular Ranges).

Although the San Andreas Fault proper ends along the eastern side of the Salton Sea, the greater San Andreas transform system continues southward. Southwest of where the San Andreas Fault ends, another strike-slip fault, the Imperial Valley Fault (27), appears in the Imperial Valley south of the Salton Sea. This fault represents the southward continuation of the San Andreas Fault system. Farther south the Imperial Valley Fault is part of a series of strike-slip fault segments that extend through the Gulf of California where they connect small, divergent spreading ridges within which new oceanic lithosphere forms (Figure 5.6). The San Andreas system ends near the southern end of the Gulf of California at the Rivera triple junction where the Pacific, North American, and Cocos tectonic plates meet. With continued displacement, all land areas west of the San Andreas Fault, including Baja California and the city of Los Angeles, will be displaced northward along the California coast. As Baja California is displaced northward, the Gulf of California will, over time, widen to become part of the open Pacific Ocean.

Figure 17.4 is a night Google Earth image that shows all the faults active in the past 1.6 million years. Several observations are of note in this figure: (1) the incipient development of faults in the Central Valley, (2) the absence of faults across a sharp line in the southeastern California Basin and Range, (3) the sharp change in fault geometry in western Nevada, and (4) the abundance of offshore faults, especially in southwestern California and along the Juan de Fuca trench.

FIGURE 17.4 A Google Earth image looking northward across California, Nevada, and Oregon. The various colored lines are faults active over the past 1.6 million years as shown in the Quaternary fault and fold database for the United States, accessed January 18, 2012 from USGS Website: http://earthquake.usgs.gov/regional/qfaults/.

DISPLACEMENT ALONG THE SAN ANDREAS FAULT

To understand earthquakes, one must first understand how rocks react to stresses that are imposed on them. *Stress* is force per unit area. If you take a small dry stick between your hands and push on it until it buckles, you are applying a stress to that stick. *Strain* is the distortion (the bending) that results from stress. If you release the stick so that it is no longer under any stress, the stick will instantly (elastically) snap back to its original shape. If you apply enough stress to the stick, it will break into two pieces. Once broken, the stick is no longer under any stress, and all of the strain (the bending prior to breaking) is recovered. The two broken pieces are no longer bent and could, theoretically, fit back together again. The type of strain that is present when the stick is under stress but absent when stress is removed is known as *elastic* (or *recoverable*) *strain*. It is akin to stretching an elastic band. A larger (stronger) stick will require a greater amount of stress to break, but it will snap back with greater force due to a greater amount of stored elastic energy.

Rocks behave the same way. Earthquakes form on the San Andreas Fault (and on any other fault for that matter) when rocks at depth on either side of the fault surface are so tightly stuck together that they bend (and, therefore, accumulate elastic strain) prior to breaking. All (or most) of the stress is relieved when the rock finally breaks and, therefore, all of the elastic strain is recovered. The broken pieces of rock snap back to their original shape. The difference between a rock and a stick is that rocks are buried deep within the Earth, and when they snap back upon breaking, they snap into adjacent rocks. The force of the snap applied to adjacent rocks creates seismic waves that, in turn, create earthquakes. The more stress that builds up prior to breaking (that is, the stronger the rock), the greater the distortion (bending) and therefore the larger the earthquake once the rock breaks and snaps back. Thus, the size of an earthquake is roughly a function of the amount of stress a rock can withstand before breaking and the length and width of the fault surface that breaks.

Rocks along the San Andreas Fault are stuck tightly together in the vicinity of San Francisco, site of the devastating 1906 earthquake as well as large earthquakes in 1838 and 1989. Rocks are also stuck tightly together near Los Angeles, site of a large 1994 earthquake. However, not all of the fault trace is subject to dangerous earthquakes. Part of the fault zone moves by a process known as *creep*. As the name implies, creep is very slow, almost constant movement. In these areas, the fault zone is either lubricated, or the rocks are so weak that they break after only a small amount of stress is applied. Stress never gets a chance to build up, so the rocks do not bend very much before they break. The result is an earthquake of such low magnitude that it is rarely felt. The main part of the San Andreas Fault that moves via creep is the section that extends approximately from a few miles south of San Juan Bautista (near Salinas), southward to about 6 miles northwest of Parkfield (south of Coalinga). This section is shown as a slightly darker line along the San Andreas Fault in Figure 17.3. Parkfield is located just south of the area of creep and has been the site of a sequence of large earthquakes in 1857, 1881, 1901, 1922, 1934, 1966, and 2004. Creep has been reported along the Hayward Fault although apparently not enough to completely relieve the possibility of another large earthquake.

THE HISTORY OF THE SAN ANDREAS FAULT SYSTEM

For more than 200 million years prior to formation of the San Andreas Fault, the entire California coast was a subduction boundary much like the present-day Oregon-Washington coast. This subduction boundary separated the North American plate from a plate that no longer exists, the Farallon plate, which at that time extended most of the way down the North American coastline. During subduction, the Farallon plate was moving obliquely eastward toward the North American plate which was moving westward over the Farallon plate. The Pacific plate was in existence at this time, separated from the Farallon plate by a divergent plate boundary. The relationships are shown in Figure 17.5a.

The San Andreas Fault began to form between 29 and 27 million years ago when the divergent plate boundary that separated the Pacific from the Farallon plate was itself subducted. It was at this time that the Pacific plate first came into direct contact with the North American plate, and because the Pacific plate was moving north-northwestward at high velocity relative to the slow-moving North American plate, the contact evolved into a transform (strike-slip) boundary as shown in Figure 17.5b. The initial transform boundary some 27 million years ago was probably at the location of the subduction zone trench. Since 27 million years ago, the San Andreas Fault system has jumped eastward several times and has grown in length in both a northward and southward direction as more and more of the Farallon plate was subducted. This implies that the Mendocino and Riviera triple junctions have migrated northward and southward, respectively, thus lengthening the San Andreas system (Figures 17.5b and 17.5c). It also implies that the San Andreas Fault is oldest and has its greatest displacement near its center and that it gets progressively younger toward the triple junctions. The eastward jump of the San Andreas implies that part of what used to be the North American plate has been transferred to the Pacific plate. Several faults currently on the Pacific plate, such as the Santa Lucia Bank (12), San Gregorio-Hosgri (8, 13), the Reliz–Rinconada–East Huasna (14), the San Clemente

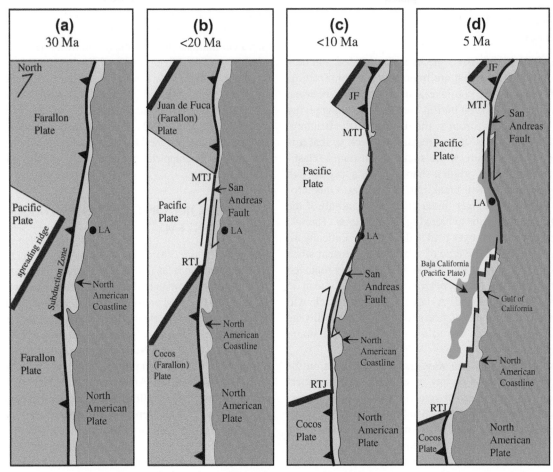

FIGURE 17.5   Sketch maps that show the sequence of development of the San Andreas Fault. Based on Atwater (1998).

(21), and the Elsinore (25) faults, may have marked the Pacific-North American plate boundary in the past.

A major eastward jump in the southern part of the San Andreas Fault at about 5 million years ago caused initial opening of the Gulf of California and resulted in shortening of the fault proper (Figure 17.5d). Since that time, Baja California has moved northward about 150 miles away from mainland Mexico at a rate of about 15.8 feet per 100 yrs. The eastward jump created the Big Bend in the San Andreas Fault, which, in turn, has strongly influenced landscape development in the Transverse Ranges as described later in this chapter.

Total displacement on the San Andreas Fault is not well constrained and, for reasons already discussed, varies along its length. Offset features across the fault suggest that minimum displacement in central California is on the order of 200 miles and could be much more. Today, remnants of the once larger and continuous Farallon plate still exist in the form of the Juan de Fuca plate north of the Mendocino triple junction, and the Cocos plate south of the Riviera triple junction. Both of these plates will shrink as the triple junctions migrate northward and southward, respectively.

A RELICT SUBDUCTION ZONE LANDSCAPE

Figure 17.6a is a schematic cross-section that shows the relationship between topography and tectonics prior to development of the San Andreas Fault. In this figure we can see that the Coast Ranges were host to the accretionary prism while, at the same time, the Central Valley formed the forearc basin and the Sierra Nevada formed the now vanished volcanic arc. As shown in Figure 17.6b, the development of the San Andreas Fault has effectively fossilized the topographic expression of this ancient subduction zone but not the actual landscape that existed during subduction. The present-day landscape is younger than even the birth of the San Andreas fault.

The Coast Ranges are characterized by a variety of rock, the oldest of which consists of accretionary prism rocks that were scraped off the down-going Farallon subduction slab and added to the North American plate during subduction. These rocks are partly buried beneath younger rock in the Central and Southern Coast Ranges. However, the Northern Coast Ranges north of San Francisco and east of the San Andreas Fault consist almost entirely of these

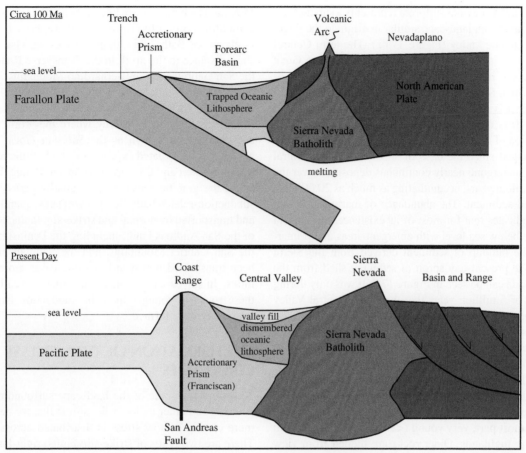

FIGURE 17.6 Cross-section sketches that depict the formation of the relict California subduction landscape. The upper figure shows the subduction landscape that existed circa 100 Ma. Note the inferred existence of the Nevadaplano. The lower figure shows the present-day topographic expression of the relic subduction landscape. Oceanic lithosphere (the Coast Range ophiolite) was incorporated into the accretionary prism during the subduction process.

rocks which are known as the *Franciscan complex*. The Northern Coast Ranges are a series of rugged, dissected, forested, semilinear mountain ranges that, with the exception of the Yallo Bally Mountain area north of Clear Lake, rarely reach 6,000 feet in elevation. Outcrop is poor within the heart of the mountain range due to extensive vegetation cover; however, linear mountain ridges and valleys north of Point Arena intersect the shoreline at a moderate angle creating a coastline of alternating headlands and inlets (Figure 17.1). The Franciscan complex is beautifully exposed in the mountain headlands, but if you cannot get that far north, you can also view the rocks in San Francisco on both sides of the Golden Gate Bridge. There are also a few exposures of Franciscan complex in the central and southern Coast Ranges, most notably in the Diablo Range southeast of San Jose and along the coast north of San Luis Obispo. The rocks form a typical mélange assemblage composed of deformed, fine-grained, oceanic sedimentary rock with embedded lenses of basalt, metamorphosed basalt (greenschist, blueschist), serpentinite, and other parts of the Farallon oceanic lithosphere (see Chapter 5). A single lens can be the size of a house or even larger. The lenses

tend to be randomly distributed throughout the Franciscan complex and are harder than the rocks in which they are embedded. As such, the lenses form resistant knobs in the mountains and sea stacks along the shoreline.

Unlike a normal sedimentary rock unit, there is little in the way of consistent layering in the Franciscan complex. Given the absence of layering, one might expect a random distribution of mountains and valleys. Instead, the mountains show a fairly strong linear landscape, and on this basis alone, we could suggest that the mountains we see today are unrelated to the subduction process that occurred prior to development of the San Andreas Fault. A clue to the age of the present-day landscape is the presence of very young marine (oceanic) sedimentary layers that lie unconformably above Franciscan rocks in many parts of the Coast Ranges. Some of these rocks are less than 4 million years old. Their presence indicates that the land area now occupied by the Coast Ranges was below sea level less than 4 million years ago. The Coast Ranges we see today must therefore be younger than 4 million years old. Later in this chapter we will discuss the origin of the Coast Ranges with particular reference to the Central-Southern ranges.

The Central Valley of California represents a subduction-related forearc basin landscape similar in origin to the Puget Sound-Willamette valley (Figure 17.6a). The term Central Valley refers to its location and topography. The term Great Valley refers to its geological form, which is that of a syncline (the Great Synclinal Valley). At 400 miles long and 50 miles wide, it is one of the major synclinal depressions of the world. It is a relict landscape that was fossilized during formation of the San Andreas Fault and preserved virtually unchanged since the days of subduction. The Central Valley has undergone nearly continuous deposition over the past 155 million years, accumulating as much as 20,000 feet of rock and sediment. The abundance of marine (oceanic) deposits indicates that for most of its existence, the Central Valley was below sea level with emergent areas due primarily from the buildup of sediment derived from the Sierra Nevada. The present-day apron of sediment shed from the Sierra Nevada can be seen in Figure 17.2. It was only during the past 4 or 5 million years that the entire Central Valley emerged permanently above sea level. Today it is a flat landscape, mostly between 50 and 300 feet in elevation, drained by the Sacramento River in the north and the San Joaquin River in the south. These two rivers join and drain into San Francisco Bay. The southern third of the Central Valley south of Fresno is a closed basin with internal drainage.

The flat-lying sediment that covers the Central Valley is, for the most part, very young (<2 Ma) and derived from surrounding highlands. Older rocks are hidden from view in the central part of the valley but are turned up on edge and exposed along the western margin adjacent to the Coast Ranges. Here, the entire sedimentary sequence can be examined. Surprisingly, oceanic lithosphere (the Coast Range ophiolite) can be found at the base of the sedimentary sequence. Apparently, part of the Farallon oceanic lithosphere became trapped in the forearc basin during the subduction process as depicted in Figure 17.6b.

The Sierra Nevada consists primarily of a large batholith of granitic rock that intruded between about 220 and 80 million years ago with a magmatic lull between 140 and 120 Ma and strong magmatism between 100 and 80 Ma. The batholith consists of more than 200 individual intrusions (plutons) emplaced at depths between 5 and 20 miles. These rocks represent the feeder batholith to the now vanished overlying, subduction-related volcanic arc. Thus, the mountains we see today are nowhere near the same as the volcanic mountains that existed 80 million years ago. The area likely was a volcanic highland 80 million years ago in close proximity to an ocean (Figure 17.6a). The origin of the present-day landscape is unsettled in the geological literature and is discussed later in this chapter.

The Sierra Nevada is not the only location in California where 220- to 80-million-year-old batholithic rocks exist. These same rocks are widely exposed in the Peninsular Ranges and particularly in the Laguna, Santa Rosa, and Jacinto Mountains where they are known as the Southern California batholith. These are high mountains that approach 11,000 feet in elevation. Batholithic granitic rocks are also exposed a short distance to the north in the Transverse Ranges within the San Gabriel and San Bernardino Mountains. These also are impressive mountains with peaks above 11,000 feet. Batholithic rocks are widely exposed in the Santa Lucia Mountains south of Monterey where they form part of an accreted terrane known as the *Salinian block*. There are also numerous isolated exposures of batholithic rocks in the Mojave Desert and California Basin and Range. In all these areas, the granitic rocks were originally part of the same subduction-related batholith that was subsequently chopped and transported by normal and strike-slip faults. Rocks west of the San Andreas Fault, including the Peninsular Ranges, the San Gabriel Mountains, and the Salinian block, have been transported northward in some cases more than 200 miles. In all cases, the landscape that currently overlies these rocks is younger than the subduction landscape in which the rocks formed.

THE FORMATION OF TRANSPRESSIONAL STRUCTURES

To understand some of the landscape surrounding the San Andreas and other major strike-slip faults, we need to look more closely at how stress is distributed across the fault. There are two types of strike-slip faults: right-lateral faults and left-lateral faults. From the perspective of someone standing on the west side of a right-lateral strike-slip fault and looking across to the east side, it would appear that the east side is moving to the right. Conversely, if someone were standing on the east side of the fault and looking across to the west side, it would appear that the west side is moving to the right; thus, the name *right-lateral*. A left-lateral strike-slip fault shows the opposite relationship. The San Andreas is a right-lateral strike-slip fault in which the east side (the North American plate) moves southward relative to the northward-moving Pacific plate.

Unlike normal and reverse faults where maximum compression and extension are approximately perpendicular to the strike of the fault, the maximum compression direction for strike-slip faults is at an angle of roughly 30° to the strike of the fault (Figure 4.3h). Figure 17.7 is a sketch of some of the features that are expected along a right-lateral strike-slip fault such as the San Andreas. The inferred maximum compression direction is shown with full arrows, and the relative movement along three fault segments is shown with half-arrows. Two of the fault segments have bends.

Under ideal conditions, opposite sides of a straight, properly oriented fault segment should slip horizontally past each other without any up or down displacement. However, even under these conditions, anticlinal ridges, synclinal valleys, thrust faults, and reverse faults are expected to

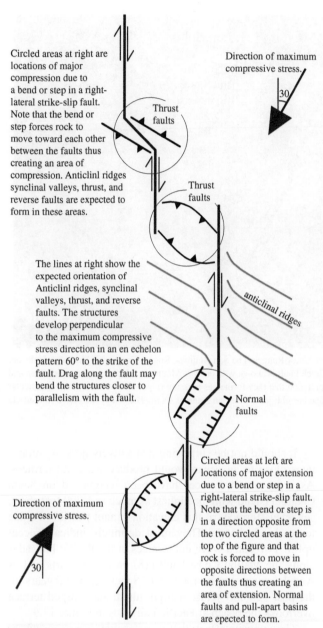

Circled areas at right are locations of major compression due to a bend or step in a right-lateral strike-slip fault. Note that the bend or step forces rock to move toward each other between the faults thus creating an area of compression. Anticlinl ridges synclinal valleys, thrust, and reverse faults are expected to form in these areas.

Direction of maximum compressive stress.

30

Thrust faults

Thrust faults

The lines at right show the expected orientation of Anticlinl ridges, synclinal valleys, thrust, and reverse faults. The structures develop perpendicular to the maximum compressive stress direction in an en echelon pattern 60° to the strike of the fault. Drag along the fault may bend the structures closer to parallelism with the fault.

anticlinal ridges

Normal faults

Direction of maximum compressive stress.

30

Circled areas at left are locations of major extension due to a bend or step in a right-lateral strike-slip fault. Note that the bend or step is in a direction opposite from the two circled areas at the top of the figure and that rock is forced to move in opposite directions between the faults thus creating an area of extension. Normal faults and pull-apart basins are epected to form.

FIGURE 17.7 A sketch map showing stress direction and areas of compression and extension expected along a right-lateral strike-slip fault such as the San Andreas Fault.

develop in an en echelon (in-step) pattern perpendicular to the compression direction and, therefore, 60° to the strike of the fault as depicted in the center part of Figure 17.7. Anticlinal ridges that form in this manner are sometimes referred to as *pressure ridges*. With displacement along the fault, it is possible for the anticlinal ridge to bend toward parallelism with the fault.

With few exceptions, the Coast Ranges are approximately parallel with the San Andreas Fault and therefore, are not at the expected orientation of a pressure ridge as described above. There are only a few areas, primarily in

the Northern Coast Ranges, where mountain ridges appear to be close to the proper orientation to have developed in this manner. One location is just north of Point Arena where the San Andreas Fault trends offshore parallel with the shoreline and the mountains intersect the shoreline at an angle (Figure 17.3). Another location, also evident in Figure 17.3, is just east of Point Arena between Clear Lake and the town of Santa Rosa, where ridges are oriented at an angle to the Maacama Fault (2). A Google Earth image of this area is shown in Figure 17.8. However, even if these particular ridges did form via compression associated with strike-slip displacement, the Coast Ranges as a whole require a different explanation. We will explore other possibilities shortly. First we need to look in more detail at landscape evolution in areas where strike-slip faults bend or jump from one segment to another. These areas include the Transverse Ranges and the Salton Sea.

THE TRANSVERSE RANGES AND THE SALTON SEA

Whereas most of the mountains in the Coast Ranges have formed along straight segments of the San Andreas Fault, the Transverse Ranges, in particular, have formed at the location where the San Andreas makes the Big Bend. As shown in the upper part of Figure 17.7, a bend to the left along a right-lateral strike-slip fault results in a zone of unusually strong compression where the two sides of the fault move directly toward each other. The upper part of Figure 17.7 shows development of thrust faults in this area. A similar type of compression in the Big Bend area of the San Andreas Fault produces nearly east-west-trending anticlinal and thrust fault mountains expressed in the landscape as the Transverse Ranges and Channel Islands (which include Santa Rosa and Santa Cruz Island). Given this origin for the Transverse Ranges, the mountains must be younger than the final eastward jump of the San Andreas Fault and the formation of the Big Bend, both of which occurred approximately 5 million years ago.

The San Andreas Fault is active, therefore compression in the Big Bend area is also active which implies that at least part of the Transverse Ranges are still growing. One example of an actively growing landform is Wheeler Ridge, a faulted, anticlinal ridge located in the Big Bend area just north of the San Andreas Fault at the border between the Transverse Ranges and the Central Valley, 25 miles south of Bakersfield and just west of the junction between Interstate 5 and Golden State Highway 99. The ridge is 2,141 feet high and rises more than 1,400 feet above surrounding flatlands. A detailed study of this ridge showed that a single identifiable erosion surface is present at progressively higher elevations with a greater degree of tilt in the higher central part of the mountain than at its eastern end. This observation suggests that the eastern part of the ridge is

FIGURE 17.8 A Google Earth image looking northward along the Northern Coast Ranges. The San Andreas, Rogers Creek, and Maacama faults are drawn and labeled. An inferred approximate stress direction for the Rogers Creek Fault is shown with arrows. Mountain ridges between Santa Rosa and Clear Lake are oriented at an angle to the fault lines consistent with an interpretation that they formed as pressure ridges via compression perpendicular to the stress direction. Note also that ridges in the lower-left part of the image are oblique to the San Andreas Fault. These may also have formed via compression perpendicular to the stress direction.

younger than the central part and therefore the eastern part has not yet been elevated or tilted as much. Not only is the ridge growing taller via uplift, it is also growing longer at its eastern end. Rates of uplift were provided by radiocarbon and uranium-series dating of charcoal and carbonate in soil horizons. These dates indicate that Wheeler Ridge has been uplifting at a minimum rate of about 1 foot per 100 yrs for at least the past 1,000 years. The dates also indicate that the fold is becoming longer at a rate of about 9.8 feet per 100 years. In other words, during a 100-year period, almost 10 feet of flat ground at the east end of the ridge is uplifted to form a small anticlinal hill. Given these rates of growth, it is estimated that the Wheeler Ridge anticline began to develop about 400,000 years ago. Because Wheeler Ridge is located at the northern margin of the Transverse Ranges, its youthful development suggests that ridge growth may be spreading northward into the southern edge of the Central Valley. Perhaps we are witnessing the very beginning of the reincarnation of the Central Valley.

Wheeler Ridge is not the only documented growing structure in the Transverse Ranges. Another fold, the Ventura Avenue anticline, located near Ventura, is one of the fastest-growing structures known. Dating techniques and soil correlations suggest that this structure grew at an average rate of about 6.5 feet per 100 years between about 200,000 and 100,000 years ago and that today, it is growing at a rate of nearly 20 inches per 100 years.

It should not be surprising that actively growing mountains close to an ocean would produce wave-cut terraces. As many as 14 terraces have been recognized on Santa Catalina Island with the highest as much as 1,700 feet above sea level. Terraces at this elevation cannot possibly be due to sea-level change. They most definitely indicate recent uplift. Two dated wave-cut terraces in the Pacific Palisades along the Malibu Coast Fault (18) on the outskirts of Los Angeles suggest uplift rates of 1.18 inches per 100 years for the past 320,000 years. A high, tilted, and warped terrace overlooks the beach at Pacific Palisades in Figure 17.9.

The opposite effect, that is, extension, can occur where a right-lateral strike-slip fault bends or steps to the right. This type of structure is shown at the bottom of Figure 17.7. A right step in the San Andreas Fault system occurs at the southern termination of the San Andreas Fault along the east side of the Salton Sea. As noted earlier, displacement picks up again along the Imperial Valley Fault (27) on the west side of the Salton Sea, thus creating a right-step, extensional geometry (Figure 17.3). The resultant landscape is a normal-fault basin in the form of the Salton Sea, which is 227 feet below sea level. The Salton Sea is California's largest lake at about 381 square miles but has a maximum depth of only 51 feet. Like the Great Salt Lake, the Salton Sea has no outlet. Water is lost through evaporation such that salinity in the lake is about 25% greater than ocean water. A basin that forms by normal faulting along a bend or jump in

FIGURE 17.9 A Google Earth image looking northward at a high terrace at Pacific Palisades west of Los Angeles.

a strike-slip fault is referred to as a *pull-apart basin*. Pressure ridges, pull-apart basins, and narrow, linear fault-valleys are characteristic features along the San Andreas and other strike-slip faults in California producing some of its distinctive physiography.

ROTATION OF THE TRANSVERSE BLOCK

Earlier we noted that the Pacific-North American plate boundary (e.g., the San Andreas Fault) has jumped eastward several times since 27 million years ago such that rocks that were originally part of the North American plate have been transferred to the Pacific plate and displaced northward. The result of this process, particularly in southern California, has been the development of several fault-bound blocks and pull-apart basins that have moved semi-independently of each other. The present-day distribution of these tectonic features west of the San Andreas Fault is shown in Figure 17.10. Significant among these are four blocks, the Santa Lucia Bank (*SL*), the Outer Borderland (*O*), the Transverse (*T*), and the Salinian (*S*) blocks, and two pull-apart basins, the Santa Maria Basin (*SM*) and the Inner Borderland Basin (*I*), which includes the Los Angeles Basin. Note that some of these features are currently offshore. Note also that the Transverse block is oriented approximately perpendicular to the trend of the other blocks.

One method by which to understand the origin of the Transverse block is paleomagnetism. As everybody knows, iron is magnetic and can become oriented in a magnetic field. This simple relationship is used in geology to determine whether iron-bearing rocks, particularly volcanic rocks, have been rotated since the time they were deposited.

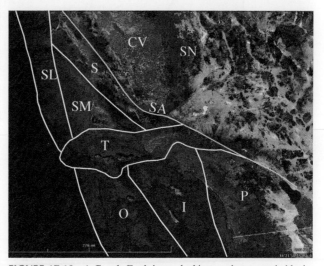

FIGURE 17.10 A Google Earth image looking north at tectonic blocks and pull-apart basins along the southern California coastline. *CV* = Central Valley, *I* = Inner Borderland Basin, *P* = Peninsular Ranges and Valleys, *O* = Outer Borderland block, *S* = Salinian block, *SA* = San Andreas Fault, *SL* = Santa Lucia Bank block, *SM* = Santa Maria Basin, *SN* = Sierra Nevada, *T* = Transverse block.

As a volcanic rock cools, the iron in the rock becomes oriented parallel with the Earth's magnetic field. Once the rock is cold, the magnetic orientation is locked in place. If a rock is rotated after the magnetic orientation is locked in place, a geologist can determine the amount of rotation. When the various blocks in southern California were analyzed, it was discovered that rocks older than 16 million years in the Transverse block had been rotated 80 to 110 degrees clockwise and that progressively younger rocks in the block were rotated progressively lesser amounts. These relationships imply that clockwise rotation began soon after 16 million years ago and has continued, possibly to the present day. It thus appears that the Transverse Ranges were first rotated into the orientation we see today and then, beginning about 5 million years ago, underwent very rapid active uplift.

Figure 17.11 is a schematic set of sketches that show how and why this rotation occurred as well as the origin of the two pull-apart basins. Twenty million years ago, the San Andreas Fault likely was positioned offshore and west of all four blocks. Sometime around 16 million years ago the fault jumped eastward, transferring all four of the tectonic blocks onto the Pacific plate. Note also that the eastward jump created a transtensional bend in the fault. The Santa Lucia Bank, Salinian, and the Outer Borderland blocks were all displaced northward. The northern (now eastern) part of the Transverse block apparently snagged against the continent

while the southern (now western) side swung seaward. Much of this rotation likely occurred while the Transverse block was below sea level. The geometry of the rotating block resulted in the opening of the Santa Maria and Inner Borderland basins. Additional room was created by the plate kinematics in which the Pacific plate was not only sliding past the North American plate but also diverging slightly. A final eastward jump beginning about 5 million years ago created the Big Bend and a transpressional plate boundary, leading to rapid uplift of the Transverse Ranges.

The smaller bend east of Monterey Bay produces compression and uplift (transpression) in the Santa Cruz Mountains south of San Francisco. As shown in Figure 17.12, the San Andreas Fault actually cuts through part of the Santa Cruz Mountains and the mountains themselves appear to be highest in the vicinity of the bend.

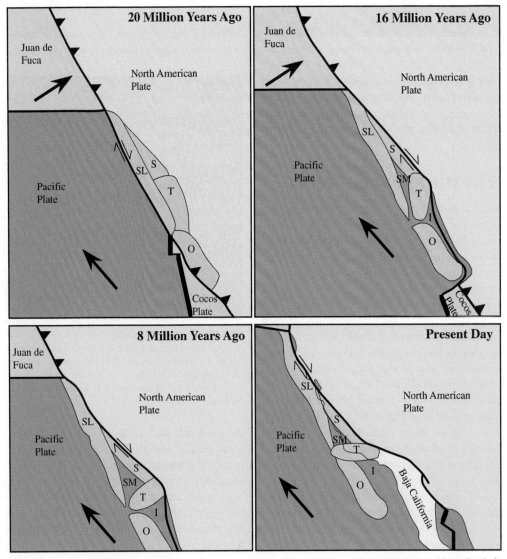

FIGURE 17.11 Sketches that show progressive rotation of the Transverse block and development of the Santa Maria and Inner Borderland basins. Based on Atwater (1998).

THE EASTERN CALIFORNIA–WALKER LANE BELT

If you look closely at Figures 17.1 and 17.2 you will notice that the trend of mountain ranges in the Basin and Range directly east of the Sierra Nevada is different from mountain ranges in central Nevada. For example, the Stillwater, Shoshone, and Toquima ranges in central Nevada all trend north to northeast, as do surrounding ranges. This trend changes to a more variable but dominantly northwest direction along a line that is most clearly seen in the vicinity of Pyramid and Walker lakes. The line is shown in Figure 17.3 as extending from the vicinity of Alturas, California, southward to the Salton Sea. The line represents the approximate eastern limit of a zone of active right-lateral strike-slip faults and normal faults known as the Eastern California–Walker Lane Belt. The belt is not defined by a single through-going fault but rather by perhaps a dozen or more separate, individual faults. The system of faults diverges from the San Andreas fault in the vicinity of the Salton Sea and extends discontinuously along the east side of the Sierra Nevada to the vicinity of Alturas, where the fault system appears to die out. The complex interaction of strike-slip and normal faulting has produced some of the most recognizable landforms in the region. Death Valley, Owens Valley, Mono Lake, and Lake Tahoe are all essentially pull-apart basins rimmed with normal faults that have formed in a manner similar to the Salton Sea. Some of the major faults associated with this belt from south to north include faults in the Mojave Desert that collectively define the Eastern California fault system (29), the Sierra Nevada

Frontal Fault System (31), the Panamint Valley (32), Fish Lake Valley-Furnace Creek-Death Valley (33), Owens Valley (34, site of major earthquakes in 1872), the White Mountains (35), Carson Lineament (40), Honey Lake (42), and Pyramid Lake (43) faults. Total displacement on all of the faults combined is on the order of 30 to 60 miles in southern California, 35 to 50 miles in the Lake Tahoe region, and essentially zero in the north, where the fault zone dies out.

The time for the initiation of faulting can be deduced from field evidence. There is an abundance of sedimentary and volcanic rock as well as unconsolidated and semiconsolidated sediment that spans the age range from about 35 million years to the present. Each has a slightly different structural history depending on its age. In very simple terms, if the rock or sediment is offset by a fault, faulting must be younger than the age of the rock. If rock or sediment cuts across a fault without offset, faulting must have occurred before deposition of the rock or sediment. A second line of evidence can be found in the many fault-generated basins. Sediment and volcanic rocks deposited in a basin must be younger than the initiation of faulting that created the basin. A third line of evidence can be found in the progressive, fault-activated tilting of rock and sediment. When these relationships are applied to the many faults in the Eastern California–Walker Lane Belt, they suggest that faulting has been active since about 11 million years ago along much of the belt except for the very northern part where faulting may have begun only 3.5 million years ago. Faulting, beginning 11 million years ago, began to restrict rivers from crossing the Sierra Nevada thus signalling initial

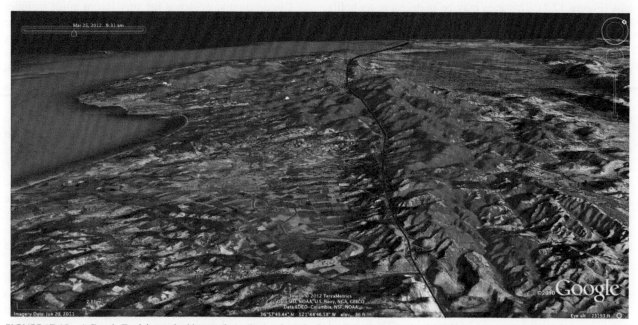

FIGURE 17.12   A Google Earth image looking north-northwest along the Santa Cruz Mountains south of San Francisco. The thick line and thin red lines represent the San Andreas Fault, which cuts obliquely through the mountains.

development of the great Sierra Nevada escarpment. Rivers were permanently cut off by 7 million years ago.

The Eastern California–Walker Lane Belt is significant because satellite measurements indicate that the area between the belt and the San Andreas Fault is moving as a single block separate from the North American plate. In other words, this area, which includes the Sierra Nevada, Central Valley, Mojave Desert, part of the Coast Ranges, and the eastern part of the Transverse Ranges, is a new tectonic plate that is in the process of forming. It is a microplate known as the *Sierran plate*. Satellite measurements indicate that the Sierran plate is moving northward and rotating counterclockwise relative to the North American plate at a rate of approximately 3 feet per 100 yrs. It is this movement that has created the diffuse zone of strike-slip and normal faults along the eastern margin of the microplate. It appears that the Sierran plate is literally being ripped off the North American plate.

Rather than becoming a separate microplate, another possibility is that the Eastern California–Walker Lane Belt is in the process of becoming the next San Andreas Fault. If so, this would constitute a major eastward jump in the San Andreas Fault system, the largest in its history, and would transfer the Sierran plate to the Pacific plate. Such a possibility is for the future. In the present, the Sierran plate is rotating toward the Klamath Mountains, producing compression that may be responsible for some of the rapid active uplift of the Klamaths as described in Chapter 16. Compression associated with rotation of the Sierran plate may also be responsible for active development of small anticlinal ridges located just north of Reading, at the northern edge of the Central Valley and southeast of the Klamaths (Figure 17.1). This area may be undergoing an uplift history similar to that described for Wheeler Ridge at the southern edge of the Central Valley.

The western, eastern, and southern boundaries of the Sierran microplate are fairly well defined by the San Andreas and Eastern California-Walker Lane fault zones. The northern boundary, however, is poorly defined. Recall that the Mendocino triple junction is slowly migrating northward. One possibility is that the Walker Lane Belt is growing (lengthwise) in a northwestward direction with migration of the triple junction. If this is the case, the Walker Lane should meet up with the triple junction somewhere along the Oregon coast. When this occurs, the eastward jump will be complete, the Sierran microplate would transfer to the Pacific plate, and the Eastern California–Walker Lane Belt would become the new San Andreas transform plate boundary.

THE COAST RANGES

Taken as a whole, the Coast Ranges could be the youngest structure-controlled landscape in the country. They extend from the Klamath Mountains in the north to an area just south of Point Sal at Santa Maria where the mountains turn abruptly east-west to form the Transverse Ranges. The Coast Ranges have strong linear trends, but a glance at Figure 17.2 reveals that the linear landscape is more pronounced in the Central-Southern Coast Ranges south of San Francisco. Here the ranges are drier and somewhat less rugged than in the north. Most of the peaks are less than 4,000 feet except for the area northeast of Santa Barbara where a few peaks top 8,000 feet. The dominant structural feature is the active San Andreas Fault, which cuts diagonally across the region at a slight angle to the coastline, intersecting the coast at San Francisco. The Coast Ranges (and valleys) are approximately parallel with the fault and therefore also intersect the coast at a slight angle.

The Central-Southern Coast Ranges are bisected by the San Andreas Fault, and as such, one would expect uplift and landscape evolution to be closely tied to fault displacement. Such an interpretation would imply that landscape could be as old as 27 million years. Instead, the evidence suggests that the ranges we see today are very young, certainly less than 4 million years old and perhaps no more than about 400,000 years old. There is abundant evidence that the mountains are still in the process of forming and that they are influenced by tectonics in addition to the San Andreas Fault. In this section, we will concentrate on the origin of this very young landscape. We noted earlier that rocks in the Northern Coast Ranges north of San Francisco consist largely of Franciscan complex. Southward, the rocks are more variable. Sedimentary rocks of various ages are most common, but Franciscan complex and granitic and metamorphic rocks of the Salinian block are also present, as are a few minor volcanic rocks.

Circumstantial evidence for very recent uplift includes the nature of the rocks from which the mountains are made. The sedimentary rock, with few exceptions, is poorly consolidated, fractured, and subject to landslides. Granitic and metamorphic rocks in the Salinian block south of Monterey are somewhat stronger, but even these rocks have been weakened by intense fracture. It is unusual, to say the least, for such weak rock to form mountains. The only explanation is that the mountains are young and that they were uplifted quickly, at a higher rate than the rather rapid pace of erosion. Clearly, these are structure-controlled, not erosional, mountains. Even the valleys appear to be structure-controlled. We know this because the valleys are synclinal in nature. They are not erosional features. They are, in fact, currently being filled with sediment rather than deepened by erosion. Evidence for recent uplift can also be seen along the shoreline. The Santa Cruz Mountains south of San Francisco, the Santa Lucia Mountains south of Monterey, and the San Luis Range west of San Luis Obispo all drop directly into the ocean or form a series of wave-cut terraces that step down to the ocean such as those already mentioned in the southern Santa Cruz Range (Figure 9.18) or those at Pacific Palisades (Figure 17.9). The combination of weak rock, steep slopes, uplift, strong wave action, and infrequent storms results in numerous landslides, both along the coastline and inland.

In addition to the hardness of rock, another indication that the mountains are young is the age of some of the rocks and where they were deposited. Marine rocks as young as 4 million years old are present in the Coast Ranges. Recall that the term *marine* indicates that the sediment was deposited in an ocean setting at or below sea level. The implication is that the present-day landscape must be younger than 4 million years. Assuming the land surface was approximately at sea level 4 million years ago, and given the average height of the mountains, we can calculate an average surface uplift rate (that is, the rate at which the ground surface was elevated into a mountain range since 4 Ma) of about 0.79 inches/100 yrs.

In spite of the presence of 4-million-year-old marine rocks, there is evidence in the rock record in the form of unconformities that indicate part of the Coast Ranges were deformed and elevated above sea level (possibly more than once) beginning about 29 million years ago at the initiation of the San Andreas Fault. These deformations, however, apparently did not result in widespread or lasting emergence of the region above sea level. This perhaps should not be surprising. The San Andreas is, after all, a strike-slip fault in which rocks are displaced horizontally, ideally with no uplift or subsidence.

What is surprising, or at least unusual, is that nearly all the Central and Southern Coast Ranges have internal folds and faults that are oriented parallel with or at a slight angle to the San Andreas Fault. Recall from our earlier discussion that the maximum stress direction associated with a strike-slip fault is approximately 30 degrees from the strike of the fault and that anticlinal mountains, synclinal valleys, thrust, and reverse faults would be expected to develop in an en echelon pattern perpendicular to the compression direction and, therefore, 60° to the strike-slip fault (Figure 17.7). We clearly do not see this orientation in the Central-Southern Coast Ranges. So, how do we explain the near parallelism of folds and faults with the San Andreas Fault?

First, we have to realize that the San Andreas Fault was probably close to but never perfectly aligned with the displacement direction of the Pacific plate relative to the North American plate. Let us look at the consequences of two hypothetical examples. Figure 17.13a assumes that the Pacific plate is approaching a stationary North American plate at a convergence direction of 30° east of north and that the plate boundary is oriented due north. Under these conditions, the plate boundary will experience pure strike-slip displacement with no additional component of compression or extension. Now let us rotate the convergence direction 40 degrees clockwise, as shown in Figure 17.13b, but keep the plate boundary in its previous due-north position. Under these conditions the plate boundary will experience compression in the form of folds, reverse, and thrust faults as

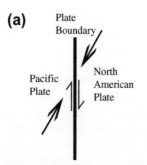
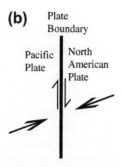
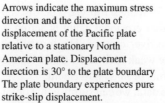

Arrows indicate the maximum stress direction and the direction of displacement of the Pacific plate relative to a stationary North American plate. Displacement direction is 30° to the plate boundary. The plate boundary experiences pure strike-slip displacement.

The displacement direction of the Pacific plate has rotated 40° clockwise so that the plate boundary now experiences a strong component of compression in addition to strike-slip displacement.

FIGURE 17.13 Sketch maps that show (a) ideal strike-slip faulting along a plate boundary and (b) transpression along a plate boundary caused by a change in relative plate convergence.

well as a component of strike-slip faulting (that is, transpression). Note also that the folds and faults will develop nearly parallel with the plate boundary.

Studies of plate motions suggest that the direction of relative motion between the Pacific and North American plates rotated clockwise about 8 million years ago but that the orientation of the plate boundary (the San Andreas Fault) did not change. The new geometry resulted in compression and strong deformation in the Coast Ranges akin to Figure 17.13b (although the amount of rotation was considerably less than that shown in Figure 17.13b). Thus, it was not movement along the San Andreas Fault that created the structure and landscape in the Coast Ranges, it was compression caused by a misfit between the orientation of the fault and the relative direction of convergence between the two plates. But wait! There is more to the story. The presence of circa 4 Ma marine rocks in the Coast Ranges indicates that 4 million years would pass following the change in convergence direction before the Coast Ranges finally and permanently emerged from below sea level. If deformation began soon after 8 million years ago, it would appear that deformation of the rocks preceded emergence of land areas by about 4 million years.

Some insight to the uplift history of the Coast Ranges may be gained by an apatite (U-Th)/He study of the Santa Lucia Mountains located south of Monterey. Recall that this dating system has a low closure temperature, which, in this case, allows for the determination of the erosional exhumation history over the past 10 million years. The data suggest slow cooling between about 10 and 6.1 million years ago, followed by rapid cooling from 6.1 to 2.3 million years ago, followed by even more rapid cooling beginning sometime after 2.3 million years ago. Geological evidence suggests that the cooling history can be correlated directly with the erosional exhumation history. The data suggest that following a period of little or no exhumation from 10 to 6.1 million years ago, the rocks were exhumed at a rather steady rate of 1.14 to 1.61 inches per 100 yrs between 6.1 and 2.3 million years ago and that this rate increased to 3.54 inches/100 yrs or higher sometime after 2.3 million years ago. The data do not constrain exactly when (after 2.3 million years ago) the exhumation rate increased to 3.54 in/100 yrs. In any case, the 3.54 in/100 yr rate is considerably higher than estimated rates of river erosion which are on the order of only 0.39 inches/100 yrs. The discrepancy suggests that some other erosional agent is acting on the mountain in addition to river erosion. The most likely and visible candidates in the weak rock of the Coast Ranges are landslides and other forms of mass movement.

In summary, given these data, we can surmise that deformation in the Santa Lucia Range began as early as 8 million years ago and that uplift began by around 6.1 million years ago, although the landscape was not widely elevated above sea level until 4 million years ago. Rapid uplift and development of present-day landscape were delayed until sometime after 2.3 million years ago.

Let us now examine the conclusion that deformation preceded initial uplift by as much as 1.9 million years. There is field evidence to support this contention. The first line of evidence is based on the trend of internal folds and faults. Earlier it was stated that internal folds and faults within mountain ranges are oriented close to parallel with the San Andreas Fault. What is significant about this is that, in some cases, the folds and faults within a particular mountain range are oriented at a slight angle to the trend of the mountain range itself. In other words, the morphology of a particular range seems to have no direct correlation with the internal structures that are present within the range. We

have seen this before. It is the same type of relationship seen in mountain blocks of the Basin and Range where the internal structure of the foreland fold-and-thrust belt has no correlation with the landscape of the range itself, which instead, is controlled by the much younger range-front normal faults. The second line of evidence is the fact that all of the internal structures are truncated (cut) by an erosion surface. Apparently any mountains or hills that did form during initial emergence 4 million years ago were quickly removed by erosion, leaving behind a relatively flat lowland erosion surface. Today this gentle lowland surface is preserved at or near the summit of many of the ranges. From this, we can surmise that the Coast Ranges existed as a lowland for an indefinite period prior to uplift into a mountain range. We can also surmise that the uplift must have been relatively rapid and recent given the weak nature of the rocks in the mountains and the preservation of the erosion surface at high elevation. Given the exhumation data on the Santa Lucia Range, one logical conclusion is that the lowland was elevated into a mountain sometime after 2.3 million years ago. There are some who would argue, based on field data, that rapid surface uplift did not begin until as late as 400,000 years ago. Thus, the mountains we see today are very likely less than 2.3 million years old and possibly as young as 400,000 years old.

We need to ask one final question. How did uplift occur? Individual mountain blocks in the Coast Ranges are bound not by normal faults as in the Basin and Range but by range-front reverse, thrust, and strike-slip faults. The mode of uplift was compression, not extension. An interesting hypothesis is illustrated in a cross-section of the Santa Cruz Mountains in Figure 17.14. The cross-section extends from the Pacific coastline across the northern Santa Cruz Mountains to the San Francisco Bay south of San Francisco. The interpretation

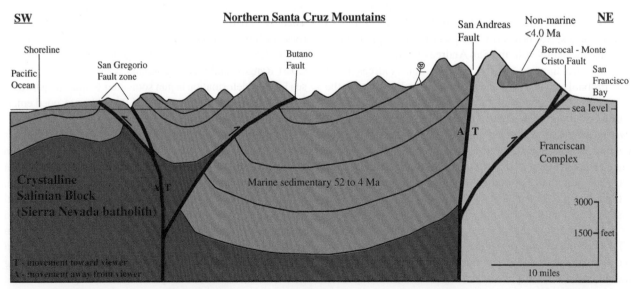

FIGURE 17.14 Southwest-to-northeast interpretive cross-section sketch across the Santa Cruz Mountains. Based on Page (1998).

is as follows. Initial compression caused deformation, which thickened and strengthened the crust beneath the Coast Ranges. Later compression then caused some of the strike-slip faults to reactivate as reverse faults resulting in mountain uplift. In some cases a preexisting strike-slip fault has splayed out into a flower-like structure of reverse and thrust faults, along which the mountains were uplifted. In Figure 17.14 the San Gregorio Fault is shown as a single fault at depth that has splayed out toward the surface, creating uplift along the Butano thrust within the Santa Cruz Mountains as well as range-front uplift along reverse faults on the west side of the mountain. The figure also shows the Berrocal-Monte Cristo reverse fault splaying out from the San Andreas Fault to form a range-front reverse fault on the east side of the Santa Cruz Mountains. Some of these faults have now reverted to strike-slip displacement which suggests that present-day compression and the rate of uplift have slowed. The sequence, therefore, was dominantly strike-slip faulting followed by compression, reverse faulting and uplift, and then back to dominantly strike-slip faulting.

Thus, the evidence suggests that uplift of the central-southern Coast Ranges is recent, that it post-dates strong deformation in the rocks, and that the cause is transpression across the San Andreas system due to a misfit between relative plate convergence and the orientation of strike-slip faults.

Recall that the southern part of the San Andreas system jumped to its final location in the Gulf of California about 5 million years ago, prior to uplift of the Central-Southern Coast Ranges. This jump, however, occurred south of Los Angeles. It certainly affected the Transverse Ranges, but whether or not it affected the Coast Ranges to the north is not clear. What we can say is that earlier eastward jumps on the San Andreas Fault probably resulted from an attempt by the plate boundary to reorganize itself following a change in the convergence direction of the Pacific plate relative to the North American plate. The final jump 5 million years ago may have been in response to the change in plate motion that occurred 3 million years earlier. Changes in relative plate motion must be one of the major driving forces causing transpression, the development of new strike-slip faults, the development of bends in existing strike-slip faults, and causing the repositioning of the San Andreas Fault itself.

PENINSULAR RANGES

The Peninsular Ranges are located south of Los Angeles and west of the San Andreas Fault. Most of the ranges, including the Santa Ana, Santa Rosa, San Jacinto, and Laguna mountains, form a separate, southward extension of the Sierra Nevada batholith. Metamorphic rocks are present within the granitic batholith in many areas. Sedimentary rocks are also present, especially near the coast. The ranges tilt westward and are separated by active faults that include the Rose Canyon (24), Elsinore (25), and San Jacinto (26; Figure 17.3). The San Jacinto Mountains form a steep escarpment on their eastern and northern sides that overlooks Palm Springs and Snow Creek more than 10,000 feet below. Figure 17.15 is a view looking east-southeastward at the steep escarpment and at the high San Bernardino Mountains, which overlook Yuciapa.

The landscape history of the Peninsular Ranges is not well understood and probably varies across the region. The general increase in elevation inland from the coastline, coupled with the character of sedimentary rocks and soil horizons, suggests that inland areas were uplifted earlier

FIGURE 17.15 A Google Earth image looking southeastward at part of the Peninsular Ranges. The steep escarpment of the San Jacinto Mountains overlooks Palm Springs (*PS*). The San Bernardino Mountains overlook Yucaipa (*Y*). The thin, barely visible lines are faults active in the past 1.6 million years, including the Eastern California shear zone (*EC*), the Santa Anna fault zone (*STA*), the San Andreas fault zone (*SA*), the San Gorgonio Pass fault zone (*SGP*), and the San Jacinto fault zone (*SJ*). The Salton Sea (*SS*) is visible at top-center. Earthquake lines are from USGS Website: http://earthquake.usgs.gov/regional/qfaults/.

than coastal areas and that they were above sea level (but not necessarily mountainous) by 40 million years ago. The timing for significant mountain uplift is poorly constrained but likely occurred prior to 3.8 million years ago. This is in contrast to marine sedimentary rocks along the coastline that are between 5 and 2 million years old. These rocks indicate that coastal areas did not emerge from below sea level until after about 2 million years ago. Wave-cut terraces are common along the coastline, including Oceanside north of San Diego and the Palos Verde Hills south of Los Angeles, implying recent uplift.

THE SIERRA NEVADA

At 400 miles long and 40 to 60 miles wide, the Sierra Nevada is the largest single continuous mountain block in the US (Figure 17.1). No rivers cross this block. It is composed primarily of massive batholithic granitic rock, with metamorphic rocks increasingly more abundant toward the north. The mountain tilts gently westward and northward such that the highest peaks are in the southeastern part of the range, including Mt. Whitney (14,495 feet), the highest peak in the contiguous US. There are at least eight additional 14,000-foot peaks in the range, all within about 40 miles of Mt. Whitney. The westward tilt of the range is visible in Figures 14.19 and 17.16. Elevation decreases sharply south of Mt. Whitney, where the Sierra Nevada ends abruptly against the Garlock Fault and the Mojave Desert. Northward, the mountain range loses elevation gradually as it plunges beneath volcanic rocks of the Cascade Mountains and Modoc Plateau.

The strong westward tilt produces a very steep eastern escarpment that is cut heavily by glaciers, creating a jagged, rugged appearance, with steep rock faces and the greatest relief anywhere in the contiguous US. There is well over 10,000 feet of elevation change within 10 miles between Mt. Whitney and Owens Valley. This great escarpment is visible in Figure 15.9, which is a Google Earth image of the southern part of the range, and in a photograph of Mt Whitney shown in Figure 17.17. In contrast to the steep eastern face, the western slope is a gently tilted erosion surface that rises gradually from the Central Valley. Deep glacial canyons, including Toulumne, Yosemite, Kings, and Kern canyons, are cut into the tilted erosion surface producing spectacular scenery, particularly at Yosemite National Park. Glaciers have excavated a fault/fracture zone in Kern Canyon south of Mt. Whitney, producing a straight, 70-mile-long gorge up to 6,000 feet deep that cuts southward through the heart of the highest peaks in the Sierras. Today this canyon is occupied by the Kern River and is visible in Figure 17.16.

The Sierra Nevada is, in a sense, a lynchpin between the older subduction-related geology of California and the younger normal fault geology of the Basin and Range. Superimposed on both of these are still younger tectonic

features associated with the Walker Lane Belt. Needless to say, the landscape history is complex and not fully understood. A more complete interpretation has emerged in only the past few years based both on detailed fieldwork and on sophisticated methods of obtaining elevation, paleoclimate, and age data, as well as methods that can reveal relative rates and recency of uplift. Such methods include (U-Th)/He dating, oxygen and hydrogen isotope analysis, fossils, and incision rate analysis. In a nutshell, the data suggest that the Sierra Nevada has been a topographic highland for considerably more than 80 million years, but has not always been a mountain range with a steep eastern escarpment physiographically distinct from the Basin and Range.

The western slope of the Sierra Nevada is cut by river valleys whose shape reveals something about the landscape history. In general, a steep river valley that is also narrow with high relief reflects more rapid or sustained down-cutting relative to one of similar size without these characteristics. The cause of down-cutting could be more rapid or sustained uplift, but it could also be due to climate or rock type variations. In the case of the Sierra Nevada, we can, to a certain extent, discount climate and rock type variations because the climate at a particular elevation is fairly uniform and because the rocks are uniformly crystalline. When this type of analysis is applied to the Sierra Nevada, we find that areas of most recent uplift are in the southern Sierra (that is, the Mt. Whitney area south of Bishop) and in the northern Sierra (Lake Tahoe area and points north). A quick gander at Figure 17.3 shows landscape evidence for recent uplift in the southern Sierras. This area is topographically highest. It is bound by several active faults, including the Sierra Nevada frontal fault system (31), the Kern Canyon Fault (30), and the Garlock Fault (20). These faults have contributed to uplift and to both the westward and northward tilt of the range. Beryllium-10 cosmogenic radionuclide surface-exposure dating along the Sierra Nevada frontal fault system (31) suggests that total vertical (up and down) displacement over the past several 100,000 years has occurred at a rate of 0.79 to 1.18 inches/100 yrs. Physiographic evidence for relatively recent uplift in the northern Sierras, on the other hand, is not as obvious. Uplift in this region could be very recent and due to the migration of the Walker Lane Belt into the area.

In order to understand the landscape history of the Sierra Nevada, we need to begin at a time 80 million years ago. Subduction-related volcanism ended in the Sierra Nevada 80 million years ago, even though subduction off the west coast of California continued uninterrupted until initiation of the San Andreas Fault 29 to 27 million years ago. We know that volcanism ended in the Sierra Nevada because there is a large gap in the age of volcanic rocks beginning 80 million years ago and ending about 16 million years ago, when volcanism returned. At this point you should be asking

FIGURE 17.16 A Google Earth image looking northward along the southern crest of the Sierra Nevada. The westward tilt of the range is visible. The major northerly trending scar through the middle of the range is Kern Canyon. Part of the relict erosion surface (*r*) is visible above the canyon walls.

FIGURE 17.17 A photograph looking westward at Mt. Whitney and the great eastern escarpment of the Sierra Nevada.

yourself, if subduction off the West Coast continued uninterrupted, why did volcanism end in the Sierra Nevada, and where did it go? We will address these questions shortly. First let us trace the landscape history of the Sierra Nevada since 80 million years ago.

Recall from Chapter 15 that by 43 to 24 million years ago the Sierra Nevada formed the western margin of the Nevadaplano (Figure 15.22). The great eastern escarpment that today separates the Sierra Nevada from the Basin and Range was not yet in existence as evidenced by ancient west-flowing stream channels that crossed this boundary on their way to the paleo-Pacific ocean. Apatite and zircon (U-Th)/He ages from west of Lake Tahoe and south of Yosemite Valley are consistent with relatively high exhumation rates on the order of 0.78 to 3.14 inches per 100 yrs between 90 and 60 million years ago, followed by much slower rates of 0.08 to 0.20 inches per 100 yrs from 60 million years to present. The apparent rapid exhumation during the 90 to 60 million-year time span could be due to uplift and concomitant erosion, or it could possibly be due to rapid erosion of a highland composed of weak volcanic rock. In either case, slow exhumation rates since 60 million years ago suggest limited uplift over that time span which, in turn, implies that the Sierra Nevada, and very likely the Nevadaplano, were already in existence and at high elevation by 60 million years ago and probably prior to 80 million years ago. The geological evidence, therefore, suggests that the Sierra Nevada 80 million years ago was a westward-sloping, rolling upland, probably with embedded volcanoes and a surface covered in volcanic material. The landscape at that time may have resembled the present-day volcanic landscape of the Central and Southern Cascades, the major differences being the probable existence of the Nevadaplano to the east and the presence of an ocean to the immediate west at the present-day location of the Central Valley. The exact elevation of the Sierra Nevada during the 80 to 60 million-year interval is unknown but likely was more than 7,000 feet and possibly 10,000 feet or more. The presence of granitic debris in sediment surrounding the Sierra Nevada indicates that by 65 million years ago the volcanic carapace had largely been eroded and batholithic rock was widely exposed.

There is little geological evidence for widespread tectonic activity in the Sierra Nevada from 65 million years ago to possibly as late as 16 million years ago, when volcanism returned to the area. The region appears to have undergone a lengthy period of slow erosion, as suggested by the previously mentioned apatite and zircon (U-Th)/He ages. The erosion surface that formed during this time interval is still well preserved along the lower western slope in the northern Sierra Nevada as well as in the southern part of the range, beyond the confines of the glacially carved Kern River canyon. This ancient erosion surface was recognized as early as 1904 as a relict surface of abnormally low relief relative to surrounding areas. A small remnant of this erosion surface is visible in Figure 17.16 above Kern Canyon. The erosion surface, although of low relief, certainly was not flat. Embedded in the erosion surface are ancient, now abandoned, river valleys filled with circa 40- to 50 million-year-old gold-bearing river gravels that were exploited during the 19th-century Gold Rush. Also present among these ancient, now abandoned river deposits are 22 million-year-old (and older) volcanic ignimbrite deposits derived from Central Nevada and used to show that rivers once flowed westward from central Nevada, across the eastern Sierra Nevada escarpment to the ocean, as described in Chapter 15. These river deposits are best-preserved west and northwest of Lake Tahoe.

Given the slow exhumation rates in the Sierra Nevada and the existence of the Nevadaplano, there is little evidence for landscape change in the Sierra Nevada until about 16 million years ago. Recall that 16 to 18 million years ago began a time of profound change in the Cordillera. This is the approximate time for initiation of the Yellowstone hot spot and the Columbia River flood basalts, for the return of volcanism to the Sierra Nevada, and for widespread normal faulting concomitant with destruction of Nevadaplano landscape in Nevada and initiation of Basin and Range landscape. We can now address our earlier question as to why volcanism ended in the Sierra Nevada 80 million years ago and where it went. The lithospheric plate that was subducting beneath California 80 million years ago was the Farallon plate. Recall that a sinking plate must reach a depth of 62 miles for melting to begin. Recall also that many of the 70- to 20-million-year old intrusive and volcanic rocks in the central Cordillera east of the Sierra Nevada (discussed in Chapter 14) are andesitic in composition with a chemical makeup that indicates they were derived from melting generated by subduction, presumably of the Farallon plate. Andesitic rocks are present in the Challis magmatic belt from northern Washington to Yellowstone and in younger volcanic areas, including the San Juan and Mogollon-Datil fields. The evidence, therefore, suggests that subduction-related volcanism shifted to areas to the east of the Sierra Nevada. The question then becomes, why did volcanism shift eastward, and why did it return 16 million years ago?

The eastward shift in volcanism is understood to be the result of shallowing of the Farallon subducting slab. Instead of sinking into the mantle within a few hundred miles of the trench, the Farallon plate appears to have pushed itself horizontally beneath the North American lithosphere, a process known as *underplating*, as shown schematically in Figure 17.18. Because of the shallow angle of subduction, the Farallon slab could not generate magma until it was well to the east of the Sierra Nevada. By about 55 million years ago, the underplated Farallon slab was creating part of the Idaho Batholith and the Challis volcanic belt.

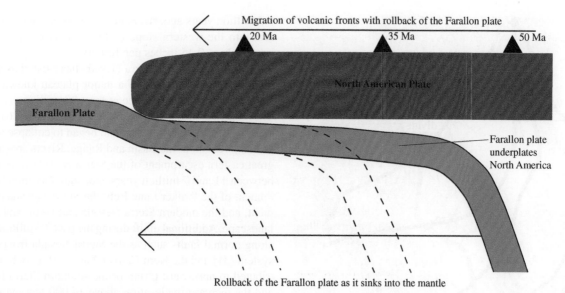

FIGURE 17.18   A cross-section sketch that shows the Farallon plate underplating North America, followed by rollback of the Farallon plate, and the migration of volcanic front with rollback.

The ignimbrite flare-up between 43 and 24 million years ago, and the eventual return of volcanism to the Sierra Nevada by 16 million years ago, are also associated with subduction of the Farallon plate. A critical piece of evidence is the progression in the age of andesite and rhyolite volcanism across the Cordillera. Figure 17.19 shows a set of lines that delineate the successive location of volcanic fronts between 50 and 20 million years ago. The figure shows that volcanism was present in eastern Washington and northern Idaho 50 million years ago and that by 20 million years ago, volcanism had spread progressively southward across Nevada. The figure also shows the migration of volcanism northwestward from the New Mexico-Mexico border region 43 million years ago to southern Nevada 21 million years ago. One way to explain these features is to suggest that the eastern part of the Farallon slab began to sink into the mantle, which, in turn, caused the plate to roll back on itself as shown in Figure 17.18. As the plate rolled back and sank, it left a void below the North American lithosphere that was filled with hot, rising mantle rock, which caused melting of the North American crust, thus generating explosive silicic volcanism (the ignimbrite flare-up).

By about 16 million years ago the sinking slab was again generating volcanism in the Sierra Nevada. The San Andreas Fault was in existence by this time but only in southernmost California. Sixteen million years ago the Mendocino triple junction, which separates the San Andreas Fault from the Juan de Fuca plate, was located in the vicinity of San Luis Obispo, well to the south of its present location. Thus, by 16 million years ago, nearly the entire Sierra Nevada was above the northern relict of the Farallon plate (which today is known as the subducting Juan de Fuca plate). 40Ar/39Ar ages indicate that volcanism lasted from

about 16 to 6 million years ago, ending first in the southern and then in the northern Sierra Nevada, as the Mendocino triple junction migrated northward toward its present location. However, rather than forming large volcanic cones such as modern-day Mt. Hood or Mt. Rainier, the field evidence suggests that this phase of Sierra Nevada volcanism was in the form of numerous small andesitic and rhyolitic domes. In some cases, the domes collapsed into blocky lava flows that entered ancient river channels. Amazingly, these are some of the same river valleys that earlier were filled with gold-bearing gravels and later with volcanic ignimbrite deposits derived from central Nevada. An analysis of these now abandoned river valley deposits provides insight to the uplift history of the Sierra Nevada.

Part of the analysis assumes that periods of rapid incision in the ancient river channels record periods of relatively rapid uplift, and that periods of deposition record periods of little or no uplift. Normally it is difficult to date periods of rapid incision; however, in this case, the nearly continuous volcanism from about 16 to 6 million years ago provides easily dated material. On the basis of field data derived from analysis of these ancient, now abandoned river channels, there is evidence for uplift at circa 16, 11, and 7.5 million years ago. If you have been paying attention you will realize that two of these dates (16 and 11 million years ago) have major significance elsewhere in the Cordillera. Earlier in this discussion we mentioned the significance of the 16 million-year date. The 11 million-year date is also significant in that it corresponds with initiation of the Walker Lane Belt, with the beginning of present-day landscape development in the Basin and Range, and with initial development of the great eastern escarpment of the Sierra Nevada. Field evidence suggests that

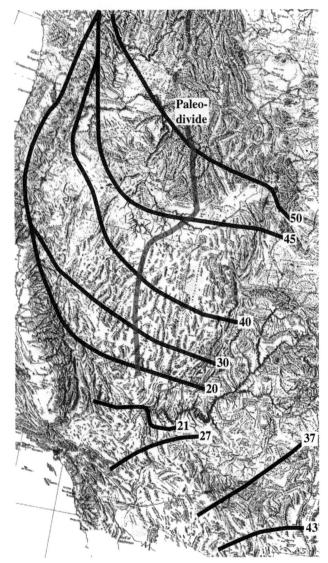

FIGURE 17.19 A landscape map that shows the migration of successive volcanic fronts from eastern Washington-northern Idaho southward to southern Nevada, and from the New Mexico-Mexico border northwestward to southern Nevada. Ages (numbers) are in millions of years. Based on Dickinson (2006) and Humphreys (1995). Also shown is the inferred location of the Nevadaplano paleodivide. Note that volcanic fronts are superimposed on present-day landscape and not on the more narrow pre-Basin and Range landscape in existence at the time of volcanism.

3.5 million years ago, rivers have incised as much as 1,300 feet into the western slope of the mountain—an incision rate of about 0.45 inches per 100 yrs.

To summarize, the Sierra Nevada has existed as a highland at the western edge of a major plateau known as the Nevadaplano, probably since at least 80 million years ago. The modern landscape began to develop as early as 16 million years ago, when the Nevadaplano began to collapse to form an early version of the Basin and Range. Rivers crossing the great eastern escarpment of the Sierra Nevada were cut off between 11 and 7 million years ago, signaling initial development of the Walker Lane Belt, the Sierra Nevada escarpment, and the modern Sierra Nevada and Basin and Range landscape. Additional uplift during the past 3.5 million years along normal faults such as the Sierra Nevada frontal fault system (31) and the Kern Canyon Fault (30) may have contributed to northward tilting of the southern Sierra Nevada and the increase in elevation above 14,000 feet, which put the final touches on the present-day landscape. An ancient erosion surface with now abandoned river channels is preserved on the western slope of the range, particularly west and northwest of Lake Tahoe. Study of these ancient river channels along with a variety of dating and paleo-elevation techniques has contributed to an understanding of the recent landscape history of the Sierra Nevada that is consistent with Cordilleran history as a whole.

QUESTIONS

1. Draw the trace of the San Andreas Fault onto Figure 17.2 or a copy of the figure. Where is the trace of the San Andreas most obvious, and where is the trace least obvious?

2. What are transpression and transtension?

3. What types of structures (extensional or compressional) should be expected between the Maacama (2) and Rogers Creek (7) faults and between the Eaton Roughs (3) and Bartlett Springs (6) faults? All four are right-lateral strike-slip faults. What type of structures (extensional or compressional) should be expected between the two segments of the left-lateral Garlock Fault (20)?

4. Why is it that strike-slip fault valleys, such as those that mark the San Andreas Fault, tend to be long, linear, and straight?

5. If we assume that the Santa Lucia Mountains are approaching a steady state, then surface uplift should be about zero. Explain how there could be an average surface uplift rate of 0.79 inches/100 yrs over the past 4 million years if, indeed, the Santa Lucia Mountains are in a steady state.

6. What rock, climate, and geographic conditions make coastal California particularly susceptible to landslides?

the 7.5 million-year event corresponds with additional Sierra Nevada range-front faulting that permanently cut off any river that previously was able to cross the developing great eastern escarpment. It also correlates with the circa 8 million-year date for the change in plate motion that eventually triggered development of the Coast Ranges. In addition to these three phases of uplift, a final pulse of uplift may have begun about 3.5 million years ago in the southern Sierra Nevada. This final pulse continues today with active faulting along numerous range-front faults that includes the Sierra Nevada frontal fault system (31). Since

7. In Google Earth, fly to Portuguese Bend, California. Describe evidence for past landslides in this area. Estimate the age of the landslides and state criteria for estimating age. Do any of the surrounding areas appear to be susceptible to future landslides? Go to www.consrv. ca.gov/cgs/information/publications/Pages/LSIM_index. aspx and look for other landslide areas.

8. What is the Big Bend along the San Andreas Fault? When did it form, and what is its significance to landscape development of the surrounding area?

9. The San Andreas Fault makes a small bend in the vicinity of the Santa Cruz Mountains. Explain why the fault trends diagonally through the mountain rather than through an adjacent valley.

10. Trace the faults in Figure 17.7 and show the types of structures expected if this were a left-lateral fault system.

11. What is fault creep? How is fault creep different from other types of fault displacement?

12. Describe both rock evidence and landscape evidence that indicates a subduction zone was present along the California coast prior to development of the San Andreas Fault.

13. What is the Franciscan complex?

14. Given a surface uplift rate of 0.79 inches/100 yrs for 4 million years, calculate the average height of the mountains in both feet and meters.

15. The Salton Sea is located between the Imperial Valley and San Andreas faults. What geologic name is applied to this type of basin?

16. The Salton Sea is below sea level and very close to the coast. Explain why it has not been drowned by the ocean. *Hint:* The answer involves the Colorado River, which is best visible in Figure 17.19.

17. What is the Eastern California-Walker Lane Belt? Where is it located? When did it first begin to develop? What might be its eventual fate? What is its relationship with the Klamath Mountains?

18. Describe the evidence that suggests the Coast Ranges are young.

19. A relict landscape is present along the western slope of the Sierra Nevada. What does the landscape look like? When did it form?

20. What evidence is there to suggest that the northern and southern margins of the Central Valley are undergoing reincarnation?

21. What is the Salinian block? Where is it located?

22. How long ago did the great eastern escarpment of the Sierra Nevada begin to develop, and what is the evidence for the timing of its development?

The Story of the Grand Canyon

The Grand Canyon is without question one of the most recognized and spectacular landforms in North America. A book on landscape would hardly seem complete without at least a discussion of this, the quintessential icon of the United States. The Grand Canyon has been studied for more than 100 years, and although we have learned a great deal, its age, origin, and evolution have been difficult to unravel. At the moment, there are three possibilities regarding its age: It was carved entirely over the past 6 million years; it was carved entirely (or nearly so) prior to 6 million years ago with most of the incision occurring prior to 70 million years ago; or it was carved both prior to and after 6 million years ago. The situation is not satisfactory, but it is what it is. The Earth does not give up all its secrets so easily.

In this chapter we will not explore all the various theories on the origin of the Grand Canyon as there are entire books written on that topic. Instead, we will look at some of the most basic evidence, add some observations, and incorporate some of what we have learned in the previous 17 chapters. We will explain why one could argue that the Grand Canyon is less than 6 million years old and explain, in simplified, abbreviated form, a recent hypothesis that suggests the canyon is more than 70 million years old and that the Colorado River had nothing to do with its inception.

THE PHYSIOGRAPHIC CANYON

Before we discuss the carving of the Grand Canyon it would be wise to review some physiography and place names. Figure 18.1 is a Google Earth image with major physiographic and structural features labeled. The Grand Canyon sits at the southwestern edge of the Colorado Plateau. It begins at Lee's Ferry, located just downstream from the Glen Canyon Dam and Lake Powell, where the Colorado River crosses the Echo Cliffs. It ends at the Grand Wash Cliffs,

which forms the boundary between the Colorado Plateau and the Basin and Range just east of Lake Mead, more than 277 river miles downstream. The river is at an elevation of 3,086 feet just beyond Lee's Ferry. The Paria Plateau to the west is at approximately 6,500 feet, and the Kaibito Plateau to the east is at 5,000 feet. From Lee's Ferry southward, the Colorado River flows progressively deeper into Marble Canyon toward the Eastern (or Upper) Granite Gorge, which is the deepest part of the canyon and the most visited part of the national park. The Little Colorado River joins the Colorado just prior to entering the gorge. Within the Eastern Granite Gorge, the Colorado River crosses a broad anticlinal flexure known as the Kaibab Arch and then curves to the northwest to form a half-loop. River elevation at the southernmost part of the gorge is 2,580 feet. The Kaibab (or Walhalla) Plateau to the north is at 8,000 feet, the Coconino Plateau to the south is at about 7,000 feet. The river flows around the half-loop, turning northeastward toward the Kaibab Plateau before abruptly turning westward where Kanab Creek joins the Colorado. Farther west, the river crosses the active Toroweep normal fault and flows past cascading lava flows of the Uinkaret Volcanic field frozen against the canyon walls (Figure 14.18). The river is at an elevation of 1,677 feet just east of Toroweep fault, and plateau elevations are 6,300 to the north and 6,400 feet to the south. The river crosses the active Hurricane Fault, then turns southward to enter the Western (or Lower) Granite Gorge, which forms a second half-loop around the volcanic Shivwits Plateau. To the south lies the Hualapai Plateau with several paleovalleys (Milkweed, Hindu, Peach Springs) and associated "rim gravels" that preserve a relict landscape that could be more than 50 million years old. The paleovalleys are associated with ancient rivers that once flowed northward and eastward through the Grand Canyon area, not southward or westward as the Colorado

River flows today. The Hualapai Plateau and these ancient paleovalleys with their northeastward river flow are an important part of the Grand Canyon story. The straight-line distance between the southernmost point in the Eastern Granite Gorge and that in the Western Granite Gorge is about 90 miles. The river is at an elevation of 1,290 feet at the southern apex of Western Granite Gorge. The Shivwits Plateau to the north is at 6,100 feet, and the Hualapai Plateau (in the vicinity of the paleocanyons) is mostly between 4,000 and 4,750 feet. Once beyond the Western Gorge, the river continues a northwesterly flow until it crosses the Grand Wash Cliffs where it drops abruptly off the plateau and out of the canyon into the Basin and Range province and Lake Mead. River elevation as it leaves the Grand Wash Cliffs is 1,205 feet. Elevation on both sides of the river in the vicinity of the Grand Wash Cliffs is 4,600 feet. Although not the deepest canyon in the world, the Grand Canyon is impressively long and up to 18 miles wide.

Both the Hurricane and Toroweep faults show active normal displacement down-dropped to the west. The Toroweep Fault has been active since probably about 2 to 3 million years ago. It has about 197 feet (60 m) of displacement in the past 600,000 years (0.39 in/100 yrs) where it crosses the

Grand Canyon. The Hurricane Fault has a far more complex history, beginning with reverse (west-side-up) faulting more than 50 million years ago during mountain-building events that affected the Colorado Plateau and the Middle-Southern Rocky Mountains. It has probably been active as a normal fault for the past 3 to 4 million years with 98 to 115 feet (30 to 35 m) of offset in the past 200,000 to 320,000 years. This is equivalent to a displacement rate of 0.37 to 0.69 inches per 100 yrs. The presence of these faults suggests that Basin and Range normal faulting is extending slowly but progressively deeper into the Colorado Plateau.

WHY LESS THAN SIX MILLION YEARS?

The prevailing view for many years has been that the Grand Canyon formed mostly or entirely within the past 6 million years and this view still holds sway with some researchers. The logic and reasoning, in their simplest form, are (1) the floor of the Grand Canyon is currently occupied by the Colorado River and (2) the Colorado River did not flow through the Grand Canyon until about 6 million years ago. Thus, if the Colorado River carved the Grand Canyon, the canyon must be no older than the river that created it.

FIGURE 18.1 A Google Earth image of the Grand Canyon from Lee's Ferry to Lake Mead. North is to the top of the image. *CP* = Coconino Plateau, *EC* = Echo Cliffs, *EGG* = Eastern Granite Gorge, = Grand Wash Cliffs (edge of Colorado Plateau), *HF* = Hurricane Fault, *HP* = Hualapai Plateau, *KA* = Kiabab Arch and Plateau, *KC* = Kanab Creek, *KP* = Kaibito Plateau, *LC* = Little Colorado River, *LF* = Lees Ferry, *LM* = Lake Mead, *MC* = Marble Canyon, *MH* = Milkweed-Hindu paleocanyons, *PC* = Peach Springs Paleocanyon, *PD* = Painted Desert, *PP* = Paria Plateau, *RG* = rim gravels, *SGW* = Southern Grand Wash Cliffs (edge of Colorado Plateau), *SV* = Shivwits Plateau and volcanic field, *TF* = Toroweep Fault, *UV* = Uinkaret Volcanic field, *VC* = Vermilion Cliffs, *WGG* = Western Granite Gorge.

The Colorado River of today flows from the Colorado Rockies through the Grand Canyon into Lake Mead and then all the way to the Gulf of California. Defined as such, the river is only about 5.36 million years old. How do we know this? Here again we rely on geological principles. The Colorado River flows out of the Rocky Mountains carrying rock fragments that are found only in the Rocky Mountains. These rock fragments, when found in river sediment, are like a fingerprint that indicates how long ago the Colorado River flowed across a certain area. Rock fragments indicative of the Rocky Mountains are present in river gravels near Grand Junction, Colorado, below the Grand Mesa basalt, which is dated at 9 to 11 million years old. These same rock fragments, however, are absent in the Lake Mead area (at the western end of the Grand Canyon) within loosely consolidated rock known as the Muddy Creek formation, dated between 13 and 6 million years old. At one location in the Lake Mead area, known as Sandy Point, river gravels that contain Rocky Mountain rock fragments are present below a basalt dated at 4.4 million years old and above the Hualapai Limestone, which forms the uppermost rock unit in the Muddy Creek Formation and which is dated at 5.97 million years old. These same river gravels do not appear at the Gulf of California until about 5.36 million years ago. What these relationships indicate is that there was an ancestral *upper* Colorado River that flowed across Colorado more than 11 million years ago, but this same river did not flow through the Grand Canyon to Lake Mead until at least 5.97 million years ago. It then took approximately 600,000 years for this river to reach the Gulf of California to become the modern Colorado River we know today. All researchers agree with this part of the story.

Given this information, a simple conclusion would be that the Grand Canyon could not have formed in its entirety until the Colorado River began to flow through Lake Mead some 6 million years ago. If we make this assumption, and we assume further that there was almost no canyon in existence prior to 6 million years ago, then the average incision rate to cut the canyon over the entire 6-million-year period would be between 1.0 to 1.1 inches per 100 yrs. This is rather fast for an extended period of time, but it is possible.

MEANDERS

The Colorado River and its tributaries, which include the Green and San Juan rivers, are well known for their incised meanders. As shown in Figure 18.2, meanders on the Colorado River are particularly strong north of the confluence with the Green River and again near the confluence with the San Juan River. Perhaps most impressive is the incredibly complex array of meanders along the Escalante River, a relatively small tributary near Lake Powell. A close-up view of these meanders is shown in Figure 18.3. Meanders continue through Lake Powell to the Echo Cliffs, where the river straightens and begins to cut progressively downward into Marble Canyon. Marble Canyon is less than 500 feet deep just south of Echo Cliffs, more than 2,300 feet deep in its central portion, and about 2,900 feet deep where it forms a few meander bends just north of the confluence with the Little Colorado River. Even the Colorado River within the Grand Canyon meanders somewhat, as shown in Figure 18.1.

A common method by which incised meanders form is to first establish themselves on relatively flat, stable ground, and

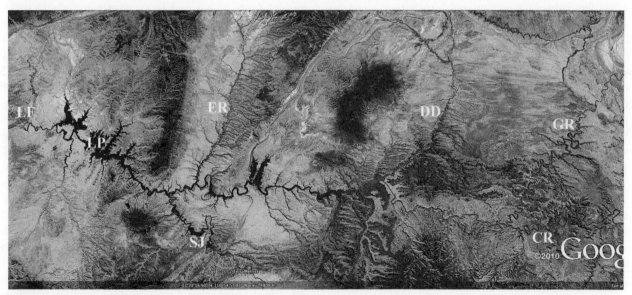

FIGURE 18.2   A Google Earth image of the canyon region along the Colorado River north of the Grand Canyon. West is toward the top of the image. Lee's Ferry (*LF*) is at extreme left. *CR* = Colorado River, *DD* = Dirty Devil River, *ER* = Escalante River, *GR* = Green River, *LP* = Lake Powell, *SJ* = San Juan River. Split Mountain (Ladore Canyon) is located just off the right side of the figure along the Green River.

FIGURE 18.3   A Google Earth image of the Escalante River and tributaries. West is toward the top of the image. The Colorado River (*CR*) is at right.

then maintain their channel during either regional uplift or a drop in base level. The Mississippi and Ohio rivers meander incessantly as they traverse the flat midcontinent. The implication is that the Colorado River and its tributaries were, at one time, also meandering across stable, flat ground. Something caused them to deeply incise their channels. That something could have been uplift of the Colorado Plateau. In Chapter 11 we presented evidence for between 1,640 and 3,280 feet of uplift within the past 10 million years centered in the Colorado Rockies. There is little doubt that at least some of that uplift affected the Great Plains and the Colorado Plateau.

Incised meanders will also form if there is a sudden drop in base level. Base level for the Colorado River dropped 1,200 feet between 6 million and 5.36 million years ago as the Colorado River slowly cut its way to the Gulf of California (based on the present-day difference in elevation between sea level and the elevation of Lake Mead). There may have been an additional drop in base level when the upper Colorado River first made its way through the Grand Canyon around 6 million years ago. The rather small Escalante River and some of its even smaller tributaries have cut canyons 500 to 1,000 feet deep into the plateau surface. All of this rather circumstantial evidence would seem to indicate incision on the order of 500 to 3,000 feet. The question then becomes, when did this incision occur? There is no clear answer, although the meandering of even small tributaries along the Colorado River in Utah is at least suggestive of incision within the past 6 to 10 million years.

THE GREAT DEFORMATION AND PALEOELEVATION

We know that at least part of the Colorado plateau was at sea level 80 million years ago, because marine rock of

that age is exposed on the plateau surface (Figure 7.1). We also know that the plateau is now, on average, about 6,200 feet above sea level, with elevations that range from 1,200 feet at the bottom of the Grand Canyon, where it crosses the Grand Wash Cliffs, to nearly 11,000 feet in the High Plateaus region. Thus, we can say with relative certainty that the Colorado Plateau, as a whole, was permanently elevated above sea level less than 80 million years ago. Clearly, the Grand Canyon cannot be older than 80 million years.

As noted in Chapter 10, there are plenty of mildly deformed sedimentary rocks on the Colorado Plateau in the form of monoclines, flat-topped anticlines, and high-angle reverse and normal faults, some of which have reactivated ancient scars within underlying crystalline shield rocks. Nearly all of this deformation (with the exception of young normal faults) is ascribed to the same mountain-building event that created the Middle and Southern Rocky Mountains. This mountain-building event began about 75 million years ago and ended between 55 and 40 million years ago. It is known as the *Laramide orogeny* and is discussed in more detail in Part 3 of this book. It was this deformation that resulted in uplift and permanent elevation of the Colorado Plateau above sea level, but exactly how much elevation was gained at this time is subject to debate. Some researchers argue, based on measurements of temperature-dependent carbon and oxygen isotopes in carbonate rocks, that present-day elevation was gained between 80 and 60 million years ago. However, even these data have enough built-in uncertainty to allow for up to 1,500 feet of elevation gain within the past 6 million years.

One way to get at paleoelevation is to look for paleo-relief in the landscape. Elevation must be at least as great as relief if we assume that the lowest elevations were at or above sea level. Recall from Chapters 15 and 17 that eastern

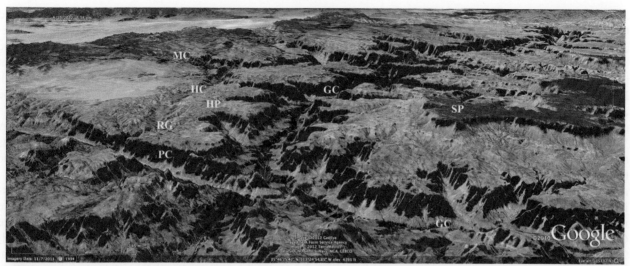

FIGURE 18.4   A Google Earth image looking west across the Western Granite Gorge. Peach Springs (*PC*), Milkweed (*MC*), and Hindu (*HC*) paleocanyons are visible. *GC* = Grand Canyon, *HP* = Hualapai Plateau, *RG* = Rim Gravels, *SP* = Shivwits Plateau.

California and much of the Great Basin were a highland from at least 80 million years ago to as late as 16 or 17 million years ago (the Nevadaplano and accompanying foreland fold-and-thrust belt). A glance at Figure 17.19 shows that the inferred location of the Nevadaplano drainage divide (that is, the paleodivide) was somewhere west of Lake Mead. This implies that rivers were not flowing westward or southwestward across the Colorado Plateau as they do today but were flowing eastward or northeastward, away from the drainage divide in a direction nearly opposite that of today's Colorado River. Field evidence that rivers did indeed flow eastward and northeastward is present in the form of imbricated pebbles derived from the southwest and from the slope of paleochannels. Because the streams were flowing northeastward, the highest elevations on the Colorado Plateau prior to and during at least part of the 75 to 40 million-year mountain-building event, were likely to the south and southwest.

Exactly when the rivers reversed themselves and began flowing south and southwestward as they do today is not known with certainty, but it probably did not occur all at once. It is entirely possible that the drainage divide migrated slowly toward its present configuration through progressive stream capture associated with the erosion of elevation in the west and uplift of the Rocky Mountains in the northeast. Reversal could have begun or been completed as early as 55 million years ago, or it could have occurred 16 or 17 million years ago during collapse of the Nevadaplano and development of Basin and Range landscape, or at around 6 million years ago when the Colorado River began to flow through the Grand Canyon. In any case, the presence of ancient (>50 Ma) paleocanyons on the Hualapai Plateau indicates that relief (and therefore elevation) was present in the Grand Canyon by the end of the Laramide orogeny.

We can gain some insight into how much elevation (or at least relief) was present by the study of these paleocanyons and their deposits. The best-preserved paleochannels are located south of the western gorge on the Hualapai Plateau. They include the Peach Springs, Milkweed, and Hindu canyons, all three of which are in existence today as shown in Figure 18.4. Associated with these canyons are paleo-gravels (known as *rim gravels*) that partially fill the Milkweed and Hindu canyons. The Peach Springs Canyon has been excavated. The dating of rim-gravel detritus, and the presence of a 19 million-year basalt flow near the top of the section, suggest the gravels were deposited between 50 and 19 million years ago. The fact that these rim gravels have not been eroded suggests that the Hualapai Plateau surface and the paleocanyons are quite old (more than 50 million years old; that is, they must have been in existence for gravels to accumulate within them). The rim gravels lie on the relatively flat, undulating Hualapai Plateau surface at elevations as high as 4,600 feet. The excavated Peach Springs paleocanyon extends down to elevations of less than 2,200 feet. This suggests a minimum paleorelief of 2,400 feet between the plateau surface and the bottom of Peach Springs Canyon prior to 50 million years ago. More detailed measurements suggest that paleorelief could have been in excess of 3,900 feet, which is nearly the entire present-day relief of the area.

RECENT INCISION RATES

Basalt is an easily datable rock and thus is a handy tool when available. Basalt flows emanating from the Uinkaret field between the Hurricane and Toroweep faults spilled into the Grand Canyon as shown in Figure 14.19. These basalts have been used to determine Colorado River incision rates. The Uinkaret field is between about 3.7 million and 1,000 years old, but basalt that flowed into the Grand Canyon is less than 723,000 years old. Incision rates east of the Toroweep fault were calculated to be between 0.59 and 0.69 inches per 100 yrs (150 and 175 m/my) over the past 500,000 years. Incision rates west of the Hurricane Fault are less due to faulting along both the Hurricane and Toroweep faults, thereby lowering base level. In this region, incision rates are between 0.20 and 0.30 inches per 100 yrs (50 and 75 m/my) over the past 720,000 years. At these rates, between 2,950 and 3,450 feet of canyon could be cut in the eastern gorge (east of the Toroweep fault) over a 6-million-year interval, and between 1,000 and 1,500 feet in the western gorge (west of the Hurricane Fault). These numbers amount to between 55% and 64% of the Eastern Granite Gorge (5,400 feet present-day relief) and between 21% and 43% of the Western Granite Gorge (4,800 to 3,500 feet present-day relief).

These numbers do not imply that the canyon could not have been cut within 6 million years. We could note, for example, that the very instruments used to calculate incision rates (the lava flows) would have obstructed incision simply by flowing into the river and creating dams or partial dams. Additionally, we could suggest that the amount of water flowing through the canyon was greater in the past and thus argue for greater cutting power. Finally, the dampening effect of the Toroweep and Hurricane faults would not have been present during the first half of the 6-million-year interval. The alternative argument is that we really don't know how much water has flowed through the canyon. If it was less in the past, then the Colorado River could not have cut the canyon in 6 million years.

EXHUMATION AGES

Perhaps some of the more thought-provoking findings in recent years have been the results of apatite fission track and (U-Th)/He analyses from the Grand Canyon area. Recall from Chapter 4 that both techniques record the time that a buried rock first reaches to within a few miles of the Earth's surface. Apatite fission tracks are preserved at temperatures below about 110°C (equivalent to less than 2.7 miles depth) and helium in apatite closes at about 30°C (equivalent to less than 4,000 feet depth). Data were obtained from crystalline shield rocks at river level and from sedimentary rocks on the plateau surface as much as 5,000 feet above river level.

Let us look first at the deeper Eastern Granite Gorge. Apatite fission track dates from crystalline shield rocks exposed at river level are between 80 and 70 million years old. Apatite fission track dates from sedimentary rocks on the plateau surface 5,000 feet above the canyon floor are as much as 50 million years older. An older age for rocks on the plateau would be expected because these rocks were not initially buried as deeply as the crystalline rocks, as shown in Figure 18.5a. If we assume that layers are stripped from

both locations at the same rate, the sedimentary rock should cool earlier than the crystalline rock, which is what we see. But this is not what we see when we look at the apatite (U-Th/He) data. Instead we find that crystalline rocks at river level and the sedimentary rocks on the plateau both record ages of about 20 million years. The data imply that the crystalline rocks were at the same depth below the Earth's surface 20 million years ago as the sedimentary rocks. A simple explanation is that the Eastern Granite Gorge was already in existence 20 million years ago, directly above the crystalline rocks, as shown in Figure 18.5b.

There is actually more to the story regarding the Eastern Granite Gorge. Recall again from Chapter 4 that there is a temperature range between which fission tracks are shortened but not destroyed and where some (but not all) of the trapped helium is able to escape from the crystal structure. Sophisticated modeling of the data through these temperatures suggests that rocks at river level and rocks on the plateau were already at about the same depth as early as 70 Ma. The modeling therefore suggests that an ancestral Eastern Granite Gorge was already in existence 70 million years ago. Recall that the Colorado River did not flow through the canyon until 6 million years ago. If the modeling is correct and the Eastern Granite Gorge is more than 70 million years old, then the Eastern Granite Gorge was cut by a river other than the Colorado. Rim gravels in

(a)

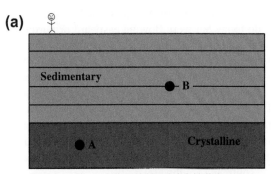

Rock A, within crystalline shield rock, is initially buried much deeper than sedimentary rock B. If layers are stripped from both locations at the same rate, then rock B would cool earlier than rock A.

(b)

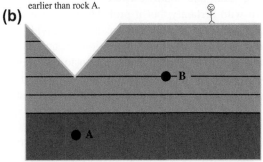

The existence of a canyon above the crystalline rock allows the crystalline rock to be at about the same depth below the surface as the sedimentary rock consistent with apatite data.

FIGURE 18.5 Cross-sections that show relative location and depth of crystalline and sedimentary rock on the plateau. Based on Wernicke (2011).

the Western Grand Canyon suggest further that this river, 70 million years ago, was flowing east-northeastward in a direction opposite that of the Colorado River.

Now let us look at the smaller but still very impressive Western Granite Gorge. Data from this area are just as surprising. Both apatite fission track and the apatite (U-Th)/He dates from river level are between 90 and 70 million years and one apatite (U-Th)/He date from the plateau is 68 million years. These data suggest that river-level crystalline rocks cooled rather quickly from above 110°C to below 30°C prior to about 70 million years ago and that, by about 68 million years ago, they were at about the same depth below the surface as sedimentary rocks on the plateau. Here again, the simplest interpretation is that the Western Granite Gorge was cut, at the location where it is today, during the period of rapid cooling between 90 and 70 million years ago (Figure 18.5b). The apatite data therefore suggest that the entire Grand Canyon was cut prior to 70 million years ago by a river other than the Colorado.

A PRE-SIX-MILLION-YEAR THEORY ON THE CUTTING OF THE CANYON

A leading theory on the cutting of the canyon takes into account all available apatite data as well as field data gathered over the course of more than 140 years, beginning with the legendary three-month 1869 expedition through the Grand Canyon led by John Wesley Powell, a one-armed professor of geology at Illinois Wesleyan University (his arm was lost in the 1862 Civil War battle of Shiloh) and founder and curator of the Illinois Museum of Natural History. Powell led additional expeditions in 1871 and 1872, producing photographs, a map, and scientific publications. Powell would ride the lead boat perched in a secured armchair so that he could see upcoming rapids and signal the other boats. His boat, named the *Emma Dean* after his wife, is shown in Figure 18.6. Beginning in 1881 Powell became the second director of the US Geological Survey, a position he held for 13 years. He also served as director of the Smithsonian Bureau of Ethnology beginning in 1880 until his death in 1902.

The story of the Grand Canyon begins 80 million years ago when a highland existed in California and Nevada (the volcanic arc and Nevadaplano) and an inland sea existed at the present-day location of the Middle-Southern Rocky Mountains (e.g. Figures 17.19 and 7.1). If you were to visit the Grand Canyon today, you would see about 1,000 vertical feet of crystalline shield rock exposed at river level at the bottom of the Eastern Granite Gorge and about 900 feet at the bottom of Western Granite Gorge. Along the walls of the Canyon you would see 4,000 to 5,000 feet of flat-lying sedimentary layers. Eighty million years ago, before the

FIGURE 18.6 A photograph of the boat (the Emma Dean) and the armchair used by John Wesley Powell to explore the Grand Canyon. Photo by E. O. Beaman, August 20, 1872, in Marble Canyon (cropped). Downloaded from the US Geological Survey photographic library, ID Hillers, J.K. 445, http://libraryphoto.cr.usgs.gov/cgibin/show_picture.cgi?ID=ID.%20Hillers,%20J.K.%20%20445&SIZE=large.

Grand Canyon came into existence, the sedimentary section above the crystalline shield rocks was two to three times as thick (almost as thick as the sedimentary section below the High Plateaus area shown in Figure 10.15). The area that was to become the Grand Canyon was a lowland that had recently emerged from below sea level, underlain by this thick sequence of sedimentary rock. The situation in the Western and Eastern Granite Gorges prior to canyon cutting is shown in a schematic cross-section in Figure 18.7a. At that time, an ancient river flowed from southern California eastward and northeastward across the Grand Canyon area toward the inland sea. This river is referred to as the California River and is superimposed across the present-day landscape in Figure 18.8a. Notice in this figure that the California River is shown flowing through the Grand Canyon in a direction opposite that of the present-day Colorado River. Neither the Grand Canyon nor the Colorado River were in existence at this time.

Beginning prior to 70 million years ago, uplift migrated from California into the Colorado Plateau and Grand Canyon region just prior to or coincident with the initiation of the rise of the Middle-Southern Rocky Mountains. The uplift produced mild deformation on the Plateau that may have included initial development of the Kaibab Arch. The California River was undeterred by this uplift and maintained its grade on its way to the inland sea, which had receded and shrunk since 80 million years ago. As uplift occurred in the Western Granite Gorge area, the California River and its tributaries stripped away layers of sedimentary rock (via bench-and-slope retreat) and at the same time cut a canyon nearly as deep as the present-day canyon down almost to the top of the crystalline shield. This scenario is shown in Figure 18.7b. Such a scenario would bring crystalline rock close enough to the surface to record circa 70 to 80 million-year-old apatite fission track and (U-Th)/He ages. The Western Granite Gorge at this point was already both visually and geologically similar to the modern-day canyon. The Hualapai Plateau was either already in existence at this time on the canyon rim or, with some additional stripping, would be in existence by 55 million years ago.

In the Eastern Granite Gorge, the California River cut across an uplifting Kaibab Arch, but the overlying sedimentary layers were not completely stripped off. This scenario, shown in Figure 18.7b, allows the sedimentary rocks to be close enough the surface to record old (>100 Ma) apatite fission track ages but not close enough for the closure of the apatite (U-Th)/He system (recall that this area recorded 20 Ma (U-Th)/He closure ages). Thus, the Eastern Granite Gorge was in existence 70 million years ago at the same location and at about the same depth as it is today, but it was not exactly the same canyon we see today, at least from a visual perspective. Instead, the Eastern Granite Gorge was cut entirely into sedimentary rock layers. Crystalline

shield rocks remained buried 3,000 to 6,000 feet below the California River. Still, according to this theory, essentially the entire Grand Canyon was in existence by 70 million years ago.

By 55 million years ago minor additional down-cutting in the Western Granite Gorge, and simultaneous bench-and-slope stripping of layers off the plateau surface, had lowered the river into crystalline rock. The present-day plateau surface above the canyon walls (that is, the Hualapai Plateau) was in existence and deeply incised by paleochannels. Also, by about 55 million years ago the area to the northeast that was once an inland sea had risen to form the Middle and Southern Rocky Mountains. The rise of the Rocky Mountains caused the Colorado Plateau to tilt gently westward. At the same time, the highland in southern California was effectively eroded along with the drainage divide that had separated this highland from the ocean. There was now a clear path from the Colorado Plateau to the paleo-Pacific Ocean. Rivers in the Grand Canyon area reversed direction and began flowing toward the ocean. The California River was destroyed. Given the preexistence of a well-developed east-northeast-flowing California river network, one would expect ponding and choking of river channels and much deposition as water first stagnated and then began to reverse direction. Such a scenario may have been the cause of deposition of some of the rim gravels on the Hualapai Plateau, which filled many of the preexisting channels. Coincident with stream reversal was development of a rather small river system originating this time on the Colorado Plateau south of the present-day Echo Cliffs. This river, known as the Arizona River, flowed west and southwestward toward the paleo-Pacific Ocean in a direction opposite that of the California River but probably utilizing many of the preexisting river channels, including the Grand Canyon. The Arizona River at this time did not have the cutting power to further incise the Grand Canyon to any substantial degree.

By 40 million years ago the Middle-Southern Rockies would reach their maximum pre-present-day elevation, but rivers exiting the mountains were trapped in large closed basins similar in some respects to the present-day Great Basin. The Rocky Mountains underwent a period of erosion and burial so that by about 34 million years ago the mountains were largely buried in their own debris (see Chapter 11). The ancestral Colorado River may have been in existence at this time, but it would have flowed a short distance out of the mountains and into one of the many closed basins. It did not cross the Echo Cliffs, which formed a drainage divide that separated the internally drained Colorado River system from the Arizona River system. The situation is depicted in Figure 18.8b. By 24 million years ago large areas of the west, including the Nevada, Utah, Colorado, and perhaps the Grand Canyon regions, were blanketed in volcanic ash hundreds of feet thick in some

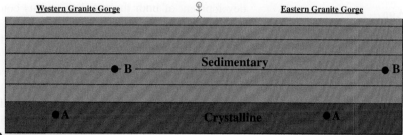

(a) <u>Circa 80 Million Years Ago</u>

At 80 million years ago, prior to the cutting of the canyon, rocks A and B are deeply buried. Rock A in both the Western and Eastern Granite Gorge is located within the crystalline shield. Rock B in both gorges is located higher in the sequence within nearly flat-lying sedimentary rock. Rocks A are 35 to 45 °C hotter than rocks B because they are more deeply buried.

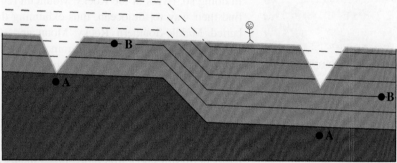

(b) <u>Circa 70 Million Years Ago</u>

Prior to 70 million years ago, rising mountains in the west resulted in greater uplift in the Western Granite Gorge and eastward tilting. Apatite fission track and (U-Th)/He data suggest that rocks A in both the Western and Eastern Granite Gorges were exhumed to the same depth below the surface as rocks B (e.g. both reached the same temperature at the same time). This constraint requires the presence of a canyon, 4000 to 5000 feet deep, above rocks A at both locations. Also during this period, sedimentary layers in the Western Granite Gorge were stripped (eroded) approximately to the present-day plateau surface allowing rocks A and B to rapidly cool through both the apatite fission track and apatite (U-Th)/He closure temperatures. By 70 million years ago the Western Granite Gorge was cut to within a few hundred feet of its present-day level. Sedimentary layers on the plateau surface in the Eastern Granite Gorge were not eroded but rock A cooled through the apatite fission track closure temperature (approximately 110 °C) during excavation of the canyon and was exhumed to the same depth below the surface as rock B. At 70 million years ago the Eastern Granite Gorge was cut entirely within sedimentary rock.

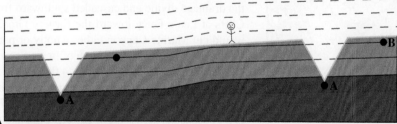

(c) <u>Circa 20 Million Years Ago</u>

By 55 million years ago, uplift of the Rocky Mountains had tilted the Grand Canyon region westward. At about 20 million years ago a pulse of uplift centered in the Eastern Granite Gorge stripped (eroded) sedimentary layers from the plateau surface while, at the same time, maintained a 5000-foot deep canyon. Rocks A and B in the Eastern Granite Gorge were both exhumed close to the surface and both rocks cooled through the apatite (U-Th)/He closure temperature at the same time. The Eastern Granite Gorge was now cut to within a few hundred feet of its present-day level. Note that the location of the Grand Canyon has not moved significantly in 70 million years.

FIGURE 18.7   Schematic cross-sections that show development of the Grand Canyon. Based on Flowers et al. (2008) and Wernicke (2011).

areas (Chapter 14). The Colorado River and its tributaries likely developed into lazy, meandering rivers as they flowed over the volcanic debris across some of the many flat basins.

Sometime around 20 million years ago a pulse of broad uplift, perhaps associated with the sinking of the Farallon slab and with volcanic-induced thermal isostasy (Chapter 17), stripped several thousand feet of sedimentary rock from the plateau surface above the Eastern Granite Gorge down to the present-day Kaibab surface. At the same time, the Arizona River incised into crystalline rock, thus maintaining

(a) Circa 75 Million Years Ago

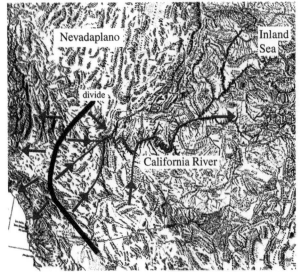

(b) Circa 40 Million Years Ago

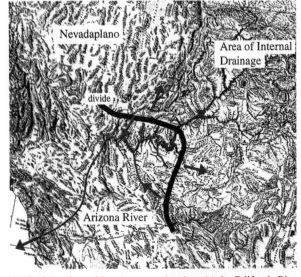

FIGURE 18.8 General location maps that show (a) the California River circa 75 million years ago and (b) the Arizona River 40 million years ago, superimposed on present-day landscape. At 6 million years ago the drainage divide shown at Echo Cliffs (Lee's Ferry) in the lower diagram would be toppled and the flow direction in the area of internal drainage would reverse from northeast to southwest thus creating the modern Colorado River.

the depth of the canyon and producing the modern Eastern Granite Gorge as we see it today. The end result is shown in Figure 18.7c. This circa 20 Ma pulse of uplift, stripping, and incision accounts for the 20 million-year-old apatite (U-Th)/He ages in both the crystalline rocks at river level and the sedimentary rocks at plateau level. This event could correlate with reactivation and uplift of the Kaibab Arch and with reactivation of a complementary syncline below the High Plateaus as shown in Figure 10.15. However,

development of both the anticline and its complementary syncline have more commonly been attributed to the older 75 to 40 million-year Laramide mountain-building event.

By about 15 million years ago, normal faulting in the Basin and Range west of the Grand Canyon began to dismember and destroy the Arizona River, leaving a small river system with internal drainage and very little cutting power. The canyon and surrounding plateau underwent little if any change until about 6 million years ago.

Some time after 10 million years ago, the Middle-Southern Rocky Mountains began a period of broad warping that eventually elevated the Rockies above 14,000 feet. In doing so, rivers in the Rocky Mountain region began to find their way to the ocean, thus exhuming the heretofore buried Middle-Southern Rocky Mountains (Chapter 11). By 6 million years ago the drainage divide along the Echo Cliffs at Lee's Ferry was toppled (removed) such that north-flowing tributary rivers on the north side of the drainage divide (that is, the area in Figure 18.8b labeled "area of internal drainage") reversed direction and began flowing southward across the Echo Cliffs toward the Grand Canyon. With the opening of an outlet at Echo Cliffs, rivers north of Echo Cliffs reorganized themselves, and the Colorado River, which previously flowed into closed basins north of the Echo Cliffs, now flowed through the breached divide and into the remnants of the Arizona River system. The modern-day Colorado River had taken shape.

The integration of the Colorado River with the Arizona River did at least two things: It stripped the Middle-Southern Rockies of its basin fill, thus exhuming the mountains, and it added a whole lot of water to the nearly dead and dismembered Arizona drainage basin. As water overtopped the drainage divide and cascaded southward from the Echo Cliffs at Lee's Ferry toward the Eastern Granite Gorge, it would cut Marble Canyon either wholly or partly. Perhaps more important, base level north of Lee's Ferry would have almost instantly been lowered by 4,000 feet or more, causing all rivers north of Lee's Ferry to incise their channels. Such a scenario could easily create the crazy meanders and goosenecks across all of eastern Utah and northeastern Arizona. As the newly integrated Colorado River passed through the canyon and out into the Basin and Range, the water would flow into one of several closed, down-dropped normal-fault basins that had earlier disrupted and destroyed the Arizona River. However, this time the additional water would be enough to fill and eventually overtop each basin as shown in Figure 18.9. Thus, between about 6 and 5.36 million years ago, the Colorado River would lengthen itself and slowly make its way to the newly opened Gulf of California by filling and overtopping basins, possibly following the old Arizona River drainage.

So, how much cutting of the Grand Canyon can be attributed to the Colorado River beginning 6 million years ago? As it stands, there is room for at least some down-cutting

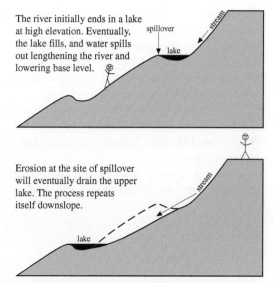

The river initially ends in a lake at high elevation. Eventually, the lake fills, and water spills out lengthening the river and lowering base level.

spillover

stream

lake

Erosion at the site of spillover will eventually drain the upper lake. The process repeats itself downslope.

stream

lake

FIGURE 18.9 A schematic cross-section that shows how a river can lengthen itself by filling, overtopping, and eventually destroying basins along its route.

to have occurred. Within the Eastern Granite Gorge there is good evidence for about 250 feet of incision in the past 500,000 years based on the dating of basalt that flowed into the canyon. This figure is about 1/20th the present-day canyon depth. As previously noted, we can extrapolate this to about 3,000 feet (more than 50% of the canyon) by simply maintaining this pace over the 6 million-year interval. Three thousand feet, however, is about twice as much as inferred by the apatite modeling data, leaving a maximum of between 1,000 and 1,500 feet of possible post-6-million-year incision. As for the Western Granite Gorge, we have noted that the Hualapai Plateau (on which the rim gravels were deposited) is approximately 3,400 feet above the canyon floor. One could speculate that the Colorado River cut the entire 3,400 feet of canyon below the Hualapai Plateau within the past 6 million years, but again, such a possibility is inconsistent with the apatite data and with data obtained from paleocanyons and rim gravels. Present-day incision rates are much lower in the Western Granite Gorge, less than 200 feet in the past 720,000 years, and careful interpretation of the data suggests that less than 600 feet of incision is reasonably possible over the past 6 million years. Thus, for now, and with the available evidence, we can suggest that there really was a Grand Canyon in existence 70 million years ago and that perhaps no more than between 500 and 1500 feet of that canyon in any one place was cut by the Colorado River in the past 6 million years. Alternatively, perhaps there is more to the story of the Grand Canyon. We need to keep in mind that the apatite data are subject to error depending on the models used to interpret the data. Additionally, the number of samples used in the analysis, thus far, are few. We will need additional data to fully vet this interesting theory because there are those who continue to argue that

two-thirds (or more) of the canyon was cut by the Colorado River in the past 6 million years. What we can say is that, with new surface and near-surface dating techniques, better understanding and accuracy of existing dating techniques, additional fieldwork, and more thorough sampling coverage, geologists, in time, will surely learn a great deal more and finally get to the bottom of the Grand Canyon (so to speak).

An interesting aspect to the story described here is how the age of landscape changes along the Colorado River. At its origin in the Front Range of the Colorado Rockies, the river flows across anticlinal mountains cored with crystalline rock and flanked by steeply dipping sedimentary rock. The structures are between 75 and 40 million years old, but the high, mountainous landscape is probably no older than about 10 million years. Westward, the river drops onto nearly flat-lying layers of the Colorado Plateau with deeply incised, meandering canyons younger than 6 million years. As the river drops into the Eastern Granite Gorge, the surrounding landscape is about 20 million years old. The river then cuts its way through cascading lava flows less than a million years old into an ancient landscape surrounding the Western Granite Gorge, perhaps no younger than 50 million years. Finally, the river spills out onto the Basin and Range, which likely began to develop less than 17 or 10 million years ago.

WATER GAPS

This interpretation of the cutting of the Grand Canyon rather neatly explains the presence of deeply incised meanders along the Colorado, Green, and tributary rivers north of Lee's Ferry, including, for example, Canyonlands National Park. If this is the case, these canyons would have had to have been cut in the past 6 million years after the drainage divide at Lee's Ferry was toppled. Thus, the interpretation for the cutting of the Grand Canyon provides two entirely different mechanisms for creating water gaps. The water gap through the Kaibab Arch in the Grand Canyon formed as early as 70 to 80 million years ago, when the California River cut down into an actively developing and uplifting structure. The California River was antecedent to the structure. In other words, the river was there first, and the structure grew up around it. Water gaps in the Middle and Southern Rockies north of Lee's Ferry, and probably including Split Mountain (Ladore Canyon; Figure 3.16), could have formed within the past 6 million years when rivers flowing over flat-lying basin sediment simply maintained their channel as they cut downward, into and across a preexisting anticline. In this case, the younger river is superposed across a pre-existing exhumed structure.

QUESTIONS

1. How long would the Toroweep fault have to remain active in order to accumulate 20,000 additional feet

of displacement? If such displacement did occur, how would it affect the look of the Grand Canyon?

2. What is the total vertical drop of the Colorado River from Lee's Ferry to Lake Mead?

3. What are the two dark areas in Figure 18.2? One is located west of Dirty Devil River, and the other is located west of the San Juan River.

4. What are rim gravels?

5. Many rivers on the Colorado Plateau consist of incised meanders. What are incised meanders? Suggest a process by which they could form.

6. Stated very succinctly, what is the primary argument in favor of a Grand Canyon that is less than 6 million years old?

7. Describe in one paragraph the incision history of the Eastern Granite Gorge based on apatite data.

8. Describe in one paragraph the incision history of the Western Granite Gorge based on apatite data.

9. Research the early exploration history of the Grand Canyon by John Wesley Powell.

10. Why is the incision rate of the Colorado River higher east of the Toroweep fault than west of the fault?

11. What is the difference between an antecedent and a superposed river?

12. In individual essays, describe the history of the California River, the Arizona River, and the Colorado River.

Part III

Mountain Building

Chapter 19

Early Theories on the Origin of Mountain Belts

Before we examine the modern theory of mountain building, it is appropriate to look first at a few of the many early ideas. Theories on the origin of mountain systems go back at least as far as Pythagoras (≈580 B.C.) who reasoned that wind trapped within the Earth with no outlet to the surface would inflate the ground producing mountains just like closing our mouth and puffing our cheeks. Aristotle (384–322 B.C.) thought that by mixing dry air with moisture, the loose earth would be converted to stone, swell, and produce mountains. These ancient writings along with writings in the Bible dominated ideas concerning the origin of mountains throughout the Middle Ages (≈476–1453). The overriding opinion was that Earth was stagnant and the Creator made the mountains which had always been there and will always be there.

However, those who studied natural processes were impressed with the erosive power of water. Avicenna (≈980–1037) considered two causes in the formation of mountains: wind and water. He suggested that earthquakes produced by wind within the Earth were capable of raising the ground. Wind and water, particularly floods, could then erode the ground and form a valley. This would leave mountains where erosion did not occur. He was one of the first to realize that mountains were in a constant state of disintegration.

Ristoro d'Arezzo (≈1281) thought that mountains were fixed in their position by the stars. The distance of a star from Earth produced an exact replica on Earth directly below it. A distant star draws the land up to produce a mountain, a close star produces a valley. The mountains were then modified by stream erosion, wave action, the Noachian Deluge, earthquakes, and deposition.

Leonardo da Vinci (≈1452–1519) was a keen observer of geological phenomena and was one of the first to draw, in great detail, the fabric of rock. He reasoned that Earth was round and that all oceans were at the same level. He recognized the immensity of geologic time by calculating that it took more than 200,000 years to deposit sediment in the Po River and that this was only a fraction of geological time. He predicted the existence of the hydrologic cycle by stating that water circulates from rivers to the sea and back to rivers, and reasoned that the erosive power of rain and rivers will slowly destroy a mountain. He thought that sea level was constantly being lowered and that all water would eventually sink into the ground, leaving the Earth's surface high and dry. His idea regarding mountains was that they originated not by uplift, but rather in the depths of the sea, first by deposition of sediment and then carved by rivers as sea level slowly lowered. He considered fossils found on mountaintops as evidence of once-higher sea level.

The idea that the Earth was stagnant and that present-day mountains had existed throughout most of Earth history persisted into the 17th century. By that time, Earth features were considered to have formed more or less instantaneously during times of great catastrophes, either at the beginning of time (which some thought was about 6,000 years ago) or during Noah's flood. For example, in 1669 Nicholas Steno suggested that caves were hollowed out by water before and after the Great Flood and that overlying rock units fell into the holes, creating mountains of folded, faulted rock. He viewed mountains as forming nearly instantaneously and that nothing of consequence was happening in the present day.

By the mid-1700s scientists in Europe began to recognize that mountains were made up of several packets of rock, each of widely different age and structure. This led to the realization that mountains were formed, not in one catastrophic event, but over time during two or more widely separate periods. On this basis, Giovanni Arduino produced the first formal stratigraphic classification of rocks. His subdivision was rather simple. The oldest were crystalline rocks which he referred to as primitive, followed by secondary (sedimentary rock), tertiary (semi-consolidated sedimentary rock and sediment), and finally volcanic rocks. Arduino inferred that several periods of mountain building had occurred, all formed by repeated upheavals and contortions ("revolutions" and "metamorphoses") brought about by the injection of magma into the substratum.

Between 1787 and 1796, the most famous geologist of his time, Abraham Werner, readvocated the lowering of sea-level theory. Werner thought that the hidden inner part of Earth (the part below the surface rock layers) had an irregular form consisting of high mountains and deep valleys that had always existed, and that rock above this hidden surface (with the exception of minor volcanic rock and some late-forming river sediment) was precipitated directly from water. Werner did not necessarily believe that the Earth was 6,000 years old, nor did he address a correlation between geologic events and Scripture. His idea was that, in the beginning of time, the Earth was covered entirely by a stormy primordial ocean that held all the world's near-surface rocks in suspension. The first precipitates from this ocean produced a layer of rock that draped entirely over the inner preexisting Earth from the mountaintops to the valleys below. Structures such as folds and faults were formed in the swirling, stormy water prior to the hardening of rock. As the ocean receded, it exposed the mountains. Additional rock was then precipitated along the flanks of mountains and in valleys, the youngest rock derived from erosion of older rock. The last gasps of subsiding ocean water cut channels that accentuated relief between mountains and valleys. Werner believed that the primitive ocean settled rapidly to its present character such that nothing of consequence was happening to shape present-day Earth. Thus, Werner was an advocate of a stagnant, unchanging Earth.

It was left to a contemporary of Werner, James Hutton, who in the 1790s published a paper that revived the idea that mountains were actively being eroded. Hutton saw Earth as immensely old and constantly changing, albeit very slowly. His view of mountains was similar to that of Arduino in which uplift was attributed to internal heat that generated magma at depth. Hutton argued that magma intrusion and thermal expansion of rock would push rock upward, thereby forming mountains. Rocks were deformed during uplift, not during deposition (precipitation), as Werner had argued. Hutton did not believe that all rocks precipitated from water. He recognized the geological differences and significance of igneous, metamorphic, and sedimentary rocks. Hutton concluded that erosion would slowly wear a mountain down and return the sediment to the sea. This sediment would then be consolidated, deformed, and metamorphosed, and then rise again to form a mountain as depicted in Figure 19.1. Hutton saw a continuous rock cycle of uplift, erosion, deposition, and uplift, with no sign of a beginning and no prospect of an end. He eventually found proof of his theory when he discovered an angular unconformity in the Scottish Highlands.

Hutton's views mirror our present-day understanding of Earth. He reasoned that processes happening today, such as river erosion, earthquakes, and volcanism, have been happening for all of eternity, and that collectively they will slowly modify and change landscape. Hutton began to interpret the rock record in terms of processes acting today; the present is the key to the past. Because he was one of the first and perhaps the most visible to apply these concepts, he is often mentioned as the father of modern geology.

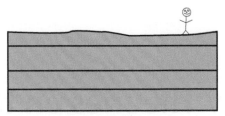

Sediment is deposited in the sea, compressed, and lithified to form sedimentary rock.

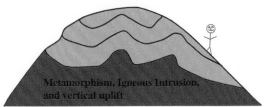

Rocks are deformed, metamorphosed, intruded, and uplifted to form a mountain.

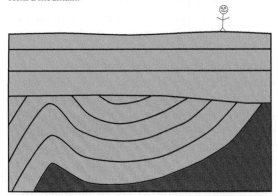

The mountain is worn down by rivers and covered by an ocean. Sediment is deposited in the sea above the remains of the eroded mountain to create an angular unconformity and to begin the cycle anew.

FIGURE 19.1 James Hutton's continuous cycle of deposition, deformation, uplift, erosion, and deposition.

Up to this time, the overriding principle in all these mountain-building theories was vertical tectonics (vertical uplift). However, by the 1830s Arnold Escher and other geologists working in the Swiss Alps had recognized large thrust faults. Seemingly the only possible explanation for these structures was horizontal compression. The problem was that there was no mechanism to cause such massive horizontal stress. Enter the shrinking Earth theory! This theory assumed that the Earth was continuously cooling from an originally molten state and that cooling caused contraction, which resulted in the shrinking of Earth's crust. This, in turn, produced horizontal stresses and mountains, much like wrinkles on a dried apple.

Meanwhile, geologists in the United States were mystified, not by large thrust faults, but by the incredible thickness of folded strata that made up the Appalachian Mountains. More than 40,000 feet of rock are present in the Appalachian Valley and Ridge, all deposited in shallow water.

Geologists of the day wondered how such a thick sequence of rock could be deposited in shallow water. They were under the assumption that sediment initially filled a deep basin. Thus, they reasoned that sediment at the bottom of the pile would be deposited in deep water and only the sediment at the top in shallow water. In 1859, James Hall suggested a solution. He proposed that sediment at the bottom of the pile was indeed deposited in shallow water and that there was never a deep basin. Hall suggested that the weight of newly deposited sediment would depress the crust slightly, allowing additional shallow-water sediment to be deposited. This process would continue until so much sediment was deposited that the lowest layers became metamorphosed. At some point, the sediment thickness would reach a critical value and the crust would rebound to form mountains of faulted, folded sedimentary and metamorphic rock. Thus, Hall's view of mountain building was without a significant horizontal driving force.

Another American, James Dana, believed, like the Europeans, that mountains were built not by vertical forces but by horizontal forces. Dana subscribed to the shrinking Earth theory. He figured that the interface between the continent and ocean would be the place where Earth would crumple the most during shrinkage. In 1873 Dana modified Hall's hypothesis to be consistent with horizontal compression due to a shrinking Earth. As the continent-ocean interface slowly buckled, it would fill with shallow-water sediment. This process continued until the rocks were compressed so much that they were uplifted to produce the Appalachian Mountains. Dana also proposed a solution to another intriguing problem with Appalachian mountain building. The character of the sediment in the adjacent Blue Ridge and Piedmont indicated the existence of an offshore (eastern) volcanic or cratonic source area referred to as a *borderland*. This borderland, however, was no longer present off the East Coast of North America. Dana proposed that the borderland must have been present during mountain building but, either sank like mythical Atlantis during or prior to uplift, or was incorporated into the mountain system. The geological setup of Dana's hypothesis is shown in Figure 19.2. Dana coined the term *geosyncline* (meaning *earth syncline*) for the down-warped basin, and his and Hall's theory became known as the *geosynclinal theory* of mountain building.

There were variations on this theory but the general theme persisted into the 1970s. One important variation from an historical perspective was the division of the geosyncline into a miogeosyncline composed of unmetamorphosed shallow-water limestones, sandstones, and shale deposited on the continent side of the proposed borderland, and a eugeosyncline composed of variably metamorphosed shale, silty sandstones, chert, and volcanic material deposited close to the borderland. In the Appalachians the miogeosyncline was represented by the Valley and Ridge.

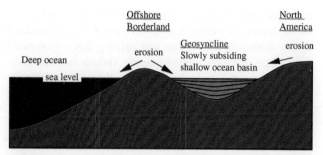

FIGURE 19.2   Hall and Dana's geosynclinal theory of mountain building. As the Earth shrinks and contracts, a basin (geosyncline) forms, subsides, and widens over time. Sediment deposition keeps up with subsidence, producing a thick shallow-water sequence. The crust fails and the basin is deformed and uplifted into a mountain. The borderland either sinks below sea level or is incorporated into the mountain.

The eugeosyncline was represented by the Blue Ridge and the Piedmont Plateau. Today, the term *miogeosyncline* has been shortened to *miogeocline* and is used to refer to a passive continental shelf, such as the present-day Atlantic shelf, where thick sequences of shallow-water limestone, sandstone, and shale are deposited. Within the Appalachian and Cordilleran mountain belts, these rocks form the foreland fold-and-thrust belt structural province in the Valley and Ridge and Northern Rockies. The term *eugeosyncline* can refer to almost anything oceanward of the miogeocline including deep oceanic rock, accreted terranes, and volcanic arc terranes. Because its meaning is nebulous, the term is avoided in favor of a more specific description.

The big debate in the late 1800s was whether or not the Earth was actually shrinking. A majority favored Dana's argument for horizontal tectonics due to shrinking, but there were physical arguments suggesting that the Earth could not shrink enough to produce all the folds and faults. Adherents would step around this discrepancy by suggesting that large quantities of water had been squeezed out of Earth through volcanoes allowing for additional contraction. Others did not believe in the shrinking Earth and instead championed vertical tectonics. These geologists explained large thrust faults as gravity slides (essentially super-large landslides) that formed during uplift. The bottom line was that nobody had a better idea. With reference to the shrinking Earth theory, Joseph Le Conte, in 1883, stated the following: "We all dearly love our own intellectual children, especially if born of much labor and thought; but I am sure that I am willing, like Jephtha of old, to sacrifice, if need be, this my fairest daughter on the sacred alter of truth."

In 1896, radioactivity was discovered, and soon thereafter it was recognized as a heat-producing reaction. With this came the realization that Earth was not steadily cooling from a primordial magma. It could not be shrinking. Thus began a slow end to the shrinking Earth theory, although it left no alternative in its wake. The geosynclinal theory was still the leading theory on the origin of mountains, and gravity sliding the leading theory on thrust faults, but there was no driving force to produce uplift. Some favored a vertical driving force generated by injection of deep-seated magma. Others favored horizontal forces, this time due to an expanding (yes, expanding!) Earth. Many clung to the idea of a shrinking Earth well into the 20th century.

The geographic fit between the continents of South America and Africa was noticed as early as 1620 by Francis Bacon, but he did not suggest that continents had moved. However, throughout the 1700s and 1800s, various forms of continental drift were proposed with driving forces that varied from Noah's Flood, the catastrophe of the Moon separating from Earth, an expanding Earth, and a shrinking Earth. An American, F. B. Taylor, in 1909, was one of the first post-shrinking Earth geologists to suggest that a form of continental drift had produced the world's mountains. He suggested that the movement of a continent would depress oceanic crust ahead of it to form a geosyncline. Further movement upheaved the geosyncline to form a mountain. Taylor did not think continents were still moving. Instead he thought that the moon was captured in Cretaceous time (circa 65 Ma) producing a great tidal wave that moved the continents and also presumably killed off the dinosaurs.

In 1912, and more thoroughly in 1915, Alfred Wegener presented a comprehensive body of evidence in support of the idea that continents had indeed drifted apart from what was once a single landmass consisting of all the world's (now separated) continents. He named this landmass Pangea, meaning "all lands" (Figure 8.5). Unfortunately, his idea was soundly rejected because he could not come up with a mechanism to move the continents. Wegener's mistake was that he initially thought the continents either slid across or plowed their way across a stationary oceanic crust—a proposal that proved to be physically impossible. However, in 1928 Arthur Holmes suggested a mechanism for continental displacement that is much closer to our present-day understanding. He suggested that continents ride on a convecting layer of lower basaltic crust and upper mantle. He considered the Mid-Atlantic ridge as marking the emergence of an upward convection system and to be the origin of continental displacement in the Atlantic region. He also postulated that the basaltic layer was carried downward into the mantle, leaving the continent at the surface. This was an early reference to what later would become known as *subduction*. Wegener quickly accepted Holmes view of crustal movement, but the idea was largely ignored by most of the geological community. Wegener died in 1930 and geology marched into the 1960s without a universally accepted driving force for mountain building!

It wasn't until the late 1960s that the plate tectonic theory began to take hold. The United States invented a variety of tools during World War II to look for German submarines beneath the ocean. These tools, and a few decommissioned ships, were later utilized by universities to explore the ocean bottom. In 1959, a detailed map of part of the ocean floor (the North Atlantic Ocean) was produced for the first time. This map ushered in a decade of oceanic discovery that by 1968 proved to most geologists that tectonic plates were indeed in motion. The plate tectonic theory is different from Wegener's original idea of continental drift in that the moving plate is the entire lithosphere (continental crust, oceanic crust, and solid upper mantle) and not just the continental crust. We now recognize that the movement of tectonic plates provides the horizontal driving force for mountain building.

QUESTIONS

1. Who was one of the first to recognize the erosive power of streams?
2. How did Leonardo da Vinci explain the existence of fossils on mountaintops?
3. What was Leonardo da Vinci's idea of how mountains formed?
4. How did Aristotle explain mountain building?
5. What was the prevailing view on the age of mountains prior to the 1700s?
6. Who was one of the first to produce a formal stratigraphic classification of rocks, and what did the classification consist of?
7. How did Giovanni Arduino explain mountain building?
8. How did Abraham Werner explain the origin of most near-surface rocks?
9. How did Werner explain mountain building (or is this a trick question)?
10. What was James Hutton's view of mountain building?
11. Why is Hutton often referred to as the father of modern geology?
12. How old did Hutton consider the Earth to be?
13. What evidence did Hutton find in the Scottish Highlands to confirm his idea of a never-ending cycle of uplift, erosion, deposition, and uplift? How does this evidence prove his hypothesis?
14. One initially well-received theory on mountain building was the shrinking Earth theory. When was this theory in vogue? What discovery in the late 1800s doomed the shrinking Earth theory, and what theory took its place as a cause of deformation?
15. How did James Hall explain the more than 40,000 feet of shallow-water sedimentary rocks in the Appalachian Valley and Ridge?

16. What specific type of structure is present in the Swiss Alps that caused geologists to look for a mechanism that could produce horizontal compression? What type of mechanism became favored?

17. Define the terms *miogeosyncline* and *eugeosyncline*. Are these terms in use today, and if so, how are they used?

18. How did nonbelievers of horizontal compression explain thrust faults?

19. How did F. B. Taylor explain geosynclines and mountain formation?

20. Who was Alfred Wegener?

21. How did Arthur Holmes solve Wegener's dilemma regarding how continents move?

Chapter 20

Keys to the Interpretation of Geological History

Part 3 of this book is concerned not so much with the evolution of landscape as with the geologic and tectonic sequence of events that, sometime in the past, produced a series of now vanished landscapes. Specifically, we want to look at how compressional mountain systems form and evolve over time. Compressional mountain systems form over a long period of time by multiple deformation pulses due usually to the convergence of tectonic plates. The primary mountain-building forces are compression and isostasy and, together, these forces produce *orogeny* which is the process of mountain building. In the United States we are blessed with an old, eroded Appalachian-Ouachita compressional mountain system and a young, actively evolving Cordilleran system, both of which are referred to geologically as *orogenic belts*. Although compressional mountain systems produce distinctive landscape, they are not defined by their landscape. In fact, they incorporate all nine of the previously described structural provinces. Compressional mountain systems are defined by their geology and are understood based on an interpretation of the rock record. Before we discuss the geology of mountain systems, it is wise to first outline some of the basic principles of how rocks are interpreted, how rocks record time, and some of the nomenclature involved in understanding the process of mountain building. This information sets the stage for a more advanced look at mountain building and can be used as a starting point to delve deeper into the geological history of the United States.

GEOLOGIC FIELD MAPPING

It is not possible to even begin to understand the history of a mountain belt without first knowing the distribution of rock types, the degree and style of deformation, and the degree to which the rocks are metamorphosed. This can only be accomplished through painstakingly slow fieldwork. Such work involves the recognition of specific rock units that can be followed across an area; the development of a *stratigraphy* (that is, the order in which rocks are stacked, one above another); the recognition of faults, folds, and other structural relationships; the measurement of all planar and linear surfaces (such as bedding and fault surfaces); and the collection of rock samples from known localities. This work is the most important step in understanding geologic history. It requires that a geologist traverse (walk across) an area and record all geologic data obtained from any rock he or she encounters. The geologist then must make additional traverses and correlate rocks found in the first traverse with subsequent traverses, eventually creating a geologic map that shows the distribution of rock types, geologic contacts, and structures, usually on a topographic base map. The procedure is a little bit like completing a jigsaw puzzle except that this puzzle requires that you go out and find the pieces which are scattered across several tens of miles. The pieces are already in place, but you must find enough of them to create a picture (a geologic map) that makes sense. Once completed, this map forms the basis of all subsequent work and interpretation. There is nothing better than a good geologic map when deciphering Earth history.

In the course of making a geologic map, the geologist must employ spatial and geological reasoning to understand what the rocks are willing to reveal about Earth history. Depositional basins, for example, hold the erosional residue of now vanished landscapes. Fossils, radiometric dates, and unconformities provide a means of separating and ordering discrete intervals of time. Texture, composition, and stacking tell us how the rock was made. The structure of a rock indicates the type of stresses that once existed. In this adventure, the rocks that make up the Earth are our only means of understanding Earth. Previous chapters referred to field data as a basis for some conclusion or interpretation. This chapter describes the most basic forms of field data and what they tell us about Earth history. It is an introduction to the techniques, methods, and some of the terminology required to study mountain belts.

HOW ROCKS REVEAL HISTORY

A thorough analysis of present-day landscape will reveal a great deal about the recent geological history of the United States. But Earth is much older than even the oldest landform. This fact implies that, before the present-day landscape existed, there must have been a progression of preexisting landscapes that have since vanished from the face of the Earth. What were these now vanished landscapes? What happened to them? How do we even know they existed? The answers to these questions are in the rocks! Just as one must learn to read the landscape to understand how a certain mountain or plateau may have formed, one must also learn to read the rocks to understand what the United States may have looked like millions or even billions of years ago. If you are a collector of antiques, then perhaps you should collect rocks because rocks are, by far, the oldest objects on Earth. Some are nearly 4 billion years old. Rocks are the only objects of such great antiquity and, as such, our knowledge of the past 4.5 billion years of Earth history is grounded in the observation and interpretation of them. Rocks represent both the recorders and the time-keepers of Earth. For example, even if the Grand Canyon formed 70-80 million years ago, this is still only the most recent history of the region revealed to us by the present-day landform, the Grand Canyon. However, the rocks that form the canyon walls reveal a much longer history—one that extends back almost 2 billion years. The rocks reveal a progression of now vanished landscapes that existed before the Grand Canyon. They tell us that, at different times in the past, the area that is now the Grand Canyon was once a mountain range, a desert, a beach, and many times covered by shallow oceanic seas.

Each of the three rock groups (sedimentary, crystalline, and volcanic) tells us something about Earth history but from different perspectives. Sedimentary rocks (as well as unconsolidated sediment) form right before our eyes usually by deposition from a well-known medium such as water, ice, or air. The sediments in a swamp, a river, a beach, or a deep ocean are all sedimentary rocks that are in the process of forming. These and many others, such as a glacier, a lake, a desert, and a reef, are all depositional environments.

Each of these environments has its own set of processes that produce a distinctive type of sediment. The study of these environments provides insight into ancient environments. Simple observation, for example, indicates that a river typically deposits sand and gravel. If we were to stumble across a 200-million-year-old bed consisting of sandstone and conglomerate (that is, lithified sand and gravel), we could recognize it as an ancient river deposit. In other words, the rock looks like a river deposit. We call the look of any sedimentary rock its *facies*. We can interpret this rock as a river facies. We also know, by observation, that when a mountain is eroded, it produces an abundance of sand and gravel which is washed into rivers that flow away from the mountain. We know that both the thickness of the sand and gravel layer and the size of individual particles increase generally with proximity to the mountain. Thus, we can gauge our proximity to an ancient mountain front by noting changes in sediment thickness and particle size, even though the mountain itself has long ago disappeared. Here we see an important characteristic of sedimentary rocks: They represent the residue of a vanished landscape.

Now, if we look at rock layers both below and above the ancient river facies, we can then see how the depositional environment at that location has changed over a period of time. What does it mean if we find layers of black shale, coal, and abundant plant fossils above the river facies? Well, when we study present-day rivers, we notice that they change location by migrating (meandering) back and forth across a valley. In doing so they carve out a floodplain. As the river migrates away from one area, it might leave behind a swamp. What do you see in a swamp? Do you see black, smelly mud and lots of organic matter? This is a swamp facies. When this sediment is compacted, dewatered, and cemented, it will consist of black shale, coal, and abundant plant fossils—exactly what we see in the rock. We can now say that over the course of time some 200 million years ago, a river flowing from a distant mountain had migrated away from this spot and left a swamp.

But what about the landscape? What did the surrounding area look like when the river facies was being deposited? Where did the river go when the swamp muds were being deposited? To answer these questions we need to realize that sedimentary layers are approximately horizontal when first deposited. We know this not because we think it is true but because we can look at present-day environments such as a lake or a river and see that the sediment is approximately horizontal. In a present-day environment we can easily walk laterally (horizontally) away from the swamp and eventually encounter the river. We can do the same thing with ancient rock layers. If we walk laterally along the black shale and coal beds (the ancient swamp facies), the rocks might change facies (that is, change their look) from a shale to a (river) sandstone. If so, we have found the location of the ancient river at that moment in time! This is great, but in most instances rocks are not continuously exposed across a wide area. Instead, a section of rock might be covered by grass or trees such that we cannot walk laterally from one rock to the next. Under these circumstances we must time-correlate two or more vertical sequences of rock from different areas. In other words, we must look at isolated exposures of vertical rock sequences from several different locations. If we determine the depositional environment (facies) of each rock layer in each vertical sequence and also determine which layers were deposited at the same time, we can then reconstruct a landscape from the ancient past that existed over a certain interval of time. We can even determine where highland areas were because these areas would have been undergoing erosion and, therefore, would have no rocks at all of that particular age. The procedure is outlined in Figure 20.1.

I know what you are thinking! How are we supposed to know if two rocks in different places are approximately the same age? The answer is fossils. From studying the fossil record for over 200 years, we can say two things with certainty: Fossil organisms evolve (some slowly or not at all, others less slowly or even quickly) and fossil organisms go extinct. This conclusion, which is based on simple observation, is extremely important because it allows us to time-correlate two rocks from widely separated areas, even if the two rocks are of different composition such as sandstone and limestone. If we find fossils from the same evolutionary period in both of these rocks, we can safely infer that the rocks are approximately the same age. With

this information, the depositional environment of the two rocks can be interpreted to reveal the landscape that existed at the time the rocks were deposited. We could then look at the layers of rock both above and below this reconstructed landscape and interpret how the landscape has changed (evolved) through time (Figure 20.1).

Sedimentary rocks are thus important geologically for at least four reasons: (1) They represent the residue of now vanished landscapes. (2) They contain fossils that allow us to determine whether a rock layer at one location is approximately the same age as a rock layer at a distant location. (3) A vertical sequence of sedimentary rock layers records the progression of vanished landscapes that existed at that particular location during the time interval those rocks were deposited. (4) Correlation of rocks of the same age at different locations allows us to reconstruct what an ancient landscape may have looked like at a single instant in time. There is an old saying in geology: Every sedimentary layer represents a page of Earth history.

Crystalline (igneous plutonic and metamorphic) rocks are produced deep within the Earth, well out of sight. They are born of different means in a different place and, for that reason, they give us information that sedimentary rocks cannot. In contrast to sedimentary rocks, the exposure of crystalline rock at the Earth's surface often requires considerable erosional exhumation. Observation tells us that deep erosion and exposure of crystalline rocks at the Earth's surface are characteristic features of mountain belts. Thus, if we find a present-day, nonmountainous area composed of crystalline

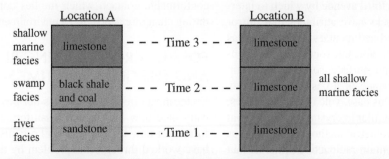

Two vertical sequences of rock and their inferred depositional environment are shown above for locations A and B on the maps below. Fossils allow the two sequences to be time-correlated as indicated by dashed lines. The maps below show the inferred landscape at times 1, 2 and 3. The rocks indicate that location B remained in a shallow sea environment over the entire time span whereas location A evolved from a river, to a swamp, and finally to shallow marine. The rocks indicate a landward transgression of the sea probably due to a rise of sea level or a lowering of land.

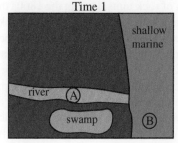
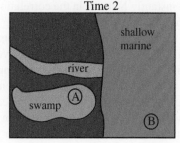
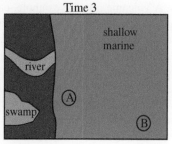

FIGURE 20.1 Depositional environment of sedimentary rocks.

rock, such as the Superior Upland, we can assume that the area was once mountainous a long, long time ago.

Crystalline rocks rarely contain fossils, but they do contain radioactive minerals. When certain minerals crystallize from magma as part of an igneous rock, or crystallize from a reaction while still in the solid state as part of a metamorphic rock, radioactive elements such as uranium and potassium become trapped in minerals and the radiometric clock begins. Radiometric dating is described in more detail later. For now we will generalize and say that a radiometric date can record one of three possibilities: how long ago (in years) an igneous rock crystallized, how long ago a metamorphic rock was metamorphosed, or how long ago either rock underwent cooling following crystallization or metamorphism. These dates often coincide with the approximate time that mountain building is in progress. Earlier we inferred that mountains must have existed a long time ago in the Superior Upland province based on the widespread presence of crystalline rocks at the surface. Radiometric dates obtained from the crystalline rocks are mostly 1.8 billion years or older. This gives us a pretty good idea as to when those mountains existed. Here we see a major difference between the information recorded in sedimentary rocks and that of crystalline rocks. Sedimentary rocks record various aspects of deposition on a time scale based on fossils. They record the residue of an eroding mountain landscape both during and following uplift. Crystalline rocks, on the other hand, record periods of metamorphism and magmatism on a time scale based on radiometric dating. These time periods are often associated directly with uplift of large mountain systems.

Volcanic rocks provide a third avenue by which to interpret Earth history. These rocks have attributes that mirror those of both crystalline and sedimentary rock as well as unique attributes of their own. Volcanic rocks, like sedimentary rocks, form at the Earth's surface, sometimes contain fossils, and record the existence of a past or present-day landscape composed of, in this case, volcanoes, lava flows, and volcanic ash. They are similar to crystalline rock in that they can be dated using radiometric methods. Fossiliferous sedimentary rocks rarely contain radioactive minerals that will date the time of deposition. Thus, an important aspect of volcanic rocks is that they offer a bridge between fossil and numerical dating. Additionally, volcanic rocks are unique in that they can be deposited in a matter of days or weeks. A volcanic ash, for example, can instantly blanket a large geographic area and, if preserved, will provide a snapshot of the landscape and geology at the time of the volcanic eruption. A radiometric date of this ash will provide a precise age constraint on the underlying ground surface and its fossils.

Thus, the three major rock types each reveal their own unique part of Earth history. Taken collectively, they mesh beautifully to bring us toward a greater understanding of Earth and its ancient landscapes.

There is yet a fourth avenue by which Earth history can be pursued. As alluded to earlier, it is, in a manner of speaking, the absence of rock. If a mountain is undergoing erosion, then the very rocks that are required to interpret Earth history are, at that moment in time, being removed from their place of origin. An erosion surface can be thought of as time not represented by rock. In geology we refer to such a surface as an *unconformity*, a term we introduced in Chapter 10. The presence of an unconformity in a vertical sequence of rock not only tells us when that particular area existed as land, it also helps us separate periods of deformation and mountain building from periods of inactivity and deposition. The various types of unconformities and some aspects of how they are interpreted are given in Figure 20.2. Angular unconformities in particular are used to bracket time periods during which mountain building may have occurred. If rocks below an unconformity are folded, faulted, metamorphosed, or intruded, and rocks above the unconformity are not, then deformation (and associated mountain building) must have occurred after deposition of the youngest rock below the unconformity but before deposition of the oldest undeformed or less deformed rock above the unconformity. Angular unconformities are used worldwide to date deformation associated with mountain building. Recall that James Hutton's view of Earth history was crystallized when he discovered an angular unconformity in Scotland. In addition to the unconformable contacts, Figure 20.2 also shows three other possible contacts between rocks layers: a fault across which adjacent beds are offset, an intrusive contact in which the younger granite intrudes older rock, and a conformable contact, which implies continuous deposition during changing depositional environments.

FOSSILS, CROSS-CUTTING RELATIONSHIPS, AND THE GEOLOGIC TIME SCALE

Under most circumstances, it is not possible to radiometrically date how long ago a fossiliferous sedimentary rock formed or how long ago a fossil organism died. Geologists have worked through this problem by developing a formal *geologic time scale* and by using cross-cutting relationships that involve datable intrusive and volcanic rocks.

In the late 1700s, geologists in Western Europe began to recognize the importance of fossil evolution and extinction in the correlation of rock units. During the 1800s they went about mapping and describing the layers of rock, which they then grouped into formations. They described a *formation* as a layered sequence of beds related to each other by similar composition, texture, depositional environment, or fossil content. Formations were named after particular geographic locations where the rocks are well exposed. Each formation was assigned a stratigraphic position above or below other formations. Eventually, these early geologists produced a geologic map of Western Europe that

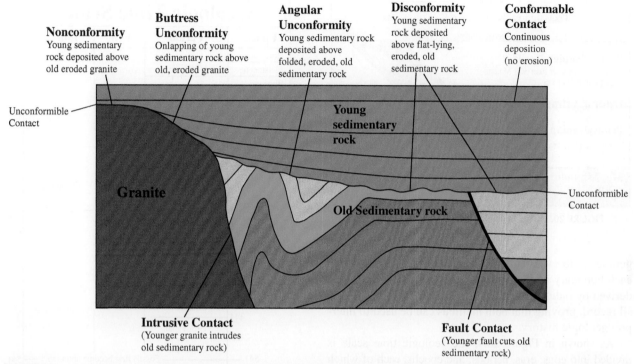

FIGURE 20.2   A schematic cross-section that shows the four types of geologic contacts (conformable, unconformable, faulted, intrusive), including the various types of unconformities. The angular unconformity provides the best evidence of mountain building. Mountains likely were in existence sometime after deposition of the youngest deformed bed below the unconformity and before deposition of the oldest bed above the unconformity.

showed the location of each rock formation and its position, above or below, other rock formations. This formal subdivision of rock is still in use today. For example, the Bright Angel formation is described in the Grand Canyon as a shale unit that overlies the Tapeats formation, which is composed mostly of sandstone. In Indiana the West Franklin Limestone is well exposed at its type locality near the town of West Franklin.

As the geologic map of Western Europe began to emerge, it became clear that there was an orderly and evolutionary arrangement of fossils in the stratigraphic sequence. Because similar fossils appear worldwide, it was recognized that this arrangement could be used as a correlation tool to determine the relative age of rock units anywhere in the world. By the 1830s there was a conscious effort to produce a formal stratigraphic hierarchy based solely on fossil content. A set of fossils within the stratigraphy was assigned a name and a type location. Boundaries were placed at major extinctions or where new fossils appeared. Because each set of fossils represents a certain period of time, each distinctive assemblage was referred to as a *period*. For example, the Devonian Period is represented by fossil assemblages in rocks located at Devon, England. Fossils in Devon are Devonian by definition. All rocks that contain similar fossils (anywhere on Earth) are Devonian (that is, approximately the same age) by comparison. The stratigraphic correlation

chart, which is now referred to as the geologic time scale, was largely completed by 1880.

Fossils tell us whether one rock is older, younger, or approximately the same age as another rock, but they tell us nothing about how old (in years) the rock actually is. This requires radiometric dating. In 1880, nobody knew how long (in absolute numbers) any of the geologic time periods lasted because radioactivity (and the radiometric dating of rocks) had not yet been discovered. Henri Becquerel discovered radioactivity in 1896, and by 1911 Arthur Holmes, Ernest Rutherford, Bertram Boltwood, and a few other scientists developed methods by which to date rocks using radioactivity. Many scientists of that period thought the Earth was no more than 100 million years old. The early experiments proved for the first time that Earth was billions of years old. These methods were then applied indirectly to date fossiliferous sedimentary rocks using the principle of crosscutting relationships, as outlined in Figure 20.3. Note the presence of volcanic ash in Figure 20.3. Ash and lava flows are particularly valuable because they are deposited directly onto the land surface in a matter of hours or days. They provide a snapshot of the land surface and an accurate depositional age of any unconsolidated sediment that had been deposited directly below the ash or lava layer.

Radiometric dating came into wide use after World War II, and using the principle of cross-cutting relationships, it allowed

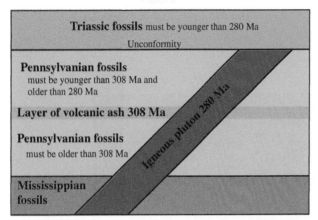

FIGURE 20.3   The principle of cross-cutting relationships.

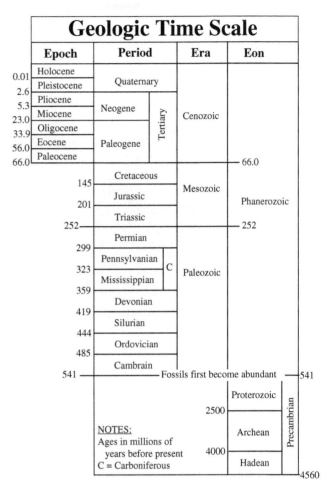

FIGURE 20.4   The geologic time scale. Based on Geological Society of America (2012) http://www.geosociety.org/science/timescale/.

geologists to eventually add absolute ages (numbers) to each boundary on the geologic time scale. Absolute ages derived by radiometric means are consistent with the fossil record, proving that both methods can be used to interpret geologic history.

As shown in Figure 20.4, the geologic time scale is divided into eons, eras, periods, and epochs, each of which represents a progressively smaller interval of time defined by a certain set of fossils. All of geologic time is divided into the Precambrian and Phanerozoic eons, the boundary of which separates highly fossiliferous rock above from rocks nearly devoid of fossils below. Note the length of the Precambrian eon and that it is divided as far back as 4,000 million years ago into the Archean and Proterozoic eons. *Archean* is from the Greek *arkhaios,* meaning ancient. *Proterozoic* literally means earliest (or former) life. The *Phanerozoic* eon (meaning evident life) began 541 million years ago and marks the beginning of time when fossils first become abundant in the rock record. It represents only one-ninth of the total age of Earth! The Phanerozoic eon is separated into the *Paleozoic, Mesozoic,* and *Cenozoic* eras, the boundaries of which are marked by massive extinction events. The mass extinction at the Paleozoic-Mesozoic boundary terminated a majority of known marine (ocean-dwelling) species. The extinction event at the Mesozoic-Cenozoic boundary terminated 25% of all known families, including the dinosaurs. A distinctively different set of animals thrived before and following these extinction events such that it is fairly easy to distinguish between Paleozoic, Mesozoic, and Cenozoic fossils. The uniqueness of the fossils from each era is reflected in the names; Paleozoic means old life, Mesozoic means middle life, and Cenozoic means new life. The periods for each of the three Phanerozoic eras and the epochs for the Cenozoic era are shown in Figure 20.4.

Each eon, era, period, and epoch on the geologic time scale represents an interval of time. Therefore, when we refer to an Ordovician mountain-building event we are implying that mountain building occurred during the interval of

time that Ordovician fossils were alive. Based on radiometric dating, this interval occurred between 485 and 444 million years ago.

In addition to time correlation, fossils also tell us a great deal about ancient environments and about the geographic location of a landmass during the time the fossils were alive. Fossils tell us whether sediment was deposited in fresh water, marine (oceanic) water, deep water, or shallow. For example, we know that reefs can survive only in relatively shallow, warm, oceanic water. Today all reefs are located within about 30° latitude north or south of the equator. The presence of Silurian and Devonian reef deposits in and around the state of Michigan indicates that Michigan was below sea level and possibly closer to the equator during the Silurian and Devonian periods. Plate tectonic movements have since transported Michigan to its present location. The alternative explanation is that the Earth was much warmer in the Silurian-Devonian than it is today. This possibility must be considered and can be checked by comparing other fossils of that time period from around the world.

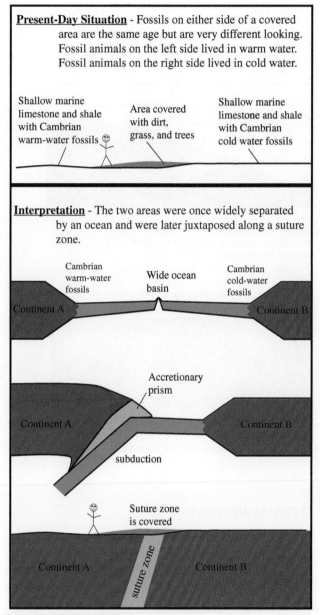

Present-Day Situation - Fossils on either side of a covered area are the same age but are very different looking. Fossil animals on the left side lived in warm water. Fossil animals on the right side lived in cold water.

Shallow marine limestone and shale with Cambrian warm-water fossils ☺

Area covered with dirt, grass, and trees

Shallow marine limestone and shale with Cambrian cold water fossils

Interpretation - The two areas were once widely separated by an ocean and were later juxtaposed along a suture zone.

Cambrian warm-water fossils

Wide ocean basin

Cambrian cold-water fossils

Continent A

Continent B

Accretionary prism

Continent A

Continent B

subduction

Suture zone is covered

Continent A

Continent B

suture zone

FIGURE 20.5 How different fossils of the same age indicate the existence of accreted terranes separated by a suture zone.

It is well known that land plants and crawling animals evolve into unique species if they become isolated from other land plants and crawling animals. Unique species are a characteristic of many of today's islands. This observation is used in geology when we examine ancient rock deposits. If rocks in one area contain land fossils that are indicative of continent A, and rocks right next to them contain a different set of land fossils of the same age that are indicative of continent B, one could assume that the two areas were separated by a body of water at the time the fossil organisms were alive. The two land areas must have been brought into contact by tectonic movements and faulting after the fossil animals died.

Similarly, we can distinguish between warm water and cold water fossils of the same age. Figure 20.5 shows a situation in which two very different sets of fossils of the same age are juxtaposed next to each other. In this case, the foreign rock (continent B) could not originally have been part of continent A. It must have accreted (that is, attached) to continent A along a suture zone when plate tectonic movements closed an intervening ocean basin and caused the collision of continent A and continent B. If a geologist stumbles across these two distinct sets of fossils, he or she should immediately set out to locate the suture zone that now separates them. This same concept can also be used to determine when two separate landmasses began to approach each other. If younger fossils in both terranes appear to evolve toward each other, the terranes by that time were no longer completely isolated. The two terranes must have been approaching each other. An analysis of this type does not provide a time when the two terranes collided; it only provides a time when the two now adjacent terranes were apart. We can only say that collision occurred sometime after the fossils were alive. Still, in summary, we can say that fossils provide constraints on the age of rocks, the depositional environment, the approximate latitude at which they lived (cold versus warm water), and tectonic plate movements.

Let us return for one moment to cross-cutting relationships, because these can be used in many ways to decipher Earth history. Figure 20.6 provides a few examples. Figure 20.6a is perhaps the most simplistic. In this example the stacking order indicates that rock unit 3 is younger than units 1 and 2. The fault cuts unit 3; therefore, faulting occurred after deposition of unit 3. Rock unit 4 cuts across the fault without offset and, therefore, must be younger than faulting. If we date rock units 3 and 4 via fossils or radioactive isotopes, we will bracket the time of faulting, uplift, and erosion. For example, if unit 3 was deposited 250 million years ago and unit 4 deposited 230 million years ago, then faulting (and uplift and erosion) occurred sometime between 250 and 230 million years ago.

Figure 20.6b is a situation in which a fault is present at the margin of an existing highland of crystalline rock. Units 2 and 3 consist of unconsolidated sediment derived from erosion of the mountain. Unconsolidated unit 2 is cut by the fault, and unconsolidated unit 3 cuts across the fault. What does this tell us about Earth history? First, the presence of a fault at the base of the mountain suggests that uplift occurred as a result of displacement on the fault. Cross-cutting relationships coupled with the presence of sediment derived from the mountain indicate that the fault was active during deposition of unit 2 but became inactive during deposition of unit 3. Thus, we can infer that the mountain came into existence and underwent tectonic uplift during deposition of unit 2, and that uplift ceased and the mountain underwent (and continues to undergo) erosional lowering during deposition of unit 3.

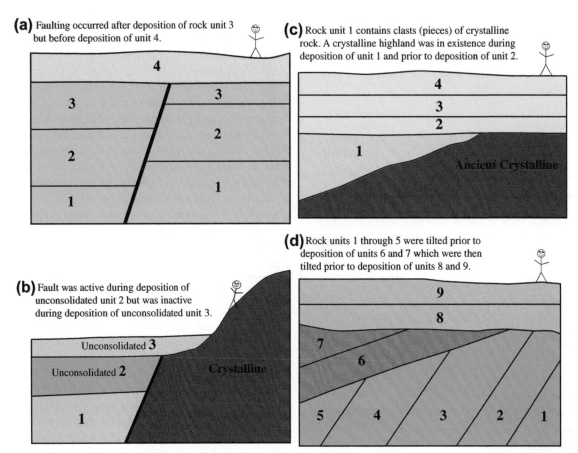

(a) Faulting occurred after deposition of rock unit 3 but before deposition of unit 4.

(b) Fault was active during deposition of unconsolidated unit 2 but was inactive during deposition of unconsolidated unit 3.

(c) Rock unit 1 contains clasts (pieces) of crystalline rock. A crystalline highland was in existence during deposition of unit 1 and prior to deposition of unit 2.

(d) Rock units 1 through 5 were tilted prior to deposition of units 6 and 7 which were then tilted prior to deposition of units 8 and 9.

FIGURE 20.6   Various aspects of cross-cutting relationships and how they are used to determine age of deformation and mountain building. Numbers in all diagrams denote sedimentary rock units.

Figure 20.6c is a situation where old crystalline rock, and a wedge of young sedimentary rock (unit 1) derived from erosion of the crystalline rock, are covered with younger, flat-lying, marine, sedimentary rock layers (units 2, 3, and 4). In this situation we can infer that a crystalline mountain (or at least a highland) was in existence during deposition of unit 1 and that the highland had eroded to a flatland and sank below sea level prior to deposition of unit 2.

Finally, Figure 20.6d shows a sequence of tilted layers in which units 1 through 5 were tilted prior to deposition of units 6 and 7, which were then tilted prior to deposition of units 8 through 10. Notice that in all of these examples, we can determine the sequence of events without fossils or radiometric dating. However, if we date the rock units via fossils or radioactive isotopes, we would then be able to determine specifically how long ago, and over what time interval, each event such as faulting, uplift, and erosion, occurred. We would also be able to determine when mountains were in existence and where they were located. It is these types of analyses, carried out through painstakingly detailed and field studies, that allow us to decipher and order Earth history.

METAMORPHISM

The *geothermal gradient* is defined as the increase in temperature with depth in the Earth. In normal continental crust a typical geothermal gradient within the first 3 to 5 kilometers (2 or 3 miles) of Earth's surface is about 25°C/km. This gradient, however, is not sustained but decreases to no more than about 16°C/km at a depth of 40 km. A geothermal gradient for continental crust that accounts for changes at depth is shown in Figure 20.7. Mountain-building processes have the effect of increasing the geothermal gradient, which results in the metamorphism of crustal rocks.

Abnormally high geothermal gradients and concomitant metamorphism result from a number of factors associated with subduction and continental collision. These factors include crustal thickening, the concentration of heat sources such as radioactive elements, the introduction of rising magma bodies, and the presence below the crust of hot mantle rock. The added heat weakens the crust, allowing it to deform more readily. Under these conditions the crust thickens via several mechanisms that include folding, magma injection, thrust faulting, and simple compression. Thickening of relatively low-density continental rock will

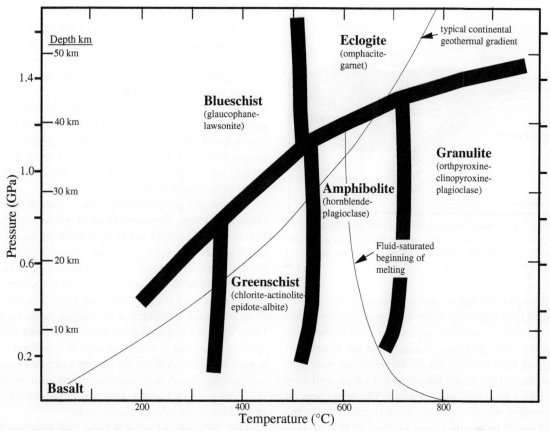

FIGURE 20.7 The five metamorphic facies and their diagnostic mineral assemblages. Also shown is a typical continental geothermal gradient and a curve that marks the beginning of fluid-saturated melting. Based on Winter (2001).

cause the crust to rise isostatically, producing mountains. This, in turn, increases the rate of erosion. A feedback mechanism is established whereby additional compression produces greater uplift and greater rates of erosion such that warm rock at depth is quickly exhumed toward the surface. A high geothermal gradient is established below the uplifting mountain and metamorphism occurs. What this means is that, in an ancient mountain belt, we can use the temperature and areal extent of metamorphism as an indirect measure of the intensity of mountain building (that is, the intensity of uplift/erosion).

A common description for the intensity of metamorphism is the metamorphic facies concept (not to be confused with sedimentary facies; the two are entirely different). Developed in 1915, the facies concept is based on the metamorphism of basalt. The basic idea states that, for a given pressure-temperature condition, the resulting mineralogy is controlled only by the chemical composition of the rock which does not change during metamorphism except for the loss of fluids. When a rock with the chemical composition of basalt is metamorphosed, it will recrystallize into one of at least five rocks, each with a diagnostic mineral assemblage depending on the prevailing temperature and pressure. The five rocks are blueschist,

greenschist, amphibolite, granulite, and eclogite. The temperature and pressure conditions under which each rock occurs, and the rocks' diagnostic mineral assemblages, are shown in Figure 20.7. Thus, a rock of basaltic composition, metamorphosed to amphibolite facies, will recrystallize to form an amphibolite. Implicit in this concept is that any surrounding rock, such as a shale metamorphosed to schist, or a limestone metamorphosed to marble, is metamorphosed within a similar temperature-pressure range as the nearby amphibolite. All of the rocks are at amphibolite facies. The concept implies that by simply looking at a rock and noting its mineral assemblage, a geologist can determine the temperature and pressure range under which the rock was metamorphosed.

The most common metamorphic facies in mountain belts are the greenschist and the higher-temperature amphibolite facies. *Greenschists*, as their name implies, tend to be green. *Amphibolites* tend to be dark green to black. Neither is difficult to recognize in the field. The widespread distribution of these facies indicates that this is the temperature-pressure range under which most mountain-building occurs (Figure 20.7). Granulite facies rocks represent extreme temperature conditions. When present they form the deep-seated core of mountain belts. Blueschist and eclogite

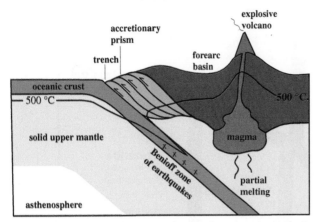

FIGURE 20.8 A schematic cross-section that shows how the depth of a 500 °C isotherm (a line of constant temperature) varies across a subduction zone boundary. Note that the isotherm is subducted with the subducting plate and rises over the volcanic arc.

facies, although relatively uncommon, are extremely important because they form only under high-pressure conditions at relatively low temperature. When rocks are pushed into the Earth they, like all things, instantly feel the pressure of the overlying rock column. Rocks, however, are poor conductors of heat. They take a long time to heat up, and once hot they take a long time to cool down. The only common tectonic setting where rocks are pushed rapidly into the Earth and buried before they have a chance to heat up is in the subducting slab of a subduction zone. A subducting slab will remain cool as it is pushed deep into the Earth, even though surrounding rocks are much hotter. Figure 20.8 is a schematic diagram that shows the depth at which 500°C is reached across a subduction zone boundary. The 500°C line is referred to as an *isotherm* because it is a line of constant temperature. Note how the isotherm dips to great depths within the subducting slab. It is this high pressure–low temperature condition that produces blueschist and eclogite facies rocks. Note also in this figure how the isotherm rises within the thickened crust of the mountain belt.

The fate of nearly all blueschist and eclogite facies rocks is that they will subduct deeper into the mantle and eventually melt. In order for the mineral assemblage to be preserved, blueschist and eclogite facies rocks must first be sliced off the subducting slab and then returned to the Earth's surface. If by chance they are sliced off the subducting slab but remain at depth, they will continue to heat and metamorphose until they reach the same temperature as surrounding rocks. Under these conditions, the blueschist facies mineral assemblage will be destroyed in favor of an amphibolite facies mineral assemblage. This possibility is shown as pressure-temperature Path A in Figure 20.9. In this example the rock is sliced off the subducting slab but remains more or less stationary at a depth of about 25 miles (40 km), where it undergoes static heating. It continues to heat as it undergoes exhumation to a depth of 12 miles

(20 km). Exhumation could be the result of a thrusting event or rapid erosion. Under most circumstances, a rock will retain the metamorphic assemblage that forms at the highest temperature attained, regardless of the pressure. The temperature and the corresponding pressure at which the highest temperature is attained are referred to as *peak metamorphism*. In the case of Path A, the mineral assemblage associated with peak metamorphism is that of the amphibolite facies. If the diligent geologist suspects such a pressure-temperature path, he or she will look in great detail at the rock to see if any evidence remains of the earlier blueschist facies mineral assemblage.

A possible pressure-temperature path that preserves blueschist facies metamorphism is shown as Path B in Figure 20.9. This rock could have been sliced off the subducting slab soon after burial, incorporated into the accretionary prism, and almost immediately carried to the surface by thrust faults. Blueschists are easy to recognize because, unlike most rocks, they really are blue. The significance of an exposure of blueschist or eclogite facies rocks at the Earth's surface is that they are a diagnostic rock type within a suture zone (that is, within an ancient fossilized accretionary prism). They indicate the presence or past presence of a subduction zone. A now vanished ocean basin must have at one time separated now juxtaposed terranes on either side of the blueschist-eclogite facies rocks.

One may ask, why doesn't the rock continue to metamorphose as it cools? There are two reasons. The increase in temperature is the primary driving force for reaction. Reactions are less likely to occur if this driving force is increasingly dissipated during cooling. The second reason is the general absence of fluid (usually H_2O or CO_2 in the form of a gas-liquid known as a *supercritical fluid* that can diffuse through solid like a gas and dissolve material like a liquid). Fluids are driven from the rock during metamorphism. If these fluids are absent when the rock returns to lower temperature then minerals such as chlorite and actinolite, which require fluids to form, will be unable to crystallize and the rock will retain its high-temperature mineral assemblage.

Another descriptor for metamorphism is grade, which is a measure of metamorphic pressure-temperature (and fluid) intensity. Both pressure and fluids are capable of influencing metamorphism. However, the dominant influence in many instances is temperature. We will therefore consider grade to be roughly synonymous with temperature. A low-grade rock implies that it was metamorphosed at relatively low temperature such as greenschist or blueschist facies. High-grade implies high-temperature metamorphism consistent with amphibolite, granulite, and eclogite facies.

With the major exception of the Valley and Ridge, nearly the entire US Appalachian Mountain belt was subjected to at least greenschist facies metamorphism, with more than half reaching high-grade amphibolite-granulite facies conditions. Metamorphism in the Cordillera, by contrast, is

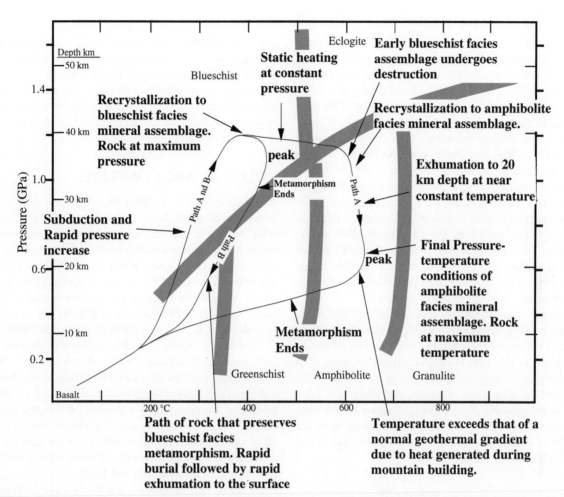

FIGURE 20.9   Hypothetical pressure-temperature paths A and B. Path A reaches blueschist facies but remains at depth, where it continues to heat up, reaching amphibolite facies. Peak metamorphic conditions are about 0.7 GPa (21 km depth) and 640°C. Path B reaches blueschist facies and is then immediately returned to the surface. Metamorphism ends before the rock leaves the blueschist facies pressure-temperature field. Peak metamorphic conditions are about 1.1 gigapascals (GPa), or 37 km depth and about 440°C.

spotty, mainly because the mountain system is young and has not yet undergone deep erosion to expose metamorphic rock. There is, however, an abundance of blueschist and some eclogite facies rocks in the Coast Ranges as well as isolated blueschist rocks in the Klamath Mountains and Sierra Nevada. Areas of greenschist facies rocks are present in mountain blocks across the Basin and Range. High-grade rocks are present in the North Cascades and in a few areas in the Coast Ranges, Klamath Mountains, Northern Rockies, and Basin and Range. Metamorphism in both the Cordilleran and Appalachian Mountain belts is complicated by multiple heating events that in some cases are superimposed on one another.

PLUTONISM

Plutonism is the process by which magma rises through the crust and crystallizes as an intrusive igneous rock beneath the Earth's surface. *Pluton* is a generic word for

any igneous intrusive rock body. Exactly what causes melting is complicated and variable. Assuming the rock is already under metamorphic temperature-pressure conditions, melting could result from an influx of fluids, a decrease in pressure as hot rocks are brought closer to the Earth's surface, or any number of factors and combinations that cause an increase in temperature, such as a concentration of radioactive elements, or additional heat flow from the mantle or from an underlying magma body. For example, melting of crustal rock will occur under amphibolite facies conditions if the rock is saturated with fluid as implied in Figure 20.7. It will not melt if fluids are absent. If melting does occur, then in nearly all situations the rock only partially melts. In other words, only selected minerals in the rock melt; other minerals remain solid. Buoyancy contrasts allow the liquid to rise to create a magma body.

There are two primary locations where magma is generated during mountain building. The first is directly within thickened continental crust where granitic magma forms during high-grade metamorphism. The second and more widespread location is in the mantle wedge directly above a subduction zone, in which case the magma is typically basaltic in composition (Figure 20.10). Basaltic magma generated in the mantle wedge above a subducting slab often rises only as far as the base of the continental crust because liquid basalt it is heavier than the warm, solid continental crust. The addition of heat into the crust as the basaltic magma cools contributes to metamorphism and could potentially melt part of the overlying continental crust producing an area of granitic magma. As the basaltic magma cools, igneous processes that include partial crystallization and magma mixing cause the liquid to evolve into a more silicic andesite composition that is light enough to rise through the crust and create explosive volcanoes. A magma chamber forms near the base of the crust that feeds the overlying volcano. The relationships are depicted in Figure 20.10.

Eventually, subduction will cease or melting will move elsewhere causing the underlying magma chamber to crystallize and the overlying volcanoes to go extinct. If mountain building in the form of uplift/erosion continues, then the underlying magma chamber will be exhumed to the Earth's surface in the form of a *batholith*, which is defined simply as an exposure of intrusive igneous rock with large areal extent. The Sierra Nevada and Idaho batholiths are examples of subduction zone magma chambers that at one time hosted volcanoes at the surface.

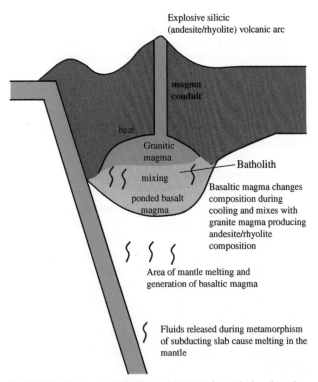

FIGURE 20.10   A schematic cross-section that shows the location where magma is generated above a subduction zone.

Granitic magma also can be generated in continental crust without subduction, particularly if continent-continent collision occurs. Under these conditions, the thickened crust simply reaches its melting point. In contrast to subduction zone magmatism where a conduit to the surface is established, the sticky granite magma normally cannot find its way to the surface and instead cools and crystallizes at depth in the form of a pluton.

VOLCANIC ARC COMPLEXES

A *volcanic arc* is a chain of volcanoes, hundreds to thousands of miles long, that forms above a subduction zone. An island volcanic arc forms in an ocean basin via ocean-ocean subduction. The Aleutian Islands off the coast of Alaska and the Lesser Antilles south of Puerto Rico are examples. A continental volcanic arc forms along the margin of a continent where oceanic crust subducts beneath continental crust. The Cascade Volcanoes are an example. In both cases, the volcanic arc is an active landform.

The term *complex* is used in geology to identify a discrete package of deformed rocks. A volcanic arc complex is a discrete package of rocks interpreted to represent the deformed remnants of an ancient, now vanished volcanic arc landform. There are many examples of volcanic arc complexes in the Appalachian and Cordilleran Mountain belts, all of which formed either offshore and were accreted to the mountain belt or, like the present-day Cascade volcanoes, formed within the mountain belt itself.

Volcanic arc complexes are fairly easy to recognize when they remain unmetamorphosed or at low metamorphic grade. Typically they contain silicic and some basaltic volcanic rock as well as volcaniclastic rock (that is, clastic sedimentary rock derived from erosion of a volcanic source). Typical volcaniclastic rocks include certain types of shale and greywacke (mud-rich sandstone). Under low-grade conditions these rocks metamorphose to greenschists. Thus, the simple presence of green-colored rock is a clue to the presence of a volcanic arc complex. Volcanic arc complexes metamorphosed to high-grade are also fairly easy to recognize. In this case, the volcanic rocks metamorphose to amphibolite facies, creating a rock known as *amphibolite*, which is dark green to blackish. Another indicator of a volcanic arc complex is the presence of intermingled plutonic rocks, which represent the ancient underground feeder chambers for the lava. The extent to which plutons are exposed depends on the level of erosion. When metamorphosed, the plutons become granitic gneisses.

There are other types of volcanic terranes that form part of an ocean basin including volcanic island chains and volcanic plateaus. Unlike arc terranes, these are unrelated to subduction. Instead they could have formed above a hot spot much like the Hawaiian Island volcanic chain. These too can be accreted to continents and deformed into rock complexes.

The Siletz volcanic complex discussed in Chapter 16 is an example. Rocks that form volcanic island chain and plateau complexes are distinct from volcanic arc complexes in that they contain large quantities of basalt and associated deep-water clastic rock. Another possible distinction is the general absence of plutonic rocks. The distinction is less obvious at high metamorphic grade because the original rock fabric is destroyed. Geologists in this case can use more sophisticated chemical techniques to decipher the origin of the rock.

OPHIOLITE, SUBDUCTION COMPLEXES, AND COLLISION

Ophiolite literally means snakestone. In modern usage it refers to an intact section of oceanic lithosphere that has been thrust onto a continental margin. A typical section of oceanic lithosphere consists of ultramafic rock and serpentinite near its base, overlain by gabbro and basalt, with bedded chert or turbidite at the top. *Turbidite* is a fine-grained sediment (or sedimentary rock) that gradually changes from coarse- to fine-grained and that was deposited by turbidity currents. It is rare for a complete, intact section of oceanic lithosphere to be thrust (obducted) onto a continent. The

only complete section in the Appalachians is the Bay of Islands Ophiolite, which is spectacularly exposed in Gros Morne National Park in the Long Range Mountains of western Newfoundland. Fairly complete sections in the Cordillera include the Trinity ophiolite in the Klamath Mountains and the Coast Range ophiolite in the Coast Range of California. The obduction of intact oceanic lithosphere onto a continental margin implies that a subduction zone once existed at that location, but it does not imply collision between two terranes. Obduction can occur during the subduction process, and following obduction the subduction process could continue. Complications arise if collision with another terrane occurs long after obduction as depicted in Figure 20.11. In this case, a geologist must differentiate between the obduction event and the later collision event. This can be an especially difficult task if rocks representing the two events are intermingled.

As noted in Chapter 5, an *accretionary prism* consists of dismembered pieces of oceanic lithosphere that have been scraped off the down-going slab of a subduction zone, mixed with sediment, and added to the underside of an overriding plate. The package is churned, tumbled, sheared, and faulted during the subduction process to create a truly

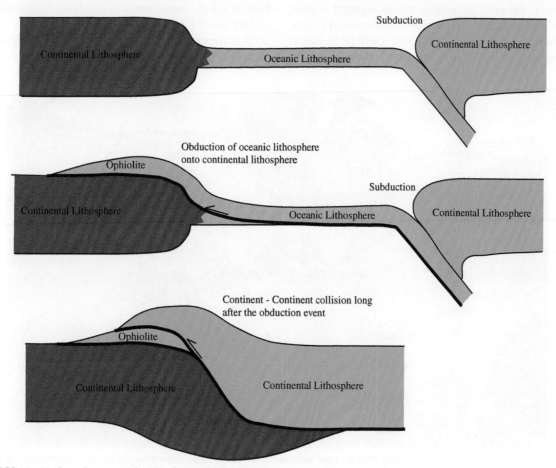

FIGURE 20.11   A schematic cross-section that shows ophiolite obduction onto a continent followed, sometime later, by continent-continent collision.

chaotic mixture of material. The resulting rock body consists of lenses, pieces, and blocks of oceanic lithosphere (ophiolite), including chert, argillite, greywacke, serpentinite, ultramafic rock, gabbro, basalt, blueschist, greenschist, talc-carbonate, and eclogite, set in a matrix of mudstone, shale, low-grade phyllite, or schist. Individual blocks vary in size from several miles across to only a few inches. Each block is individually metamorphosed or unmetamorphosed. Blocks are accreted (added) over the course of millions or hundreds of millions of years during the life of the subduction zone and are derived from different parts of the subducted oceanic lithosphere which could have originally been hundreds or even thousands of miles wide. The accretionary prism could even include a small island volcanic arc or an island volcanic chain.

When this mixture is preserved as rock within a mountain belt, it is known as *mélange* and the rock unit is referred to as a *mélange zone*, a *subduction complex*, or a *suture zone*. These terms are used interchangeably. Figure 20.12 is a photograph of a mélange. The presence of a mélange zone (suture zone) implies that landmasses on either side were once separated by a now vanished ocean basin (Figure 5.7). The age of final suturing must be younger than any oceanic rock found in the mélange zone. In the strict sense, the mélange zone is not an accreted terrane, because even though it is composed of far-traveled blocks, it may have formed in place. However, it will be treated as a terrane because it is distinct relative to surrounding rock and because it can be separated, usually by faults, from surrounding rock. A classic example is the Franciscan complex in the northern California Coast Ranges.

Collision between two tectonic plates is defined as an event large enough and buoyant enough to change either the location of a subduction zone resulting in accretion of a landmass to the continental margin, or to cause a change in the pattern of plate motions (Figures 5.7 and 5.9). Collision thus includes the accretion of a large volcanic arc. Ongoing accretion at a subduction zone, and the obduction of intact ophiolite slabs, are not considered to be collision.

RECOGNITION OF CRYSTALLINE BASEMENT

The term *basement* refers to rocks at the base of a rock sequence. They are usually crystalline and are unconformably overlain by a cover of sedimentary or volcanic rocks. The basement and cover rocks have completely separate geologic histories. A classic example is the contact between the North American crystalline shield that forms most of the Superior Uplands, and the unconformably overlying sedimentary rock of the Central Lowlands. A second example is the Fall Line between crystalline rock of the Piedmont Plateau and the unconformably overlying sedimentary rock of the Coastal Plain.

It is important to recognize the type of basement rock in an area because this can show whether or not the rock is

FIGURE 20.12   A photograph of a mélange zone. Note the large lenticular-shaped lens at upper-center below the tree and several smaller lenses set in a sheared shale and mudstone matrix.

exotic (derived from far away) or native. In the Appalachians, the native North American crystalline shield was metamorphosed circa 1 billion years ago during a pre-Appalachian mountain-building event known as the *Grenville orogeny*. Crystalline basement in the western part of the Blue Ridge conforms to this age; therefore it is considered to have been part of North America prior to Appalachian mountain building. Crystalline basement in the Piedmont is not Grenville in age, implying that it is composed of accreted terranes that were added to North America during Appalachian orogenisis. There must be a suture zone that separates the two areas. A geologist could conceivably correlate the accreted basement rock with basement rock from other continents to determine where the accreted terrane originated. This, in turn, provides constraints on ancient plate movements.

RADIOMETRIC DATING (GEOCHRONOLOGY)

Recall from Chapter 4 that the closure temperature of a mineral is the temperature at which the radiometric clock begins to record the age of the mineral. For example, a mineral with a closure temperature of 200°C and an age of 40 million years indicates that this particular mineral cooled through a temperature of 200°C 40 million years ago. It also indicates that this mineral was never heated above 200°C within the past 40 million years because otherwise the isotopic clock would reset itself such that the 40-million-year-old date would be lost from the rock record and the mineral would record the age that corresponds with the second cooling below 200°C.

Also in Chapter 4 we introduced methods such as fission track and (U-Th)/He geochronology to interpret final exhumation to the present-day land surface. The methods employed are ideal for this purpose because the minerals have closure temperatures less than 310°C. They record final cooling as a rock approaches the temperature at the Earth's surface. An understanding of how vanished landscapes evolved will require radiometric systems with much higher closure temperatures, ones that more closely approximate the temperature of plutonism, peak metamorphism, and initial cooling. The two most widely used are the U-Pb and argon-argon ($^{40}Ar/^{39}Ar$) methods. The argon method was discussed in Chapter 4 with respect to its role in landscape evolution.

The geologic significance of a mineral date depends on the closure temperature and the environment in which the mineral crystallized. A mineral date can represent either the time of crystallization in an igneous magma (plutonism), the time of crystallization during metamorphism, or the time the already existing mineral cooled through a closure temperature within a preexisting rock. A mineral will record an igneous or metamorphic crystallization age if it crystallizes at a temperature less than its closure temperature. As

an example, let us examine a mineral with a closure temperature of 600°C. If this mineral crystallizes in a metamorphic rock when the temperature is 500°C, then the date is locked in immediately upon crystallization. If the same mineral crystallizes when the temperature is 700°C, then the radiometric clock will not begin until the mineral temperature cools to 600°C. The beauty of this system is that it is possible to date a variety of minerals with different closure temperatures from the same rock (or from nearby rocks) to obtain the crystallization age and a range of cooling ages. This will allow a geologist to decipher the rock's temperature at various times throughout the mountain-building process. An analysis that utilizes both crystallization ages and cooling ages will allow a geologist to create pressure-temperature paths such as Paths A and B in Figure 20.9 but with time constraints. Creating a pressure-temperature-time path and understanding how a mountain formed, evolved, and eventually disappeared is what makes geology so incredibly interesting!

In addition to dating a single metamorphism, it is often possible to date more than one period of metamorphism within a rock or to date the time of crystallization associated with displacement along a fault zone. To do this, the geologist must be able to associate crystallization of specific minerals with each geologic event. For example, it might be possible to date a relict (leftover) mineral from the earlier blueschist facies metamorphism of a rock that followed Path A in Figure 20.9. To properly interpret any radiometric date, the geologist must know every detail surrounding the mineral dated and the rock from which the mineral came. He or she must have an understanding of the geologic history of the rock, its structure, its contact relationships, and its composition. The geologist needs to create a geologic map of the area or field check an existing geologic map and know exactly where on the map and from what rock unit the sample was collected. Here again, field geology is of paramount importance.

The closure temperature of any mineral will vary by 25°C or more depending on the size of the mineral grain, the speed at which ions move through the mineral (a process known as *diffusion*), the mineral purity, and the rate of cooling. The first three variables can be determined in the laboratory; however, the rate of cooling must be inferred based on geological conditions. Thus, the best minerals to date are those where these parameters are well understood or where the parameters do not highly affect the closure temperature.

The mineral zircon (zirconium silicate) is by far the most common mineral used in U-Pb geochronology. It is an accessory mineral that crystallizes during the late stages of granite intrusion and, less commonly, during metamorphism. It is highly resistant to weathering and is found as a detrital mineral (a fragment) in many sedimentary rocks, including sandstone. Zircon has a very high closure temperature, on the order of 900°C. It almost always records a crystallization age, usually an igneous crystallization age. Zircon is relatively immune to diffusion, which means that it retains its igneous crystallization age even if later subjected to high-grade metamorphism or to weathering and redeposition in a clastic sedimentary rock.

A U-Pb zircon age from a plutonic rock will provide the age of magma crystallization following intrusion. Such dates can be used with cross-cutting relationships to decipher Earth history. If the pluton is undeformed and the surrounding rocks are strongly deformed and metamorphosed, then we know that crystallization occurred after deformation and metamorphism. If the pluton and the surrounding rocks are equally deformed and metamorphosed, then crystallization occurred prior to deformation and metamorphism. If we date both the undeformed and deformed plutons (assuming both are present), we can bracket the time interval during which deformation and metamorphism occurred.

Zircon is not the only mineral used in U-Pb geochronology. Others include titanite, rutile, apatite, and monazite, all with different closure temperatures. We can increase our understanding of a particular orogeny (mountain-building event) if we also date minerals that grew during metamorphism. Titanite is a good candidate. It has a U-Pb closure temperature of around 660°C. Such a high temperature is well into the amphibolite facies (Figure 20.7) and potentially close to or above the highest temperature attained during metamorphism. Thus, a U-Pb titanite date will record a crystallization age or at least a high-temperature cooling age that is very close to the time when the highest temperature of metamorphism occurred. This time would likely correspond with the time that mountain building was active.

We can complete our understanding of a particular orogeny if we date minerals with lower closure temperatures. These minerals will record the time and the rate of cooling as igneous or metamorphic temperatures wane. For example, $^{40}Ar/^{39}Ar$ dates on hornblende, muscovite, and biotite will record the time when the rock cooled through approximately 500°C, 425°C, and 330°C, respectively.

PRE-, SYN-, POST-, AND INTRA-

Geologists use terms like *pre-orogenic, syn-orogenic,* and *post-orogenic* (as well as more generic terms such as *pre-, syn-,* and *post-deformational*) to refer to events (deposition, intrusion, faulting, metamorphism, etc.) that occurred prior to (*pre-*), during (*syn-*), or following (*post-*) another event or between (*intra-*) two discrete events. For example, a pluton that intrudes after a certain deformation is a post-deformational pluton. One that intrudes before deformation is a pre-deformational pluton, and one that intrudes during deformation is syn-deformational. We have to be careful in using these terms to indicate specifically what the terms relate to. If there are multiple deformational events, we should not say that the pluton is post-deformational without

explicitly specifying the deformational event in question. This brings us to the prefix *intra-,* as in *intra-orogenic.* Such a term implies deposition or intrusion following one orogeny but prior to another. Here again, we must specify exactly what orogeny or deformational event we are referring to so that the meaning of the term is clear.

DETRITAL ZIRCON GEOCHRONOLOGY

A major problem encountered in mountain belts is that large areas of sedimentary and metamorphic rock contain no fossils or are so strongly metamorphosed that no fossils have survived. Without fossils, it is difficult to determine when the rocks were deposited. The dating of detrital zircon helps alleviate this problem. A geologist collects a sample of sedimentary or metamorphic rock. He or she separates perhaps 100 zircon grains from the rock and then obtains an age on each one of them. In the past, such a procedure would have been prohibitively expensive, but with techniques that involve laser and ion beams, it is now inexpensive and fairly routine. Recall from our previous discussion that zircon has a very high closure temperature and is relatively immune to later events. It almost always records the time of crystallization from a magmatic body (a pluton). The dates obtained from detrital zircons, therefore, do not indicate how long ago the sedimentary or metamorphic rock was deposited. They do not indicate how long ago the rock was metamorphosed. The dates indicate only how long ago each zircon grain crystallized from magma in an intrusive rock. The history of each zircon grain following crystallization is likely to be variable and difficult to determine. But what they all have in common is that they were all exhumed to the surface, eroded from an ancient landscape, and deposited into the same basin at the same time. The magmatic age of each zircon grain could vary considerably, by a billion years or more. Given a group of 100 dates, it has been found that most will group into clusters of similar age. This makes sense because there can only be a limited number of source areas at any one time. A similar age likely reflects a similar source area and, therefore, a similar geologic history for those zircons.

So, what do the detrital zircons tell us? First, it should be obvious that the age of deposition of the rock sample must be younger than the youngest cluster of zircon ages. The zircons do not provide an exact date for deposition, but they limit the time of deposition and provide information on source area. For example, in the Appalachians, if the youngest cluster of zircon dates is in the range of 450 Ma, the rock must have been deposited at some time after this date. The information can then be combined with other geologic clues to interpret orogenic history. If, for example, the rock sample is overlain or intruded by rocks that are known to be 380 million years old, then the rock sample must have been deposited between 450 and 380 million years ago. If the rock also contains a zircon age cluster in the 1-billion-year range, then a possible source area for these zircons (but not for the circa 450 Ma zircons) is the 1-billion-year-old Grenville (North American) basement. Alternatively, if a majority of the dates are older or younger than 1 billion years, then the source area could not have been Grenville basement and, instead, could have been an accreted terrane. We could then look for an accreted terrane with magmatic zircons of that age as a potential source area.

FAULT AND BELT TERMINOLOGY

We end this section with a few paragraphs on terminology. The term *allochthon*, from *allo* (other) and the Greek *khthon* (Earth), is used to signify a large body of rock that has moved a great distance from its original location, usually along thrust faults, and that is distinctive from underlying rocks. Three well-known allochthons in the US are the Taconic allochthons in Vermont and New York, and the Roberts Mountains and Golconda allochthons in Nevada. In all three examples, the allochthons were transported more than 50 miles along thrust faults. The rocks in each allochthon are distinct from the rocks on which they were thrust, and all three are internally deformed with additional thrust faults and folds. *Root zone* is a field term that refers to the location from which thrust faults originate. Thrust faults underlying all three allochthons likely root into a subduction zone. The Taconic allochthons are *klippen* in the sense that later episodes of folding and erosion have separated the allochthon from its root zone. A *klippe* and its root zone are shown in Figure 20.13.

The term *belt* (which we have used throughout this book) can be applied to any distinctive package of rock that is longer than it is wide. A belt may contain faults but is not necessarily bound by faults. A *fault slice* is any package of rocks that is bound by faults. If a fault slice is longer than it is wide, it could also be referred to as a belt. A *terrane* is a fault-bound package of rock that is markedly distinct from surrounding rocks. More specifically, a terrane consists of any combination of oceanic lithosphere, volcanic arc, or continental lithosphere, and has a stratigraphy or a deformational, metamorphic, or intrusive history different from surrounding rocks. A terrane could be composed of several fault slices and could be in the shape of a belt. One should also note that faults that define the boundary of a terrane could be covered or masked by later deformation, intrusion, or deposition and thus may not be readily visible. As noted in Chapter 5, the geologic term *terrane* is different from a geographic *terrain,* which is a physical stretch of land.

A *microcontinent* is a terrane composed mostly of continental lithosphere. An *accreted terrane* is one that formed elsewhere and later become attached (sutured) to a continent. A *superterrane* is a grouping of individual fault-bound terranes, all of which share a similar or distinctive tectonic history that is different from surrounding rocks. Superterranes are assembled (amalgamated) elsewhere, usually during an orogeny. Sometime later the superterrane collides with and is accreted to a large continent, creating a second orogeny. The first orogeny is absent in the large continent (which constitutes the distinctive tectonic history within the superterrane), yet both the continent and superterrane share the deformation and metamorphism associated with the second orogeny. The term *zone* can be used interchangeably with supercontinent. The term *realm* implies a characteristic set of fossils within any body of rock.

An *anticlinorium* is a giant anticlinal structure, usually on the scale of an entire mountain range, composed of many smaller anticlines and synclines. The Colorado Rockies, as a whole, could be considered an anticlinorium. Likewise, a *synclinorium* is a giant composite synclinal structure.

Two terms that are used quite often in Part III are *foreland* and *hinterland*. In their most simplistic form, foreland refers to sedimentary rocks at the front of a mountain belt. The foreland includes the foreland fold-and-thrust belt structural province and flat-lying rocks in the US interior (Chapters 10, 12). The hinterland refers to crystalline rocks in the core of a mountain belt such as rocks in the Blue Ridge and the Northern Rocky Mountains (Chapter 13).

Finally, we should be familiar by now with terms such as *clastic, detritus, detrital, carbonate,* and *marine,* all of which are defined in Chapter 2.

So, there you have it. This chapter does nothing more than provide some of the necessary background and terminology to understand mountain building from the perspective of a geologist. We are now in a position to describe the characteristic features of a mountain belt and provide some insight as to how a mountain belt forms and evolves.

Klippe (singular) and **Klippen** (plural) refer to a remnant outlier of an eroded thrust sheet. A large klippe, composed of several thrust sheets, could also be termed an allochthon.

Root zone - The location where the klippe connects back to the fault. It is the location where the fault originates.

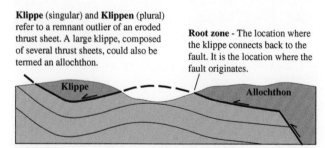

FIGURE 20.13 A cross-section that shows a klippe that roots into an allochthon. An allochthon is a large body of rock that has moved a great distance from its original location, usually along thrust faults, and that is distinctive from underlying rocks. A large klippe composed of several thrust sheets could be termed an allochthon.

QUESTIONS

1. Explain how the three rock groups reveal different aspects of geological history.
2. What is a sedimentary facies?

3. Describe the type of sediment and the resulting sedimentary rock expected in the following depositional environments: deep marine, shallow marine, beach, river, desert, and lake.

4. In viewing a vertical sequence of rock, how do we tell when the area was a highland rather than covered in water?

5. What does it mean to time-correlate two vertical sequences of rock?

6. How do we know when rocks in two different places are the same age?

7. Describe four reasons that sedimentary rocks are important in the interpretation of ancient, now vanished landscapes.

8. How are crystalline rocks used to determine periods of ancient mountain building?

9. In what way are volcanic rocks similar to sedimentary rocks, and how are they similar to crystalline rocks in terms of understanding ancient landscapes?

10. What is the significance of an angular unconformity?

11. Describe each of the various types of unconformities shown in Figure 18.2.

12. What is a formation? Give several examples of formations, including one located close to where you live.

13. What is the geologic time scale? What is it used for?

14. On what basis are boundaries between periods of the geologic time scale defined?

15. Why is volcanic ash important in the understanding of ancient landscapes?

16. Name the eons, eras, and periods represented by the geologic time scale.

17. Sketch a simple cross-section showing cross-cutting relationships. Describe what the cross-cutting relationship in your drawing indicates.

18. What does it mean if two fossils of the same age—one warm-water, one cold-water—are found within a mile of each other? What type of contact might exist between these two fossils?

19. There have been numerous extinction events throughout geologic history. However, only two resulted in a nearly complete change in the look and type of fossils. Between which two eras and which two periods did these extinction events occur, and how long ago (in years) did they occur?

20. Explain why the duration of each geologic time period is different. For example, the Triassic lasted 51 million years, whereas the Silurian lasted only 28 million years.

21. Describe the metamorphic facies concept. What rock type is it based on, and how is it used?

22. What is metamorphic grade? How is the term applied in this book?

23. What is an ophiolite?

24. Describe how a subduction complex forms.

25. What is a pluton? What is a batholith?

26. Suggest an example of a basement-cover rock sequence.

27. A radiometric age can represent one of at least three things. What are they, and how does one differentiate between a crystallization age and a cooling age?

28. Why is zircon the most widely used mineral for geochronology?

29. What can a geologist infer (and not infer) from a date of a detrital zircon obtained from a low-grade, metamorphosed sandstone?

30. What are anticlinoriums and synclinoriums?

Chapter 21

Tectonic Style, Rock Successions, and Tectonic Provinces

Although no two collisional orogenic belts are exactly the same, many, including the Appalachian and Cordilleran systems, have characteristics in common. Most obvious is the physical length of the mountain belt which is greater than its width. The Appalachian range extends from Newfoundland to Alabama but is rarely more than 150 miles wide. The Cordillera is over 1,000 miles wide but arguably extends from Alaska to the Andes. The length of a mountain belt is a function of the length of the original subduction/collision zone which could extend for thousands of miles. Width is a function of the strength of colliding plates and the angle of subduction. It can be quite variable along the length of the mountain belt.

Many orogenic systems, including the Appalachian and Cordilleran systems, are bordered on one side by a relatively stable, weakly deformed region that grades into the compressional mountain belt. This area, known as the *craton*, corresponds roughly but not exactly with the physiographic Interior Plains and Plateaus region. The craton is at the front of an orogenic system beyond which lies the mountain belt. The overall shape of crustal rocks in a fully developed compressional mountain belt is that of a tectonic wedge that thickens with distance from the craton into the mountain belt. Normal thickness of continental crust on the US craton is 26 miles. Crustal thickness at the heel of a tectonic wedge can be more than two times this amount.

Parts of both the Appalachian and Cordilleran systems reach crustal thicknesses of about 34 miles. The tectonic wedge is one of the most important concepts in the origin of mountains. It is the tectonic wedge that is pushed over the craton during mountain building. A well-defined thrust fault, known as the *basal décollement*, typically separates rocks within the tectonic wedge from weakly deformed or undeformed cratonic rocks underneath. The direction that rocks are transported within the tectonic wedge is known in geology as *vergence*. Thrust faults in the Appalachians mostly dip east and verge west, whereas those in the Cordillera dip west and verge east. It can be inferred from this geometry that the tectonic wedge in both mountain belts has been pushed toward the craton and that there has been a great deal of crustal shortening (compression).

An important characteristic of orogenic belts throughout the world is that they show changes in rock type, structure, and, topography across the tectonic wedge. The most important changes are illustrated in a schematic cross-section shown in Figure 22.1. From the front of the mountain belt to the rear, we can define a craton, a foreland basin, a foreland fold and thrust belt, a hinterland thrust belt, and a belt of accreted terranes. The craton is the stable interior part of a continent that was not involved in mountain building. The foreland basin is actually part of the craton. It develops directly in front of the

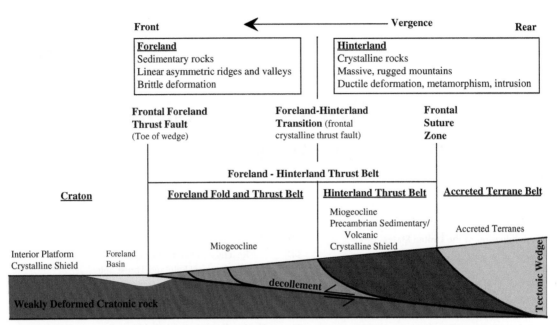

FIGURE 21.1   A schematic cross-section that shows the nomenclature and structural characteristics of the tectonic wedge.

mountain belt and receives detritus (erosional debris) from eroding highlands. The foreland fold and thrust belt forms the sedimentary toe of the tectonic wedge. The hinterland thrust belt forms at the rear of the foreland fold and thrust belt where thrust faults and erosion have exposed crystalline rock. Collectively, the two are referred to as the foreland–hinterland thrust belt. A belt of accreted terranes forms the rear of the tectonic wedge separated from the hinterland thrust belt by a suture zone. This sequence of changes can often be followed for hundreds or thousands of miles along the length of an orogenic belt and is related directly with the building of the tectonic wedge. In this chapter we show where each of these features are present in the Appalachian and Cordilleran Mountains. We divide the US into six tectonic provinces. Each province is defined by a certain tectonic style and by one or more rock successions stacked upon one another. We begin by defining tectonic style and rock successions, and then discuss their distribution within tectonic provinces.

Geologic history, particularly Phanerozoic geologic history, is described relative to the lifespan of fossils as defined in the geologic time scale (Figure 20.4). It is advisable to have a copy of the geologic time scale handy when reading this and subsequent chapters so that you can place events in their proper order.

A TECTONIC SUBDIVISION OF THE UNITED STATES

Changes in rock type and structure across nearly all wedge-shaped, orogenic mountain belts allow us to define tectonic provinces. Previously we divided the United States into physiographic and structural provinces based on topography,

rock/sediment type, and style of rock deformation. We did this to gain a better understanding of how landscape forms and evolves. In this section we ignore the landscape aspect of geology and focus exclusively on bedrock geology. Our goal is to better understand, for example, why crystalline rock is found in some areas but not in other areas, how rocks are stacked on top of each other, and what the rocks tell us regarding mountain building and the geological history of the US. Rather than simply looking at combinations of rock type and style of rock deformation as we did with structural provinces, we will look instead at combinations of tectonic style and rock successions to divide the US into six *tectonic provinces*. The tectonic provinces described here are different from structural provinces in that they are based specifically on geology and geologic history, not on the present-day landscape. Our goal is to explore the rock record in order to better understand the sequence of events, and the sequence of now vanished landscapes, that have led to the present-day landscape. The difference between looking at rock succession/tectonic style (tectonic provinces) versus rock type/style of deformation (structural provinces) is something like looking at a movie versus a photograph. The movie allows us to see a succession of events and how each event is related to the one before and the one after. By looking at the rock succession/tectonic style, we see an interval of history rather than a snapshot.

The *tectonic style* is the overall style of deformation which can be described as orogenic, epeirogenic, or post-orogenic. Each is listed in Table 21.1. Tectonic style is different from the previously described landscape-forming style of deformation because it can include a variety of structures as well as other geologic phenomena such as igneous intrusion, metamorphism, earthquakes, and volcanism.

TABLE 21.1 Tectonic Style

Orogenic

Epeirogenic

Post-orogenic

TABLE 21.2 Rock Successions

North American (Precambrian) crystalline shield

Precambrian sedimentary/volcanic

Interior platform

Miogeocline

Accreted terrane rock succession

Atlantic marginal basin rock succession

TABLE 21.3 Tectonic Provinces

Craton

Reactivated Western Craton

Foreland Fold and Thrust Belt

Hinterland Thrust Belt

Accreted Terrane Belt

Atlantic Marginal Basin

A *rock succession* (or *rock sequence*) is a stack (succession) of rocks, be they sedimentary, crystalline, volcanic, or some combination, that are related to each other by their age, thickness, and geologic interpretation. They are different from the previously described landscape-forming rock/sediment types in the sense that we will apply a genetic interpretation to the succession of rocks. In the United States only five rock successions form a majority of rock within the epeirogenic and orogenic tectonic domains, and all five were deposited either prior to or during mountain building (that is, they are pre- and syn-orogenic). They are the North American crystalline shield, the Precambrian sedimentary/volcanic rock succession, the interior platform, the miogeocline, and accreted terranes. The first four developed entirely on the North American plate and, therefore, are referred to as native (or North American) rock successions. The accreted terrane succession developed initially offshore. This rock succession is exotic to North America and, as the name implies, was added (accreted) to the North American continent during orogeny. A sixth rock succession forms the post-orogenic Atlantic marginal basin. Each rock succession is listed in Table 21.2.

The combination of tectonic style and rock succession produces six tectonic provinces. They are the craton, the reactivated western craton, the foreland fold-and-thrust belt, the hinterland thrust belt, the accreted terrane belt, and the Atlantic marginal basin. The foreland fold-and-thrust belt and the hinterland thrust belt together form the foreland-hinterland thrust belt. The Cordilleran and Appalachian Mountains share a single craton, but contain separate foreland-hinterland thrust belts and separate accreted terrane belts. The reactivated western craton is unique to the Cordilleran Mountain belt. The Ouachita Mountains-Marathon Basin consist only

of a foreland fold-and-thrust belt. The tectonic provinces are listed in Table 21.3. Most of the Ouachita-Marathon orogenic belt and part of the Appalachian accreted terrane belt are hidden beneath younger, post-orogenic rocks of the Atlantic marginal basin.

TECTONIC STYLE

Figure 21.2 shows the distribution of the orogenic, epeirogenic, and post-orogenic tectonic styles across the United States. From this figure, we can see that the Appalachian, Ouachita, and Cordilleran Mountain systems are orogenic and that the central US is epeirogenic. The physiographic Coastal Plain forms the post-orogenic Atlantic marginal basin which also shows epeirogenic deformation. Note that most of the active and recently active landscape-forming features of the Cordillera are not shown. In Part III of this book we will, for the most part, ignore the relatively young, landscape-forming normal faults and volcanism in order to concentrate on the mountain-building aspects of orogeny.

Orogenesis

In Chapter 20 we introduced the term *orogeny* as the process of mountain building. What follows is a more formal description. *Orogenesis* derives from the Greek *oros*, which means mountain, and *genesis*, which means the origin or mode of formation. The term *mountain building* implies that the rate of surface uplift is greater than the rate of erosion such that, over time, a lowland area evolves into a mountain system. *Orogeny* and *orogenic tectonic style* refer specifically to deformation imposed during mountain building. Although mountains form in a variety of ways, most geologists associate orogeny with continental-size mountain systems that develop as a result of the convergence and accretion of two or more tectonic plates. Such a mountain system is known geologically as an *orogenic belt* (or *orogenic system*). An orogenic belt is a deformational belt. The energy for orogeny is derived from horizontal compression, gravity, heat, and climate, particularly climate-driven erosion. Together, these energy sources produce folds, thrust faults, strike-slip

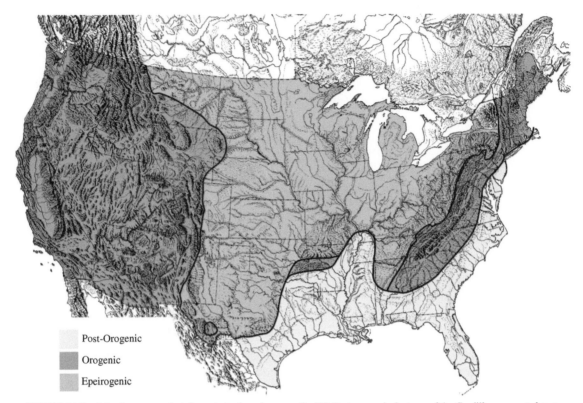

Post-Orogenic

Orogenic

Epeirogenic

FIGURE 21.2 A landscape map that shows tectonic style across the US. Post-orogenic features of the Cordillera are not shown.

faults, normal faults, metamorphism, granitic intrusion, iso-static adjustment, and, if one plate is pushed (subducted) beneath the other, explosive silicic volcanism.

The Appalachian and Cordilleran Mountain belts have both undergone several orogenic events during the Phanerozoic eon, each lasting 30 million years, more or less. Deformation associated with an orogenic event does not occur all at once within a particular mountain range but rather may affect only one part of a mountain belt or may propagate from one area to the next to affect different parts of a mountain at different times. Deformation in any one area and during a single orogeny can occur in a series of relatively short pulses separated by periods of quiescence that can last several million to several tens of millions of years, during which time the rocks could undergo metamorphism, isostatic adjustment, volcanism, intrusion, and erosion. The rocks can be folded or deformed multiple times during a single orogeny.

Epeirogeny

An *epeirogenic tectonic style* (or *epeirogeny*) is a mild (or weak) deformation characterized by regional uplift, tilting, or subsidence of an area many hundreds of miles wide. Whereas orogenic deformation occurs in pulses primarily along plate boundaries at the margins of continents, epeirogenic deformation typically affects continental interior regions and can occur over a time span of

hundreds of millions of years. The primary structures are broad anticlines (or domes) and synclines (basins) in which sedimentary rock is tilted but remains nearly flat-lying. Igneous intrusive activity and volcanism are rare, and metamorphism is absent. Epeirogenic deformation of this type likely results from vertical movement such as thermal and isostatic adjustment or from mild compressional stress. An area may warp upward over a warm mantle plume or downward in an area of rapid deposition. An orogenic pulse along a continental margin could conceivably propagate to the continental interior to produce weak far-field compressional stresses that warp rocks into broad anticlines or synclines. In the US, an epeirogenic region refers specifically to one that has undergone mostly mild deformation during the Phanerozoic eon. These areas escaped strong deformation associated with Appalachian and Cordilleran orogeny and are present today in the continental interior.

Post-Orogenic

The term *post-orogenic* refers to volcanism, faulting, intrusion, and deposition that occurs after the final compressional orogenic push within a mountain system. Post-orogenic events tend to reshape and cover both the rocks and the structures that formed during earlier orogenic phases. In doing so, they reincarnate the landscape such that the older orogenic landscape is no longer recognizable. In the Appalachians, post-orogenic elements include sedimentary rocks of the Coastal Plain (the Atlantic marginal basin rock succession) and rocks that form the Triassic Lowlands. In the Cordillera, they include volcanism and normal faulting, particularly associated with the Columbia River Plateau, Snake River Plain, and Basin and Range.

ROCK SUCCESSIONS

In this section, we discuss the origin of the six rock successions, four of which are pre- and syn-orogenic and native to North America. They are the Precambrian crystalline shield, the Precambrian sedimentary/volcanic, the interior platform, and the miogeocline. The accreted terrane rock succession is pre- and syn-orogenic and exotic to North America. The Atlantic marginal basin is post-orogenic.

North American (Laurentian) Rock Successions

Figure 21.3 is a conceptual west-to-east cross-section of the United States that shows the original location and stacking order of each of the five pre- and syn-orogenic rock successions. This cross-section is provided so that you can gain a visual understanding of where each rock succession originated and how the rocks were originally stacked above one another. Note that the Cordilleran and Appalachian sides are near mirror images of each other. We will discover that the pre- and syn-oroginic distribution of rock successions, although modified, is preserved in the present-day arrangement.

Notice in Figure 21.3 that the Precambrian (or North American) crystalline shield formed the substratum of the entire continent prior to orogenesis. The shield rocks are overlain primarily by the interior platform succession within the continental interior, and by the Precambrian sedimentary/volcanic and miogeoclinal rock successions at the continental margins. Deep-water slope/rise sedimentary rocks, and other terranes, lay farther offshore underlain by oceanic lithosphere. These rocks will later be added to the continental margin in the form of accreted terranes.

The crystalline shield rock succession developed as a result of several orogenic events, the strongest of which

occurred roughly 2.5, 1.8, and 1.0 billion years ago. Each of these orogenic events culminated in the formation of a giant landmass (a supercontinent) that included all the Earth's landmasses in existence at the time. Rifting and separation of continents followed each orogenic event. The last major Precambrian orogeny, from about 1.1 to 1.0 billion years ago, is named the *Grenville orogeny*. This event resulted in continental collision primarily in the eastern part of North America that produced the supercontinent Rodinia. Because Rodinia formed previous to Pangea, it is also known as proto-Pangea. Figure 21.4a is a cross-section that depicts the Rodinian supercontinent. During the latter part of the Precambrian, beginning less than 800 million years ago, Rodinia began to break apart, and the part that would become North America eventually became isolated from other continental fragments. The isolated North American continental fragment is known as *Laurentia*. It is the original (or native) North America that existed prior to Appalachian and Cordilleran tectonic accretion. It is shown in Figure 21.4b. Laurentia would continue to grow over time by the addition of the Precambrian sedimentary/volcanic rock succession, the interior platform, and the miogeocline as shown in Figure 21.4c. The Precambrian shield rock succession thus forms the nucleus of North America, the nucleus of Laurentia. It is a 15- to 34-mile-thick slab of Precambrian crystalline rock, nearly all of which is between 1 and 4 billion years old.

The Precambrian sedimentary/volcanic rock succession was deposited in basins primarily along the margins of the crystalline shield at various times, mostly between about 1.6 and 0.6 billion years ago. These are the oldest unmetamorphosed sedimentary and volcanic rocks in the US. The oldest part of this succession, including the Belt Supergroup (Chapter 12), was deposited prior to breakup (rifting) of Rodinia in what were likely extensional basins as depicted in Figure 21.4a. The younger rocks form a rift-related clastic sequence that is directly associated with the breakup of Rodinia beginning some time after 800 million years ago (Figure 21.4b). The 1.1-billion-year-old Keweenawan rocks in the middle of the continent (Chapter 13) represent a failed attempt at rifting apparently just prior to the climax of the Grenville orogeny!

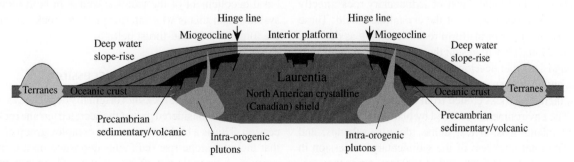

FIGURE 21.3   A conceptual west-to-east cross-section across the US with Appalachian and Cordilleran orogenesis removed. The cross-section shows the original distribution and stacking of pre- and syn-orogenic rock successions.

(a)

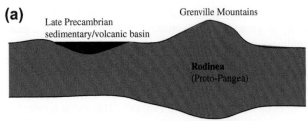

Rodinea forms during Grenville orogeny about 1.0 billion years ago. Deposition of Late Precambrian sedimentary/volcanic rock occurs in the west.

(b)

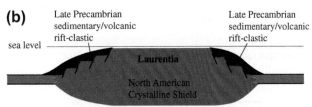

By circa 542 million years ago Rodinea had rifted apart and the North American fragment (Laurentia) was eroded to sea level exposing crystalline shield rock at the surface.

(c)

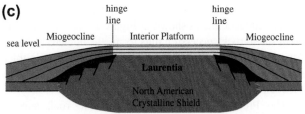

By 480 million years ago the crystalline shield was receiving interior platform and miogeoclinal deposition prior to Appalachian orogenesis.

FIGURE 21.4 A series of cross-sections that show the breakup of Rodinea and deposition of the Precambrian sedimentary/volcanic, Interior Platform, and miogeoclinal rock successions to form Laurentia.

By the beginning of the Phanerozoic, all mountains in the United States had eroded to a relatively flat surface (a platform) bordered on all sides by a passive continental margin. Shield rocks were exposed across the continental interior, and Precambrian sedimentary/volcanic rocks were either exposed at the surface or hidden beneath a shallow ocean along the continental margin as shown in Figure 21.4b. Throughout the Phanerozoic, the craton would periodically subside below sea level and be covered by a shallow ocean that deposited layer after layer of sedimentary rock directly on top of the eroded stump of the crystalline shield. These rocks form the interior platform rock succession as depicted in Figure 21.4c. They represent the progression of landforms and inland seas that developed on top of the shield prior to, during, and following Appalachian and Cordilleran orogenic events, including rocks eroded from adjacent orogenic highlands. The environments recorded by platform rocks include shallow inland seas, rivers, deltas, deserts, swamps, and lakes. The total thickness of the sedimentary succession in most areas is between 5,000 and 15,000 feet. Areas that continually subsided throughout most of the Paleozoic, such as

the Michigan and Illinois Basin, received a thick succession of platform rock. Areas that underwent periods of uplift, such as the Transcontinental and Cincinnati Arch, received a much thinner succession (Figure 10.2). The defining characteristics of the Interior Platform succession are that they are sedimentary, they are Phanerozoic in age, and they are thin relative to rocks of the miogeocline.

Rocks of the miogeocline represent a continuation of interior platform deposition along a passive continental margin where the crystalline shield begins to thin and merge with oceanic lithosphere (Figure 21.4c). The rocks are the same age as those of the interior platform, but because they were deposited along the continental margin, there are some differences. A major difference is the thickness of the miogeoclinal succession which can exceed 50,000 feet. Another difference is that miogeoclinal rocks are, almost everywhere, underlain by the Precambrian sedimentary/volcanic rock succession rather than directly by shield rocks. A final difference is that miogeoclinal rocks were folded, thrust-faulted, and, in some areas, metamorphosed, during Appalachian and Cordilleran orogeny. These rocks form the primary rock succession of the foreland fold-and-thrust belt tectonic (and structural) province.

One important similarity with the interior platform is that miogeoclinal rocks, in spite of their great thickness, were deposited in shallow water. These are the rocks that confounded early geologists in the Appalachians that led to the geosynclinal theory. The rocks represent sediment shed from the continental interior and deposited along the continental shelf, much like the present-day Atlantic marginal basin. Constant deposition in a slowly subsiding basin produced a rapid transition between the thin sedimentary platform succession of the continental interior and the much thicker miogeoclinal succession. This original transition is referred to as a *hinge line* (or *hinge zone*; Figures 21.3, 21.4c). The hinge line forms an abrupt transition because it also marks the approximate craton-ward limit of significant rifting associated with the breakup of Rodinia. Major thrust faults in both the Appalachian and Cordilleran foreland fold-and-thrust belts would later take advantage of this hinge line by pushing the miogeoclinal rock succession (the toe of the tectonic wedge) over the interior platform succession. The hinge line forms a natural structural break for the basal décollement of the tectonic wedge in both mountain systems, and this is why interior platform rocks are rare in the foreland fold-and-thrust belt.

Accreted Terrane Rock Successions

Pre- and syn-orogenic rocks not originally deposited as part of Laurentia are considered part of the accreted terrane rock succession. This is a highly diverse and complex group of rocks that includes slope-rise sediments deposited on transitional oceanic-continental crust off the coastline of Laurentia (that is, the deep-water continuation of the Miogeocline, Figure 21.5),

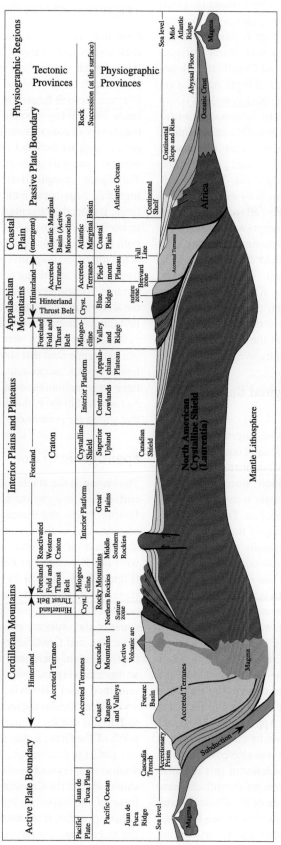

FIGURE 21.5 A schematic cross-section that shows the present-day distribution of tectonic provinces, rock successions, and physiographic provinces.

island volcanic arcs, volcanic chains and plateaus, small continental masses (microcontinents), pieces of oceanic lithosphere (ophiolite), and transported plutonic rock. There are dozens of distinct terranes accreted at different times across both the Appalachian and Cordilleran Mountain belts. All of these rocks became part of North America via subduction, collision, and tectonic accretion and, in doing so, have undergone variable amounts of deformation, metamorphism, and intrusion. With the possible exception of offshore slope-rise sediments, these rocks are surrounded by suture zones and have a stratigraphy and geologic history that is different from surrounding rocks. They are terranes in the truest sense. Terranes in both mountain belts are overwhelmingly oceanic. Nearly all of the rock in the Appalachians consists of metamorphosed sedimentary and volcanic rocks, much of it metamorphosed to high grade. Sedimentary and volcanic rocks are widespread in the Cordilleran accreted terrane belt, with metamorphic rocks present in the Coast Ranges, Klamath Mountains, the Blue Mountain region of Oregon, the North Cascades, the northern Sierra Nevada, and the northwestern Northern Rocky Mountains (Figure 2.9). Both the Appalachian and Cordilleran belts contain abundant intrusive rock.

Post-Orogenic Atlantic Marginal Basin Rock Succession

The Atlantic marginal basin is a post-orogenic basin that incorporates the entire Coastal Plain and continental shelf. It is an actively subsiding passive continental shelf that has undergone continuous deposition since initial rifting of Pangea beginning in the Triassic some 230 million years ago. It is a modern-day miogeocline. The rocks completely bury and hide the eroded stump of the Appalachian-Ouachita Mountain belt, which originally extended to the edge of the continental shelf. The Atlantic marginal basin is epeirogenic in the sense that it has subsided and continues to subside isostatically due to both thermal cooling and the accumulated weight of deposited sediment. Subsidence has been greater on the seaward side creating an overall seaward tilt of both the rock units and the Coastal Plain landscape. Nearly the entire Coastal Plain was below sea level during the latter part of the Cretaceous when the oldest exposed rocks were deposited close to the Fall Line (Figure 7.1). The presence of these rocks at the Fall Line implies that about one-half of the Appalachian Mountain system had already eroded to sea level by the Cretaceous. It is possible that part of the Piedmont was also covered by an ocean at this time although little evidence of this remains. Since the Cretaceous, there has been an overall decrease in sea level, exposing more and more of the Coastal Plain. As we have noted previously, the lowering of sea level has been subject to short-term fluctuations, particularly during the most recent glaciation.

TECTONIC PROVINCES

Now that we have an understanding of tectonic style and rock successions, we can look more closely at the distribution of rock successions and the structural characteristics within each of the pre- and syn-orogenic tectonic provinces. For now we will ignore the post-orogenic Atlantic marginal basin. We will begin on the craton and work our way toward the interior of both mountain belts. Figure 21.5 is a schematic present-day, west-to-east cross-section of the United States, with post-orogenic normal faulting and volcanism removed. This figure attempts to show the connection between physiographic provinces, rock successions, and tectonic provinces. It extends conceptually from San Francisco to Denver to northern Wisconsin to Charleston. Tectonic provinces in both the Cordilleran and Appalachian Mountains are shown at the top of the diagram (craton, reactivated western craton, foreland fold-and-thrust belt, hinterland thrust belt, accreted terranes, Atlantic marginal basin). Rock successions that crop out at the surface are listed below the tectonic provinces. The first thing to notice is that the rock successions are somewhat in the same order they were prior to deformation. Figure 21.6 is a landscape map that shows the present-day surface distribution of rock successions. Also shown in this figure are the three large batholiths of the Cordillera: the Peninsular, Sierra Nevada, and Idaho batholiths. Figure 21.7 is a landscape map that shows the present-day distribution of tectonic provinces.

Epeirogenic Craton

The craton is the stable interior part of North America that corresponds roughly but not exactly with the physiographic Interior Plains and Plateaus region. The primary rock successions across the craton are the Precambrian crystalline shield and the unconformably overlying interior platform, with the Keweenawan rocks as the only major exception (Figures 21.5, 21.6, 21.7). The craton has, for the most part, remained undeformed and has received marine deposition periodically throughout the Phanerozoic eon even during some of the orogenic pulses that formed the Appalachian, Ouachita, and Cordilleran mountain systems. Deformation on the craton, therefore, is epeirogenic in the sense that sedimentary layers, although warped into broad folds, have remained nearly flat-lying. A few of the up-warps expose crystalline shield rock, including the St. Francois, Arbuckle, and Llano domes. A major up-warp centered in Canada north of the Great Lakes has completely removed the interior platform sedimentary rock succession, thus exposing a wide area of crystalline shield rocks known as the Canadian shield. Part of the Canadian shield extends southward in the form of the Superior Upland and Adirondack provinces. Other up-warps, most notably the Cincinnati and Transcontinental arches, still retain their sedimentary platform cover. Major down-warps include the Michigan and Illinois basins

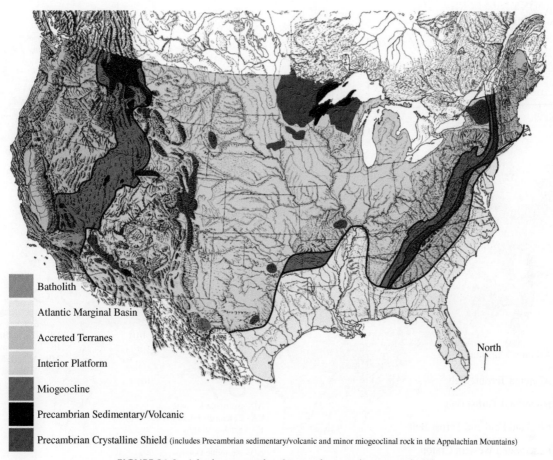

Batholith

Atlantic Marginal Basin

Accreted Terranes

Interior Platform

Miogeocline

Precambrian Sedimentary/Volcanic

Precambrian Crystalline Shield (includes Precambrian sedimentary/volcanic and minor miogeoclinal rock in the Appalachian Mountains)

North

FIGURE 21.6 A landscape map that shows rock successions across the US.

and the Appalachian syncline. Down-warped areas close to orogenic belts, such as the Denver Basin, received sediment from the adjacent eroding orogenic mountains.

Many of the domes and basins on the craton developed slowly over time periods of several hundred million years, primarily during the Paleozoic era. The craton was below sea level during much of that time. Simultaneous folding and deposition are inferred to have occurred because individual sedimentary layers are thick in synclinal areas such as the Michigan and Illinois Basins (>15,000 feet) and thin in adjacent anticlinal areas such as the Cincinnati and Transcontinental arches (<5,000 feet). This is interpreted to indicate steady subsidence and deposition in synclinal areas versus less subsidence or mild uplift and erosion in anticlinal areas. The timing of subsidence or uplift can be inferred based on thickness variations as outlined in Figure 21.8. Erosion along the crest of an arch will produce a disconformity or a mild angular unconformity in which rocks below the unconformity are tilted very slightly relative to overlying rock layers. This type of unconformity, along with thickness variations in sedimentary layers across domes and basins, forms some of the best evidence for epeirogeny.

The craton is not without local areas of strong, even orogenic, Phanerozoic deformation. The most pronounced area extends from Wyoming to Texas. Rocks across this region were deformed during a Late Pennsylvanian-Early Permian cycle, understood to be related to the Ouachita orogeny. The extent of these cratonic mountains and their associated depositional basins is shown in Figure 21.9. The mountains are known as the Ancestral Rocky Mountains because they partly overlap the present-day Southern Rocky Mountains. The uplift brought ancient Precambrian shield rock to the surface along high-angle reverse faults that likely reactivated ancient Precambrian faults. Differential uplift was as much as 20,000 feet, generating mountains that may have been on the order of 10,000 feet. The proof for existence of these now vanished mountains is based on the presence of eroded residue preserved as sedimentary rock in nearby depositional basins in a manner similar to what is shown in Figure 20.6c. Relationships are best recorded in Colorado where two ancient mountain ranges cross the state. The two ranges are known as the Ancestral Front Range and the Ancestral Uncompahgre Mountains due to their coincidence with the present-day Front Range and Uncompahgre

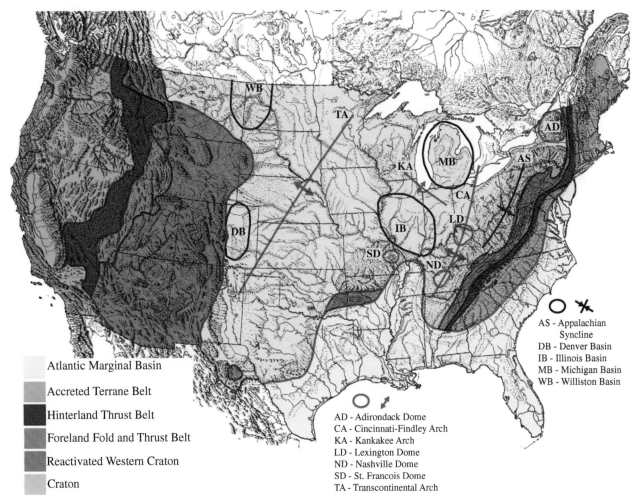

Atlantic Marginal Basin

Accreted Terrane Belt

Hinterland Thrust Belt

Foreland Fold and Thrust Belt

Reactivated Western Craton

Craton

AS - Appalachian
         Syncline
DB - Denver Basin
IB - Illinois Basin
MB - Michigan Basin
WB - Williston Basin

AD - Adirondack Dome
CA - Cincinnati-Findley Arch
KA - Kankakee Arch
LD - Lexington Dome
ND - Nashville Dome
SD - St. Francois Dome
TA - Transcontinental Arch

FIGURE 21.7   A landscape map that shows tectonic provinces across the US.

Plateau. A deep trough between these ancient mountains collected almost 13,000 feet (4,000 m) of sediment, some of which, including red beds of the Fountain formation, are exposed in present-day Rocky Mountain uplifts. The trough received fine-grained detritus during the Late Permian, suggesting that the mountains had eroded to lowlands by that time. Small remnants of this ancient high topography exist today as the Wichita and Arbuckle Mountains.

Deformed rocks are also present on the craton in the western part of the Interior Low Plateaus where several periods of high-angle normal and reverse faults are associated with the Rough Creek-Pennyrile fault zone. The location is shown in Figure 21.9. This deformation tilted sedimentary layers to angles well beyond horizontal but did not create high mountains. The Rough Creek-Pennyrile fault zone is part of the Reelfoot rift fault system that includes the New Madrid-Wabash Valley earthquake zone, part of which is hidden beneath sedimentary rock of the Coastal Plain. As noted in Chapter 10, the Reelfoot rift system likely originated within the Precambrian crystalline shield between

800 and 500 million years ago during a failed rift attempt. Part of the area was reactivated in the early Paleozoic and again in the Early Tertiary with faults cutting into the younger, overlying interior platform succession. The New Madrid and Wabash Valley areas are active today based on earthquake seismicity, but there are no obvious active surface faults.

Foreland Basins

The foreland basin develops on the craton directly adjacent to the mountain belt. It is largely a product of the isostatic response of the crust to the weight of the adjacent mountain. The process is similar to placing heavy books in the middle of a shelf. As the shelf sags, a low area is created in the shape of a wedge which is deep next to the books but becomes shallow toward the edge of the shelf. A foreland basin has a similar shape. It is deep at the mountain front, tapering to more shallow levels with distance from the mountain. There is often a frontal foreland thrust fault that separates the basin

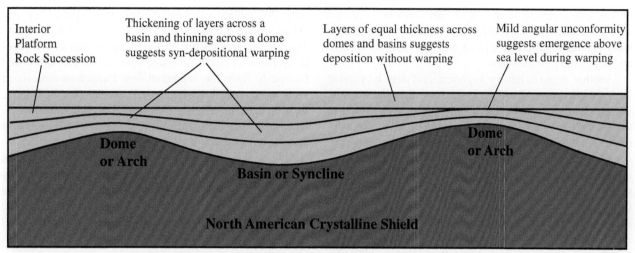

Interior Platform Rock Succession

Thickening of layers across a basin and thinning across a dome suggests syn-depositional warping

Layers of equal thickness across domes and basins suggests deposition without warping

Mild angular unconformity suggests emergence above sea level during warping

Dome or Arch

Basin or Syncline

Dome or Arch

North American Crystalline Shield

FIGURE 21.8   Deposition and erosion across basins and arches on the craton.

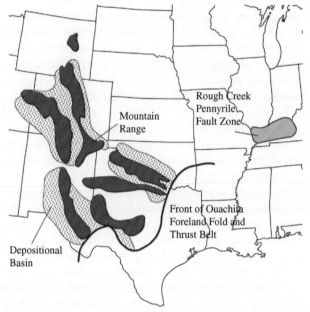

Rough Creek Pennyrile Fault Zone

Mountain Range

Front of Ouachita Foreland Fold and Thrust Belt

Depositional Basin

FIGURE 21.9   Extent Ancestral Rocky Mountains and associated depositional basins in the southwest, and the Rough Creek–Pennyrile fault zone.

from the mountain front as shown in Figure 21.1. This fault represents the leading edge (the toe) of the foreland fold-and-thrust belt and the leading edge of the tectonic wedge.

The size and depth of a basin are a function of the strength (the stiffness) of the underlying lithosphere and the size (weight) of the adjacent mountain. A stiff lithosphere is analogous to a stiff board. It will not bend easily and will produce a long, shallow basin. A weak lithosphere or a large mountain mass will favor a small but deep basin. In either case, sediment derived from erosion of the mountain fills the basin. Sediment is coarse-grained (conglomerate and sandstone), and can be more than 5 miles thick

close to the mountain front, becoming thinner and generally finer-grained with distance, thus forming the wedge shape. Most of the sediment is alluvial, meaning it was deposited by rivers.

A foreland basin, such as the one just described, is also variably referred to as a *foreland clastic wedge*, a *molasse basin*, or a *foredeep*. It is composed of detritus (clastic material) eroded directly from the mountain region and carried into the basin by rivers. The location, composition, and shape of the basin result in the term *foreland clastic wedge*. The term *molasse* was originally defined in the Swiss Alps as a clastic rock sequence consisting mostly of nonmarine conglomerate, sandstone, and shale produced from erosion of an adjacent mountain belt. The term *foredeep* refers to the location of the deep (the basin) at the fore (front) of the mountain belt. These terms are used interchangeably.

Technically, the foreland is the undeformed region at the front of a mountain belt closest to the craton. A developing tectonic wedge, however, is not static. It grows wider over time beginning in the hinterland and, eventually, it propagates into the undeformed foreland. Thus, today's foreland fold-and-thrust belt was once part of the underformed foreland. Depositional basins too may be deformed and incorporated into a mountain belt. A molasse basin will develop during and following development of a foreland fold-and-thrust belt. As the thrust belt grows wider and propagates onto the craton, part of the molasse basin will be overrun by thrust faults, uplifted, and inverted into a highland. As this occurs, the active molasse basin will migrate farther onto the undeformed craton.

In an ancient eroded orogenic mountain system, the age of the molasse deposit dates the existence of a mountain belt. Molasse can also date the timing and progression of unroofing (exhumation). If sediment in the lower part of a molasse basin consists entirely of eroded sedimentary rock debris, and sediment in the upper part consists of eroded crystalline rock debris, it may be possible to date the time when crystalline (hinterland) rock was first exposed, eroded, and deposited into the basin.

Another deposit, known as *flysch*, may also be present in the foredeep, usually in the lower part of the basin below younger molasse. More commonly, flysch is deposited in the vicinity of a subduction zone trench. Also originally defined in the Swiss Alps, flysch is considered a deep-water, oceanic deposit composed dominantly of turbidites (shale and mudstone deposited from oceanic turbidity currents), greywacke (silt-rich sandstone), and fine-grained volcanic sediment. The rocks are eroded from emergent landmasses such as the accretionary prism and the adjacent volcanic arc. Flysch deposits are very commonly deformed, metamorphosed, and incorporated into the mountain belt. The age of these rocks can potentially date the earliest phases of mountain building (that is, subduction) as well as ongoing phases. Flysch and molasse basins in both the Appalachians and Cordillera date the former existence of several generations of mountain belts as discussed in later chapters. Molasse basins in the Appalachian Mountains include much of the rock deposited on the present-day Appalachian Plateau. Although today it is an uplifted plateau, this area acted as a depositional basin (a foredeep) during the final stages of Appalachian Mountain building. In the Cordillera, the Denver Basin forms a foredeep to the adjacent Rocky Mountains (Figure 21.7).

Orogenic Provinces

Both the Appalachian-Ouachita and Cordilleran Mountain systems have undergone several orogenic events during the Phanerozoic eon, each affecting a different part of the mountain belt at different times. Each distinct orogenic event is given a proper name so that it can be distinguished and discussed. For example, earlier in this chapter we mentioned the Grenville orogeny as the final orogeny to affect the eastern US prior to Appalachian orogenisis. The classic interpretation is that the Appalachian Mountains were affected by three orogenic episodes, the Cordillera by five, and the Ouachita-Marathon region by one. We shall see that this is an oversimplification, however, it is an accurate description to use as an introduction.

In the Appalachians, the Taconic orogeny occurred in the Middle to Late Ordovician, the Acadian orogeny during the Early and Middle Devonian, and the Alleghany orogeny from Middle Mississippian to Middle Permian. Between each orogeny, the mountains in some areas of the Appalachians were eroded to sea level and reclaimed by the sea. Only the Alleghany orogeny is well represented in the Ouachita (and Marathon) belt, and because of its location, it is referred to as the *Ouachita orogeny*. Notice that the final orogenic episode ended during the Paleozoic era

some 265 million years ago. Clearly the Appalachians and Ouachitas are old mountain belts that are no longer active. The Alleghany-Ouachita orogeny culminated in continental collision of North America, South America, Africa, and Europe to form the supercontinent Pangea as envisioned long ago by Wegener (Figure 8.5). At the time of Pangea, the Appalachian-Ouachita belt formed a continuous mountain range that extended eastward across the present-day Coastal Plain and Atlantic continental shelf, and southwestward across Louisiana and Texas. It even extended to the British Isles and Scandinavia. The mountains may have reached elevations of 20,000 feet or more. Because the Alleghany-Ouachita orogeny was the last major orogeny to affect the Appalachian-Ouachita belt, we will consider all deposition and deformation that followed the Alleghany-Ouachita orogeny to be post-orogenic.

The Cordilleran orogenic belt has had a long history of tectonic accretion but has never experienced major continent-continent collision. Orogenic activity began in the Paleozoic, was strongest in the Mesozoic and early Cenozoic, and remains active today. Orogenic episodes overlap in different parts of the mountain belt. The major events are the Antler orogeny in Late Devonian-Early Mississippian, the Sonoma orogeny at the Permian-Triassic boundary, the Nevadan orogeny during late Jurassic-Early Cretaceous, the Sevier orogeny in the mid- to late Cretaceous, and the Laramide orogeny from late Cretaceous to Eocene. Present-day orogenic activity is confined to the Pacific Coast in the form of the Cascadia volcanic arc system (Chapter 16) and the California transpressional system (Chapter 17). The Sevier orogeny is primarily responsible for development of the Cordilleran (Sevier) foreland-hinterland thrust belt that extends from Montana to Arizona (Figure 21.4). The timing of the Laramide orogeny partly overlaps the Sevier orogeny. The Laramide orogeny is primarily responsible for orogenic features on the reactivated western craton. This is the orogeny that produced the faulted crystalline-cored anticlines of the Middle and Southern Rocky Mountains and Black Hills. It also produced the monoclines and flat-topped anticlines on the Colorado Plateau. The Laramide orogeny ended between 40 and 55 million years ago and was last major orogenic event to affect the interior US Cordilleran region. For that reason, and for the purpose of discussion, we will consider all deposition and deformation that followed the Laramide orogeny to be post-orogenic. Post-orogenic deposition and deformation therefore includes Columbia River, Snake River Plain, and Basin and Range normal faulting and volcanism as well as ongoing orogenic activity along the Pacific Coast.

Reactivated Western Craton

The reactivated western craton represents an area where rocks normally restricted to the craton (specifically, the interior platform rock succession) are involved in orogenic deformation. Widespread involvement of the craton in orogenic deformation is somewhat of an aberration not seen in the Appalachian belt (but seen in the Ouachita belt in the form of the Ancestral Rocky Mountains). Perhaps the most significant result is that the foreland fold-and-thrust belt (the toe of the tectonic wedge) in the Cordillera is not everywhere at the eastern margin of the orogen. Deformation of the reactivated western craton is largely the result of the Laramide orogeny and involves the Precambrian crystalline shield, the interior platform, and, in a few areas, the Precambrian sedimentary/volcanic rock succession. None of the rocks were metamorphosed. Most of the area is underlain by the interior platform rock succession. Shield rocks are widely exposed primarily in two belts as shown in Figure 21.6. One belt stretches from the Southern Rocky Mountains northward and northwestward across the Middle Rockies and includes the Black Hills. This belt is shown as a faulted anticlinal mountain of exposed shield rocks in Figure 21.5. The second belt extends westward along the southern margin of the Colorado Plateau and across the foreland-hinterland thrust belt (Figure 21.6). The Precambrian sedimentary/volcanic rock succession forms nearly all of the anticlinal Uinta Mountains and locally forms a tilted series in the Grand Canyon below the interior platform and above the crystalline shield as shown in Figure 21.10.

The Foreland Fold-and-Thrust Belt

The Cordilleran and Appalachian foreland-hinterland thrust belt described here includes both the sedimentary foreland fold-and-thrust belt and the variably metamorphosed (crystalline) hinterland thrust belt (Figure 21.7). Although the crystalline shield and Precambrian sedimentary/volcanic rock successions are present, the most characteristic rock succession across the thrust belt is the miogeocline. The miogeocline is, in fact, restricted to the foreland-hinterland thrust belt in both the Cordillera and Appalachians. This section provides a brief description of the frontal half of the thrust belt, the foreland fold-and-thrust belt.

As will be described in Chapter 22, the foreland fold-and-thrust belt typically represents the final installment in the building of the tectonic wedge. It represents a part of the foreland that had remained undeformed during the early stages of mountain building but which eventually became incorporated into the orogenic belt as the tectonic wedge widened. It forms the toe of the tectonic wedge in both mountain belts where the miogeocline, along with, in some cases, the Precambrian sedimentary/volcanic rock succession and early foredeep deposits, have been pushed toward and above the craton (Figures 21.5 and 21.7). The Valley and Ridge forms a classic foreland fold-and-thrust belt, especially in the Southern Appalachians. Similar rocks and structures form the Ouachita Mountains and Marathon Basin. A

foreland fold-and-thrust belt extends most of the length of the Cordillera although much of it has been reincarnated via Basin and Range extension as explained in Chapters 12 and 15. The northern part of the Cordilleran belt is unusual due to the widespread presence of Precambrian sedimentary/volcanic rocks in the form of the Belt supergroup (Figure 21.6).

The geometry of a foreland fold-and-thrust belt is such that all thrust faults merge (or root) into a basal thrust known as the *basal décollement* or *sole thrust*, below which the rocks remain largely undeformed (Figure 21.1). The basal décollement represents the base of the tectonic wedge, which characteristically tilts gently toward the hinterland. Individual thrust faults emanate upward from the basal décollement with typical flat-ramp-flat geometry, thereby producing anticlines, particularly above ramps (Figure 3.3g). The basal décollement could conceivably remain at a gently inclined angle and extend for miles within sedimentary layers below the foreland region. A fold-and-thrust belt with this geometry is referred to as a *thin-skinned thrust belt* because the basal thrust fault lies entirely within sedimentary rock and rocks below the basal thrust remain largely undeformed. The basal thrust will eventually cut downward and carry crystalline rock toward the surface but this occurs at the rear of a wide foreland (sedimentary) fold-and-thrust belt as shown in Figure 21.11a.

Alternately, the basal thrust could turn abruptly downward and cut into the underlying crystalline rocks, which are then carried to the surface. In this case, the deformed sedimentary foreland region is narrow and the thrust belt is referred to as *thick-skinned* as shown in Figure 21.11b.

Many factors govern the geometry of a foreland thrust belt. Primary among these is the amount of resistance that is present along the décollement at the base of the tectonic wedge. If rocks are strong, or if preexisting steeply oriented weak zones are present in the rock succession, they will resist horizontal sliding, causing thrust faults to steepen at depth, resulting in a short, thick-skinned thrust stack with crystalline rock close to the front. Conversely, a thin-skinned thrust belt forms if there is a weak, horizontal to gently inclined sedimentary layer in the rock succession along which rocks can easily slide with little resistance. Sliding occurs whenever stress is applied, thus creating a long, thin, tapered tectonic wedge. Nowhere in the southern Valley and Ridge are crystalline rocks brought to the surface. Thrust faults instead flatten at depth along weak sedimentary layers, forming a thin-skinned thrust belt. The Sevier foreland fold-and-thrust belt in the Cordillera is also considered thin-skinned along most of its extent.

Another factor that governs the geometry of a thrust belt is the thickness of the sedimentary pile above crystalline basement. Although the structure in the Middle-Southern Rockies is anticlinal, many of the anticlines are bordered by reverse and thrust faults. If we consider the volume of crystalline rock brought to the surface during Laramide

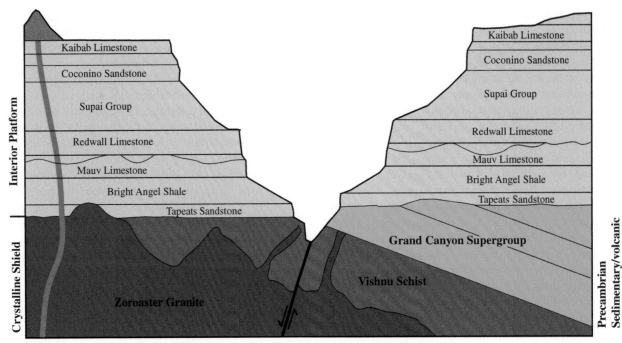

FIGURE 21.10   A cross-section of the Grand Canyon area that shows the Precambrian shield, the Precambrian sedimentary/volcanic rock succession, and the Interior Platform rock succession. Note the buttress unconformity between the crystalline shield and Tapeats Sandstone. Based on O'Dunn and Sill (1988, p. 175).

orogeny, we would have to consider this deformation to be thick-skinned.

Foreland-Hinterland Transition

Metamorphism within both the Appalachian and Cordilleran mountain belts, intrusions, and exposure of crystalline shield rock at the rear of the foreland fold-and-thrust belt allow us to define a foreland-hinterland boundary. In an idealized orogenic system the foreland-hinterland boundary separates sedimentary from crystalline rock (Figure 21.1). The change in rock type creates a topographic step that coincides with a marked change in landscape. Such a step is seen in the Appalachians between the sedimentary Valley and Ridge and the higher, more resistant crystalline topography of the Blue Ridge. The topographic step will retreat over time if the contact is close to horizontal but will remain relatively stationary if the contact is close to vertical as outlined long ago in Figure 3.13. The most distinctive foreland-hinterland boundary is one marked by a major thrust fault. Again we can cite the southern Appalachians as an example where a series of thrust faults, collectively referred to as the Blue Ridge thrust, forms the foreland-hinterland boundary. The boundary, however, can also be marked by an intrusive contact or by a nonconformity. All three are present in the Appalachians.

In the Cordillera, in order to define a foreland-hinterland boundary we need to: (1) ignore the fact that crystalline rock

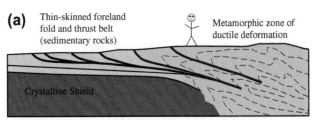

Thin-skinned thrust belt that dies into a metamorphic zone of ductile deformation.

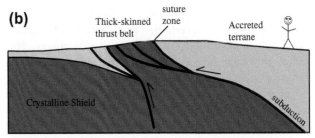

Thick-skinned thrust belt in which foreland faults steepen into preexisting, vertical fault/fracture zones in the crystalline shield. Hinterland faults root into a subduction zone.

FIGURE 21.11   Cross-sections that show thin-skinned and thick-skinned thrust belts. Based on Twiss and Moores (1995).

is widespread across the reactivated western craton in front of the foreland fold-and-thrust belt, and (2) ignore the fact that sedimentary rocks are present across the thrust belt all the way to the boundary with accreted terranes and beyond. Crystalline rocks crop out within the hinterland thrust belt as isolated anticlinal culminations, granitic intrusions, and normal fault exhumed crystalline core complexes surrounded by miogeoclinal sedimentary rock. We can, therefore, define a foreland-hinterland boundary as the first appearance of crystalline rock west of the Sevier foreland fold-and-thrust belt. Given the presence of sedimentary rock in the hinterland, we can conclude that the Cordillera does not display a fully developed, mature foreland-hinterland transition.

There are marked differences in both the landscape and the style of deformation across the foreland-hinterland thrust belt. If you were to look at rocks along the side of the road within a foreland fold-and-thrust belt, they probably would not look terribly deformed. Perhaps all you would see are tilted layers of sedimentary rock. If you are lucky enough to see a fault, you would notice that the rocks are clearly broken across the fault surface. If you are really lucky and have a grand view of a mountainside, you might see giant folds that broadly deform thick sedimentary layers. What you likely would not see is any evidence of metamorphism, granitic intrusion, or abundant small-scale folds. The reason for this is that rocks in a foreland fold-and-thrust belt are deformed under brittle conditions where temperatures are too low to cause obvious metamorphism and where rocks tend to break rather than bend. The foreland fold-and-thrust belt landscape is one of asymmetrical, linear mountains that are steep on the cratonic side and tilted with a more gentle dip on the oceanward side. This type of landscape was described in Chapter 12 (Figure 12.6). The thrust faults themselves impart the asymmetry to the mountain belt because they transport rock layers mostly in one direction: toward the craton.

Mountains in the hinterland are higher, massive, and more rugged, primarily because of the presence of hard, poorly layered, crystalline rock. Here, if you were to look at rocks along the side of the road, you would likely see multiple small-scale folds, granitic intrusions, and evidence of metamorphism. These rocks reached temperatures hot enough to bend without breaking. They were sheared, folded, and intruded under warm, ductile, metamorphic conditions. The difference between brittle and ductile deformation is a little bit like the difference between a stick of cold wax and a stick of warm wax. Cold wax will snap and break. Warm wax will bend but not break. The change from cold, brittle-deformed sedimentary rock at the front of a mountain belt, to warm, ductile-deformed crystalline rock at the rear, can be expressed in terms of a foreland–hinterland transition.

The Hinterland Thrust Belt

The hinterland thrust belt represents the part of the orogenic belt where crystalline rocks are brought to the surface. As used here, all of the crystalline rocks are native North American rock successions (Figures 21.5, 21.7). They represent miogeoclinal and Precambrian sedimentary/volcanic rocks metamorphosed during Appalachian and Cordilleran mountain building, intrusive rocks, and outliers of variably deformed and remetamorphosed Precambrian shield rocks. The primary mode of deformation is thrust faulting; however, strike-slip faulting, multiple fold generations, metamorphism, and intrusion are locally significant. The Appalachian hinterland thrust belt is continuous from the western Blue Ridge to the Green Mountains of Vermont. This hinterland belt consists dominantly of metamorphosed Precambrian sedimentary/volcanic rocks and remetamorphosed crystalline shield rocks, but a few metamorphic equivalents of miogeoclinal rocks are also present. We have already noted that the Cordilleran hinterland thrust belt is markedly different from its counterpart in the Appalachians in that it is dominated by unmetamorphised or low-metamorphic-grade miogeoclinal rocks that extend westward to the boundary with accreted terranes. Rather than a continuous belt of crystalline rocks, the first appearance of isolated exposures of high-grade metamorphic rocks, intrusive rocks, and crystalline shield rocks define the hinterland region.

Accreted Terrane Belt

The accreted terrane belt in both the Cordilleran and Appalachian Mountains is a part of the hinterland that consists of material added to the continent during orogeny. As such, the rocks are separated from North American rock successions by suture zones. Structures are complex and, in addition to suture zones, may include multiple fold generations, metamorphism, igneous intrusion, strike-slip faults, normal faults, and thrust faults. These rocks make up the entire western half of the Cordillera including most of California, Oregon, and Washington. In the Appalachians, about one-half of the Blue Ridge, the entire Piedmont, and all but the western (Green Mountain) part of New England consists of accreted terranes. Figure 21.5 suggests that crystalline shield rocks are present at depth below parts of both accreted terrane belts. Figures 21.5 and 21.6 show that part of the Appalachian accreted terrane belt, including a part of Africa, is buried beneath younger rock of the Atlantic marginal basin. Accreted terrane rocks, and North American hinterland rocks in general, are not exposed in either the Ouachita or Marathon belts but drilling and geophysical data indicate they are present at depth below younger rocks of the Atlantic marginal basin. These data also clearly indicate that the Appalachian, Ouachita, and Marathon regions once formed a continuous orogenic belt in the US that stretched from New England to Texas prior to erosion and burial beneath the Atlantic marginal basin.

THE IDEALIZED OROGENIC BELT

The idealized orogenic belt discussed above is best exemplified by the Appalachian Mountain system (Figure 21.1).

Many of the rocks in the now uplifted Appalachian Plateau were deposited in foreland basins adjacent to rising mountains in the east. The Valley and Ridge forms a classic foreland fold-and-thrust belt, especially in the Southern Appalachians. The western half of the Blue Ridge in the southern Appalachians, and the Green Mountains in the New England Highlands, consist of reworked, variably metamorphosed rocks that include parts of the North American crystalline shield. These rocks form part of the hinterland thrust belt. The eastern Blue Ridge, the Piedmont, and most of New England consist of accreted terranes and suture zones also metamorphosed at various grade. The US Cordilleran Mountain system contains the requisite parts of an orogenic belt but with modifications and additions. The northernmost US Cordillera is least modified. In this region, the reactivated western craton is absent (Figure 21.7). The Cordilleran (Sevier) fold-and-thrust belt, including the Montana Disturbed Belt and the Belt Supergroup (Chapter 12), is followed to the west by the Northern Rocky Mountain–North Cascades crystalline deformation belt which includes both a North American crystalline thrust belt and a belt of accreted terranes (Chapter 13).

Conventional wisdom suggests that North America (the US in particular) was narrower prior to Appalachian and Cordilleran orogenesis than it is today such that it did not extend far beyond the accreted terrane boundaries shown in Figure 21.5. The reasoning for this is that landmass has been added to the margins of North America throughout the Phanerozoic eon via tectonic accretion such that accreted terranes now account for nearly one-half of both the Appalachian and the Cordilleran landscape. Additionally, there has been a great deal of crustal extension particularly during the past 17 million years. However, we must also remember that the continent has undergone a considerable amount of shortening during orogeny with at least some of the shortening affecting the crystalline shield. It is not known with certainty how far crystalline shield rocks extend eastward below the Appalachian belt. Shield rocks are likely present below part of the Coastal Plain and could, in some areas, extend below the continental shelf. However, even given these considerations, it remains possible that the US portion of Laurentia was indeed narrower than present-day North America as suggested in Figure 21.5, but one must not assume that the present-day accreted terrane boundaries define its original extent.

TERMINATION OF DEFORMATION AT THE MARGINS OF AN OROGENIC SYSTEM

We now turn our attention to the foreland and hinterland termination of deformation and mountain building. In other words, what does the transition from deformed foreland to undeformed craton look like, and what happens at the rear of an orogenic belt? Termination of foreland deformation, in theory, is straightforward. The wedge will push only so far onto the craton during plate tectonic collision. Deformation either ends abruptly at the toe of the wedge along a thrust fault that separates the foreland fold-and-thrust belt from the foreland basin, or the deformational front ends in a series of folds that become weaker or more widely spaced toward the craton.

In the Appalachians, the western termination of deformation is often drawn at the physiographic boundary between the Valley and Ridge and Appalachian Plateau. This is a rather abrupt boundary in the Southern Appalachians marked along most of its extent by a thrust fault. However, the deformational front is more diffuse in the central (Pennsylvania) Appalachians where thrust faults are less common in the Valley and Ridge and where folds extend into the Appalachian Plateau. In the Cordillera, we can draw the eastern termination of deformation at the eastern boundary of the reactivated western craton. This boundary is sharp in some areas (for example, the Colorado Front Range) and rather obscure in other areas (for example, surrounding the Black Hills).

Termination in the hinterland is complex and variable. What happens, for example, to thrust faults that dip toward the interior hinterland? What is the nature of the transition from continental lithosphere to oceanic lithosphere? Regarding thrust faults, there are at least two viable possibilities. The faults could root into an active or ancient subduction zone, or they could be lost in a metamorphic zone of ductile deformation. Both possibilities are shown in Figure 21.11. The Appalachian side of Figure 21.5 shows the basal decollement fault rooting into an ancient, no longer existent subduction zone.

To understand the continent-ocean transition more fully, we need only to look at how the Cordillera and Appalachians terminate today. The southern US Cordilleran mountain belt terminates along a series of strike-slip faults that includes (but is not limited to) the active San Andreas Fault. In the northern US Cordillera there is active and ongoing subduction of the Juan de Fuca plate. In both situations, the transition from continental to oceanic lithosphere occurs along a fault zone that marks a tectonic plate boundary. The continental shelf is less than 30 miles wide and drops very rapidly to deep ocean as shown in Figure 1.7. In most areas along the Pacific Coast, sea level drops from less than 500 feet below sea level near the edge of the continental shelf to more than 9,000 feet below sea level across a distance of 20 to 40 miles. The Appalachians terminate at a rifted margin along which the colliding plates (North America, Africa, Europe) have separated. Under these circumstances, the transition from continental to oceanic lithosphere is not a fault and not a plate boundary. A continental shelf, mostly between 50 and 200 miles wide, overlies the rifted margin. Here, in most areas, depths of less than 500 feet below sea level near the edge of the continental shelf drop to more than 9,000 feet below sea level across a distance of 30 to 60 miles.

Finally, there is the possibility of a doubly vergent mountain belt, as shown in Figure 21.12, that could arise by the collision of two continents of roughly equal strength. The Northern Appalachians may have originally formed a doubly vergent mountain belt at the height of orogeny following continental collision with Europe but prior to rifting.

INTRA-OROGENIC DEPOSITION, PLUTONS, AND SUTURE ZONES

Intra-orogenic elements are those rocks that were either deposited above, or were intruded into, rocks that had previously undergone orogenic deformation, but were later themselves deformed during a second orogeny. In other words, deposition or intrusion of these rocks followed one orogeny but occurred prior to another.

In the early Paleozoic, following the breakup of Rodinia but prior to Cordilleran and Appalachian orogenisis, the rocks of the crystalline shield, the Precambrian sedimentary/volcanic sequence, and the earliest deposits of the interior platform and miogeocline together formed the continent of Laurentia. Over time, compression associated with Appalachian and Cordilleran orogenisis would shorten Laurentia while, at the same time, tectonic accretion at the continental margins would add mass. Interior platform rocks on the craton would continue to be deposited throughout most of the orogenic cycles. Miogeoclinal rocks, on the other hand, were deformed and uplifted into a mountainous highland. In some instances, the mountainous highland was eroded and eventually subsided below sea level. Under these circumstances, a new passive continental margin could establish itself and deposit a younger miogeoclinal sequence above the eroded stump of the old mountain belt. This miogeocline could, in turn, be deformed and uplifted during a later orogenic event in a scenario not unlike the never-ending cycle envisioned by James Hutton (Figure 19.1).

The younger miogeocline could be referred to as an *intra-orogenic deposit*, a *successor deposit*, or an *overlap deposit* (we will use these terms interchangeably). These are sedimentary rocks that unconformably overlie older, more strongly deformed rock units, but are themselves deformed. They were deposited between discrete orogenic episodes.

The Appalachian and Cordilleran mountains both developed over the course of several hundred million years throughout the Paleozoic and, in the case of the Cordillera, throughout the Mesozoic and Cenozoic. During that time there were several major orogenic events separated by periods of erosion that resulted in deposition of successor assemblages. Thus, the portrayal of a single pre-orogenic succession of miogeoclinal and slope-rise rocks as suggested in Figure 21.3 is a simplification.

A successor basin need not be a miogeocline but rather is defined as any depositional basin that unconformably overlaps one cycle of orogeny and erosion but is deformed by a later cycle. Intra-orogenic sedimentary deposits do not fit very well into any of the previously listed categories of North American rock successions, nor are they entirely post-orogenic. They were not part of the original Laurentian continent, and although they may depositionally overlie accreted terranes, they were not themselves accreted to North America. They are North American rocks, but they are in a special category of their own.

Plutons can intrude anywhere in a mountain belt prior to, during, or between orogenic episodes, and therefore can also be intra-orogenic. If they intrude into an accreted terrane that was later transported and accreted to the margin of a continent, they would be considered part of that accreted terrane. If they intrude into a terrane that was previously accreted to Laurentia or if they intrude into Laurentia directly, they would have to be considered North American rocks but in a special category of their own. The largest intra-orogenic plutons in the US form the Sierra Nevada, Peninsular, and Idaho batholiths. These rocks are shown in Figure 21.6.

Suture zones are fault zones that form during tectonic accretion and thus are syn- and intra-orogenic in the strict sense. Because they form weak, linear zones within an orogenic belt, they are susceptible to deformation during subsequent orogenic cycles. It is not uncommon for suture zones to be partly or wholly reactivated as a strike-slip, thrust, or normal fault, to be partly obliterated by younger granitic intrusions, or to be strongly folded or metamorphosed. Suture zones can also be wholly or partly covered by younger (overlap) sedimentary rock layers. It is not always

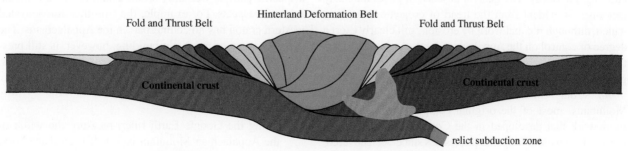

FIGURE 21.12   A cross-section sketch of a doubly vergent mountain system.

easy to determine how long ago a suture zone formed or even where it is located. As we have heard before, fieldwork in this case is of paramount importance.

POST-OROGENIC REINCARNATION

The geologic history of the Appalachians following the Alleghany orogeny is relatively simple. Pangea began to rift apart beginning in the Triassic period, a process that culminated with opening of the Atlantic Ocean and formation of the Atlantic marginal basin. However, the only surface evidence of the rifting event are the Triassic Lowlands in the New England Highlands and the Piedmont Plateau where Upper Triassic red sandstone and basalt crop out extensively (Figure 15.28). The Triassic Lowlands are similar to the Basin and Range in the sense that both formed via extension, normal faulting, and volcanism. However, a major difference is that the Triassic Lowlands represent only a small amount of extension, perhaps a few tens of miles. These areas do not affect topography to the extent that Basin and Range extension affects the Cordillera. The Triassic Lowlands reincarnate only small, isolated parts of the Appalachian landscape. Most of the extension, deposition, and volcanism associated with rifting occurred farther east and is now buried beneath rocks of the Atlantic marginal basin.

With the exception of the Atlantic marginal basin and Triassic rift valleys, the Appalachian Mountain system still retains the same compressional structural framework that was in existence immediately following the Alleghany orogeny. In other words, the landscape has endured 265 million years of erosional modification, sea-level changes, and isostatic adjustment, but at no time has it undergone tectonic modification and reincarnation. The landscape that exists today is an erosional reflection of the landscape that was in existence 265 million years ago. Deep erosion is the primary reason for the widespread surface exposure of crystalline rock in the Blue Ridge, Piedmont, and New England provinces.

From an historical viewpoint, the Appalachians are a classic compressional mountain belt built by continent-to-continent collision and split apart by continental rifting. These mountains have been studied for over 200 years and, in spite of poor-quality outcrop particularly in the southern half, they serve as a model for the general theory of mountain building and decay. The general absence of post-Alleghany tectonism is evident from the trend of orogenic structures, which, although old and worn by erosion, still clearly exert a degree of control on the Appalachian landscape as shown by the general parallelism between geologic trends and physiographic provinces (compare Figures 21.7 and 1.6).

In strong contrast to the much older Appalachian Mountains, most of the original compressional structural framework that developed in the Cordillera as a result of Laramide and older orogenic phases has since been covered, modified, or destroyed by younger, post-Laramide tectonic modification that includes faulting, volcanism, intrusion, regional uplift, and deposition. Some of the tectonic modification began before the end of Laramide deformation. Much of it occurred only during the past 17 million years, and much of it is active today in the form of frequent earthquakes, continuing volcanic activity, and measurable tectonic uplift and subsidence. In addition to continuing orogenic activity along the coast in the form of the Cascadia volcanic arc system and the California transpressional system, post-Laramide volcanic accumulations (primarily basalt) cover nearly all of the Columbia River and Snake River physiographic provinces as well as a large part of the Basin and Range (Figure 14.1). Normal faults associated with Basin and Range extension permeate not only the Basin and Range physiographic province but also extend into the Rocky Mountain, Colorado Plateau, and Cascade provinces. Post-Laramide regional uplift and river incision are at least partly responsible for the present-day landscape of the Middle-Southern Rocky Mountains and Colorado Plateau.

The result of these post-Laramide tectonic episodes has been an almost complete reincarnation of the original orogenic landscape. The scale of reincarnation is evident when one notices that the trend of tectonic elements in Figure 21.7 cuts directly across the trend of physiographic provinces shown in Figure 1.6. Unlike the Appalachian landscape, most of the Cordilleran landscape does not reflect the orogenic episodes that built it in the first place. Instead, it is the post-Laramide structures that dominate a decidedly tectonic landscape and that are responsible for the present-day distribution of physiographic and structural provinces. Overall, the tectonics are far more recent than Appalachian geology, and as a result, the mountains are bigger and much of the landscape mimics the youngest structure imposed on the rock. Although erosion is active and intense, it has not yet removed very much of the overlying sedimentary rock; as a consequence, crystalline rock is not nearly as widespread as in the Appalachians.

An important observation that can be seen in Figures 21.3, 21.5, 21.6, and 21.7 is that, by removing the effects of young normal faulting and volcanism from the Cordillera, we discover an orogenic belt that is a near mirror image of the Appalachian belt. What this implies is that mountain building is not a random process, but rather that it follows a specific pattern. Certainly all mountain belts are different in some respects. For example, the Cordilleran reactivated western craton has no counterpart in the Appalachians. The overall pattern of mountain building, however, is still present, and it is this pattern that will be discussed in Chapter 22.

QUESTIONS

1. Using the Google Earth ruler, measure the width of the Appalachian Mountain belt at Boston, New York, Washington, DC, Charlotte, and Atlanta. How did you

define the eastern and western limits of the orogen? Why should these values be considered absolute minimum values. and why do they not reflect the true width of the orogenic belt?

2. What does the term *reactivated* mean?

3. How is a rock succession different from the rock/sediment type as used in this book?

4. How is a tectonic province different from a structural province as used in this book?

5. What is epeirogenic deformation, and where in the US are its effects best seen?

6. Describe each of the four North American rock successions.

7. Why is the Cordilleran reactivated western craton something of an unusual feature within a compressional mountain belt?

8. What are the Ancestral Rocky Mountains? Where did they form? How long ago? What is the evidence for their existence? What orogeny are they likely related to?

9. Why are the Ancestral Rocky Mountains not included within the reactivated western craton?

10. Why is the foreland basin at the mountain front in the shape of a wedge, and what controls its shape?

11. Define the following terms: foredeep, clastic, and detritus.

12. Explain how detritus in a molasse basin can be used to better understand the exhumation history of a now vanished mountain belt.

13. Describe differences between molasse and flysch.

14. With respect to timing of a mountain belt, what do flysch and molasse date?

15. What factors contribute to the development of a deep foreland basin?

16. Describe differences between the foreland and the hinterland of a mountain belt in terms of rock type and style of deformation.

17. How is the Appalachian hinterland different from the Cordilleran hinterland?

18. Sketch examples of a thin-skinned and thick-skinned fold-and-thrust belt. What are the primary differences between the two?

19. What are the parts of an idealized orogenic belt?

20. Draw a cross-section of a deformed intra-orogenic deposit that unconformably overlies older, more strongly deformed sedimentary rocks.

21. Label and describe each tectonic unit (that is, each color) shown in Figure 21.12. For example, what type of rocks, structures, and tectonic setting are expected in the area shown as yellow?

22. The Appendix contains an uncolored version of Figure 21.6. Color this map and describe each of the rock successions.

23. The Appendix contains an uncolored version of Figure 21.12. Color this map and describe each of the tectonic provinces.

Chapter 22

Formation, Collapse, and Erosonal Decay of Mountain Systems

With the tectonic setup of the US complete, we now focus on the topic of how orogenic systems are made. The modern view of mountain building stresses interdependence among uplift, climate, and erosion. Broadly speaking, any tectonic uplift of land will affect a change in climate, which will cause rates of erosion to change. Erosion removes material from the mountain, which causes isostatic uplift. Heat below the mountain could result in additional thermal isostatic uplift. If uplift results in a rain shadow, the wet, windward side of the mountain could conceivably undergo erosion at a faster rate than the dry side. This could cause the wet side to isostatically uplift at a higher rate. A mountain will achieve steady state topography only if all variables are at or close to equilibrium. In this chapter we illustrate how an idealized orogenic system develops as a result of the convergence of tectonic plates based broadly on the classic Appalachian model. Important concepts include the development of a tectonic wedge, and isostasy which is active prior to, during, and following mountain building. We then discuss how heat plays a two-faced role by maintaining mountain elevation on one hand but potentially causing a decrease in mountain elevation via a process known as *gravitational collapse*.

SUBDUCTION AND DEVELOPMENT OF THE TECTONIC WEDGE

To show how compressional mountains are made, we will use an example in which a change in plate motion causes oceanic lithosphere to buckle and break, forming an ocean-ocean subduction zone. A passive continental margin is present on the down-going plate at some distance from the subduction zone as shown in Figure 22.1a. A trench develops, and an accretionary prism begins to build. Once the subducting slab reaches a depth of at least 62 miles (100 km) it will be hot enough for dehydration to cause melting. Water

released into the mantle wedge between the subducting slab and overlying oceanic crust generates basaltic magma that rises to the base of the crust. Because oceanic crust is thin and dense, some magma reaches the surface to form a string of undersea volcanoes. As the crust thickens, less magma reaches the surface and instead ponds and begins to cool at depth below the surface. Igneous processes that include the preferential crystallization and removal of low-silica minerals will change the magma to a less dense, more silicic, andesite composition. This lighter magma reaches the surface and begins to build an explosive and dangerous oceanic island volcanic arc above sea level located some 90 to 200 miles from, and parallel with, the subduction zone. The situation just prior to arrival of the passive continental margin at the trench is shown in Figure 22.1b.

With continued subduction, the accretionary prism grows wider and higher but remains below sea level. It is possible for the subduction zone trench to migrate away from the volcanic arc as more and more material is added to the accretionary prism. At the same time, the angle at which the subducting slab sinks into the mantle may change. If the slab begins to subduct at a steeper angle, active arc volcanism will migrate toward the trench. If the slab begins to subduct at a more gentle angle, active arc volcanism will migrate away from the trench. The arc-trench gap (the distance between the trench and the active volcanic arc) will grow narrower or wider with time.

As the passive continental margin (the miogeocline) of the continent begins to approach the subduction zone, it may flex upward above sea level and undergo erosion. The upward flexing results from the downward bending of the lithosphere as it approaches the subduction zone and is shown in Figure 22.1b as an outer bulge. The process is similar to bending the end of a stiff 2-by-4 board. As you push the end downward, the central part of the board

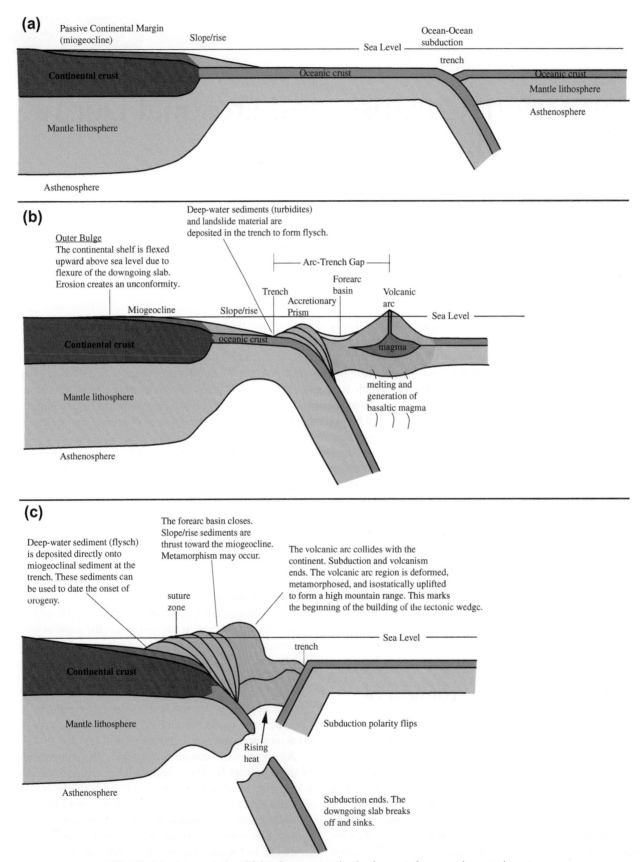

FIGURE 22.1   A conceptual model that shows progressive development of an orogenic mountain system.

(d)

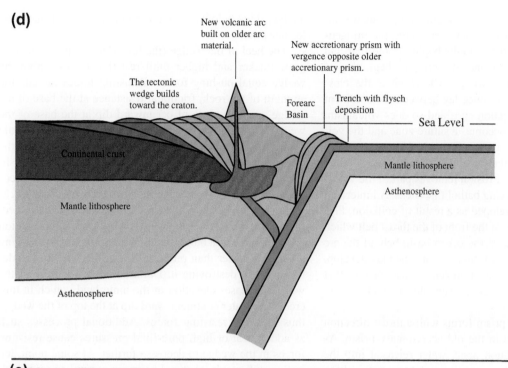

New volcanic arc
built on older arc
material.

The tectonic
wedge builds
toward the craton.

New accretionary prism with
vergence opposite older
accretionary prism.

Forearc
Basin

Trench with flysch
deposition

Sea Level

Continental crust

Mantle lithosphere

Mantle lithosphere

Asthenosphere

Mantle lithosphere

Asthenosphere

(e) Mature collisional system

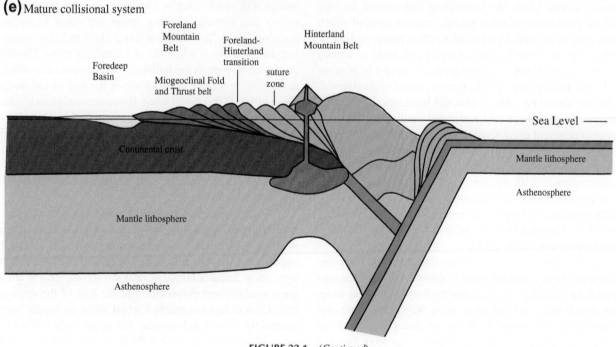

Foreland
Mountain
Belt

Foreland-
Hinterland
transition

Hinterland
Mountain Belt

Foredeep
Basin

Miogeoclinal Fold
and Thrust belt

suture
zone

Sea Level

Continental crust

Mantle lithosphere

Asthenosphere

Mantle lithosphere

Asthenosphere

FIGURE 22.1   (*Continued*)

will flex upward. In the case of the continental margin, the upward flex will be a short-lived event. Soon thereafter, the continental margin is pulled down into the deep water of the trench and flysch is deposited directly on top of the miogeocline. The preservation of deep-water flysch above miogeoclinal rocks (possibly separated by an unconformity) is one of the best ways to date the beginning of orogeny within an ancient orogenic belt.

As the passive continental margin is pulled down into the trench, collision ensues between the continental margin and the volcanic arc. The volcanic arc is a large mass of material that acts like a bulldozer. In geology we refer to it as a *backstop*. The size of the arc, and the force of its crunch, cause thrust faults to develop directly in front of the arc within the forearc basin. The forearc basin closes and is thrust toward the passive continental margin along with the accretionary prism. Thrust faults propagate toward the continental rise which is then thrust onto the miogeocline. This stacking of thrust faults begins in the rear and progresses toward the foreland. It marks the beginning of the building of a tectonic wedge. The result is the closing of the ocean basin that once lay between the continent and volcanic arc. Subduction ends and so does volcanism. The accretionary prism becomes a suture zone and the volcanic arc is accreted to the continental plate as shown in Figure 22.1c. Rocks in the vicinity of the volcanic arc may undergo metamorphism. Erosion removes part of the volcanic arc, exposing underlying batholithic rocks. A hinterland mountain range has developed as a result of collision. Flysch deposition continues at the front of the thrust belt which remains below sea level. If the ocean basin behind the arc continues to converge, a new subduction zone may develop. In Figure 22.1c, a new subduction zone forms opposite that of the earlier subduction zone. The old subducting slab breaks off and sinks.

A new accretionary prism forms with a thrust direction opposite that developed in the old accretionary prism. As with the earlier subduction zone, water released into the mantle wedge above the subducting slab causes melting and the production of dense basaltic magma, most of which rises only to the base of the crust where it ponds and cools. Igneous processes will again change the magma to a more silicic andesite-rhyolite composition, allowing it to isostatically rise through the crust to form a second explosive volcanic arc above the older collisional boundary. Rising heat causes metamorphism and additional melting in the overlying continental crust. Compression of warm crust causes folding and faulting, which, along with rising magma, cause the crust in the vicinity of the volcanic arc to thicken. The heat combined with thickening of relatively light crustal rock causes thermal and isostatic uplift which could potentially increase the height and width of the mountain range. Flysch is deposited in the vicinity of the trench. The situation is shown in Figure 22.1d.

In our scenario we will assume that the entire orogenic belt remains in a general state of compression throughout mountain building. As with the earlier subduction zone, the trench may migrate away from the arc as more and more material is added to the accretionary prism. At the same time, the dip of the subducting slab may become less steep, causing the volcanic arc to migrate toward the craton. The arc-trench gap becomes wider. Areas of old volcanic activity in the hinterland undergo thermal and isostatic uplift. Erosion exposes metamorphic rock and more of the underlying batholithic roots of the volcano. A major mountain range of crystalline rock, along with a volcanic mountain range, is developed in the hinterland, but much of the foreland remains close to sea level and undeformed as shown in Figure 22.1d. The orogenic belt (above the mantle lithosphere) is in the shape of a tectonic wedge: thick below the mountain range, normal thickness farther inland.

The heel of the wedge (the hinterland) will continue to grow thicker and higher until resisting forces within the wedge equal pushing forces. Resisting forces include the strength of the rock, frictional resistance at the base of the wedge, and the angle of oceanward dip at the base of the wedge. Pushing forces include the magnitude of tectonic compressional stress, and the continent-ward dip at the top of the wedge which gravitationally aids tectonic pushing forces and lowers resisting forces. If tectonic forces required to push the wedge are less than resisting forces within the wedge, only the back of the wedge is deformed. In a sense, the back of the wedge is crushed. This situation is comparable to a bulldozer plowing through and crushing a building rather than pushing the building off its foundation without destroying it. Deformation at the back of the wedge increases elevation in the hinterland, which in turn creates a greater continent-ward dip at the top of the wedge, thus lowering resisting forces. Additional processes such as added heat or high pore-fluid pressures cause resisting forces in the wedge to decrease further. At some point, the wedge will reach what is known as *critical taper*, which means that tectonic pushing forces are equal to tectonic resisting forces. The wedge at this point is in a state of equilibrium such that it will slide as a coherent mass. The bulldozer can now push the building off its foundation without destroying it. Rather than deform at the rear of the wedge, any additional force will move the entire wedge as a coherent mass forward toward the craton such that deformation is transferred to the front (the toe) of the wedge within sedimentary rock of the miogeocline. A foreland fold-and-thrust belt develops beginning first at the rear of the miogeocline (closest to the volcanic arc) and progressing toward the craton as shown in Figure 22.1e.

As a side note, the process of achieving critical taper is similar to pushing a thin layer of snow to the side of your driveway. The process is illustrated in Figure 22.2. Keep the angle of the shovel constant as you push it across the driveway. Snow initially builds against the shovel. This is analogous to hinterland deformation at the rear of the orogenic belt. Critical taper is reached when snow no longer builds against the shovel and, instead, the entire wedge slides forward, breaking fresh snow at the front of the wedge. This is analogous to development of a foreland fold-and-thrust belt.

As the front of the tectonic wedge grows, foreland thrust faults will eventually reach the hinge line that separates the miogeocline from thinner layers of the interior platform. At this point the hinge line becomes a thrust fault that carries miogeoclinal rock above rocks of the interior platform. A foreland mountain belt, composed of Precambrian and miogeoclinal sedimentary rock, is now fully developed at the toe of the tectonic wedge. A foreland molasse basin is created on the craton in front of the rising

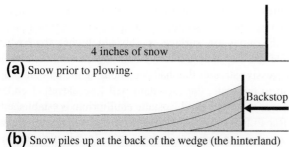

(a) Snow prior to plowing.

(b) Snow piles up at the back of the wedge (the hinterland) until the wedge achieves critical taper.

(c) Once critical taper is achieved, the wedge moves as a single rigid mass, breaking snow at the front of the wedge (the foreland). Snow at the front of the wedge breaks in sequence from oldest to youngest as shown.

(d) Faulting in the foreland causes the wedge to change shape. If the wedge loses critical taper, then faulting may occur at the rear of the wedge in order to reestablish critical taper.

FIGURE 22.2   Formation of the tectonic wedge using snow as an example.

foreland mountains. We now have a fully developed tectonic wedge with two distinctive mountain belts as shown in Figure 22.1e. It is important to reiterate that foreland mountains begin to form well after the beginning of hinterland deformation and mountain building at the rear of the wedge.

The resulting orogenic belt is similar to what is seen in the northern Washington-Montana area near the Canadian border. The eastern part of the Northern Rocky Mountains forms the sedimentary fold-and-thrust belt, and the North Cascades and the western part of the Northern Rocky Mountains form the volcanic, batholithic, and metamorphic hinterland. In the southern Appalachians, the Valley and Ridge forms the foreland fold-and-thrust-belt and the Blue Ridge and Piedmont form the hinterland.

Once critical taper is achieved, thrusting and erosion will alter the shape of the wedge. A change in shape could potentially cause the wedge to lose critical taper. If this should happen, deformation transfers back to the hinterland until the wedge builds to critical taper whereupon deformation moves to the front of the wedge. It is possible for a seesaw effect to be created within a mature orogenic belt in which deformation continuously shifts from foreland to hinterland.

Mountain building will be especially intense if an additional large buoyant mass enters the subduction zone such as an offshore volcanic island, a small continental fragment, or a thick section of oceanic crust. The mass will resist subduction and will accrete to the edge of the continent thus changing the shape of the wedge and creating an extra bulldozer effect. Higher or additional mountain ranges will develop in the hinterland until critical taper is achieved. Once critical taper is achieved and deformation shifts to the foreland, the wedge will push farther inland toward the craton potentially increasing the size and height of foreland mountains. The mountain belt will continue to grow as long as the wedge remains in a state of compression and subduction and accretion occur at the rear. An especially large mountain belt can develop if two continents collide, possibly creating a doubly vergent mountain system as shown in Figure 21.12.

GRAVITATIONAL COLLAPSE OF A MOUNTAIN

It would seem counterintuitive to suggest that crustal extension is a common phenomenon in compressional mountain belts, but, indeed, it is present in all mountain belts of the world, including the Cordillera and Appalachians. In some areas, extensional structures (that is, normal faults) develop during orogeny at the same time as compressional structures. In other areas, extensional structures follow orogeny. For example, Basin and Range extension is currently active in the Cordillera at the same time as orogenic subduction and strike-slip faulting along the Pacific Coast. Basin and Range extension was also active in northeastern Washington and in the Rocky Mountain Basin and Range 40 to 55 million years ago during the final stages of the Sevier and Laramide orogenies. It should be obvious that any form of extension within a mountain belt, for any reason, can result in the development of normal faults such that the mountain as a whole loses elevation. We will use the term *gravitational collapse* for any process, except erosion, by which a mountain range loses elevation.

Extension in a mountain belt can occur due to any one of several reasons. One possibility is a change in plate motion. In Chapters 15 and 17 we presented evidence that, as late as 17 million years ago, the Basin and Range area existed as a high plateau (the Nevadaplano) behind a Cordilleran foreland fold-and-thrust mountain belt. This plateau probably had a crustal thickness on the order of 25 to 40 miles and must have been higher than today's average elevation of the Basin and Range (about 4,600 feet). Recall that between 29 and 27 million years ago, plate interaction along the Pacific Coast changed from subduction to transform, leading to development of the San Andreas fault. Extension and collapse of the Nevadaplano may ultimately be associated with this change in plate motion even though it occurred as much as 12 million years before the onset of collapse.

We noted previously that bends or jumps in strike-slip faults can produce extensional (normal fault) basins such as the Salton Sea along the San Andreas Fault. Strike-slip faults have been active in both the Appalachian and Cordilleran mountain systems, and there is no doubt that they have produced extensional basins similar to that of the Salton Sea.

Another process associated with strike-slip faulting is known as *lateral extrusion* (or *lateral escape*). Lateral extrusion occurs where two or more major strike-slip faults combine to displace crustal blocks toward the margin of an orogen. The extrusion of a crustal block can potentially create room for extension and gravitational collapse as suggested in Figure 22.3.

Gravitational collapse could occur even if the orogenic wedge has achieved critical taper and the mountain is in isostatic equilibrium. Any change in variables that created critical taper could potentially produce gravitational collapse, even if the magnitude of stress across the mountain range remains unchanged. One possibility is the addition of heat into the tectonic wedge. Heat can be introduced by rapid exhumation which drives deep-seated warm rock toward the surface, or by igneous intrusion. The added heat softens and weakens the crust such that it can no longer hold the weight of the mountain. The tectonic wedge will adjust to the new, warmer rock conditions by losing elevation in the hinterland. It will collapse under its own weight. The process is not unlike a large mound of wax that is slowly heated. However, in the case of a mountain range, the situation can become very complex. The addition of heat lowers the density of rock, which results in thermal isostatic uplift. It is possible, therefore, for part of the hinterland to lose elevation through collapse, while at the same time, average elevation across the entire hinterland region is increased. Further complications arise if the mountain remains under strong compression during heating. In this case, the now warmer hinterland may undergo compressional deformation, crustal thickening, and uplift, rather than collapse.

As a fifth example of gravitational collapse, keep in mind that a single orogenic cycle can involve several short-lived compressional pulses. It is possible for rocks below a mountain to be squeezed so much during a compressional pulse that the mountain surface is elevated higher than what simple isostasy should allow. If, for any reason, the strong compressive stresses that had been holding the mountain up begin to dissipate, the mountain will lose elevation under the force of gravity until isostatic equilibrium is established.

Recall from Chapter 15 that extension and normal faulting result in an increase in land area. In this case, extension and normal faulting imply an increase in the width of a mountain belt in order to create room for gravitational collapse. Any one of the above five mentioned mechanisms for gravitational collapse can result in an increase in the width of the orogen. Figure 22.4a is a schematic cross-section of a mountain under tectonic compressional stress. Figure 22.4b is a situation whereby gravitational collapse occurs via normal faulting in the hinterland such that the mountain belt becomes wider. The hinterland losses elevation commensurate with an increase in mountain width.

It is possible, however, for collapse to occur without increasing the width of a mountain belt. In Figure 22.4c, the width of the mountain does not expand. Mass is instead transferred from high areas in the hinterland to low areas in the foreland. The loss of elevation in this case might be manifested in the hinterland by normal faulting at the surface and ductile flow of rock material below the surface toward the margins of the orogenic belt. Room for collapse is created in the foreland by thrust-faulting and crustal thickening. Mass is transferred from high areas in the hinterland to low areas in the foreland without necessarily expanding the width of the mountain. Elevation loss in the hinterland is commensurate with elevation gain in the foreland such that elevation becomes more even across the mountain belt. This situation is analogous to a mound of wax that is confined by the sides of a pan. The wax mound will flatten out in the pan as it is heated.

There is also the possibility that gravitational collapse can occur as a result of changing conditions in the lower crust without necessarily creating normal faults at the surface. In extreme cases, the lower or middle crust can get so hot that it melts or partially melts and begins to flow laterally, away from high-pressure (high-elevation) areas below hinterland mountain ranges and toward adjacent lowland foreland areas. Essentially, what happens is that the partially melted middle crust tunnels its way away from the hinterland and toward the margin of the orogenic belt. This process, shown in Figure 22.4d, tends to even out topography by increasing elevation in the foreland while decreasing elevation in the hinterland without necessarily creating any surface effect (that is, no faulting at the surface). Such a process can create a wide, flat plateau with or without an increase in the width of the orogen.

Regardless of the magnitude of compression, the processes of erosion and gravitational collapse seem to suggest that a mountain cannot grow indefinitely to great

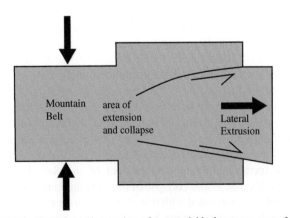

FIGURE 22.3 Lateral extrusion of a crustal block creates room for mountain collapse.

heights. Elevation is limited by the weight of the mountain mass, by the temperature of rock below the mountain surface, by the rate of erosion (particularly glacial erosion), and by any change in structural or tectonic conditions whereby compressive stress, or the strength of rock in the

mountain, changes. The process of gravitational collapse is not restricted to the hinterland but can occur elsewhere in an orogenic belt, such as within the accretionary prism of a subduction zone.

LITHOSPHERIC DELAMINATION

Continental lithosphere is composed approximately of a 20- to 35-mile layer of light granitic crust underlain by a 60- to 80-mile layer of heavy mantle peridotite (an olivine-, pyroxene-rich rock; Figure 5.1). Recall that these layers together form a tectonic plate. *Lithospheric delamination* refers to any process whereby the mantle layer peels away from the crustal layer and sinks, on its own, into the mantle. The process occurs where the mantle lithosphere is heavier than the deep mantle rocks below it. The delaminated layer is replaced by an upwelling of warm mantle rock, which, in turn, warms the overlying crust resulting in thermal and isostatic uplift and possibly crustal extension and gravitational

(a) A mountain is under active tectonic stress. The width of the mountain is defined by the two vertical lines.

(b) Gravitational collapse occurs via normal faulting in the hinterland such that the mountain belt becomes wider. The hinterland losses elevation commensurate with an increase in mountain width.

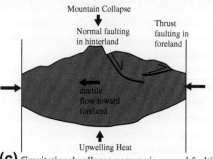

(c) Gravitational collapse occurs via normal faulting and ductile flow in the hinterland commensurate with thrust faulting and crustal thickening in the foreland. The hinterland losses elevation and the foreland gains elevation. Hinterland collapse occurs without increasing the width of the mountain belt. The mountain belt, in this example, remains under compressive stress. Collapse occurs due to addition of heat.

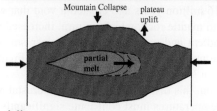

(d) The middle crust partially melts and begins to flow laterally away from high-pressure, high mountain areas in the hinterland and toward adjacent lowland foreland areas. The process increases elevation in the foreland potentially creating a high plateau. Mountain width does not necessarily expand.

FIGURE 22.4 Examples of gravitational collapse; based, in part, on Winter (2010; Figure 18.8).

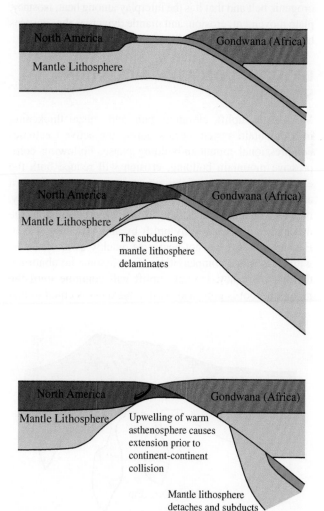

FIGURE 22.5 A series of cross-sections that show lithospheric delamination and normal faulting just prior to the Allegheny orogeny. Based on Sacks and Secor (1990; Figure 2).

collapse. Delamination may have produced normal faults in the southern Appalachians just prior to collision of North America and Gondwana (Africa) during Alleghany orogeny as shown in Figure 22.5.

A form of lithospheric delamination known as *lithospheric drip* is a process by which the heavy lower lithosphere becomes unstable and sinks into the mantle in the shape of teardrops as depicted in Figure 22.6. This type of process is believed to have occurred below the Sierra Nevada between 11 and 3.5 million years ago, allowing hot asthenosphere to rise, causing volcanism and possibly as much as than 3,000 feet of surface uplift. In this case, the addition of heat coupled with the loss of a high-density lithospheric root resulted in a pulse of uplift. However, at about the same time, extension (normal faulting) occurred in Owens Valley to the immediate east of the Sierra Nevada, and compression occurred to the west in the Coast Ranges. From this example alone, we can clearly understand that the processes involved in mountain building and collapse are complex across an orogenic belt and that it is the interplay among heat, isostasy, plate movement, erosion, and mantle dynamics that governs how the Earth's surface will respond.

EROSIONAL DECAY OF MOUNTAIN SYSTEMS

Most of the uplift, elevation gain, and crustal thickening in a mountain system occurs during the active (tectonic) compressional mountain-building phase. Following compressive mountain building, erosion will reduce both the elevation and weight of the mountain mass, which in turn causes isostatic uplift of the thickened crust. Generally, if 5 feet of mountain height are removed by erosion, the mountain will isostatically uplift by approximately 4 feet. Elevation is lost but not nearly as quickly as it would be without isostatic compensation. If we assume no abnormal thermal buoyancy, isostatic uplift will continue until the mountain root is gone and crustal thickness is equal to that

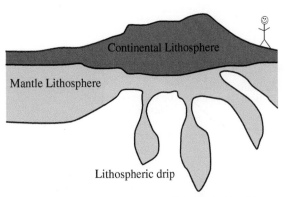

FIGURE 22.6 A cross-section that shows lithospheric drip. This process may have occurred beneath the Sierra Nevada. Based on Winter (2010; Figure 18.8).

of the craton. At that point, the mountain will have been reduced to a flat plane at the elevation of the craton and crystalline rock will be exposed at the surface. In other words, the mountain will have evolved into a craton as shown in Figure 7.7. The Appalachian Piedmont is close to becoming part of the craton, but the Blue Ridge still has a long way to go.

A scenario that involves erosion to a flat plain following compressive mountain building seems to work generally for the Appalachians. But it does not work for the Cordillera. There are wide areas in the Cordillera with thin crust and high elevation. Crustal thickness below the Basin and Range, for example, is only about 20 miles, yet there are several mountain peaks above 12,000 feet and the province as a whole has an average elevation of 4,600 feet. These areas would be at a lower elevation if isostatic equilibrium were dependent solely on crustal thickness. The high elevation, in this case, is due to thermal isostasy. Both the crust and underlying mantle beneath the Cordillera are warm, particularly when compared with cratonic areas to the east. We have noted repeatedly that when rocks within a mountain belt become hot, they expand, become less dense, and rise isostatically. Mantle-generated heat is at least partly responsible for the present-day elevation of the Southern Rocky Mountains and Colorado Plateau. A similar process was alluded to in Chapter 11 when we discussed uplift of the Adirondack Mountains. We have even suggested that warm rock at depth is responsible for the present-day elevation of the Appalachians (Chapter 13).

High geothermal gradients across the Cordillera result from a combination of both active and recently active causes. Recent and active normal faults, especially horizontal (listric) normal faults, have the effect of thinning the crust, allowing warm mantle rock to rise toward the surface (for example, crystalline core complexes). Active Cascade subduction beneath the northern Cordillera perturbs the underlying mantle allowing for high heat flow. Rapid exhumation of any form will have the effect of driving warm rock to the surface. There is evidence that the Cordilleran lithosphere has been warm for at least the past 50 million years. Recall that progressive sinking of the Farallon slab between about 50 and 16 million years ago created a void that was filled with warm mantle rock, which, in turn, increased the geothermal gradient and created explosive volcanism (the ignimbrite flare-up) across the Basin and Range (Figures 17.18, 17.19).

It is clear from this discussion that elevated crustal and upper mantle temperatures must contribute significantly to both the production and maintenance of elevation regardless of crustal thickness and regardless of whether or not the rocks are under compressive stress. Such heat was probably significant during the early stages of erosion of the Appalachians, but after 265 million years, much of this latent heat of orogenesis has dissipated.

Now that we have a basic understanding of how compressional mountain systems form and evolve, we can take a detailed look at the classic Appalachian orogenic belt, followed by a somewhat less detailed overview of the Cordilleran orogenic belt.

QUESTIONS

1. Mountain belts have several characteristics in common. List these characteristics.
2. Why are mountain belts wedge-shaped?
3. What is one of the best ways to date the beginning of orogeny?
4. Why does the hinterland thrust belt develop before the foreland thrust belt?
5. What is critical taper?
6. Name the resisting forces, and the pushing forces, within a tectonic wedge.
6. Given that resisting forces are greater than pushing forces, what is expected to happen within a tectonic wedge?
8. What is meant by a backstop, and why is it important in compressional mountain building?
9. What factors might create a wide tectonic wedge of low mountains versus a narrow tectonic wedge of high mountains?
10. What is gravitational collapse? Why does it occur?
11. Suggest some ways by which extension can occur in an active mountain belt.
12. Given a constant compressional stress, can a mountain grow higher indefinitely? Explain.
13. What is lithospheric delamination and lithospheric drip? Why do they occur? Where have they been proposed to occur?
14. Describe the role of heat in both mountain building and collapse.
15. Describe what could happen to a mountain belt as warm rock is exhumed toward the surface.

The Appalachian Orogenic Belt: An Example of Compressional Mountain Building

As an example of compressional mountain building we will take a detailed look at the geology and tectonics of the Appalachian orogenic system. This is a classic orogenic belt that, with the exception of erosion, has not changed appreciably since its final amalgamation some 265 million years ago. The overall structure remains wedge-shaped; thin in the foreland with a well-developed sedimentary fold-and-thrust belt, thicker toward the interior with North American and accreted crystalline rock. The geology has been studied and scrutinized for more than 200 years and, as a result, much of our basic understanding of mountain belts has been derived through examples found in the Appalachians. Such examples include foreland-hinterland relationships, the overall wedge-shape geometry, the flat-ramp-flat geometry of foreland thrust faults, the classic geosynclinal theory, and even the sequential development of a mountain system as outlined in Figure 22.1. The primary purpose of this chapter is to show how geologists interpret features they see in the field into a comprehensive understanding of geologic history. Similar to landscape analysis, the type of analysis and conclusions outlined here can be applied to other compressional mountain systems because they have all formed via similar mountain-building processes.

SETTING THE STAGE

The story of the Appalachians begins south of the equator about 600 million years ago when the supercontinent Rodinia began to break apart. By 540 million years ago there were at least three large continental fragments separated by an ocean basin known as the Iapetus Ocean. The situation at 540 Ma is shown in Figure 23.1. In Greek mythology, Iapetus was the father of Atlas, from whom the name Atlantic Ocean is derived. The North American fragment, known as Laurentia, was drifting northward toward the equator. The African fragment, known as Gondwana and also consisting of parts of South America, India, and Antarctica, had drifted to the South Pole. The third fragment, known as Baltica, included most of northern Europe and part of Asia. All three continents looked different back then; consisting only of Grenville and older crystalline shield rock along with overlying Precambrian sedimentary/volcanic rift successions.

Appalachian orogenesis resulted from the closing of Iapetus and the collision of Laurentia, Gondwana, and Baltica. This took a long time. Orogeny was underway by

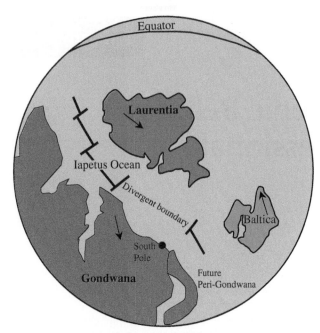

FIGURE 23.1 Reconstruction at 540 million years ago showing location of Laurentia, Gondwana, Baltica, and the Iapetus Ocean. Arrows point toward present-day north. Based on Nance and Linnemann (2008).

about 470 Ma as Iapetus began to close but did not climax until the end of the Paleozoic, about 265 Ma, when final collision formed the supercontinent Pangea (Figure 8.5). The end of the Paleozoic was when the Appalachian system was alive and at its greatest extent. Parts of it were likely more than 20,000 feet high, spanning not only eastern and southern North America but also parts of South America, Greenland, Europe, and Africa. Today we see remnants of the Appalachian chain in the Ouachita Mountains, the Marathon region, eastern Canada (including New Brunswick, Nova Scotia, and Newfoundland), the Cordillera Oriental of Mexico, the Venezuelan Andes, the Caledonides of Great Britain, Ireland, Scandinavia and Greenland, and the West African Fold Belt from Morocco to Senegal. Our discussion will focus on the US Appalachian belt.

The US Appalachians extend from central Alabama, at about 32 degrees north latitude, to the northern boundary of Maine at approximately 47 degrees north latitude. It includes the Valley and Ridge, Blue Ridge, Piedmont Plateau, and New England Highlands physiographic provinces. The Appalachian Plateau forms part of the now uplifted foredeep basin to the ancient mountains. The southern and eastern margins of the Appalachian belt are buried beneath younger rock of the Coastal Plain. The Appalachian belt continues north of Maine to Newfoundland and Greenland although it is partly submerged below sea level. Traditionally, the mountain system in the US is divided into southern, central, and northern segments with boundaries approximately at 37 degrees north (near Roanoke) and 41 degrees north

latitude (near New York City). The three segments arc broadly westward into the North American continent forming salients that are separated near Roanoke and New York City by recesses. The geometry is shown in Figure 23.2. This configuration is thought to represent the original shape of Laurentia prior to Appalachian mountain building. There will be repeated reference to the southern, central, and northern (New England) Appalachians throughout this chapter.

The classic explanation of the Appalachian Mountains is that they are the product of three distinct orogenic episodes. Each was felt along the length of the mountain belt but with different intensities and at slightly different times. These are the Middle to Late Ordovician Taconic orogeny, the Early to Middle Devonian Acadian orogeny, and the Middle Mississippian to Middle Permian Alleghany orogeny. In detail, however, the orogenic events that affected the Appalachians are more complex. A pre-Taconic Middle Ordovician event known as the Blountian orogeny is apparent in the Southern Appalachians and is sometimes considered an early part of the Taconic orogeny. Additionally, a Late Ordovician to Late Silurian event known as the Salinic orogeny fills much of the time gap between Taconic and Acadian orogeny in the Northern Appalachians. Finally, it is now known that the classic Acadian orogeny was not widely felt in the Southern Appalachians. This area was instead affected by a Late Devonian to Early Mississippian Neoacadian event that fills part of the time gap between Acadian and Alleghany orogeny. Overall, the more modern version of Appalachian mountain building is one in which one part of the belt or another was affected by nearly continuous orogeny from Ordovician to Permian. While one area underwent orogeny, other areas experienced quiet deposition, either in a flysch or molasse basin related to orogeny, or in a carbonate basin completely unaffected by orogeny.

At least two additional orogenic events are known to exist, although both occurred within tectonic terranes before they were accreted to North America. These are the Virgilinan orogeny, which affected terranes in the Southern Appalachians during Late Precambrian-Early Cambrian time, and the Penobscot orogeny of Middle Cambrian to Early Ordovician age, which is best defined in parts of Maine northward into Canada and in the Potomac Valley of Virginia and Maryland where it locally is called the Potomac orogeny.

The Northern Appalachians evolved throughout the Taconic and Salinic orogenies by accretion of numerous island volcanic arc systems and various sections of the Iapetus ocean basin, culminating in New England during Acadian and Neoacadian orogeny with collision of two microcontinents (superterranes) known as Avalon and Meguma. The Southern Appalachians experienced Blountian-Taconic orogeny followed by strong Neoacadian orogeny and then

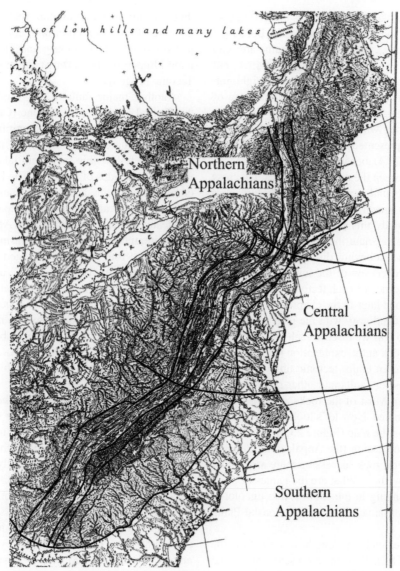

FIGURE 23.2   A landscape map that shows the northern, central, and southern Appalachians. Also shown, from west to east, are the western front of the Appalachian fold-and-thrust belt, the foreland-hinterland transition, the western limit of accreted terranes, and the Fall Line.

felt the brunt of head-on collision between Gondwana and Laurentia during Alleghany orogeny. The exposed part of the Northern Appalachians received only a glancing blow during Alleghany orogeny, with strike-slip faulting but no major collision.

The Ouachita orogenic belt is often considered a continuation of the Appalachian belt; however, the two are physically separated by the younger, overlapping, post-orogenic Atlantic marginal basin rock succession. Orogenic rocks and structures in the Ouachita Mountains and the Marathon basin consist entirely of foreland fold-and-thrust belts and foredeep basins. Hinterland rocks are not exposed. The Ouachita orogeny began during the Mississippian at about the same time as the Alleghany orogeny and is part of the final amalgamation of Pangea.

A TECTONIC MAP OF THE APPALACHIANS

The geology of the Appalachian Mountains is understood primarily through fieldwork and the creation of geologic maps. Ideally, a geological map shows no interpretation. It shows only the spatial distribution, orientation, and shape of rock units, faults, and folds relative to the Earth's surface. It is data in the purest sense, and it is through such maps that geological history is deciphered. A *tectonic map* is a variant of a geological map in which rock units and structures are grouped or separated based on an interpretive history of the rocks. Ideally, all rocks within a tectonic unit have undergone a similar geologic history that is distinct from surrounding tectonic units. Many, but certainly not all, contacts on a tectonic map are faults. A tectonic map simplifies, or in a sense, compartmentalizes the geology so that geologic history is more readily understood. Because a tectonic map is interpretive, two maps of the same area can look different depending on the author's perspective and on what the author deems important or wishes to highlight. With this in mind, and to facilitate our understanding of Appalachian geologic history, a tectonic map of the Appalachian Mountains, along with an explanation of tectonic units and symbols, is presented in Figures 23.3 and 23.4. The map itself (Figure 23.4), because of its size, covers four pages, with overlap between the pages. Thin lines that cross tectonic units are state boundaries. A version of Figure 24.4, reduced to one page, is shown in the Appendix.

It is important to study this map and to understand it so that you can fully appreciate the spatial distribution of each tectonic unit. The following discussion is meant to guide you through an overview of the map. You need not worry about the significance of each tectonic unit; that will become better understood later in the chapter. At this point you need only to gain familiarity with the map so that you can easily refer back to it as we discuss each tectonic unit.

Let us first have a look at the explanation (Figure 23.3). Notice that there are seven major tectonic units and many subunits. Abbreviations for faults and other structures are shown at bottom center. Most of the rocks are Ordovician or older except where noted. Now let's look at the distribution of tectonic units on the map (Figure 23.4). Notice that the interior platform succession (the Appalachian Plateau) is not colored. Within this rock succession there are several large foredeep clastic wedges. What are the names of these foredeep wedges beginning in the south? Also uncolored on this map are the Adirondack Mountains, Coastal Plain, and Atlantic Ocean.

A major fault that extends the length of the map is the Taconic suture zone (TSZ). This line represents the frontal suture zone. It separates North American rock successions (the Laurentian realm and transitional Laurentian realm tectonic units) to the northwest, from accreted terranes which include the internal massifs, Iapetan realm, and peri-Gondwana microcontinent tectonic units (Figure 23.3). Take a moment to follow the Taconic suture zone along the length of the map, noting the tectonic units on either side. Notice that the suture zone disappears beneath younger rock of the Triassic Lowlands near Washington, DC, and below the Coastal Plain near Princeton.

Within the Laurentian realm we have the three North American rock successions that were discussed in previous chapters. In Vermont, these rocks crop out along the Green Mountain anticlinorium (GMA). Four areas are shown as a transitional part of the Laurentian realm. These are the Hamburg (HA) and Taconic (TA) allochthons in Pennsylvania and eastern New York, respectively, the Talladega terrane (T) in Alabama, and the Westminister terrane (WT) near Washington, DC. As the name implies,

these are terranes that contain rocks deposited at the distal edge of the Laurentian miogeocline and in deep water just offshore. The Hamburg and Taconic allochthons are giant klippen (isolated thrust sheets) emplaced during the Taconic orogeny.

Another major fault in the Southern Appalachians is the Central Piedmont shear zone (CPSZ). This fault separates the Iapetan realm tectonic unit from the Carolina peri-Gondwana microcontinent tectonic unit. Take a moment to locate and follow this fault, and take note of the tectonic units on both sides of the fault. The Iapetan realm units are dominantly oceanic accreted terranes (that is, volcanic arc systems and ocean basins). Four separate tectonic units are delineated in the Southern-Central Appalachians, including the Laurentian Arcs and Ocean Basins unit which is further divided into four subunits indicated by letter designation. Be sure to locate all of the tectonic units within the Southern-Central Appalachian Iapetan realm. The Carolina peri-Gondwana microcontinent is a superterrane composed of many smaller tectonic units that are not shown on the map. However, the map does delineate between areas of high metamorphic grade, which includes the Charlotte terrane, and low metamorphic grade, which includes the Carolina Slate Belt.

Map relationships in the Northern Appalachians are perhaps a bit more confusing. For one thing, a younger tectonic unit, the Silurian-Devonian (Acadian) unit, covers most of the rock associated with the Iapetan and peri-Gondwana tectonic units, especially in eastern Vermont, New Hampshire, and Maine. The Silurian-Devonian rocks were deposited following the Taconic orogeny and were involved in Acadian deformation. Notice on the map that the rocks are especially common along two major synclinoriums, the Connecticut Valley-Gaspe synclinorium and the Merrimacke-Kearsarge-Central Maine-Aroostook synclinorum. These two major downfolds are separated by the Bronson Hill anticlinorium (BHA), where buried Iapetan realm and peri-Gondwana microcontinent rocks have been exhumed. Notice that the colors for tectonic units in the Northern Appalachian Iapetan realm are the same as those used for the Southern-Central Appalachians. A similar color implies a similar (but not identical) tectonic history. Here we are attempting to broadly correlate tectonic units. Notice also that rocks of the Ganderia superterrane are intermingled with the Iapetan realm Peri-Gondwana Arcs and Ocean Basins tectonic unit but not with the two other Iapetan realm tectonic units (the Taconic suture mélanges (TS) and the Laurentian Arcs and Ocean Basins). A thick dashed line labeled Red Indian Line (RIL) marks the westernmost extent of Ganderia and serves to separate the Iapetan realm Peri-Gondwana Arcs and Ocean Basins unit from the two other Iapetan units which occur only in the area to the west of the Red Indian Line.

Laurentian Realm

Paleozoic Miogeocline

Precambrian Sedimentary/Volcanic
Succession (rift clastic and slope/rise)

Precambrian Grenville
(North American) Shield
MP - Manhattan Prong
RP - Reading Prong

Transitional Laurentian Realm

T - Talladega Terrane
WT - Westminster Terrane

TA - Taconic Allochthon
HA - Hamburg Allochthon

Internal Massifs

Southern-Central Appalachians
Iapetan Realm
B - Baltimore Dome
S - Sauratown Mountain Window
TF - Tallulah Falls Dome
W - Wilmington-
    West Chester Complex
Carolina Terrane
G - Goochland Terrane
P - Pine Mountain Window

Northern Appalachians
Iapetan Realm
CL - Chain Lakes Massif
C - Chester-Athens Dome
W - Waterbury Dome

Iapetan Realm

Southern-Central Appalachians

CB — Central Blue Ridge Terranes (Iapetus West)

Laurentian Arcs and Ocean Basins (Iapetus West)
(Eastern Blue Ridge-Western Inner Piedmont)
C - Chopawamsic Terrane (Precambrian-Devonian)
M - Milton Terrane
P - Potomac Terrane
T - Tugaloo Terrane

Peri-Gondwana Arcs and Ocean Basins (Iapetus East)
S - Smith River Terrane

CS — Cat Square basin (Eastern Inner Piedmont)
(Middle Silurian-Middle Devonian)

Northern Appalachians

Ts — Taconic Suture Melanges (Iapetus West)

Laurentian Arcs and Ocean Basins (Iapetus West)
SF - Shelburne Falls arc
    also Baie Verte, Annieopsequatch, Notre Dame

Peri-Gondwana Arcs and Ocean Basins (Iapetus East)
BH - Bronson Hill arc
P - Popelogan-Victoria arc
E - Ellsworth-Penobscot ocean basin and arc
T - Tatagouche-Exploits ocean basin

Southern-Central Appalachians
BRT - Blue Ridge Thrust
BCF - Brindle Creek Fault
BF - Brevard Fault
CPSZ - Central Piedmont Shear
    Zone
EPFS - Eastern Piedmont Fault
    System
TSZ - Taconic Suture Zone

Note: All rock units are Ordovician
or older except where noted.

Northern Appalachians
BHA - Bronson Hill Anticlinorium
CT - Champlain Thrust
LO - Liberty-Orrington Line
GMA - Green Mountain Anticlinorium
HLBF - Honey Hill-Lake Char-Bloody
    Bluff Fault
HT - Hinesburg Thrust
NF - Norumbega Fault
RIL - Red Indian Line
TRZ - Taconic Allochthon Root Zone
TSZ - Taconic Suture Zone

Peri-Gondwana Microcontinent Realm

Carolina Superterrane (Southern Appalachians)

Low-Grade (Carolina Slate Belt)

High-Grade (Charlotte Terrane)

Gandaria Superterrane (Northern Appalachians)

CP — Coastal Plutonic and Volcanic Belt
(Silurian-Early Devonian)

Ganderia Passive Margin and
Basement Rocks

Avalon Superterrane (Northern Appalachians)

A

Silurian-Devonian (Acadian)
Ocean Basin, Flysch, and Plutons

With Acadian deformation. Includes the
Piscataquis Volcanic Arc (PVA) and
Upper Ordovician rocks

Alleghany and Post Orogenic

Tr — Triassic Rift Basins

MP — Mississippian-Pennsylvanian Basins

White Mountain Batholith
(Jurassic-Lower Cretaceous)

FIGURE 23.3 An explanation of the tectonic map of the Appalachian Mountains.

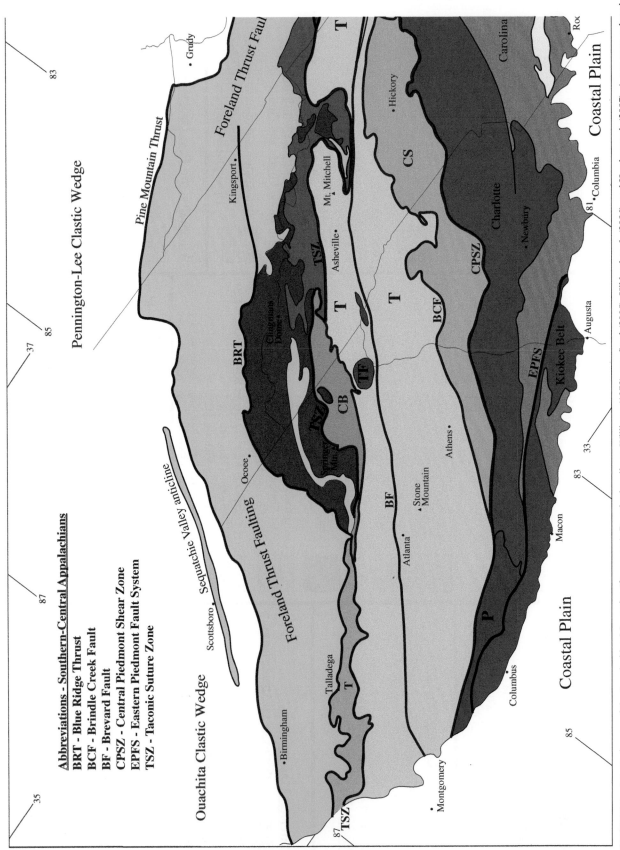

Abbreviations - Southern-Central Appalachians
BRT - Blue Ridge Thrust
BCF - Brindle Creek Fault
BF - Brevard Fault
CPSZ - Central Piedmont Shear Zone
EPFS - Eastern Piedmont Fault System
TSZ - Taconic Suture Zone

FIGURE 23.4   A tectonic map of the Appalachian Mountains (four pages). Based primarily on Williams (1978), van Staal (2005), Hibbard et al. (2006), and Hatcher et al. (2007). A one-page reduced version of this map can be found in the Appendix.

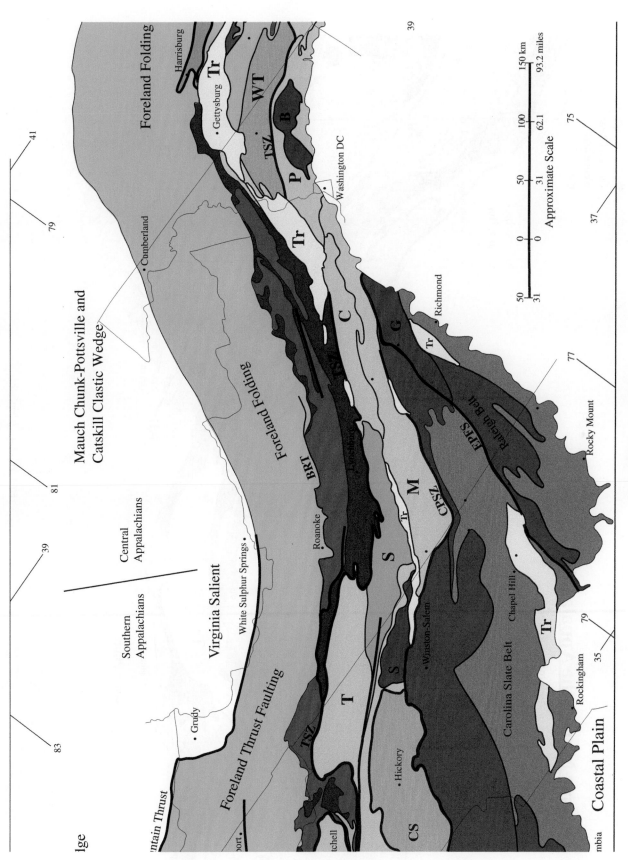

FIGURE 23.4 (Continued)

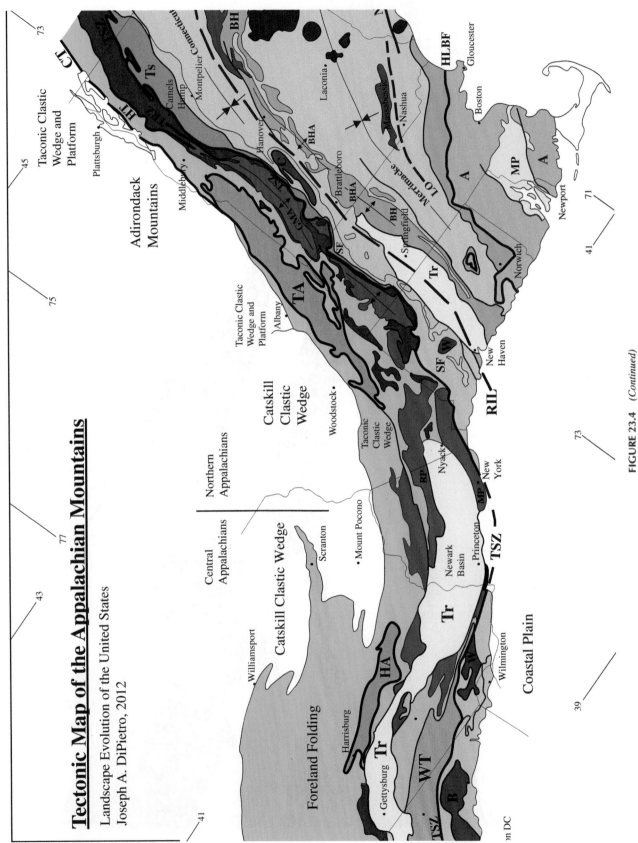

Tectonic Map of the Appalachian Mountains

Landscape Evolution of the United States
Joseph A. DiPietro, 2012

FIGURE 23.4   (*Continued*)

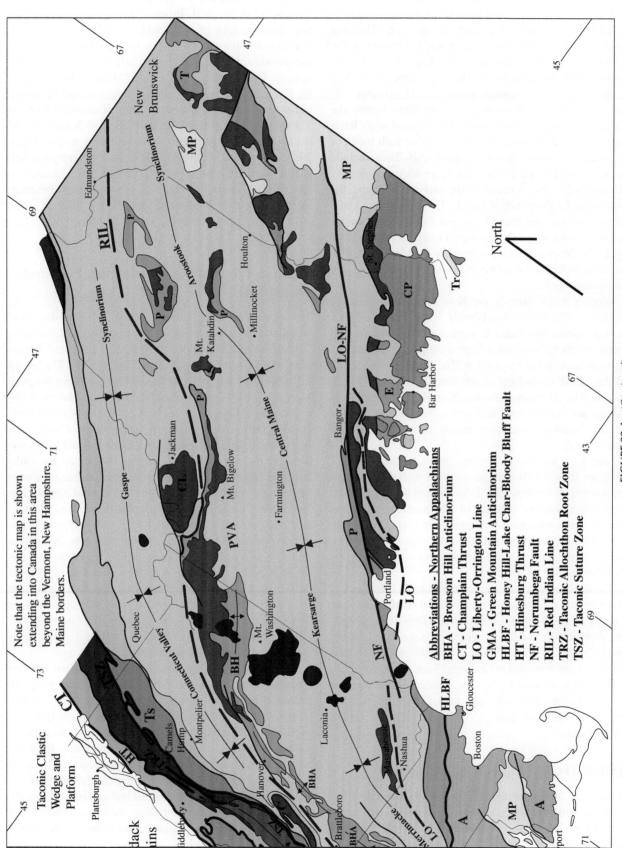

Note that the tectonic map is shown extending into Canada in this area beyond the Vermont, New Hampshire, Maine borders.

Abbreviations - Northern Appalachians
BHA - Bronson Hill Anticlinorium
CT - Champlain Thrust
LO - Liberty-Orrington Line
GMA - Green Mountain Anticlinorium
HLBF - Honey Hill-Lake Char-Bloody Bluff Fault
HT - Hinesburg Thrust
NF - Norumbega Fault
RIL - Red Indian Line
TRZ - Taconic Allochthon Root Zone
TSZ - Taconic Suture Zone

FIGURE 23.4 (Continued)

The Red Indian Line is a suture zone. It is dashed along most of its extent because it is older than, and buried by, the Silurian-Devonian (Acadian) tectonic unit. The Red Indian Line is nowhere exposed in the US except in the Jackman area just northwest of Mt. Bigelow in Maine. Can you find this location on the map? East of the Red Indian Line, several Iapetan realm peri-Gondwana volcanic arcs and ocean basins crop out from below the Silurian-Devonian tectonic unit, all indicated with letter designation and all seemingly intermingled with rocks of the Ganderia passive margin tectonic unit. They include the Bronson Hill arc (BH), Popelogan-Victoria arc (P), Ellesworth-Penobscot ocean basin and arc (E), and the Tatagouche-Exploits ocean basin (T). Notice that intermingled Iapetan realm and Ganderia rocks extend across the Liberty-Orrington-(Norumbega) fault (LO-NF) all the way to the Maine coastline where the Ganderia superterrane also includes the relatively young Coastal Plutonic and Volcanic Belt.

Another major fault in the Northern Appalachians is the Honey Hill-Lake Char-Bloody Bluff Fault (HLBF). Can you locate this fault? It separates the Ganderia and Iapetan realm tectonic units from another peri-Gondwana microcontinent known as Avalon. The Avalon superterrane and the HLBF extend out to sea near Gloucester, Massachusetts. The three superterranes (Carolina, Ganderia, and Avalon) are shown with different colors which implies that any tectonic correlation among them is suspect or controversial.

Next let's have a look at the internal massifs, all of which are shown with a single color but with different letter designation. Internal massifs are present along the entire Appalachian orogenic belt east of the Taconic suture zone. Their origin is somewhat problematic in that they may or may not be part of Laurentia. The map explanation identifies nine internal massifs by letter designation. Try to locate all of them on the tectonic map.

The final tectonic unit shown in Figure 23.3 consists of Alleghany and post-orogenic rocks. The Mississippian-Pennsylvanian basins in Rhode Island and New Brunswick, Canada were deposited during the Alleghany orogeny. The Jurassic-Lower Cretaceous White Mountain batholith and the Triassic rift basins were discussed in Chapters 13 and 15, respectively.

Hopefully, now that you have studied and are familiar with the tectonic map and explanation, our discussion of Appalachian geology will proceed with greater understanding.

MAJOR TECTONIC BOUNDARIES

On the basis of our previous discussion of mountain building, we can distinguish three major boundary zones that can be traced the length of the Appalachians. Although each is marked by distinct changes in rock type and structure, the actual boundaries are not everywhere sharp or even easily recognized in the field. In some areas, the boundaries are zones several tens of miles wide. In other areas they are hidden beneath younger rock or younger fault slices. They are introduced here because they help define the mountain system. The major boundaries are (1) the western front of the foreland fold-and-thrust belt, (2) the foreland-hinterland transition, and (3) the western limit of accreted terranes. These three boundaries, and the Fall Line boundary with the Coastal Plain, are shown on a landscape map in Figure 23.2. In Figure 23.4, the three boundaries correspond with (1) the western boundary of the Paleozoic miogeocline with the (uncolored) interior platform, (2) the eastern boundary of the Paleozoic miogeocline, and (3) the Taconic suture zone (TSZ), which forms the eastern boundary of the Laurentian realm.

Western Front of the Appalachian Fold-and-Thrust Belt

The western front of the Appalachian fold-and-thrust belt is defined as the boundary that separates thrust-faulted or folded miogeoclinal rock from weakly deformed interior platform and foreland basin rock (Figure 23.4). The boundary coincides roughly with the boundary between the physiographic Valley and Ridge and Appalachian Plateau. It also approximates the transported hinge line between miogeoclinal and platform deposition (Figure 21.3). It is a fairly sharp boundary in the Southern Appalachians marked by a series of thrust faults or tight folds such as the Sequatchie Valley anticline and Pine Mountain thrust. The line becomes more diffuse and somewhat arbitrary in the Central Appalachian Valley and Ridge where thrust faults are less common and folds extend across much of the Appalachian Plateau. Folding on the Appalachian Plateau, however, is so broad that rocks, for the most part, remain close to horizontal (that is, the deformation appears epeirogenic). Frontal deformation in both the Southern and Central Appalachians occurred during the Alleghany orogeny, placing folded and thrust-faulted Cambrian through Pennsylvanian miogeoclinal rock against or above platform and basin-fill (that is, foredeep) rock of Devonian through Permian age.

The foreland fold-and-thrust belt narrows in the Northern Appalachians and thrust faults again become prevalent. Frontal thrust faults in this region, however, are Taconian in age. The western limit of strong deformation is defined in part by the Taconic Allochthon thrust sheets, which place Late Precambrian-Ordovician rift-clastic and slope-rise rocks above miogeoclinal and basin-fill rocks of the same age. In northern Vermont, the Champlain thrust defines the western limit of the fold-and-thrust belt and is magnificently exposed at Lone Rock Point along the eastern shoreline of Lake Champlain where it places Cambrian dolomite above Middle Ordovician shale.

Northern Front of the Ouachita Fold-and-Thrust Belt

Of the three major tectonic boundaries listed, the northern front of the Ouachita fold-and-thrust belt is the only one exposed in the Ouachitas. The Ouachita thrust belt, in general, is characterized by a series of very tight folds and thrust faults that place Devonian though Pennsylvanian miogeoclinal rocks above rocks of the same age. The northern front is located at the boundary between the Ouachita Mountains and Ozark Plateau physiographic provinces, where thrust faults die out rather abruptly into a series of folds. The boundary coincides with the Choctow thrust in the western Ouachitas and the Y-City thrust in the east. Both place Mississippian-Pennsylvanian miogeoclinal rocks above platform and basin-fill rocks of the same age. The thrust front in the Marathon region is not defined. In this area, strongly folded and thrust-faulted Cambrian through Pennsylvanian miogeoclinal rocks are surrounded by unconformably overlying Cretaceous rocks. The true northern front is not exposed.

Foreland-Hinterland Transition

The foreland-hinterland transition is also the transition into the core of the mountain belt. Along the length of the Appalachian belt this contact separates unmetamorphosed or weakly metamorphosed foreland sedimentary rocks of the miogeocline from weakly to moderately metamorphosed hinterland Precambrian sedimentary/volcanic rocks and underlying crystalline shield rocks with Grenville-age deformation. The transition coincides roughly with the boundary between the physiographic Valley and Ridge and Blue Ridge-Green Mountains. The transition is gradual in the sense that there is not a great jump in metamorphism across the contact. The metamorphism instead increases eastward into the Blue Ridge and Green Mountains, eventually reaching high-grade conditions.

The foreland-hinterland transition is most obvious in the Southern Appalachians. Here the contact is a major Alleghany-age thrust fault that goes by different names in different areas but collectively can be referred to as the *Blue Ridge thrust* (BRT). From south to north, the local names are the Talladega, Cartersville, Great Smoky, Miller Cove, and Holston Mountain faults. This fault system is more or less continuous to the vicinity of Clingmans Dome where the contact is cut by a large number of small faults that imbricate Grenville basement, the Precambrian sedimentary/volcanic rock succession, and miogeoclinal rock. From this location northward, all the way to southern Vermont, the transition into the hinterland is variably a depositional contact or a fault. Where it is depositional, Late Precambrian to Ordovician miogeoclinal rocks either pass downward into the Precambrian sedimentary/volcanic rock succession

with a gradual increase in metamorphism, or they sit unconformably above high-grade Grenville-age crystalline shield rocks (granitic gneiss). The boundary is fairly distinct from Clingmans Dome to the vicinity of Gettysburg where there is a clear separation between the miogeocline to the west and metamorphosed Precambrian sedimentary/volcanic and crystalline shield rocks to the east. However, between Gettysburg and northwestern Connecticut, miogeoclinal rocks are intermingled primarily with crystalline shield rocks and both are partly buried below post-orogenic rocks of the Newark Basin (part of the Triassic Lowlands) such that a distinctive boundary cannot be easily drawn (Figure 23.4). This area corresponds with a topographic break where both the Blue Ridge Mountains and the Green Mountains are absent. The boundary is again distinctive in Massachusetts and southern Vermont along the western side of the Green Mountains where it is primarily a depositional contact. In northern Vermont the contact is marked along part of its extent by the Hinesburg thrust.

Western Limit of Accreted Terranes

Most of the exposed rock across the eastern part of the Blue Ridge province, the Piedmont Plateau, and New England Highlands east of the Green Mountains consists of accreted terranes that originally were not part of Laurentia. The contact between these rocks and North American (Laurentian) rocks is a fault zone (a suture zone) along the entire Appalachian orogenic mountain belt. The suture zone, however, is not obvious. In different areas it is variably offset across younger faults, folded, or subject to younger metamorphism. In some areas it has been reactivated as a fault with different displacement characteristics, or is buried beneath a younger thrust sheet or younger rocks and is no longer exposed. Where it is recognized, it does not show typical fault zone characteristics such as highly crushed and broken rock. The absence of typical fault-zone fabrics indicates that the suture zone was active prior to peak metamorphism. Heat from metamorphism has, in a sense, baked the fault contact, recrystallizing and destroying nearly all evidence of displacement. The fault looks like a normal depositional contact in many areas. It is recognized as a fault primarily by a sharp change in rock type and, more important, by the presence of ophiolitic lenses (remnants of oceanic lithosphere).

Individual faults that mark the suture zone, or mark the reactivated/buried boundary between North American rocks and accreted terranes are, in the Southern-Central Appalachians, the Hollins Line, Allatoona, Hayesville, Fries, Pleasant Grove, Huntingdon Valley, and Cameron's Line faults. In the Northern Appalachians, the same structure is known as the Baie Verte-Brompton Line and includes the Whitcomb Summit thrust in Vermont. Collectively, we will refer to this contact as the *Taconic suture zone* (TSZ) because it formed during the Taconic orogeny. Along its entire length the Taconic suture zone places Ordovician and older accreted terranes against the ancient Laurentian continental margin (Figure 23.4).

TECTONIC FRAMEWORK

The origin of the US part of the Appalachian Mountain system is simple in a general sense. Laurentia collided with Gondwana (the northwestern part of Africa) in the late Paleozoic, thus closing the Iapetus Ocean and producing the supercontinent Pangea (Figure 8.5). However, in detail the story becomes murky and is not fully agreed upon. The orogenic belt is old and much of it is eroded or covered with younger rock. Additionally, much of the rock that is present is poorly exposed beneath a cover of forest and soil. For these reasons, relationships between and among various rocks and structures are not straightforward. The story depicted here is a compilation of many published articles, some of which do not agree but, when taken together, give a fairly accurate first-order understanding of Appalachian geology.

Of primary importance is the physical disconnect between the Northern Appalachians and the Southern and Central Appalachians. This is evident when one notices that the belt of accreted terranes disappears below sedimentary rocks of the Coastal Plain at the orogenic recess in the vicinity of New York City (Figure 23.2, 23.4). Thus, it is not possible to continuously trace accreted terranes in the New England Appalachians along strike to the Southern Appalachians. The disconnect is also geological. Thrusting and folding associated with the Alleghany orogeny permeate the Southern and most of the Central Appalachians such that it overprints and masks deformation associated with earlier orogenic events. It is difficult in these areas to see evidence of earlier orogeny. Luckily, or perhaps fortuitously, the New England Appalachians, and particularly the Canadian Appalachians, are quite different with respect to the distribution of orogenic phases. Deformation in the foreland and in the western part of the hinterland is Taconian in age with little overprint by later events. Acadian-age deformation and metamorphism dominates central New England with little overprint from Alleghany-age deformation which is restricted to only a few areas primarily along coastal New England and Canada. This physical separation has

allowed geologists to study the effects of one orogeny without the complicating overprint of later orogeny. Without the Northern Appalachians as a guide, it would have been considerably more difficult to decipher the early orogenic events that created the Southern Appalachians.

FORMATION OF LAURENTIA

Let us begin with the rifting of Rodinia, the opening of the Iapetus Ocean, and the development of a passive continental margin along the newly formed eastern seaboard of Laurentia. Rodinia rifted in stages. The oldest rift-related rocks are in the Southern Appalachians (Grandfather Mountain and Mount Rogers formations). These rocks, however, record only a failed rift attempt at about 735 Ma. The continent was pulled apart at this time but not wide enough to create an ocean basin. Actual breakup and initial opening of Iapetus occurred between 620 and 520 Ma as is evident from an abundance of rift-related clastic rock (with or without volcanic rock and igneous intrusions) along the Laurentian continental margin. These rift clastic rocks are part of the Precambrian sedimentary/volcanic rock succession shown in Figure 23.4.

Many of the rocks associated with rifting were deposited above sea level from streams draining normal fault-block mountains. As the Iapetus Ocean continued to open, the newly formed Laurentian continental margin began to cool and isostatically sink. Normal faulting became inactive, the block mountains were eroded, and the continental margin sank below sea level. The rifted margin transitioned into a passive continental shelf. Deposition along the passive continental shelf created an unconformity that separates the eroded mountains and rift-clastic deposits from the newly formed miogeoclinal shelf deposits. This unconformity is called a *breakup unconformity* or the *rift-to-drift transition unconformity* and is shown in Figure 23.5. The unconformity is important because it marks the end of rifting and the beginning of passive continental margin (miogeoclinal) sedimentation. It can be dated using fossils. In Vermont, the underlying rift-clastic rocks include sandstones and conglomerates (now metamorphosed to schist) of the Pinnacle and Underhill formations. The Cambrian Cheshire Quartzite (a weakly metamorphosed sandstone) forms the

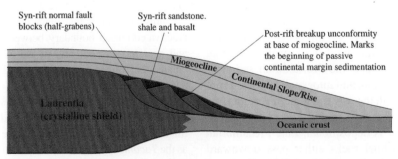

FIGURE 23.5 A schematic cross-section that shows the rift-to-drift (breakup) unconformity.

initial miogeoclinal rock unit above the breakup unconformity, followed by the Ordovician Dunham Dolomite. The transition in the Southern Appalachians occurs between weakly metamorphosed sandstone of the Late Precambrian Ocoee Supergroup, and sandstones and shale of the Cambrian Chilhowee Group. The Chilhowee Group, and the overlying Shady dolomite and Rome formation, together form the lower part of the passive continental margin.

It is important to realize that when Rodinia broke apart, it did not simply break into three large continents as depicted in Figure 23.1. In addition, there were a series of small islands, composed of Grenville basement, that were rafted off the coast of Laurentia. These islands would later collide with Laurentia as accreted terranes. In Canada, the accreted islands are known collectively as *Dashwoods*. We will refer to them as *internal massifs*. There were additional island fragments within the newly formed Iapetus Ocean as well as several subduction zones and island volcanic arcs. All of these would eventually collide with Laurentia. Our survey of the Appalachians will begin in the foreland and progress across the miogeocline to the hinterland.

FLYSCH AND MOLASSE BASINS: DATING APPALACHIAN OROGENY

Syn- and post-orogenic flysch and molasse deposits are present across the Appalachian Mountains from the hinterland to the undeformed foreland. Most of the rocks are currently in the Valley and Ridge and Appalachian Plateau, but some, particularly Acadian flysch, are in the hinterland of the Northern Appalachians. The rocks are important because they represent the erosional remnants, the residue, of now vanished mountains. They were deposited primarily in foreland basins in front of advancing thrust faults and thus record the timing of orogeny. Individual flysch or molasse deposits in the Appalachians tend to be somewhat older and more coarse-grained in the east, indicating that the mountain source was east of Laurentia or along its uplifted eastern margin. They also tend to have the shape of a clastic wedge, thick and coarse at their center and becoming thinner laterally away from their center toward the craton. Coarse-grained members contain rock fragments of the eroding mountain. These fragments offer clues to the type of rock that existed in the now vanished mountain, which in turn gives information on geological history. For example, the type of fragment could indicate whether the source is from uplifted Grenville (Laurentian) basement or from an accreted terrane. In this section we discover what the eroded residue tells us about the ancient Appalachian Mountains. The rock units discussed in this section are shown in a time stratigraphic column in Figure 23.6.

The Appalachian miogeocline shows no sign of mountain building prior to Middle Ordovician. Clean white Cambrian sandstones are overlain by Early Ordovician

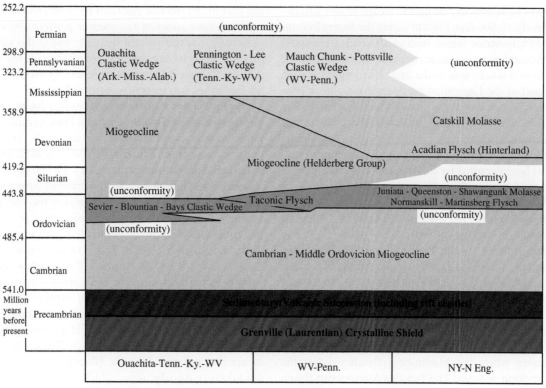

FIGURE 23.6 Time stratigraphic units of the Appalachian miogeocline and foreland basin.

carbonates (limestone and dolostone) along the entire length of the miogeoclinal foreland fold-and-thrust belt. These rocks transgress westward over the eroded Grenville highlands. They indicate the existence of a slowly subsiding passive continental margin with Grenville highlands in the distant interior. The Cambrian and Early Ordovician was a quiet time on Laurentia.

The situation began to change in the Middle Ordovician (Figure 23.6). Miogeoclinal sandstones and carbonates are overlain by an unconformity and then by deep-water black shales and greywackes (fine-grained sandstones) derived from the east. The unconformity is interpreted to have developed when the continental margin flexed upward above sea level, creating an outer bulge as it was approaching a trench as depicted in Figure 22.1b. The deep-water sediments are interpreted as flysch deposits that settled in the trench and forearc region after the miogeocline was pulled down toward the subduction zone as shown in Figure 22.1c. These rocks are known regionally as the Taconic flysch. They represent the beginning of the Taconic orogeny.

The oldest Taconic flysch and molasse deposits in the US are exposed in the southern Appalachian Valley and Ridge, particularly in Tennessee. The flysch is known as the Blockhouse and Sevier shale and greywacke and is early Middle Ordovician in age. These rocks are directly overlain by late Middle Ordovician river-derived sandstone and conglomerate (molasse) of the Bays Formation. Together these rocks form the Sevier-Blountian clastic wedge as shown in Figure 23.6. In Tennessee, the clastic wedge attains a thickness of nearly 10,000 feet. Present within the Sevier Formation are conglomerate lenses with clasts (rock fragments) of Late Precambrian rift-clastic rock and miogeoclinal rock derived from a source area to the east. This implies that an uplifted part of the Laurentian continental margin, or possibly rafted islands, lay to the east of the flysch basin. The Sevier-Blountian clastic wedge is the principal evidence for the Blountian orogeny and is the first clear indication in the United States of Appalachian orogeny. The orogeny, however, must have occurred far to the east because there is little evidence for strong Blountian (or Taconic) deformation within the flysch deposits or anywhere within the Southern Appalachian foreland. The miogeocline acted as a foredeep during Blountian-Taconic orogeny.

Classic evidence for the Taconic orogeny is present in and around the Taconic and Hamburg allochthons of western New England, New York, and Pennsylvania in the form of the Middle to Late Ordovician Normanskill and Martinsburg formations, which locally reach thicknesses of more than 9,000 feet (Figure 23.6). These flysch deposits (shale and greywacke) extend northward into Quebec and southward to Tennessee where the Martinsburg flysch overlies the Bays Formation molasse. The fact that Martinsburg flysch overlies Bays molasse is an important relationship because it implies that the Martinsburg flysch is younger and entirely separate from the Sevier-Blockhouse flysch. It tells us that orogeny and mountain building began in the south with Sevier-Blockhouse deposition (the Blountian event) and, over time, progressed northward to Quebec (Normanskill-Martinsburg deposition) to form the classic Taconian event. The orogeny also progressed westward. Fossils in New England, New York, and Quebec indicate that flysch gets progressively younger from east to west. The approximate ages are 456 Ma in the east and 446 Ma in the west (Upper Ordovician). In contrast to the Sevier-Blockhouse flysch, these flysch deposits would eventually be deformed and overrun by the Taconic and Hamburg allochthons and incorporated into the Taconic foreland fold-and-thrust belt.

In Pennsylvania and surrounding areas, the Martinsburg flysch is overlain by red-colored, nonmarine, Upper Ordovician river and delta molasse deposits of the Juniata-Queenston formation (Figure 23.6). The rocks are coarse-grained in the east and finer-grained in the west, indicating that Taconic Mountains were present east of Pennsylvania by the Late Ordovician. Molasse deposition continued into the Early Silurian as indicated by deposition of the Shawangunk sandstone and conglomerate which, today, is famous for its fabulous rock-climbing routes along the southeastern margin of the Appalachian Plateau only 90 miles north of New York City.

There was continuous carbonate deposition in western Pennsylvania far to the west of the Taconic Mountains, indicating that an interior shallow sea covered the craton throughout the Taconic orogenic cycle. These cratonic rocks are part of the interior platform rock succession. Silurian-age carbonate rocks of the interior platform become progressively younger toward the east; and at the same time, molasse deposits become less voluminous. This indicates that the Taconic Mountains were slowly eroding and that the inland sea was migrating eastward.

By Early Devonian, the mountains had disappeared completely in parts of the New York and Pennsylvania and were likely no more than low-lying hills surrounded by a shallow ocean in other areas. Erosion of these mountains produced a widespread unconformity that marks the end of the Taconic mountain-building phase. Shallow marine limestone of the Helderberg Group was deposited above the unconformity in Early Devonian, forming part of a passive continental margin (a successor miogeocline) that surrounded remnant Taconic highlands (Figure 23.6). Deformation present in the Normanskill and Martinsburg flysch below the unconformity is not present in the Helderberg Group above the unconformity. This relationship (an angular unconformity) proves that deformation seen in the flysch occurred prior to deposition of the Early Devonian Helderberg Group. The deformation must be Taconic in age. These flysch deposits are thus different from the Blockhouse-Sevier flysch in the Southern Appalachians, which were not deformed and incorporated into the Taconic orogenic belt.

Helderberg limestone deposition ended when a second major orogenic cycle, the Acadian orogeny, resulted in subsidence and a new round of flysch deposition. These flysch deposits are centered in the New England hinterland within the Kearsarge-Central Maine Basin and in the Connecticut Valley-Gaspe Basin, both of which are located in Figure 23.4. The rocks include the Waits River, Gile Mountain, Littleton, and Seboomook formations and are referred to simply as *Acadian flysch* in Figure 23.6. The rocks were derived from the east and are thickest in the east where they reach 13,000 feet. Fossils indicate that the rocks become progressively younger from east to west and from north to south. The rocks are approximately 414 My old in the east and approximately 387 Ma in the west (Early to Middle Devonian). These rocks were likely deposited in a closing forearc basin that migrated westward and southward over time with a migrating deformation/uplift front as depicted schematically in Figure 22.1d.

Originally, the Acadian flysch basin may have extended southward, but little evidence remains of its existence. There is, however, a sequence of slightly younger black shales (Upper Devonian, 385 to 362 Ma) present from southern New York to Kentucky (Genesso, Rhinestreet, Cleveland formations). These rocks also become younger from east to west and from north to south, which suggests they represent the southern continuation of the New England (Acadian) flysch basin. Whereas the Taconic orogeny apparently began in the south and migrated northward, the distribution of Devonian flysch suggests that Acadian orogeny may have begun in the north and migrated southward. The southern Appalachian foreland fold-and-thrust belt does not appear to have been greatly affected by flysch deposition and, instead, experienced miogeoclinal deposition of limestone, sandstone, and shale throughout orogeny (Figure 23.6). In other words, there is no structural or depositional evidence in the Southern Appalachian foreland for the existence of Acadian mountains to the east.

As the Acadian orogeny developed in the north, a westward-encroaching landmass deformed the flysch basin and produced a major Middle to Late Devonian foredeep molasse deposit known as the *Catskill Delta*. Similar to all molasse deposits, these rocks are primarily nonmarine river sandstones, conglomerates, and shales that overlie and interfinger with Acadian flysch deposits to the east. The Catskill Delta is thickest in the Catskill Mountains and in eastern Pennsylvania. The rocks thin southward into Virginia and westward into Ohio. This delta is larger and more coarse-grained than the Juniata-Queenston delta with a much greater volume of metamorphic and granitic pebbles, suggesting deeper erosion. The delta deposit is nearly 10,000 feet thick in the Catskill Mountains where the top of the rock unit has been eroded. Westward and southward thinning of the rock unit indicates that a major mountain range existed to the northeast during Late Devonian. The size and composition of the Catskill Delta suggests that the Acadian Mountain range was larger and higher than its forbearer, the Taconic Mountains.

With the exception of the modern Coastal Plain, a passive continental margin never reestablished itself in the northeast following Acadian orogeny. This suggests that mountains have existed in the Northern Appalachian region at least since the mid-Devonian (390 Ma) and possibly since the mid-Ordovician (470 Ma) assuming that some of the Taconic highlands persisted. Mountains likely also existed in the Southern Appalachians; however, they must have been located far to the east of the present-day Appalachian Mountains because, although there is depositional evidence of Taconic orogeny in the Southern Appalachian foreland in the form of the Sevier-Blountian clastic wedge, there is little evidence for Taconic or Acadian deformation. In other words, the Southern Appalachian foreland remained undeformed throughout Taconic and Acadian orogeny. The foreland fold-and-thrust belt had not yet formed. The Southern Appalachian miogeocline must have existed as a shallow inland seaway during at least Taconic orogeny with mountains located much farther east along the margin of Laurentia.

With the end of Acadian mountain building, and with the continued erosion of the Acadian landscape, deposition of the Catskill Delta waned. This allowed Mississippian limestones from the continental interior platform to once again transgress eastward, this time over the western part of the Catskill Delta. The limestone, however, did not transgress as far east as the Helderberg limestone following the Taconic orogeny. It was replaced, in the Middle Mississippian-Pennsylvanian, by a third cycle of molasse deposition associated with the Alleghany orogeny that accumulated across both the Valley and Ridge and Appalachian Plateau from Texas as far north as Pennsylvania. However, with few exceptions, molasse of this age is absent in New England and eastern New York. Apparently the New England-New York area did not feel the erosional effects of strong Alleghany deformation and mountain building.

South of New England there are three large Alleghanian molasse piles primarily on the Appalachian Plateau (Figures 23.4, 23.6). The Mauch Chunk-Pottsville clastic wedge is centered in Pennsylvania and West Virginia. In the north, it directly overlies the Catskill clastic wedge. The Pennington-Lee clastic wedge is centered in Tennessee, Kentucky, and West Virginia, and the Ouachita clastic wedge is centered in Arkansas, Mississippi, and Alabama. All are Middle Mississippian to Early Permian in age, and all are associated with the culminating collision of Gondwana (primarily Africa and South America) with Laurentia to form Pangea. The thickness and coarseness of these wedge deposits indicate the existence, for the third time, of high mountains to the east and south. The absence of a clastic wedge north of Pennsylvania is curious and suggests that high mountains did not exist in the Northern Appalachians at this time. It mirrors the absence of an Acadian clastic wedge in the Southern Appalachians. A favored explanation is that Alleghany collision in the Northern Appalachians was not head-on. Instead, it may have been a glancing blow with development of strike-slip faults rather than thrust faults.

There are few Paleozoic sedimentary rocks younger than Devonian anywhere in New England. One major exception is the nonmarine, molasse-like Narragansett Bay group near Newport, Rhode Island (Figure 23.4). This is a sequence of Pennsylvanian conglomerate, sandstone, coal, and shale that is more than 11,000 feet thick. It is not interpreted as a clastic wedge but more likely represents a pull-apart basin that formed during Alleghanian strike-slip faulting.

The Ouachita clastic wedge is enormous. It consists of more than 33,000 feet of deep-water flysch that grade upward into deltaic molasse deposits that merge eastward with the Pennington-Lee clastic wedge. These flysch deposits record the existence of a subduction zone south of the Ouachita Mountains that evolved into a mountain range with continental collision. The size of the wedge implies that the mountain belt was large, but unfortunately, the suture zone, and the eroded stump of the mountain range, are completely buried beneath rocks of the Atlantic marginal basin.

In summary, we can say that clastic wedge deposits in the US help define the three classic orogenic events: the Taconic, Acadian, and Alleghanian-Ouachita orogenies. In each case, the rocks indicate that a major land mass existed east of, or at the margin of, Laurentia and that the landmass propagated toward the craton. The age of each deposit suggests that landmasses appeared earlier in one part of the Appalachian belt versus another. Other indicators regarding the timing of orogeny, such as the age of volcanic and plutonic rock, the age of cross-cutting relationships, and the age of metamorphism, are found mostly in the hinterland. When these are taken into account, they collectively blur the distinction between three separate orogenic events and suggest instead that orogeny was occurring in one part of the Appalachians or another throughout the Paleozoic.

THE FORELAND FOLD-AND-THRUST BELT

The physiographic Valley and Ridge province forms the geological foreland fold-and-thrust belt in the Appalachian Mountains where Paleozoic miogeoclinal rocks have been pushed westward and northward onto the craton. The style and age of deformation vary from north to south. It is a narrow, thick-skinned, thrust belt of Taconic age in the Northern Appalachians where thrust faults continue into the hinterland. The Southern Appalachians, by contrast, host a classic, thin-skinned thrust belt that is nearly 200 million years younger than its counterpart to the north. In this area,

decollement thrusts of Alleghanian age stack sections of Cambrian through Mississippian miogeoclinal rock, one above another. The Central Appalachian foreland belt is different again in that it consists mostly of tight folds of Alleghanian age. Taconic deformation is present in the Central Appalachian foreland but is restricted to the eastern margin of the belt. Thus, it appears that Alleghanian foreland deformation dies out gradually to the north while Taconic foreland deformation dies out more abruptly to the south.

Some of the best evidence for Taconic-age foreland deformation is in the southern New England-New York area where a group of giant thrust slices, known as the Taconic Allochthons, sit like islands directly on top of deformed miogeoclinal and flysch assemblages (Figure 23.4). The allochthons extend for about 150 miles along the Vermont-New York border from just north of New York City to the area southwest of Middlebury, Vermont where they form the Taconic Mountains. A group of smaller thrust slices, known as the Hamburg Allochthons, are present in Pennsylvania in the Central Appalachian foreland. Interestingly, rocks in the allochthons are the same age as the miogeoclinal rocks on which they sit. The rocks consist of the Precambrian rift clastic succession originally deposited below the miogeocline, as well as Cambrian-Ordovician slope-rise rocks originally deposited oceanward of the miogeocline. These rocks were thrust westward a distance of more than 70 miles and placed above the deformed miogeocline. The time of thrusting is constrained to be late Middle Ordovician to Late Ordovician based on the age of associated flysch and the presence of weakly deformed Silurian-Devonian overlap assemblages (including Helderberg Group) that unconformably overlie deformed Ordovician beds.

The history of thrusting is more complicated than typical foreland thrust models that predict initial thrusting in the hinterland and, as the orogenic wedge builds, a progression to the foreland. If this were the case, slices derived from the east would be stacked above slices derived from the west (as in Figure 22.1e). Instead, the Taconic allochthons are stacked with the western slices near the top. The lowest fault slices consist of sedimentary rocks originally deposited on the continental slope and rise off the coast of Laurentia. The upper fault slices are metamorphic and composed of Precambrian rift clastic rock deposited originally on Grenville basement west of the lower slices. This implies that the eastern lower slices traveled the farthest either prior to metamorphism or in front of a wave of metamorphism. The higher western slices were thrust, out of sequence, above the eastern slices after metamorphism had already begun. Figure 23.7 shows schematically how this may have happened.

At one time the Taconic allochthons may have extended as a continuous sheet from Newfoundland southward at least as far as Pennsylvania. Today they sit as islands underlain by thrust faults surrounded by deformed flysch and miogeoclinal rock. They are separated from their root zone by the Green Mountain anticlinorium (Figure 23.4). Erosion removed both the allochthons and the underlying miogeoclinal rock from the crest of the Green Mountain anticlinorium, exposing Grenville crystalline shield rock. Part of the root zone to the Taconic allochthons appears to be along thrust faults now located on the east side of the Green Mountains within the Laurentian realm in Northern Vermont. These faults, marked *TRZ* in Figure 23.4, carry late Precambrian rift clastic and slope-rise rocks of the Underhill and Hoosac formations (now metamorphosed to schist). Farther south, in the vicinity of the allochthons, the root zone is buried below younger faults and thus is not exposed.

The influence of Taconic-age deformation in the foreland begins to diminish south of the Hamburg allochthons. In this area, Pennsylvanian-age rocks across the Valley and Ridge are deformed to the same degree as older rock. This type of relationship indicates that deformation must have occurred following Mississippian and Pennsylvanian deposition and, therefore, during Alleghany orogeny. Similar relationships farther south indicate that foreland fold-and-thrust deformation in the Southern Appalachian miogeocline is entirely Alleghanian in age.

The Southern Appalachian miogeocline is a classic foreland fold-and-thrust belt with flat-ramp-flat geometry and anticlines above ramps. There are as many as 10 major thrust faults across the foreland, none of which cut downward into crystalline rock. Instead, they flatten and root into the same basal decollement within Cambrian rock near the base of the miogeocline. This geometry implies that the thrust belt is thin-skinned in the foreland region. The foreland

fold-and-thrust belt forms the toe of a tectonic wedge that began to build far to the east during Taconic orogeny and slowly progressed westward, reaching the foreland Valley and Ridge during Alleghany orogeny. Thrust faults are well displayed in the Tennessee Valley and Ridge where hundreds of millions of years of erosion have exposed long, straight ridges of resistant rock between equally long valleys of nonresistant rock (Figure 12.10).

In Chapter 13 we discussed the existence of tectonic windows in the Blue Ridge where sedimentary layers are exposed below eroded thrust sheets of crystalline rock (Figure 13.9). These windows form some of the original evidence for large displacement along thrust faults in the Appalachians. However, nobody knew how far below the hinterland the thrust sheets extended until the late 1970s when a seismic reflection survey was conducted. The survey revealed the presence of nearly flat-lying sedimentary layers approximately 2.5 to 7 miles (4 to 10 km) below crystalline rock of the Blue Ridge and 7.5 to 11 miles (12 to 18 km) below the Piedmont. These deeply buried sedimentary rocks are understood to represent part of the Laurentian interior platform that was overthrust during Alleghany orogeny. These are the same sedimentary rocks that crop out in tectonic windows in the Blue Ridge. The significance of the seismic survey is that it indicates that the entire exposed Southern Appalachian orogenic belt (foreland and hinterland) has been thrust westward above the eastern margin of Laurentia a distance of more than 200 miles! In other words, the foreland fold and thrust belt basal décollement underlies the entire Southern Appalachian orogenic belt and probably at one time rooted directly into a subduction zone. The interpretation is depicted in a schematic cross-section shown in Figure 23.8. This figure also shows the major terranes and bounding faults in the Southern Appalachian hinterland discussed later in this chapter and can be compared with the tectonic map (Figure 23.4).

Alleghany-age thrust faults are far less numerous in the Central Appalachian miogeocline where folds form the dominant structure. This implies that thrust displacement diminishes to zero toward the north and is evidence that Alleghany-age deformation was not felt this far inland in the Northern Appalachians. It is consistent with other evidence suggesting that final collision was head-on in the south but only glancing with strike-slip faulting in the north. Alleghany-age folds are spectacularly displayed in the erosional landscape of the Pennsylvania Valley and Ridge where resistant layers form zigzag ridges that follow the limbs of folds (Figure 12.11).

The Northern Appalachians are sometimes contrasted with the Southern Appalachians as being thick-skinned in the sense that thrust faults dip steeply into the hinterland rather than extend seemingly indefinitely horizontally below the hinterland, as in the Southern Appalachians. However, prior to erosion, the Taconic allochthons would

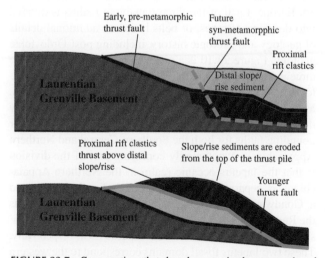

FIGURE 23.7   Cross-sections that show how proximal, metamorphosed rift-clastics are thrust above distal sedimentary slope/rise rocks in the Taconic allochthons. Note that distal slope/rise sediments have eroded from the top of the thrust pile.

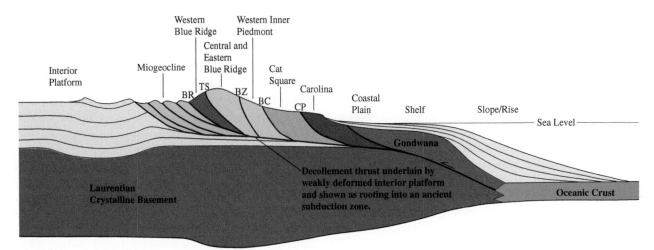

FIGURE 23.8 A cross-section of the Southern Appalachians showing a thin-skinned foreland fold-and-thrust belt and some of the major tectonic terranes and bounding faults. This figure suggests that a basal decollement lies below the entire orogenic belt and that the decollement roots into an ancient subduction zone beneath the Gondwana continent. *BR* = Blue Ridge thrust, *TS* = Taconic suture zone, *BZ* = Brevard zone, *BC* = Brindle Creek fault, *CP* = Central Piedmont shear zone.

have extended continuously back into the hinterland, possibly creating a thin-skinned situation similar to the present-day southern Appalachians. Later Acadian deformation steepened and overprinted the faults, thus destroying the thin-skinned geometry. Such a geometry may be preserved in the Southern Appalachians only because it was associated with the final phase of deformation.

THE FIVE APPALACHIAN REALMS

If one were to read older geological literature on the Appalachians, one would find this region to be divided into several tectonic zones and terranes, not all of which are applied to the entire orogenic belt. In the Northern Appalachians (primarily Canada) they include the Humber, Dunnage, Gander, Avalon, and Meguma zones. With the exception of the Meguma zone, all were named for rocks in Newfoundland. In the Southern and Central Appalachians they include the Western, Central, and Eastern Blue Ridge; the Western and Eastern Inner Piedmont; and the Carolina zone (or Outer Piedmont). Each is defined primarily based on the distribution of pre-Silurian rock, but a number of other factors, including stratigraphy, internal structure, and bounding faults, are also important. A division based on pre-Silurian geology works because it became apparent early on that many, if not most, of the Appalachian tectonic terranes consist of pre-Silurian rock. This is particularly true of terranes in the Southern Appalachians where Silurian-Devonian (Acadian) flysch is absent (Figure 23.4).

Fieldwork coupled with extensive isotopic dating in recent years have refined and, in some cases, redefined the older nomenclature, resulting in a broader genetic division applicable to the entire Appalachian belt. These new divisions are referred to as *geologic realms*. They are based on pre-Silurian

geology and therefore incorporate criteria used to define the tectonic zones of older literature. We will discuss these more recently devised realms and correlate them with tectonic zones of older literature. Throughout the discussion I will use names and definitions that seem to be in wide use among Appalachian geologists. An outline of this correlation is given in Figure 23.9.

Five geologic realms are defined based on the Late Cambrian-Early Ordovician paleogeography of Earth. They are the Laurentian continental realm, the Iapetus oceanic realm, the peri-Gondwana microcontinental realm, the Rheic oceanic realm, and the Gondwana continental realm. The Rheic and Gondwana realms are not exposed in the Appalachian Mountains and will not be discussed in detail. Both are present below sedimentary rocks of the Coastal Plain and both are exposed in Western Europe. Each of the three remaining realms is divided into domains, terranes, or belts based on additional details of geology and geologic history, including post-Ordovician geology. Figure 23.10 is a correlation chart that shows the three exposed realms in the US at the top of the figure followed by a tectonic interpretation regarding their origin. Major terranes discussed in this chapter are listed below the tectonic interpretation along with the names of major bounding faults for both the Southern-Central and Northern Appalachians. A potentially confusing part of the division is that the Iapetus oceanic realm in the Northern Appalachians is separated into a Laurentian fossil domain and a Gondwana fossil domain by a suture zone known as the Red Indian Line. For simplicity, I refer to these two fossil domains as Iapetus West and Iapetus East, respectively. These two Iapetus fossil domains correspond to the western Dunnage zone and the eastern Dunnage zone, respectively.

Note also in Figures 23.4 and 23.10 that rocks of the Ganderia passive margin, and rocks correlated with Iapetus

	Appalachian Realms				
	Laurentian Continental Realm	Iapetan (or Iapetus) Oceanic Realm	Peri-Gondwana Continental Realm	Rheic Oceanic Realm	Gondwana Continental Realm
Northern Appalachians	Humber	Dunnage	Ganderia Gander Avalon Meguma	Not Exposed	Not Exposed
Southern Appalachians	Western Blue Ridge	Central Blue Ridge Eastern Blue Ridge Western Inner Piedmont Eastern Inner Piedmont	Carolina (Outer Piedmont)	Not Exposed	Not Exposed

FIGURE 23.9 An outline of Appalachian hinterland realms and equivalent zones.

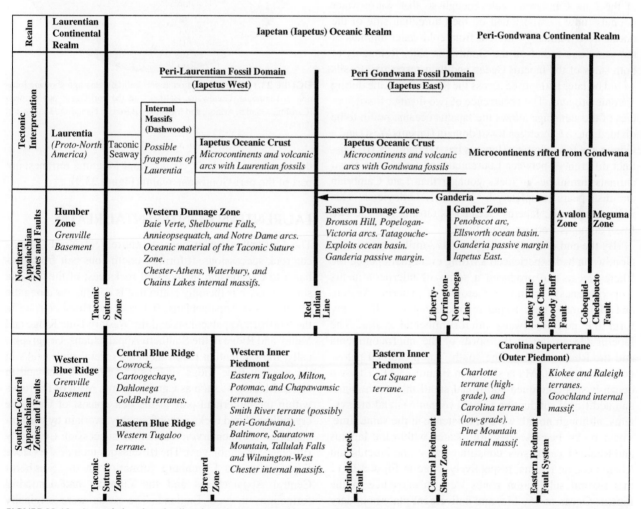

FIGURE 23.10 A correlation chart that lists the major terranes and bounding faults within the three exposed realms of the US Appalachian Mountains.

East, are present on both sides of the Liberty-Orrington-Norumbega Line (LO-NF). Rocks of the Ganderia passive margin and those of the Iapetus East are intermingled in the Northern Appalachians from the Red Indian Line to the Maine coast. I use the term Gander zone for the area

of Ganderia that lies east of the Liberty-Orrington Line, consistent with terminology in Newfoundland. Following a short discussion of Late Cambrian-Early Ordovician paleogeography, we will discuss the tectonics of each realm.

LATE CAMBRIAN-EARLY ORDOVICIAN PALEOGEOGRAPHY

By Early Cambrian, Rodinia had rifted into several continents including Laurentia and Gondwana, which were separated by the Iapetus Ocean (Figure 23.1). Rocks deposited on these continents form the Laurentian and the Gondwana continental realms, respectively. The Laurentian continent was close to the equator and oriented approximately 90 degrees clockwise from its present position such that the present-day East Coast faced southward. Gondwana was located close to the South Pole.

The Iapetus realm (also known as the Axial realm) consists of all rocks that formed within the Iapetus Ocean basin. The Iapetus Ocean was as much as 3,100 miles wide in the Late Cambrian, wide enough so that warm-water animals (now fossils) found on the Laurentian side of the ocean were markedly different from cold-water animals of the same age found on the Gondwana side. Remnants of both sides of the Iapetus Ocean and their respective fossils would be later juxtaposed across the Red Indian Line during Taconic orogeny. The occurrence of two distinct fossil species of the same age allows the Iapetus oceanic realm to be divided into a Laurentian fossil domain (Iapetus West) and a Gondwana fossil domain (Iapetus East) (Figure 23.10). Distinct differences between Laurentian and Gondwana fossil animals diminishes in rocks younger than Late Cambrian and disappears completely by late Early Ordovician. This indicates that the Iapetus Ocean was closing and the two fossil domains were intermingling.

By the end of Early Ordovician 472 million years ago, Gondwana had experienced a rifting event that opened the Rheic Ocean and produced a series of microcontinents (superterranes), including Carolina, Ganderia, Avalon, and Meguma. These island microcontinents collectively form the peri-Gondwana microcontinental realm. The Iapetus Ocean lay to the north of the microcontinents and the Rheic Ocean to the south. The presumed paleogeography at 460 Ma is shown in Figure 23.11. The two ocean basins contained additional small microcontinents and additional subduction-related volcanic island arc systems, although not all were in existence at the same time. These rocks, because they are located within the Iapetus and Rheic Ocean basins, constitute part of the Iapetus and Rheic oceanic realms, respectively. Note in Figure 23.11 that several subduction zones were in existence in the Iapetus Ocean at 460 Ma, and a divergent plate boundary was in existence in the Rheic Ocean. This was the time of Taconic orogeny.

Thus, we can summarize and say that all pre-Silurian rocks that formed on the Laurentian continent belong to the Laurentian realm. All rocks that formed within the Iapetus Ocean basin belong to the Iapetus realm, and rocks that form major tectonic terranes in the Appalachians, including

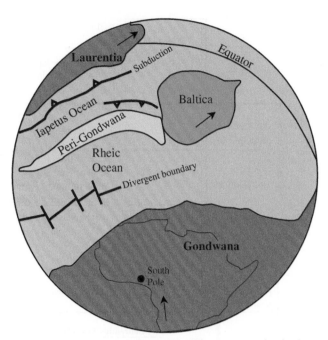

FIGURE 23.11   Reconstruction at 460 million years ago showing location of Laurentia, Gondwana, Baltica, and the peri-Gondwana terranes (Carolina, Gander, Avalon, and Meguma). Based on Nance and Linnemann (2008).

the Carolina, Ganderia, Avalon, and Meguma terranes, are part of the peri-Gondwana realm (Figure 23.9).

LAURENTIAN CONTINENTAL REALM

The Laurentian realm consists entirely of North American rock successions. It forms a continuous belt from Alabama to Canada and includes all rocks east of the Taconic suture zone. It is the only realm that is continuous along the length of the Appalachians. It forms the Humber zone in the Northern Appalachians and the Western Blue Ridge and Valley and Ridge in the Southern Appalachians. Geographically, the Laurentian realm includes several high peaks in the western Blue Ridge such as Clingmans Dome and in Green Mountains such as Camels Hump. Most of the rocks within the hinterland part of the belt consist of Grenville crystalline shield rocks unconformably overlain by the Precambrian sedimentary/volcanic rock succession (primarily rift-clastic assemblages). The rift-clastics include the Ocoee Supergroup and Lynchburg formation in the Southern-Central Appalachians and the Pinnacle and Underhill formations in New England. The rocks were originally sandstone but are now metamorphosed to schist. A few areas contain inliers of variably metamorphosed Cambrian-Ordovician miogeoclinal rock, particularly in the Southern-Central Appalachians.

The entire Laurentian hinterland experienced deformation and metamorphism circa 450 to 460 Ma during the Taconic orogeny producing an assortment of gneiss, schist, and amphibolite, particularly in the deeper-seated eastern part of the belt. Field data suggest that several of the Taconic-age thrust faults pre-date the metamorphism. This includes some of the Taconic allochthon faults in the Northern Appalachians, and the Hayesville fault in the Southern Appalachians which forms part of the Taconic suture zone. In addition to strong Taconic metamorphism, the Western Blue Ridge was also weakly affected by circa 360 Ma Neoacadian metamorphism and by Alleghany-age thrust and strike-slip displacement that permeates the Valley and Ridge. The Northern Appalachian Laurentian realm, by contrast, experienced thrusting and metamorphism during Taconic orogeny but was largely unaffected by post-Taconic events.

There are several distinctive thrust slices within the Laurentian realm west of the Taconic suture zone that are transitional between the Laurentian and Iapetus realms. They include the Talladega belt, the Westminster terrane, and the Hamburg and Taconic allochthons. We have already discussed the Hamburg and Taconic allochthons.

The Talladega belt consists of Precambrian sedimentary/volcanic rocks overlain by Ordovician to Mississippian deep-water sandstone, siltstone, conglomerate, and volcanic rocks metamorphosed to greenschist facies. It likely existed as a basin along the outermost (oceanward) part of the Laurentian miogeoclinal shelf. $^{40}Ar/^{39}Ar$ dates on muscovite indicate that the rocks were metamorphosed to low-grade in the Late Mississippian between 334 and 320 Ma. The Talladega belt, like the Southern Appalachian miogeocline, apparently was not deformed until Alleghany orogeny.

Rocks of the Westminster terrane consist of low-grade, Late Precambrian-Cambrian phyllite, schist, quartzite, metagreywacke, carbonate, and greenstone, interpreted to have been deposited in deep water on the continental slope-rise just outboard of the Laurentian miogeocline. The Westminster terrane is bound on the west by the Martic fault and on the east by the Taconic suture zone but is, itself, strongly deformed with many internal thrust faults. $^{40}Ar/^{39}Ar$ dates on biotite indicate an Early Silurian (circa 430 Ma) Salinic metamorphism in the western part and a Late Devonian (circa 370 Ma) Neoacadian metamorphism in its eastern part. It is probable that the Westminster terrane, as a whole, was emplaced against Laurentia during one or both of these metamorphic stages although bounding faults may also have been active as strike-slip faults during Alleghany orogeny.

INTERNAL MASSIFS

There are a number of exposures of strongly deformed, high-grade metamorphic Laurentian or Laurentian-like rocks east of the Taconic suture zone primarily within the Iapetus Oceanic Realm but also in the Carolina (peri-Gondwana) realm. All are surrounded by accreted terranes. Most consist of Grenville-age crystalline rock, in some cases overlain by Precambrian to Ordovician schist and gneiss (Laurentian rift clastic and miogeoclinal assemblages). Collectively, they are referred to as *internal massifs*. Nine are shown in Figure 23.4. They are the Chain Lakes massif in the Boundary Mountains on the Maine-Quebec border, the Chester-Athens dome in Vermont, the Waterbury dome in Connecticut, the Baltimore dome in Maryland, the Wilmington-West Chester complex in Delaware and Pennsylvania, the Goochland terrane in Virginia, the Sauratown Mountain Window in North Carolina, the Tallulah Falls dome in Georgia, and the Pine Mountain Window in Georgia and Alabama. Their origin is uncertain. Three possible interpretations are shown in Figure 23.12. They are as follows: (1) The internal massifs were rifted from Laurentia during breakup of Rodinia to become a string of small islands in the Iapetus Ocean that later collided with Laurentia as accreted terranes (that is, they are part of Dashwoods). (2) They were part of a Laurentian Grenville basement that was overthrust by accreted terranes. Younger thrust faults then cut through the basement to carry a sliver of Grenville (and overlying) rocks to the surface. (3) They are a part of Laurentia that was overthrust by accreted terranes and then folded and eroded, thereby exposing Grenville basement in structural windows. Part of the Canadian Dashwoods island chain is believed to have extended at least to the New England Appalachians; therefore, the first explanation may be correct for at least some of the internal massifs. The other two explanations may be valid for some of the massifs located farther south. Of course, it is also possible that some of the internal massifs were never part of Laurentia and therefore are true accreted terranes.

IAPETUS OCEANIC REALM

The Iapetus (Iapetan, or Axial) realm forms the high-grade metamorphic core of the Appalachian Mountains. In the United States it is a multiply deformed, strongly metamorphosed, intruded region that is poorly exposed relative to Canada. In New England, pre-Silurian rocks affected by Taconic orogeny are mostly hidden beneath younger rocks of the Silurian-Devonian (Acadian) tectonic unit that were themselves deformed and metamorphosed during Acadian orogeny (Figure 23.4). Extrapolation of Canadian geology to New England is straightforward in most instances but is less certain in the Southern Appalachians where differences in geologic history and correlation are more pronounced. The Iapetus realm corresponds with the Dunnage zone in the Northern Appalachians, and with the Central and Eastern Blue Ridge and Western and Eastern Inner Piedmont in the Southern-Central Appalachians (Figure 23.10). As noted previously, the Ganderia passive margin is distributed across the Iapetus realm east of the Red Indian Line in the Northern Appalachians.

(a) A fragment of Laurentian basement is rifted during breakup of Rodinea. The fragment later collides with Laurentia as an accreted terrane.

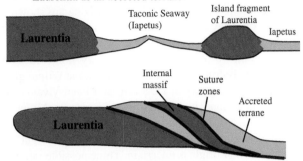

(b) Laurentia is overthrust by accreted terranes. A younger thrust fault cuts up through the accreted terrane to carry an internal massif to the surface.

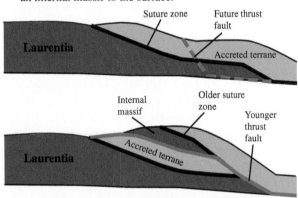

(c) Laurentia is overthrust by accreted terranes and then folded thereby exposing an internal massif within a structural window.

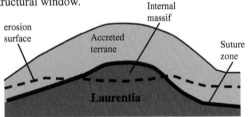

FIGURE 23.12 A schematic series of cross-sections that show three possibilities for the origin of internal massifs.

The Iapetan realm consists of dismembered sections of accretionary prism, slope-rise, and oceanic rocks (ophiolite) as well as microcontinents and volcanic arc complexes, all of which once existed within the Iapetus Ocean basin. These rocks are now metamorphosed to schist, gneiss, amphibolite, and a small amount of eclogite near Bakersville, North Carolina.

Deformation across the Iapetan realm is very complex and not completely understood. Through years of study, a number of different tectonic terranes, areas, or structures have been defined and then redefined or even abandoned. This has resulted in sometimes confusing nomenclature. Many of the tectonic terranes defined as part of this realm

have seemingly experienced different deformation, metamorphic, and intrusive histories.

In the Northern Appalachians, rocks of the Iapetus realm extend from the Taconic suture zone, across the Liberty-Orrington (Norumbega) Line, to the Maine coastline and to the Honey Hill-Lake Char-Bloody Bluff Fault in Massachusetts. It is divided by the Red Indian Line into Iapetus West (the Laurentian fossil domain or Notre Dame sub-zone) and Iapetus East (the Gondwanian fossil domain or Tetagouche-Exploits subzone). The Red Indian Line can be traced through north-central Maine to the northern tip of New Hampshire and southward along the Connecticut River Valley. It cannot be traced farther south and is absent in the Southern Appalachians (Figure 23.4).

On the basis of radiometric dating we can say that the Northern Appalachians have a more ordered arrangement of deformation and metamorphism relative to the Southern-Central Appalachians. The Northern Appalachian metamorphic core has stepped eastward during successive orogeny. Taconic-Salinic deformation and metamorphism dominate western New England along the Green Mountain anticlinorium and within the Taconic suture mélanges. Acadian events dominate central and eastern New England where a prominent area of granulite facies rocks extends from southern Maine across central New Hampshire and Massachusetts to eastern Connecticut surrounded by granitic intrusions and amphibolite facies rocks. This metamorphism, along with a wide area of greenschist facies and low-grade rocks across central and northern Maine, is attributed to the Acadian orogeny because rocks of Silurian and Devonian age are metamorphosed. These rocks form the wide Silurian-Devonian (Acadian) rock unit on the tectonic map (Figure 23.4) that buries and hides much of the evidence for Taconic-Salinic orogeny in central and eastern New England. Strong Alleghany effects are restricted to a small area in eastern Connecticut, Rhode Island, and eastern Massachusetts that includes greenshist to (rarely) granulite facies rocks.

The Iapetus realm in the Southern-Central Appalachians extends from the Taconic suture zone in the northwest to the Central Piedmont Shear zone in the southeast. This region is more difficult to decipher because nearly all the rocks are Ordovician or older, and Taconic, Neoacadian, and Alleghanian orogenic effects completely overlap. There remains much controversy regarding the timing of collision for many of the terranes. With the exception of the Central Blue Ridge terranes, the primary metamorphism within the Iapetus realm appears to be Neoacadian. This younger Acadian age for metamorphism suggests that Acadian orogeny began in the Northern Appalachians and, over time, migrated southward consistent with depositional evidence in the foreland that includes the absence of an Acadian foredeep in the south. There is evidence that Neoacadian metamorphism and deformation overprints Taconic metamorphism and deformation and, in some areas, overprints Grenville or Penobscot metamorphism and deformation. All of the rocks were later shuffled along thrust and strike-slip faults related to the Alleghany orogeny, which further complicates the situation. There is evidence in many areas for Alleghany metamorphic overprint although this is stronger within the Carolina zone than within the Iapetus realm. Taconic deformation and metamorphism appear to be dominant primarily in the Central Blue Ridge terranes. The overall result is that the Southern Appalachian Iapetan realm, together with parts of the adjacent Western Blue Ridge and Carolina zones, forms one of the largest regions of high-grade metamorphic rock in the world. We will discuss the two Iapetan realm fossil domains separately.

Laurentian Fossil Domain in New England (Iapetus West)

Cambrian-Ordovician rocks that define Iapetus West crop out in northern Vermont just east of the Taconic suture zone near Montpelier. The Taconic suture melanges (*Ts* in Figure 23.4) are part of the Rowe, Stowe, and Moretown formations and consist of schist with interspersed lenses of serpentinite, peridotite, meta-volcanics (ophiolite), and plutons thought to represent a collage of accretionary prisim, ocean basin, slope-rise, and forearc basin sediment. Rocks that form the Laurentian Arcs and Ocean Basins tectonic unit in Figure 23.4 crop out to the east of the Taconic melanges in a narrow belt that can be followed northward through Canada to the northwestern corner of Maine, and southward the length of Vermont into Massachusetts and Connecticut where it is known as the Shelburne Falls volcanic arc complex (*SF* in Figure 23.4). The only other location in the US where these rocks crop out is in the Boundary Mountains near Jackman, along the Quebec-Maine border. This is also the only location in the US where the Red Indian Line suture zone is exposed (Figure 23.4). Elsewhere, the rocks are hidden below younger Silurian-Devonian (Acadian) rocks within the Connecticut Valley-Gaspe synclinorium.

The Shelburne Falls volcanic arc complex consists of Barnard schists and meta-volcanics, and the Collinsville and Hallockville Pond granitic gneiss. Radiometric dating suggests that the arc was active between about 485 and 470 Ma (latest Cambrian-Early Ordovician) and that it collided with Laurentia between about 470 and 450 Ma during Taconic orogeny. The age and tectonic setting suggests that the Shelburne Falls volcanic arc may correlate with the Notre Dame arc exposed within the Notre Dame subzone in Newfoundland. An older volcanic arc, the Baie Verte arc (507 to 490 Ma), and a slightly younger volcanic arc/oceanic tract, the Annieopsequotch terrane (480 to 462 Ma), are also present in the Notre Dame subzone of Newfoundland with the latter situated immediately west of the Red Indian Line. Poorly exposed and mostly circumstantial evidence suggests both terranes may extend into New England.

The Boundary Mountains host the enigmatic Chain Lakes (internal) massif (*CL* in Figure 23.4), which crops out just west of the Red Indian Line. The rocks consist of strongly metamorphosed, partially melted, quartzo-feldspathic gneiss and minor amphibolite. The origin of the massif is something of a mystery, but detrital zircon analysis suggests a Laurentian source area, implying that it may have been one of the Laurentian Dashwoods microcontinents that separated during initial breakup of Rodinia. Although highly metamorphosed now, the rocks may have originally been deposited in a basin in close proximity to both a volcanic arc and to Laurentia.

Gondwana Fossil Domain in New England (Iapetus East)

Iapetus East in the New England Appalachians consists of Ordovician and older rocks located between the Red Indian Line and the Honey Hill-Lake Char-Bloody Bluff Fault. The rocks are found on both sides of the Liberty-Orrington Line (also known as the Dog Bay Line). Across part of Maine, the Liberty-Orrington Line is offset and replaced by a relatively young bedrock structure, the Norumbega Fault. The Norumbega Fault was likely active as a right-lateral strike-slip fault in mid-Devonian and has reactivated intermittently throughout the late Paleozoic and Mesozoic. Fission-track dates on the west side of the fault zone are between 113 and 89 Ma, whereas those east of the fault zone are between 159 and 140 Ma. This age discontinuity has been interpreted to suggest final reactivation during the Late Cretaceous at or soon after 89 Ma, with vertical, east-side-down displacement.

Most of the pre-Silurian rock within Iapetus East is hidden beneath younger deformed and metamorphosed Silurian-Devonian cover rocks that form the Merrimack-Kearsarge-Central Maine-Aroostook-Metapedia synclinorium. Pre-Silurian Iapetus realm rocks crop out primarily along the Bronson Hill Anticlinorium (*BHA*) just east of the Red Indian Line in central New England. From Connecticut to New Hampshire these rocks form the Bronson Hill volcanic arc and back-arc basin complex (*BH* in Figure 23.4). The volcanic arc was active between about 478 and 454 Ma and includes the Oliverian Plutonic suite (granitic gneiss), the Ammonoosuc metavolcanics (amphibolite, greenstone, and schist), and the Partridge Formation (slate, schist, metagreywacke). Radiometric dating and cross-cutting relationships suggest it collided with Laurentia during Salinic orogeny sometime after 454 Ma. It is probably correlative of the Popelogan-Victoria arc assemblage (*P*) and the Tatagouche-Exploits ocean basin (*T*), both of which crop out along strike in Maine and Canada. The Middle to Late Cambrian (513 to 486 Ma) Ellsworth-Penobscot terrane (*E*) is exposed east of the Liberty-Orrington-(Norumbega) Line. In the US, these rocks are hidden below the younger

Silurian-Devonian tectonic unit in the area west of the Liberty-Orrington-(Norumbega) Line but are exposed in Canada. We have already noted that rocks of the Ganderia passive margin are present across all of Iapetus East.

Iapetus Realm Rocks of the Southern–Central Appalachians

In the Southern and Central Appalachians the Iapetan realm is bound on the west by the Taconic suture zone and on the east by another major suture zone known as the Central Piedmont Shear zone (Figures 23.4, 23.8). The area between these two faults includes the Central and Eastern Blue Ridge, the Western Inner Piedmont, and the Eastern Inner Piedmont. The rocks represent a variety of depositional settings including slope-rise and ocean basin (ophiolite) just offshore of the Laurentian continent, island volcanic arc complexes, and possibly rafted Laurentian and Gondwana microcontinents. A remarkable aspect is that each terrane shows a rather unique deformational and metamorphic history. Radiometric dates show that Taconic-age metamorphism dominates in a few terranes, whereas Neoacadian or even Alleghanian metamorphism dominates in others. Much of the area was weakly overprinted with Alleghany metamorphism (335 to 325 Ma). Several terranes have multiple internal thrust faults across which the age and style of deformation and metamorphism are different. This feature suggests that these terranes may actually be composed of several smaller terranes that were either amalgamated into a superterrane in the Iapetus Ocean before colliding with Laurentia, or were juxtaposed following collision. The area is poorly exposed, intensely metamorphosed, and with few fossils. This makes correlation with the Northern Appalachians somewhat tenuous. Additionally, the region is imbricated along Alleghany-age thrust and strike-slip faults, a feature mostly absent from the Iapetan realm to the north. Because of a general absence of fossils, it is not possible to definitively distinguish between Iapetus West and Iapetus East fossil domains. Detrital zircon analysis and isotopic studies suggest that most of the Iapetus realm terranes had a Laurentian (Grenville basement) source and thus were close to the Laurentian continent during deposition. This would imply that they are part of Iapetus West. One possible exception is a narrow belt of rock known as the Smith River terrane which is discussed shortly (Figure 23.4).

The westernmost group of terranes in the Iapetan realm is collectively referred to as the Central Blue Ridge (*CB*, Figure 23.4). These rocks contain abundant ophiolitic lenses; therefore, they are tentatively shown as correlative with the Taconic suture mélanges in the Northern Appalachians (*TS*, Figure 23.4). They include the Cowrock, Cartoogechaye, and Dahlonega Gold Belt terranes. All three consist of Late Precambrian to Ordovician meta-sandstone, schist, gneiss, and ophiolite metamorphosed to high-grade and locally intruded by Middle Ordovician (Taconic) granitic gneiss. Detrital zircon analysis suggests that much of the source rock of all three terranes was from erosion of Grenville basement. This implies that all three originated close to the Laurentian margin. Similar to the Taconic suture mélanges, radiometric dating indicates that metamorphism and probable accretion with Laurentia is Taconic in age (470 to 450 Ma).

The Blue Ridge Escarpment forms the physiographic boundary between the Blue Ridge Mountains and the Piedmont Plateau (Chapter 13). In the Southern Appalachians, the geologic boundary between the Blue Ridge and Piedmont is often considered to be a straight, narrow lineament known as the Brevard zone, which does not coincide exactly with the Blue Ridge Escarpment. The Brevard zone lineament is clearly evident between Asheville, North Carolina, and Atlanta, Georgia in the Google Earth image shown in Figure 23.13. A single, rather large terrane known as the Tugaloo terrane (T, Figure 23.4) occupies both the physiographic Eastern Blue Ridge and the Western Inner Piedmont on both sides of the Brevard Fault and on both sides of the Blue Ridge Escarpment (Figure 23.4). Included in this terrane is the highest mountain in the Appalachians, Mt. Mitchell, located in the Eastern Blue Ridge near Asheville.

The Brevard zone is one of the most famous and enigmatic faults in the Appalachians. It is not a suture zone because similar rocks of the Tugaloo terrane are present on both sides. It may have originally been a thrust fault. There is evidence of multiple phases of both high-temperature and low-temperature fault displacement. The Brevard Fault was active as a strike-slip fault during both the Late Devonian-Early Mississippian Neoacadian orogeny and the Pennsylvanian-Early Permian Alleghany orogeny. Later it was reactivated as a high-angle fault with vertical motion (perhaps similar to strike-slip faults in the California Coast Ranges). The combination of steep dip and weak fractured rock produce the topographic lineament, which is about 465 miles long.

The Brevard Fault zone dies out north of Mt. Mitchell in the Central Appalachians. Here the Eastern Blue Ridge-Western Inner Piedmont is occupied by several terranes in addition to the Tugaloo terrane, including the Chopawamsic (*C*, Figure 23.4), Milton (*M*), Smith River (*S*), and Potomac (*P*) terranes. All of the terranes consist of complexly deformed, Late Precambrian-Ordovician high-grade schist, gneiss, amphibolite, ophiolite, and migmatite. Note in Figure 23.3 that most of these terranes are interpreted as volcanic arc/ocean basin tracts that correlate with the Northern Appalachian Shelburne Falls arc. It is difficult to say with certainty if this correlation is entirely accurate. The presence of ophiolite along the Taconic suture zone within the Tugaloo terrane in the Eastern Blue Ridge, for example, suggests that some of the rocks may correlate more directly with the Taconic suture mélanges of the northern Appalachians

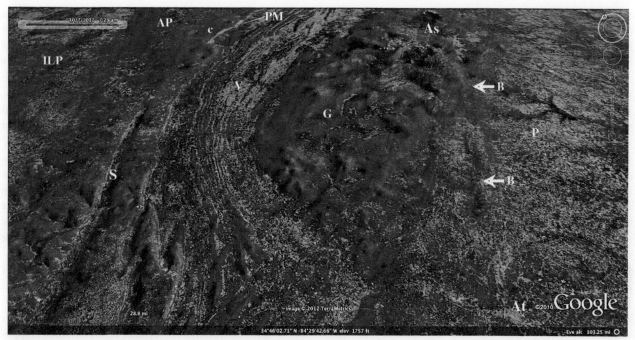

FIGURE 23.13   A Google Earth image looking northward at the southern end of the Blue Ridge Mountains. *AP* = Appalachian Plateau, *As* = Asheville, NC, *At* = Atlanta, GA, *B* = Brevard zone lineament (shown with arrows), *c* = Cumberland Mountains, *G* = Great Smoky Mountains (Blue Ridge), *ILP* = Interior Low Plateaus, *P* = Piedmont Plateau, *PM* = Pine Mountain thrust fault, *S* = Sequatchie anticlinal valley, *V* = Valley and Ridge.

rather than with the Shelburne Falls arc. However, unlike the Taconic metamorphism found in the Central Blue Ridge terranes, the age and intensity of metamorphism across the Eastern Blue Ridge-Western Inner Piedmont vary from one terrane to the next. The Tugaloo terrane contains evidence of both Taconic and Neoacadian metamorphism. The strongest and best-preserved metamorphism across several of the terranes, including the eastern Tugaloo and Milton terranes, is Neoacadian (circa 350 to 360 Ma), followed by weak Alleghany metamorphism (circa 325 Ma). A late Alleghany greenschist-amphibolite facies metamorphism seems to characterize the Chopawamsic terrane. It is clear that many of the terrane-bounding faults were active (or reactivated) during the Neoacadian and Alleghany orogeny.

All the Eastern Blue Ridge-Western Inner Piedmont terranes contain Taconic-Salinian (Ordovician-Silurian) intrusions. A few, including the Tugaloo and Milton terranes, contain Alleghanian (Mississippian-Permian) intrusions, and the Tugaloo also contains Neoacadian (Devonian-Mississippian) intrusions. None of this evidence tightly constrains the timing of collision with Laurentia. Given the widespread occurrence of Ordovician-Silurian plutons, the prevailing hypothesis suggests that all of the terranes collided with Laurentia during Taconic orogeny. If this is the case, the hypothesis suggests that the bounding faults were then reactivated and the rocks strongly metamorphosed (or re-metamorphosed) during Neoacadian orogeny as a result of collision and thrusting of the Carolina terrane above the now expanded Laurentian continental margin. Additional

shuffling of terranes along faults, as well as remetamorphism, occurred during Alleghany orogeny associated with final collision of Gondwana (Africa). The evidence, however, does not rule out the possibility that some of the terranes accreted sometime after the Taconic orogeny.

As an example of the geologic complexity in the Central-Southern Appalachians, we can take a closer look at the Smith River and Potomac terranes, both of which show unique aspects to their metamorphic history that do not fit very well into the preceding hypothesis.

Detrital zircon ages suggest that the Smith River terrane (also referred to as an *allochthon*) received sediment from erosion of the Laurentian continent; however, unlike other terranes, the metamorphic phases are dated as Early Cambrian (circa 530 Ma) and Late Cambrian-Early Ordovician (circa 480 to 490 Ma). Neoacadian metamorphism apparently is absent. The older metamorpic age may correspond with the Penobscot orogeny, whereas the younger age could be a very early stage of Taconic orogeny. Alternatively, it could be a continuous metamorphism. In any case, the early metamorphism could not have occurred as a result of collision with Laurentia because Laurentia had a passive margin until at least 480 Ma. If the early metamorphism is indeed associated with Penobscot orogeny, the Smith River terrane could have originated in the peri-Gondwana (Iapetus East) realm as suggested in Figure 23.3. Exactly when the Smith River terrane collided with Laurentia is not known. Collision could have occurred in the Late Ordovician during Taconic orogeny, or alternatively, the absence of a Neoacadian metamorphism coupled with weak Mississippian-age metamorphism (335 to 325 Ma) allows the possibility that collision could have been delayed until Alleghany orogeny.

The orogenic history of the Potomac terrane is equally unique. Schist, gneiss, and ophiolitic rock in one part of the Potomac terrane show strong Early Ordovician Taconic-age metamorphism (circa 475 Ma), whereas other areas show only Early Devonian Acadian-age metamorphism (circa 400 Ma). Here again the timing of collision is not well constrained. The entire terrane could have collided with Laurentia during Taconic orogeny; alternatively, it could have amalgamated offshore during Taconic orogeny and collided with Laurentia anytime during later orogenies.

Not included in this discussion is the Eastern Inner Piedmont terrane which is present only in the Southern Appalachians and is occupied by a single unique terrane known as the Cat Square. This unit lies between the Brindle Creek Fault and the Central Piedmont Shear zone. Detrital zircons are as young as 430 Ma, which indicates that the Cat Square was a depositional basin less than 430 million years ago. Radiometric dating indicates that the rocks were metamorphosed to high-grade schist and gneiss, and intruded by plutons, during Acadian-Neoacadean orogeny between 380 and 350 Ma. Thus, the Cat Square depositional basin must have existed between 430 and 380 Ma (similar in age to Silurian-Devonian flysch basins in the Northern Appalachians). The rocks were apparently deposited during Salinic orogeny and, therefore, are younger than rocks in the Western Inner Piedmont. Another important aspect is that a significant fraction of detrital zircons is older than the typical 1-billion-year-old Laurentian (Grenville) source terrane, implying that the Cat Square terrane may have received sediment from a peri-Gondwana source. There are two prevailing interpretations for the origin and subsequent deformation of the Cat Square terrane, both of which involve the Carolina superterrane. They will be discussed in the following section.

PERI-GONDWANA MICROCONTINENTAL REALM

There are four peri-Gondwana microcontinents: Ganderia, Avalon, and Megma in the Northern Appalachians and Carolina in the Southern Appalachians. Each is believed to represent a superterrane composed of continental, oceanic, or island arc complexes assembled prior to collision with Laurentia. In New England, the Gander zone is located between the Liberty-Orrington Line and the Honey Hill-Lake Char-Bloody Bluff Fault, but relics of the supercontinent (the Ganderia passive margin, Figure 23.4) are also found in Iapetus east. The two areas together constitute Ganderia. Avalon is restricted to the area east of the Honey Hill-Lake Char-Bloody Bluff Fault surrounding Boston. Meguma lies offshore east of Avalon. It is not exposed in the US but crops out in Nova Scotia. The Carolina superterrane occupies the entire region southeast of the Central Piedmont shear zone. Avalon was once thought to be equivalent with Carolina, but more recent thinking suggests the two may have evolved independently.

Volcanic evidence primarily in Canada suggests that the Avalon microcontinent, along with the Ganderia and Carolina superterranes, were rifted from the Gondwana mainland by Early Ordovician. This is the rifting event that opened the Rheic Ocean and created the peri-Gondwana microcontinents as shown in Figure 23.11. Laurentia at this time was a passive margin. By Late Ordovician, the Taconic orogeny had begun and Gondwana fossil animals were mixing with Laurentian fossil animals. This implies that the Iapetus Ocean was shrinking at the expense of a widening Rheic Ocean and that the peri-Gondwana microcontinents were approaching Laurentia.

Carolina Superterrane

The Carolina superterrane is the only well-defined peri-Gondwana microcontinent exposed in the Southern Appalachians. With the exception of internal massifs such as the Goochland terrane and the Pine Mountain window, it covers all of Southern Appalachia east of the Central Piedmont Shear zone. The superterrane is an amalgamation of as many as 15 smaller terranes, many with variable depositional-tectonic histories. Overall, the rocks consist of upper Precambrian to lower Ordovician volcanic arc, accretionary prism, and ocean basin assemblages with Gondwana fossils. Given that Carolina amalgamated primarily from oceanic rock, the microcontinent may have been an oceanic plateau rather than a true fragment of Gondwana. A strong metamorphism occurred in the Late Precambrian-Early Cambrian (617 to 530 Ma) associated with folding, faulting, and plutonic intrusion, and this is the dominant metamorphism in several areas. On the basis of radiometric dating and cross-cutting relationships we can suggest that much of the deformation and metamorphism occurred over a short interval between 557 and 535 Ma. This was the Virgilinan orogeny and it is responsible for final assembly of the Carolina superterrane. Assembly apparently occurred far from the Laurentian mainland while Carolina was attached to or near the larger Gondwana continent. Later orogenic events affected various parts of the Carolina superterane, as discussed below.

The Carolina superterrane is divisible into belts of low-grade and high-grade metamorphism. Three belts from west to east are the high-grade Charlotte Belt, the low-grade Carolina Slate Belt, and a belt of mixed high- and low-grade rock along the eastern margin that includes the Raleigh and Kiokee Belts (Figure 23.4). The eastern belt is partially hidden beneath younger rock of the Coastal Plain. The contact between the Charlotte Belt and the Carolina Slate Belt is depositional along part of its length and a fault along other parts. Both belts are separated from the Raleigh and Kiokee Belts by several faults collectively referred to as the Eastern Piedmont fault system. Internal deformation within each belt is complex and variable. High-grade rocks include gneiss, schist, amphibolite, and minor ophiolite (Mocksville and Burks Mountain complexes, Raleigh and Lake Murray gneisses, Battleground and Blacksburg formations) with Virgilinan, Salinic, and Alleghanian intrusions. Low-grade rocks consist of silicic volcanic, pyroclastic, and volcanoclastic rocks, basalt, sandstone, and mudstone (Virgilina, Albemarle, South Carolina, and Cary sequences) with Alleghanian intrusions.

The Virgilinan orogeny and assembly of the Carolina superterrane was followed in the Middle Cambrian-Early Ordovician by deposition of clastic rock along a passive continental margin (Asbill Pond formation). Late Ordovician-Early Silurian $^{40}Ar/^{39}Ar$ ages on micas in low-grade rocks, and on hornblende in high-grade rocks, suggest that at least part of the Carolina superterrane was affected by Taconic-Salinic metamorphism. On this basis, it has been argued that the previously assembled Carolina superterrane was accreted to Laurentia by the close of the Taconic orogeny.

The timing of accretion, however, is controversial and is one of several unsolved mysteries of the Appalachians. There is also evidence in the Carolina superterrane for minor right-lateral strike-slip faulting, deformation, and metamorphism during Middle-Upper Devonian Acadian-Neoacadian orogeny (391 to 358 Ma). On this basis, and on the basis of strong Neoacadian metamorphism in the underlying Inner Piedmont zones including the Cat Square terrane, it has been suggested that the Carolina superterrane was accreted to Laurentia during Neoacadian orogeny. The idea, as mentioned previously, is that the Carolina superterrane overthrust the Inner Piedmont during collision, causing metamorphism in the underlying rocks.

Still others have argued for an Alleghany collision. This hypothesis is based on the presence of Allegheny-age strike-slip faulting, thrust faulting, folding, metamorphism, and plutonism across all of Southern Appalachia, including the Carolina superterrane. The Alleghany orogeny was the first major event to have affected both Carolina and the Inner Piedmont. Orogeny could have resulted from Carolina-Laurentia collision or, alternatively, from final collision of Gondwana against an already accreted Carolina terrane.

Any collision scenario must account for the presence of Neoacadian and Alleghanian thrust and strike-slip faults across Southern Appalachia. These faults, perhaps more than anything else, have altered or hidden earlier collision-related structures, in large part creating the controversy. The Central Piedmont Shear zone itself is an Alleghanian (circa 330 Ma) thrust fault that has carried the Carolina zone more than 20 miles above the Inner Piedmont, thereby burying the original suture zone that once separated the two terranes. The Eastern Piedmont Fault system was active as a strike-slip fault during Alleghany orogeny.

The presence of strike-slip faults, and the absence of an Acadian foredeep, allow the suggestion that the Carolina zone, and perhaps many of the Inner Piedmont terranes, collided initially north of their present location and then sometime later were transported southward along Neoacadian and Alleghanian strike-slip faults. Two variants to the preceding collision scenarios take into account both strike-slip faulting and the presence of the Cat Square terrane, whose circa 430 Ma detrital zircons and circa 380 Ma intrusions imply that it existed as a depositional basin between 430 and 380 Ma. The first interprets the Cat Square terrane as a pull-apart basin associated with strike-slip faulting. The second interprets the Cat Square as a remnant ocean basin that closed like a zipper from north to south during collision of Carolina.

In the first scenario, Carolina collides with Laurentia by the close of the Taconic orogeny well to the north of its present location, possibly in the vicinity of New Jersey and Delaware. Following collision, the Carolina terrane was rafted away from Laurentia and transported southward thereby opening a narrow pull-apart ocean basin now recognized as the Cat Square terrane that was perhaps similar to the Gulf of California. Carolina then recollided with Laurentia in the Southern Appalachians probably during Neoacadian orogeny. The Cat Square was squeezed, metamorphosed, and deformed during Neoacadian orogeny.

In the second scenario, Carolina is oriented oblique to Laurentia such that the northern half of Carolina collides sometime after 430 Ma, leaving a small ocean basin (the Cat Square terrane) between Laurentia and the southern half of Carolina. This collision also is interpreted to have occurred in the Central Appalachians. Between 430 and 380 Ma, Carolina rotated clockwise into Laurentia, thereby closing the Cat Square ocean basin like a zipper resulting in deformation and metamorphism in the Inner Piedmont. The terranes were later shuffled and transported southward along strike-slip faults. The major difference in the two interpretations is that the first involves full collision during Taconic (or possibly early Silinian) orogeny and subsequent opening of a rift basin, whereas the second involves oblique collision entirely during Acadian-Neoacadian orogeny beginning in the north and progressing southward via clockwise rotation.

Ganderia Superterrane

There is a close association between Iapetus East and Ganderia. Both contain similar Gondwana fossils and locally they are in depositional contact rather than separated by a fault. Rocks within both realms extend from the Red Indian Line to the Honey Hill-Lake Char-Bloody Bluff Fault. Defined in this way, Ganderia overlaps with the eastern part of the Dunnage zone (Figure 23.10).

The defining tectonic unit in the Ganderia superterrane is the Ganderia passive margin (Figure 23.3). The rocks consist of Lower Cambrian to lowest Ordovician (520 to 479 Ma) quartz-rich sandstone and shale that is believed to represent the outer shelf, slope, and rise of the Ganderia passive margin prior to deformation and accretion (Albee, Baskahegan Lake, Grand Pitch, Cape Elizabeth formations). These rocks crop out from below Silurian-Devonian cover rocks across all of Ganderia (including Iapetus East) (Figure 23.4). Also present but sparsely exposed in New England are older basement rocks of the Ganderia microcontinent. These include sedimentary (circa 1230 to 750 Ma) and volcanic rocks (circa 630 to 525 Ma), most of which are now metamorphosed to gneiss. The largest exposure, the Massabessic gneiss complex, is shown in Figure 23.4 near Nashua, New Hampshire. The Ganderia passive margin, and intermingled Iapetus East rock units, were deformed and variably metamorphosed during Late Ordovician-Middle Silurian Taconic-Salinic orogeny and during Middle Silurian-Devonian Acadian-Neoacadian orogeny.

In addition to the Ganderia passive margin and basement rocks, a Silurian-Early Devonian continental volcanic arc complex known as the Coastal Plutonic and Volcanic Belt crops out along the Maine coast (Figure 23.4). The Coastal Plutonic and Volcanic Belt consists of granitic plutons, metamorphosed basalt, rhyolite, and sedimentary rocks (Cadillac Mountain granite and Eastport formation). These rocks unconformably overlie and intrude both the Penobscot-Ellsworth complex and the Ganderia passive margin. The rocks likely formed via subduction associated with the encroaching Avalon supercontinent. All of the various tectonic terranes associated with Ganderia are well exposed surrounding Acadia National Park in coastal Maine where metamorphism is generally low-grade. A tectonic map of this area is shown in Figure 23.14.

Recall from our earlier discussion that the Penobscot-Ellsworth arc complex is found in Canada west of the Liberty-Orrington-(Norumbega) Line. The presence of these rocks on both sides of the Liberty-Orrington Line is explained in the final section of this chapter.

Avalon Superterrane

In Southern New England, Ganderia is separated from Avalon by the Honey Hill-Lake Char-Bloody Bluff Fault, which extends from eastern Connecticut northward and eastward around greater Boston to just north of Gloucester where the Bloody Bluff Fault trends offshore. From this location northward, all of New England east of the Liberty-Orrington-(Norumbega) Line is part of Ganderia. Avalon reappears east of the Caledonia fault in New Brunswick and at its type locality east of the Dover Fault on the Avalon Peninsula of Newfoundland.

Rocks that form the Avalon superterrane record strong volcanic arc magmatism between 635 to 570 Ma. In southern New England, these rocks include variably deformed granite and granitic gneiss some with distinctive blue quartz (Esmond-Dedham and Milford-Ponaganset Plutonic suite), and rhyolites, andesites, volcanoclastics, pyroclastics, and minor basalt (Lynn-Mattapan Volcanic complex) some of which are intruded by granite. Magmatism of this age must have occurred on the continental margin of Gondwana far from the Laurentian mainland which was a quiet passive margin at the time. Magmatism, however, ended without evidence of continental collision. Instead, the volcanic rocks are overlain by up to 17,000 feet of Late Precambrian-Cambrian conglomerate and weakly metamorphosed mudstone and shale (argillite) known as the Boston Bay group. Rocks near the top of the sequence consist of shallow marine quartzites, shales, and limestones with Cambrian Gondwana fossils. A favored interpretation is that the region evolved from a subduction zone to a transform boundary and finally to a passive continental margin. The transition from subduction to transform without collision may have been similar to the transition in the California Cordillera from the Farallon subduction boundary to the San Andreas transform boundary (Figure 17.5). Radiometric dating suggests that all of the rocks were deformed, metamorphosed, and intruded during Acadian orogeny, an event associated with accretion of Avalon with Laurentia. Additional deformation, metamorphism, and minor intrusion occurred during Alleghany orogeny.

I have one additional note on the rocks of Avalon: In the Boston area there is a series of Ordovician-Silurian alkali (potassium-rich) granites of uncertain tectonic origin. Included in this suite is the famous Quincy granite, which was quarried for its ornamental stone from 1825 to 1963.

Meguma

The Meguma terrane is exposed only in Nova Scotia but is included here for completeness. Rocks in the Meguma terrane are unlike those in other terranes in terms of their lithology, thickness, and age of deformation. The entire section consists of sandstone with a lesser amount of intervening shale (slate) and minor volcanic rock. Basal rocks (the Meguma Supergroup) form a thick (>40,000 feet) assemblage of late Precambrian-Early Ordovician turbidites, sandstones, and shale. These rocks are interpreted to represent the continental rise, slope, and outer shelf of Gondwana (most likely the passive continental margin of northwest Africa). The Meguma Supergroup is unconformably overlain by more sandstone and shale and by interlayered basaltic and felsic volcanic rocks, a sequence that likely records rifting from Gondwana and the opening of the Rheic Ocean followed by development of a passive continental margin between Early Ordovician and Early Silurian. The entire

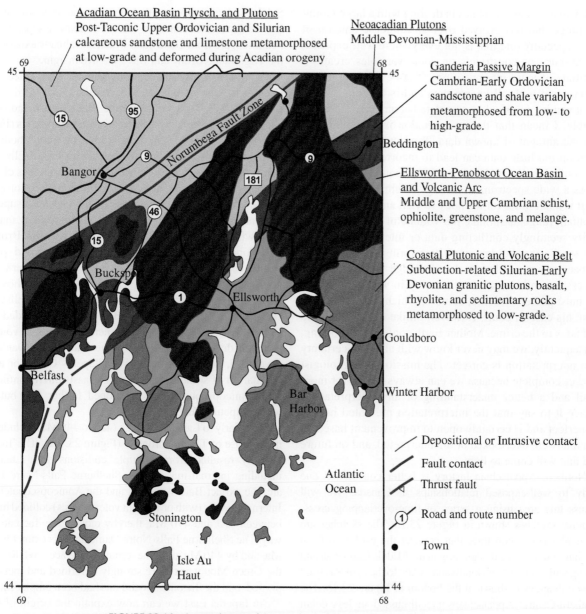

Acadian Ocean Basin Flysch, and Plutons
Post-Taconic Upper Ordovician and Silurian calcareous sandstone and limestone metamorphosed at low-grade and deformed during Acadian orogeny

Neoacadian Plutons
Middle Devonian-Mississippian

Ganderia Passive Margin
Cambrian-Early Ordovician sandsctone and shale variably metamorphosed from low- to high-grade.

Ellsworth-Penobscot Ocean Basin and Volcanic Arc
Middle and Upper Cambrian schist, ophiolite, greenstone, and melange.

Coastal Plutonic and Volcanic Belt
Subduction-related Silurian-Early Devonian granitic plutons, basalt, rhyolite, and sedimentary rocks metamorphosed to low-grade.

Depositional or Intrusive contact

Fault contact

Thrust fault

Road and route number

Town

FIGURE 23.14   A tectonic map of the area that includes Acadia National Park, Maine.

rock section is estimated to be as much as 75,000 feet thick. Beginning in earliest Middle Devonian (ca. 395 Ma) and continuing to the Early Mississippian (ca. 350 Ma), the rocks were deformed, intruded (South Mountain Batholith), and metamorphosed, mostly at low temperature. This Acadian-Neoacadian orogenisis is associated with overthrust and strike-slip collision of Meguma with Avalon (which was part of Laurentia) along the Cobequid Fault.

SEQUENCE OF APPALACHIAN COLLISION

We end our discussion with summary diagrams that depict the major events in the creation of the Appalachian orogenic belt (Figures 23.15, 23.16). I do this to give you, the reader,

an idea of how amazing it is that geologists can actually piece together a story from such fragmented and incomplete data. In my own lectures, I sometimes tell my students that geology is not rocket science; it's harder! The problem is that most of the evidence (the rock) has been removed by erosion or is buried beneath a thin soil. Geological history can be interpreted in detail only as far as the evidence will allow. Therefore, our interpretation of geological history is biased toward areas with the greatest exposure of rock. The geological history of the Northern and Canadian Appalachians can be interpreted in greater detail than their counterpart to the south simply because of a greater percentage of exposed rock. Imagine our level of understanding if we had 100% exposure virtually everywhere.

There are many variables in dealing with a large mountain range that has evolved over an immense amount of time, especially one that is not everywhere well exposed. The absence of constraints on some variables creates a multitude of valid interpretations. The best interpretation is consistent with all known data and is flexible enough that it can be modified as new data become available. By flexible, I mean that the interpretation is commensurate with the amount of known data. Too much interpretation based on too little data can lead to incorrect assumptions and confusion. A valid interpretation requires knowledge across a wide spectrum of geology, chemistry, and physics. It is a tall order for anybody to visit and map an entire mountain range, assimilate the volume of published data, rectify seemingly conflicting data or interpretations, and then synthesize the data in written form to be presented for peer review. Although the challenge is different, the process is similar to a detective story in which the detective must gather evidence, some of which is missing, and prove his case in court. However, unlike the burglar who confesses to the crime, Mother Earth confesses to nothing. Consequently, we may never know with absolute certainty if an interpretation is correct. The mission of a geologist is never complete because we can always search for more detail and a better understanding of any interpretation. Suffice it to say that the interpretation presented here is not perfect and is certainly open to improvement based on data that were overlooked or misinterpreted, and on future data that will come to light.

Northern Appalachian history is better constrained due partly to well-exposed relationships in Canada. We will discuss this area using a series of time-overlapping cross-sectional sketches shown in Figure 23.15. The sketches are self-explanatory, with lines that connect the evolution from one diagram to the next beginning in the Middle Cambrian and ending with collision of Gondwana. A key to the symbols used in each diagram is shown at the bottom of Figure 23.15. The Southern-Central Appalachians are discussed in less detail using a separate series of sketches shown in Figure 23.16. When presenting an orogeny in such a generalized fashion it is important to point out that some of the terranes and subsequent collisions shown in cross-section do not necessarily extend the length of the orogen. We begin in the Northern Appalachians.

By Middle-Late Cambrian (525 to 490 Ma), Laurentia was a passive continental margin separated from a series of island internal massifs (Dashwoods) by the narrow Taconic seaway, and from Gondwana by the Iapetus Ocean (Figure 23.15a). The island massifs likely did not extend the length of the Appalachians but rather were scattered along the coastline. The Iapetus Ocean was wide, with Laurentian fossil animals separated from Gondwana animals by what would later become the Red Indian Line. By about 500 Ma two island volcanic arcs had developed in the Iapetus Ocean: the Baie Verte arc on the Laurentian side and the Penobscot arc on the Gondwana side.

Subduction associated with both volcanic arcs was shrinking the Iapetus Ocean. The Ellsworth Seaway separated the Penobscot arc from Gondwana which had already experienced the Virgilinan orogeny and was a passive margin.

The peri-Gondwana microcontinents were rifted from Gondwana during the latest Cambrian-Early Ordovician (490 to 470 Ma), thus opening the Rheic Ocean and possibly additional seaways between the microcontinents (Figure 23.15b). Also during this time frame, between about 485 Ma and 478 Ma, the Penobscot arc and slices of Ellsworth oceanic lithosphere were accreted to Ganderia, thus closing the Ellsworth Seaway and creating the Penobscot orogeny. This event marked the beginning of the Ganderia superterrane and the beginning of the Ganderia passive margin. Following accretion of the Penobscot arc, the Bronson Hill-Popelogan-Victoria volcanic arc developed partly within oceanic crust but mostly along the northern margin of Ganderia by 478 Ma. This arc developed above the now extinct collided remnants of the Penobscot-Ellsworth complex. In West Iapetus, the Baie Verte arc collided with internal massifs and became extinct, but two new volcanic arcs had developed. The Shelburne Falls-Notre Dame arc developed, either within the West Iapetus Ocean or along the margin of some of the internal massif microcontinents. The Annieopsequatch arc developed farther east but still within Iapetus West.

The Early to Late Ordovician (480 to 446 Ma) includes the main phase of Taconic orogeny (Figure 23.15c). The Taconic orogeny resulted from multiple collisions with Laurentia including internal massifs, the Shelburne Falls-Notre Dame arc, the extinct Baie Verte arc, and the Annieopsequatch arc. Internal massifs (with or without volcanic arcs) collided mostly between 475 and 470 Ma, thereby closing the Taconic Seaway. The Shelburne Falls-Notre Dame arc was extinct by 458 Ma, and by 446 Ma all of the terranes were accreted. Rocks in the Green Mountains were strongly deformed and metamorphosed, and the Taconic allochthons had been emplaced.

In Iapetus East we can now explain the origin of Ganderia. By the latter part of the Middle Ordovician, between about 465 to 461 Ma, the Bronson Hill-Popelogan arc had rifted away from Ganderia but took part of Ganderia with it, thereby opening the Tetagouche-Exploits oceanic basin (Figure 23.15c). Ganderia was split in half and both halves contained rocks of the Ganderia passive margin and the extinct Penobscot-Ellsworth arc complex. The two halves are shown in Figure 23.15c as west Ganderia and east Ganderia. Today the two halves are separated by the Liberty-Orrington Line. The Bronson Hill-Popelogan-Victoria arc was still active in west Ganderia.

The Late Ordovician-Middle Silurian (460 to 423 Ma) marks a continuous transition from Taconic to Salinic orogeny (Figure 23.15d). The Bronson Hill-Popelogan-Victoria arc and both sides of the Gander terrane collided with Laurentia, thereby closing the Tetagouche-Exploits Seaway and

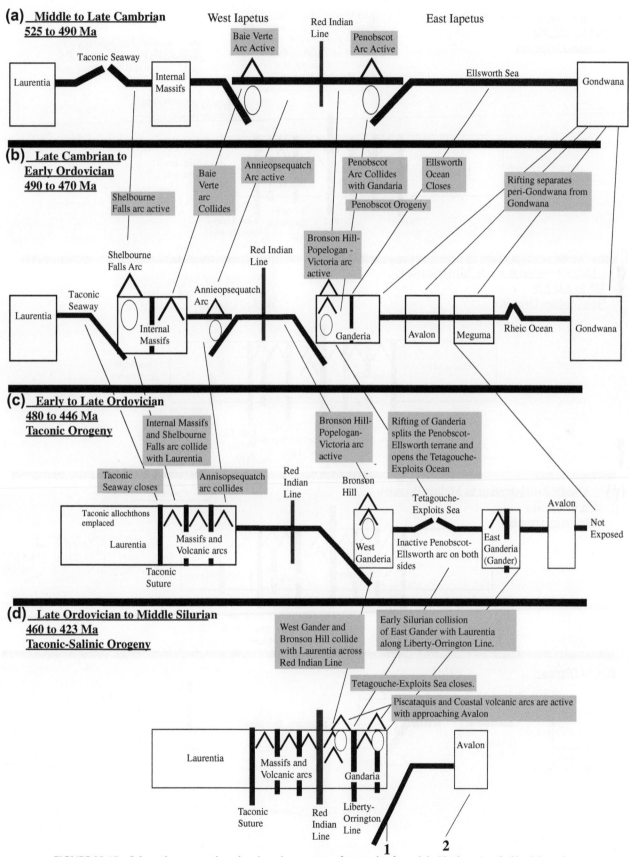

FIGURE 23.15   Schematic cross-sections that show the sequence of events that formed the Northern Appalachian Mountains.

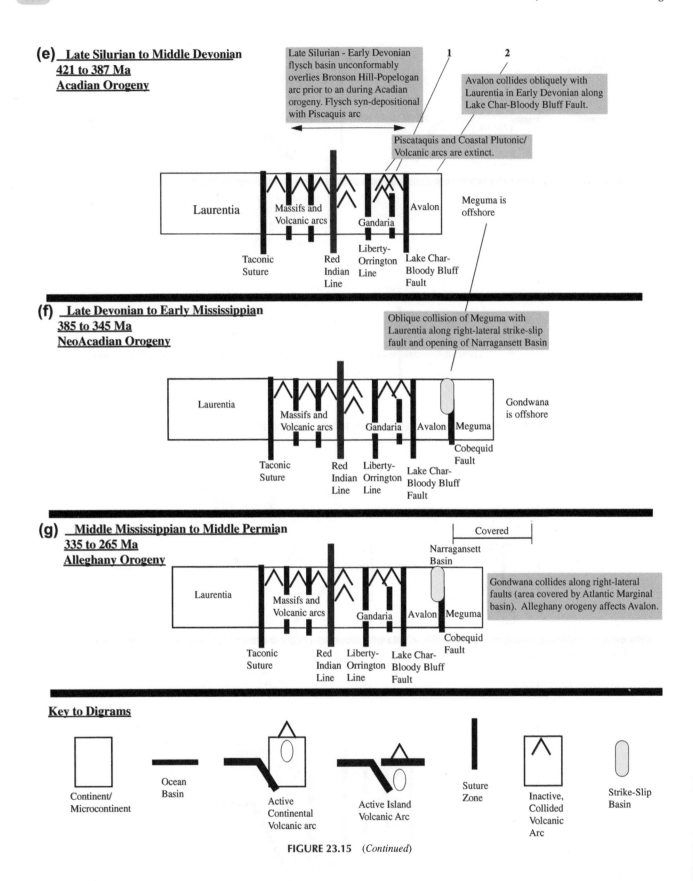

(e)   Late Silurian to Middle Devonian
421 to 387 Ma
Acadian Orogeny

Late Silurian - Early Devonian
flysch basin unconformably
overlies Bronson Hill-Popelogan
arc prior to an during Acadian
orogeny. Flysch syn-depositional
with Piscaquis arc

1   2

Avalon collides obliquely with
Laurentia in Early Devonian along
Lake Char-Bloody Bluff Fault.

Piscataquis and Coastal Plutonic/
Volcanic arcs are extinct.

Laurentia

Massifs and
Volcanic arcs

Gandaria   Avalon

Meguma is
offshore

Taconic
Suture

Red
Indian
Line

Liberty-
Orrington
Line

Lake Char-
Bloody Bluff
Fault

(f)   Late Devonian to Early Mississippian
385 to 345 Ma
NeoAcadian Orogeny

Oblique collision of Meguma with
Laurentia along right-lateral strike-slip
fault and opening of Narragansett Basin

Laurentia

Massifs and
Volcanic arcs

Gandaria

Avalon   Meguma

Gondwana
is offshore

Taconic
Suture

Red
Indian
Line

Liberty-
Orrington
Line

Lake Char-
Bloody Bluff
Fault

Cobequid
Fault

(g)   Middle Mississippian to Middle Permian
335 to 265 Ma
Alleghany Orogeny

Covered

Narragansett
Basin

Laurentia

Massifs and
Volcanic arcs

Gandaria

Avalon   Meguma

Gondwana collides along right-lateral
faults (area covered by Atlantic Marginal
basin). Alleghany orogeny affects Avalon.

Taconic
Suture

Red
Indian
Line

Liberty-
Orrington
Line

Lake Char-
Bloody Bluff
Fault

Cobequid
Fault

Key to Digrams

Continent/
Microcontinent

Ocean
Basin

Active
Continental
Volcanic arc

Active Island
Volcanic Arc

Suture
Zone

Inactive,
Collided
Volcanic
Arc

Strike-Slip
Basin

FIGURE 23.15   (*Continued*)

causing deformation and metamorphism across the entire eastern side of the Northern Appalachian collisional area. West Ganderia collided with Laurentia (and the Annieopsequatch arc) along the Red Indian suture line between about 460 and 450 Ma during Late Ordovician Taconic orogeny. East Ganderia collided along the Liberty-Orrington suture line, probably during the early part of Salinic orogeny. Immediately following collision, during the Early and Middle Silurian (445 to 420 Ma), a new subduction zone formed along the newly expanded eastern margin of Laurentia, creating the Coastal Plutonic and Volcanic Belt likely in response to subduction associated with the approaching Avalon superterrane. The Piscataquis arc was active within the Acadian flysch basin above some of the eroded remnants of the Bronson Hill arc (Figure 23.15d).

Late Silurian to Middle Devonian (421 to 387 Ma) marks the Acadian orogeny which corresponds with strong metamorphism, deformation, and plutonism across the central New England flysch basin in response to the accretion of Avalon with Laurentia (Figure 23.15e).

The Late Devonian-Early Mississippian Neoacadian orogeny (385 to 345 Ma) corresponds to deformation, intrusion, and weak metamorphism in response to accretion of Meguma (Figure 23.15f). Meguma likely collided north of its present location in Nova Scotia and was later transported southward along strike-slip faults. This event marked the beginning of collision between Laurentia and Gondwana and the closing of the northern Iapetus Ocean.

The final orogeny was the Middle Mississippian to Middle Permian Alleghany orogeny (335 to 265 Ma; Figure 23.15g). Final collision must have been associated with clockwise rotation of Gondwana relative to Laurentia because the Northern Appalachians experienced mostly strike-slip faulting, whereas the south first experienced strike-slip faulting and then strong head-on collision. The result was that Iapetus closed like a pair of scissors or like a zipper pulled shut from north to south.

We pick up the story in the Southern-Central Appalachians during the Taconic orogeny (Figure 23.16a). Volcanic arcs associated with the Central Blue Ridge were accreted to Laurentia along the Taconic suture zone (Figure 23.16a). The Eastern Blue Ridge-Western Inner Piedmont terranes (including the Tugaloo terrane) were also accreted along the Taconic suture but this collision probably occurred in the Pennsylvania-New Jersey-Maryland area well north of its present location. These terranes would later be transported southward with the Carolina terrane along strike-slip faults, possibly, in part, along the future Brevard zone. The Westminster terrane and the Hamburg allochthon were emplaced, and rocks in the Western Inner Piedmont were intruded. All of the terranes, and the Laurentian Western Blue Ridge, experienced deformation and metamorphism. The area subjected to Taconic orogeny, including the Western Blue Ridge, may have been separated from the Laurentian

(a) Early to Late Ordovician (480 to 446 Ma) Taconic Orogeny

Eastern Blue Ridge and Western Inner Piedmont collide creating strong metamorphism and deformation. The Hamburg allochthon is emplaced. A narrow sea may have separated the orogenic belt from the Laurentian mainland.

(b) Late Ordovician-Middle Devonian(460 to 385Ma) Taconic-Acadian Orogeny

Carolina collides in Central Appalachians and is transported south along strike-slip faults along with other terranes.

(c) Late Devonian to Early Mississippia (385 to 345 Ma) NeoAcadian Orogeny

The Eastern Blue Ridge and Inner Piedmont, including the Cat Square terrane, experienced strong metamorphism, plutonism, and deformation including strike-slip faulting. The rocks were shoved beneath a colliding Carolina Superterrane.

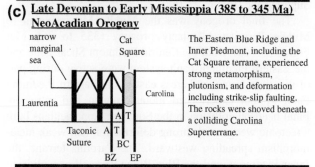

(d) Middle Mississippian to Middle Permian (335 to 265 Ma) Alleghany Orogeny   Gondwana Collides.

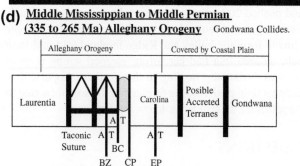

Notes: BC-Brindle Creek Fault, BZ-Brevard Zone, CP-Eastern Piedmont Shear Zone, EP-Eastern Piedmont Fault System. A-away and T-toward along strike-slip faults. Other symbols as in Figure 23.15.

FIGURE 23.16 Schematic cross-sections that show the sequence of events that formed the Southern-Central Appalachian Mountains.

mainland by a narrow sea because the foreland miogeocline (that is, the future fold-and-thrust belt) received erosional debris (flysch) but remained undeformed.

Events concerning the Carolina terrane between Late Ordovician and Middle Mississippian (460 to 330 Ma) are not well constrained. However, by the end of this time period the Carolina superterrane was accreted to Laurentia, probably north of its present location, and was in the process of strike-slip translation southward. Figure 23.16b shows a possible situation during Late Ordovician–Middle Devonian (460 to 385 Ma), a time frame that includes the Taconic, Salinic, and Acadian orogenies. Figure 23.16b leaves open the possibility that Carolina collided with Laurentia either by the close of the Taconic orogeny, circa 450 Ma, or sometime after 430 Ma, near the beginning of Acadian orogeny. In either case, the Eastern Inner Piedmont (Cat Square) terrane was in existence either as a strike-slip basin or a marginal sea. All of the terranes of the Western Inner Piedmont had accreted to Laurentia probably by the end of the Salinic orogeny. There is little evidence for strong Acadian orogeny in the Southern Appalachians.

The Eastern Blue Ridge and Inner Piedmont, including the Cat Square terrane, experienced strong Neoacadian (385 to 345 Ma) metamorphism, plutonism, and deformation, including strike-slip faulting along the Brevard zone, possibly as the rocks were shoved beneath a colliding Carolina Superterrane (Figure 23.16c).

The final orogeny was the Middle Mississippian to Middle Permian Alleghany orogeny (335 to 265 Ma; Figure 23.16d). Both the Central Piedmont Shear zone and the Eastern Piedmont Fault system were active during part of this time frame, and areas east of the Eastern Piedmont Fault system experienced metamorphism and intrusion. Final head-on collision in the Southern Appalachians built a tectonic wedge with strong deformation and weak metamorphism spreading westward. The Carolina terrane, the Inner Piedmont, and the Blue Ridge were all consolidated and pushed 200 miles or more northwestward above major flat-lying thrust faults that eventually cut into the miogeocline to form the Valley and Ridge foreland fold-and-thrust belt, an area that had heretofore escaped strong orogenic stresses. Strike-slip faulting in the Northern Appalachians, and this final push into the foreland in the south, created the Appalachian system as we know it today.

QUESTIONS

1. Name the major orogenies to affect the Northern Appalachians; the Southern-Central Appalachians.
2. What is the fundamental difference between the age of deformation in the Southern Appalachian foreland and the Northern Appalachian foreland?
3. What is the fundamental difference in the style of deformation in the Southern Appalachian foreland and the Central Appalachian foreland?
4. Refer to the Tectonic map of the Appalachians and name all of the tectonic units that lie directly west of the Taconic suture zone (TSZ) and all of the tectonic units that lie east of the Taconic suture zone.
5. Why is it difficult to correlate accreted terranes in the Northern Appalachians with accreted terranes in the Southern-Central Appalachians?
6. How does one date the time at which a continental margin transitions from a rifted margin to a passive margin?
7. Explain how foreland sedimentary rocks can be used to date the beginning of orogeny.
8. What is the age and significance of the following formations: Blockhouse, Sevier, Bays, Normanskill, Martinsburg, Juniata, Catskill?
9. What evidence suggests that the Northern Appalachian foreland was deformed during Taconic orogeny?
10. What foreland evidence suggests that the Taconic orogeny began in the south and migrated northward?
11. What evidence suggests that the Acadian orogeny began in the north and migrated southward?
12. What evidence is there in the Southern Appalachians that a shallow inland sea may have existed in the Ordovician that separated the rising Taconic Mountains from the Laurentian mainland?
13. What does the absence of Alleghanian molasse suggest regarding the Alleghany orogeny in the Northern Appalachians?
14. What is the presumed depositional environment of the Narragansett Bay group?
15. Name the three major realms in the US and their bounding faults (or rocks).
16. Explain the significance of the Red Indian Line.
17. Name all volcanic arc complexes present within the Iapetus realm.
18. How is the Smith River terrane different from other terranes in the Southern-Central Appalachians?
19. Describe the Brevard fault. Where is it located? Is it a suture zone? When was it active?
20. How is the Cat Square terrane different from other terranes in the Southern Appalachians?
21. What evidence is there that the Taconic orogeny in the Southern Appalachians occurred outboard of the Laurentian miogeocline?

The Cordilleran Orogenic Belt

We end our discourse with a general overview of the Cordilleran orogenic belt. We will not discuss the Cordillera in the same manner as we did the Appalachians. We will make no attempt to synthesize the geologic history. Instead, we will highlight and characterize some of the major features as outlined on a map of the Cordillera shown in Figure 24.1.

The Cordillera extends more or less continuously from Alaska to South America as one of the largest and most active mountain systems on Earth. It is part of the circum-Pacific belt of convergent and transform plate boundaries that encircles the Pacific Ocean extending to Japan and New Zealand. This region is known as the *ring of fire* and the *ring of shakes* because it is home to 75% of the world's volcanoes and 90% of the world's earthquakes. Similar to the Appalachians, the US Cordillera has been the recipient of numerous accreted terranes, but unlike the Appalachians, it was never involved in collision of two or more large continental masses. The history of orogenic activity is long and complex. It began about 360 million years ago with the Antler orogeny, but like the Appalachians, there is evidence for earlier orogeny that occurred offshore within terranes that are now part of North America.

The Cordilleran Mountain system has a foreland, a hinterland, and a belt of accreted terranes, but in some ways it is the antithesis of a classic orogenic belt. This is not because it formed in a markedly different manner than the Appalachians, but mostly because of what has happened during the past 70 million years and especially during the past 20 million years. An important aspect with respect to the Cordillera will be the ability to see through the young and active tectonics to reveal a slightly older classic orogenic

mountain system that in many areas is no longer reflected in the landscape. This is the part of Cordilleran history that we will emphasize in this chapter.

There are several key areas where orogenic relationships crop out from below the cover of volcanic rocks. These include the entire region south of the Snake River Plain, including most of Utah, Nevada, and California. This area, in spite of widespread normal faulting and volcanic rock, is where we have gained most of our knowledge of US Cordilleran tectonics. In addition to the Basin and Range, it includes the Sierra Nevada and the Klamath Mountains, where both Paleozoic and Mesozoic rocks are well exposed. It includes the California coastal region where Mesozoic and Cenozoic tectonics are exposed in spite of younger displacement along the San Andreas and related faults. Two other hinterland areas are important: the Blue Mountain region of Oregon and Washington, which forms an island of Paleozoic-Mesozoic tectonics surrounded by Columbia River and associated basaltic rock, and the northern Washington transect from Montana to the San Juan Islands and Olympic Mountains where Paleozoic-Cenozoic tectonics are well displayed.

Let us begin by having a quick gander at Figure 24.1. The figure is a little busy, so you have to take time to study it. Four rock units are shown along with 13 geologic boundaries. The rock units are the Precambrian crystalline shield, the Precambrian sedimentary/volcanic rock succession, areas of Cordilleran metamorphism, and areas of subduction-related plutonism. Twelve of the geologic boundaries are shown with different colors and are numbered for easy reference. You should read the label of each geologic boundary so that you can refer back to them when prompted.

410

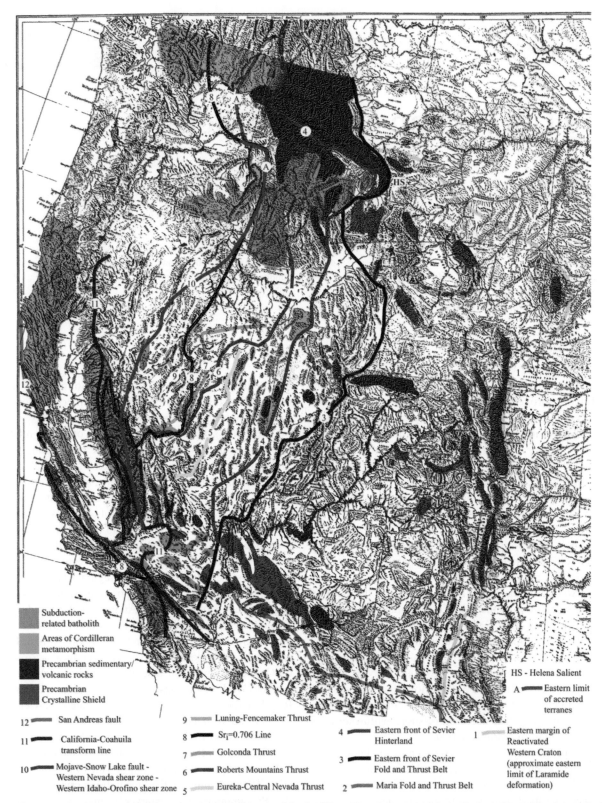

FIGURE 24.1 A landscape map showing selected tectonic features of the Cordillera. The primary source is the Geologic Map of North America. The Sevier belt, the foreland-hinterland transition, and the Eureka-Central Nevada belt are from DeCelles (2004) and DeCelles et al., (2006). The Roberts Mountains and Golconda fronts are placed at the eastern side of the Basin domain and the Golconda terrane, respectively, from Crafford (2008). The $Sr_i = 0.706$ Line is from Armstrong et al. (1987) and Kistler (1990). The Luning-Fencemaker front is based on Oldow (1984) and Wyld (2002). The Mojave-Snow Lake–Western Nevada shear zone–Western Idaho suture zone is from Schweickert and Lahren (1990), Wyld and Wright (2001), McClellend and Oldow (2007), and Dickinson (2008). The California-Coahuila Line is based on Dickinson (2006; 2008). Metamorphic areas are from Miller and Gans (1989) and Ernst (1990). Some metamorphic areas within the Precambrian sedimentary/volcanic succession in Idaho are not shown.

The line that depicts the eastern limit of accreted terranes is labeled *A*. This line is shown with a unique color in northeastWashington because it does not correspond with any of the other lines on the map. Southward, the line corresponds briefly with lines 8 and 10 (which coincide at this location). Farther south, the eastern limit of accreted terranes crosses below the Idaho batholith and corresponds with line 6 across central Nevada.

One thing to notice on the map is that several of the lines, including 3, 5, 6, 7, and 9, all end rather abruptly in southeastern California. This area, as we discovered in Chapter 18, is part of the Eastern California-Walker Lane Belt. It is also an area where numerous thrust and strike-slip faults have existed over the past nearly 300 million years. Two of these structures are shown in Figure 24.1: the California-Coahuila transform line (10) and the Mojave-Snow Lake-Western Nevada-Western Idaho-Orofino line (11), both of which are discussed later in the chapter. These fault zones, as well as recent faults associated with the San Andreas and Walker Lane systems, have severely disrupted the geology in California by shuffling and displacing rock units. The result is an intermingling of accreted and Laurentian rocks.

We will begin with a discussion of the major North American rock successions, and then discuss various aspects of the Cordillera generally from east to west. Our goal is to characterize and develop the structural framework of the Cordilleran system so that you will have the necessary background to pursue additional reading at your leisure.

THE PRECAMBRIAN SHIELD

The Grenville orogeny was not strongly felt along the western seaboard, unlike in the Appalachians. Instead, the final orogenic amalgamation and metamorphism in the western US occurred mostly between about 1.86 and 1.65 billion years ago, with a few slightly younger circa 1.65- to 1.35-billion-year events which were mostly intrusive in nature but included accretion in the southeastern Cordillera. An older, circa 2.5 Ma event, is preserved primarily in the Middle Rocky Mountain region where rocks of this age are exposed in the Wind River Range, the Teton Range, the Bighorn Mountains, the northern part of the Wyoming Front Range, and locally in the Black Hills.

PRECAMBRIAN SEDIMENTARY/VOLCANIC SUCCESSION

An abundance of Precambrian sedimentary/volcanic rock, as old as 1,500 Ma and as young as the Precambrian-Cambrian boundary, indicates that the Cordillera was undergoing deposition prior to, during, and following Grenville orogeny. Early deposits may have filled extensional basins within the continent such that only the youngest rocks are associated with final rifting and breakup of Rodinia.

Some of the oldest and most widespread Precambrian sedimentary/volcanic rocks in the US form the Belt Supergroup (1,470 to 1,370 Ma) in Montana and Idaho. These rocks have already been mentioned in relation to the Cordilleran fold-and-thrust belt (Chapter 12). *Purcell Supergroup* is the name given to Belt-equivalent rocks in Canada. The Belt Supergroup thickens from zero near the miogeocline-platform hinge line to as much as 65,000 feet (20 km) at its western margin. The extreme thickness coupled with the presence of syn-depositional normal faults suggest that the rocks were deposited in a large rift basin that evidently failed to go to completion. The Belt basin has been likened to a present-day Black Sea or Caspian Sea.

The Belt Supergroup does not extend south of Idaho, but there are other pre-rift and rift-related sedimentary/volcanic rock sequences scattered across the southwest. They include the Unita Mountain Group and the Big Cottonwood Formation (circa 975 to 775 Ma), which crop out extensively in and around the Unita Mountains and reach a thickness of 26,000 feet; the Unkar (1,220 to 1,070 Ma) and Chuar (775 to 735 Ma) Groups, which form part of the Grand Canyon Supergroup (Figure 21.7); the Apache and Troy Groups on the Colorado Plateau south of the Grand Canyon; and the Pahrump Group in the Death Valley region. All of these rocks were deposited in isolated extensional basins that were much smaller than the Belt Supergroup basin. The rocks underlie both miogeoclinal and platform sedimentary rocks.

THE MIOGEOCLINE

The Cordilleran miogeocline encompasses most of the area between the eastern front of the Cordilleran (Sevier) fold-and-thrust belt (line 3) and the belt of accreted terranes (line A). Early Cambrian miogeoclinal rocks from California to Canada form a very clean quartz-rich sandstone/quartzite known in different areas as the Flathead, Prospect Mountain, and Tapeats sandstone. These rocks are renown for their purity and are nearly identical to rocks of the same age in the Appalachians. As in the Appalachians, sandstone deposition gave way to limestone deposition by the Ordovician. The rocks imply that Laurentia, at this time, was a low-lying continent close to sea level and partly covered by a shallow continental sea. Clastic sediment, when it was deposited, was derived from the craton.

The Cordilleran miogeocline remained quiet and continued to develop throughout the Taconic orogeny and even during the main phase of Acadian orogeny. It was not until the latest Devonian-Early Mississippian (during Appalachian Neoacadian orogeny) when orogeny struck the western seaboard. In areas not affected by this orogeny, the miogeocline continued to develop into the Early Jurassic and attained a thickness in excess of 50,000 feet.

THE $Sr_i = 0.706$ LINE

One question that we addressed regarding the Appalachian Mountain belt but could not answer was how far (east) does Laurentian crystalline basement extend below the orogenic belt. We can ask a similar question with respect to the Cordilleran Mountains, but this time we would like to know how far west shield rocks are present at depth below the orogenic belt. In this case, we can provide an answer based on the geochemistry of intrusive rocks that have pierced the edge of the shield on their way to the surface. We define the western limit of shield rocks as coinciding with a boundary line known as the initial $^{87}Sr/^{86}Sr = 0.706$ line (*Sr* is the symbol for the chemical element strontium). We will refer to this boundary informally as the *706 Line*. It is line 8 in Figure 24.1. Such a line cannot be defined in the Appalachian orogenic belt because intrusions that pierce the edge of the shield do not reach the surface. Instead, they are buried beneath younger sedimentary rock of the Atlantic marginal basin.

The 706 Line is based on the chemical properties of magma generated through partial melting of mantle rock. Magma generated in the mantle will have an initial ^{87}Sr to ^{86}Sr ratio of 0.702 to 0.704. If the magma subsequently rises into old Precambrian shield rock (that is, Laurentia) it will be contaminated such that the $^{87}Sr/^{86}Sr$ ratio will increase. The reason for the increase has to do with: (1) the mineral composition of mantle rock versus continental rock, (2) the relatively old age of crystalline shield rock, and (3) the chemical properties of ^{87}Sr and ^{86}Sr.

^{87}Sr is a stable decay product of radioactive ^{87}Rb (rubidium 87). ^{86}Sr is a stable strontium isotope. ^{86}Sr cannot form by the decay of any radioactive substance. What this means is that if rubidium is present in a rock, the amount of ^{87}Sr in the rock will increase over time as ^{87}Rb undergoes radioactive decay. The amount of ^{86}Sr, on the other hand, will remain constant in the rock for all of eternity. Thus, the presence of Rb in a rock will cause the $^{87}Sr/^{86}Sr$ ratio to increase over time.

Rubidium is rarely found in abundance in igneous rocks. However, it does substitute for potassium (K), which means that it is found in small quantities in any K-bearing mineral. The Earth's mantle is composed of minerals such as olivine, pyroxene, spinel, and garnet, none of which contain potassium (and, therefore, no rubidium). In the absence of rubidium, the amount of ^{87}Sr in mantle rock will not increase over time, and because ^{86}Sr also does not increase, the $^{87}Sr/^{86}Sr$ ratio in the mantle remains constant at values between 0.702 and 0.704, irrespective of the age of the mantle rock. A magma that forms via partial melting of mantle rock will have these same values.

The Precambrian Laurentian shield, on the other hand, contains feldspar and mica, both of which are rich in potassium and, therefore, rich in rubidium. Shield rocks are also very old (between 4.0 and 1.0 By). This means that the original ^{87}Rb in shield rock has had ample time to decay such that the amount of ^{87}Sr has increased to the point where the $^{87}Sr/^{86}Sr$ ratio can be 0.710 or higher. A magma that forms initially in the mantle and then rises through these old rocks will be contaminated with enough ^{87}Sr to increase the initial $^{87}Sr/^{86}Sr$ ratio to 0.706 or higher. Magma that forms directly via partial melting of crystalline shield rock will also have a high initial $^{87}Sr/^{86}Sr$ ratio. The Cordilleran hinterland is riddled with intrusive rock, and all of it can be analyzed to determine its initial $^{87}Sr/^{86}Sr$ ratio (written as Sr_i). The line across which Sr_i increases above 0.706 thus marks the approximate location where rising magma must have encountered the buried edge of ancient crystalline shield rock on its way to the surface. In other words, the magma encountered the original but now deformed Laurentian continental margin. Recall from Figure 21.5 that the miogeocline extends to the edge of a continent. The 706 Line, therefore, also approximately coincides with the deformed and buried western limit of the Cordilleran miogeocline. Note that we are interested in the initial $^{87}Sr/^{86}Sr$ ratio of the intrusive rock when it first crystallizes from magma. The $^{87}Sr/^{86}Sr$ ratio will increase as the rock ages, but luckily there are methods by which the initial ratio can be determined.

An important piece of the story is the age and composition of accreted terranes and the fact that ^{87}Rb decays to ^{87}Sr very slowly. Accreted terranes in the Cordillera contain substantial amounts of relatively young oceanic volcanic rock derived from partial melting in the mantle. These rocks, partly because of their young age and partly because many are potassium-poor, have $^{87}Sr/^{86}Sr$ ratios less than 0.706. So, even if magma passes through these rocks, or if some of the accreted rock melts, the initial $^{87}Sr/^{86}Sr$ ratio likely would not rise above 0.706. The $^{87}Sr/^{86}Sr$ ratio will rise above 0.706 only if magma encounters ancient crystalline shield rock.

Let us trace the 706 Line beginning in the north (line 8, Figure 24.1). The 706 Line (line 8) is located in northern Washington where it trends southward through metamorphic rock and below Columbia River flood basalt before turning sharply eastward and then southward where it coincides with another line shown on the map, line 10. In this area, the east-west segment of line 10 is known as the *Orofino shear zone*. The north-south segment along the western side of the Idaho batholith is known as the *Western Idaho shear zone*. The Orofino shear zone is understood to be associated with an ancient fault zone or transform boundary between Precambrian shield and accreted rock. This boundary, incidentally, is roughly parallel with the Lewis and Clark Line, which, as you may recall from Chapter 12, forms a series of west-northwest-oriented, right-lateral strike-slip fault segments active off and on since the Precambrian (Figures 10.4 and 12.5). The Western Idaho shear zone corresponds to a near-vertical suture zone that is well exposed along the eastern flank of Seven Devils Mountains. At this location, the surface trace of the eastern limit of accreted terranes (line A) coincides with the 706 Line. The implication is that the contact between Laurentian crystalline shield and accreted terranes is vertical and remains near vertical to the base of the continental crust.

However, just south of Seven Devils Mountains, the eastern front of accreted terranes (line 6) turns eastward below the Idaho batholith and emerges on the east side of the batholith as the Roberts Mountains thrust. The 706 Line continues southward through central Nevada (Figure 24.1). Here, the implication is that rocks of the Laurentian crystalline shield must exist at depth, in southern Idaho and north-central Nevada, west of the surface trace of the Roberts Mountains thrust.

In California the 706 Line is disrupted by both young and ancient faults. It is shown looping northward to the west side of the Sierra Nevada, and then extending southward to the San Andreas Fault, which displaces the line northward almost to San Francisco. The 706 Line then trends southward along the west side of the San Andreas Fault to the California-Mexico border. Here, it is the area between the 706 Line and the San Andreas Fault where initial $^{87}Sr/^{86}Sr$ ratios are greater than 0.706 (i.e. where Laurentian shield rocks are present). In contrast to areas to the north, the Laurentian shield extends almost to the coastline in southern California.

Thus, the 706 Line provides a first-order approximation of the (now deformed) westward limit of Laurentia prior to Cordilleran accretion. On this basis it appears that Laurentia extended almost to the Pacific Ocean in Southern California but only to eastern Nevada and easternmost Washington farther north.

CRATONIC DEFORMATION: LARAMIDE AND MARIA THICK-SKINNED BELTS

The Laramide orogeny was the most recent (nonactive) orogeny to affect the Cordillera. It occurred primarily between 75 and 55 million years ago, although some structures are as young as 40 million years. It overlaps in time and also partly in space with the slightly older Sevier orogeny. The major difference between the two is tectonic style; Laramide deformation is dominantly thick-skinned, Sevier deformation is dominantly thin-skinned. Laramide deformation produced most of the structures across the reactivated western craton. The deformation is thick-skinned in the sense that it involved cratonic (crystalline shield and platform) rock and because deformation resulted in anticlinal structures with steep marginal thrust faults that dip into crystalline rock.

The western limit of Laramide deformation corresponds roughly with the eastern front of Sevier deformation (line 3). The eastern limit marks the Cordilleran deformational front and is drawn to coincide with the eastern margin of the reactivated western craton (line 1). The eastern boundary is most abrupt along the Colorado Front Range where relatively flat-lying rocks of the Great Plains are turned upward to produce crystalline-cored anticlines with or without frontal, west-dipping thrust faults. Farther north the boundary is diffuse as it swings east to include the Black Hills. North of the Black Hills the northern boundary of Laramide-style deformation coincides approximately with an eastward extension of the Lewis and Clark Line, wrapping around the Snowy and Little Belt Mountains before terminating against the Sevier fold-and-thrust belt at the southern end of the Montana Disturbed Belt (Figure 24.1).

South of Colorado Springs, the eastern limit of Laramide deformation is diffuse and segmented as it steps across the Arkansas River to the front of the Wet Mountains and then to the front of the Sangre de Cristo Mountains, both of which are structurally similar to the Front Range (that is, fault-bound anticlines cored with crystalline rock). The boundary becomes quite obscure in central New Mexico south of the Sangre de Cristo Mountains due to normal faults associated with the younger Rio Grande rift. Laramide-age folds and faults crop out sparingly in these normal fault-block mountains. The easternmost exposures of folded and thrust-faulted rocks occur along the Manzano and San Andres Mountains southward to the US-Mexico border at El Paso.

The Laramide orogeny is something of an aberration because it developed in cratonic rock and because it disrupts the tapered shape of the Sevier tectonic wedge. The deformation was likely in response to underplating of the Farallon slab (Chapter 17). The style of deformation was controlled, in part at least, by preexisting structures that include steeply dipping fault and fracture zones of Precambrian age within the crystalline shield. Some of these structures were reactivated during Pennsylvanian-Early Permian deformation associated with the Ancestral Rockies, and reactivated again during Laramide orogeny.

The Maria fold-and-thrust belt is also something of an aberration because it extends across the craton in an east-west direction. Its approximate northern extent is shown as line 2 in Figure 24.1. Thrust-faulting is thick-skinned, and vergence (thrust direction) is dominantly toward the south (although locally toward the north). The thrust belt was most active between about 70 and 90 million years ago during the early stages of Sevier deformation. Similar to Laramide deformation, it may have formed across preexisting weaknesses in the underlying crystalline basement. Today the thrust belt is severely disrupted by younger Basin and Range normal faults. Evidence for its existence is seen in the form of folds and faults within normal fault mountain blocks that include the Harquahala Mountains.

MIOGEOCLINAL DEFORMATION: THE SEVIER THRUST BELT

The Sevier orogeny is primarily responsible for development of a classic, thin-skinned foreland fold-and-thrust belt that involves miogeoclinal sedimentary rock. The eastern front of the thrust belt (line 3) forms the toe of the tectonic wedge. The western limit corresponds roughly to the belt of accreted terranes (line A) or with the Eureka-Central Nevada thrust belt (line 5). The Eureka-Central Nevada thrust belt and the Luning-Fencemaker thrust belt (line 9) also helped build the tectonic wedge and could be thought of as part of an expanded Sevier belt.

The Sevier fold-and-thrust belt extends from southeastern California well into Canada where it forms the Northern Rocky Mountain fold-and-thrust belt. The entire thrust belt has been referred to as the *Cordilleran thrust belt*. In the US, the Sevier thrust belt is disrupted by younger tectonism that includes normal faulting, volcanism, and, in a few areas, Laramide-style, thick-skinned deformation. The eastern front (line 3) is marked by a discontinuous series of thrust-fault segments that nucleated along the depositional hinge line that once separated miogeoclinal rock from interior platform rock prior to deformation (Figure 21.5). The hinge line in Utah and southern Idaho is known as the *Wasatch line*. Thrust-faulting began well to the west of the Wasatch line beginning in the Middle to Late Jurassic in the form of the Luning-Fencemaker belt (line 9), progressed eastward in the form of the Late Jurassic-Cretaceous Eureka-Central Nevada thrust belt (line 5), and finally to its present position by early Cenozoic overlapping with thick-skinned Laramide deformation. This west-to-east thrust progression over the course of more than 100 million years is what created the Cordilleran tectonic wedge. The Sevier belt proper encompasses thrust faults generally less than 100 million years old that were active in the Late Cretaceous-early Cenozoic.

A transect across the Sevier fold-and-thrust belt at almost any location would reveal between four and eight major thrust faults, all carrying mostly Cambrian to Cretaceous miogeoclinal sedimentary rock above platform and foreland basin rock of the same age. The major exception, of course, is in the north, where the thrust belt carries the Precambrian Belt Supergroup. Nearly all the thrusting occurred between 100 and 55 million years ago. All of the thrust faults are segmented, and each segment is given a different name. Thrust faults that form the eastern front (line 3) tend to be the youngest and to carry the youngest rock. Total shortening across central Nevada is as much as 150 miles (240 km).

The classic characteristics of a thin-skinned, eastward-propagating fold-and-thrust system are on display in northernmost Montana within Glacier National Park and in the Montana Disturbed Belt as previously discussed in Chapter 12. Faulting along this frontal zone is Late Cretaceous-Paleocene in age (74 to 59 Ma) based on the age of deformed rock units and cross-cutting relationships. As we have noted previously, it is the Belt Supergroup that forms the upper plate of thrust faults in Glacier National Park, including the Lewis thrust, which is well exposed both within the Park and as klippen on Chief Mountain a few miles to the east (Figure 12.7). The Lewis thrust is truly a major thrust fault. It extends northward into southern Alberta for more than 250 miles and extends southward into the interior of the US thrust belt.

The Montana Disturbed Belt displays tightly spaced thrust faults that place Paleozoic and Mesozoic miogeoclinal rock above platform and basin-fill rock of the same age. Thrusts that make up this zone are, for the most part, small. Much greater displacement occurs along thrust faults just west of the frontal belt that carry the Belt Supergroup.

Low-grade (dominantly greenschist facies) metamorphic rocks and a few high-grade rocks crop out west of the Sevier frontal thrust belt surrounded by miogeoclinal sedimentary rocks. This relationship allows us to define a foreland-hinterland boundary based on the first appearance of crystalline rocks. The boundary is shown in Figure 24.1 as line 4. It is defined by rocks metamorphosed in the Jurassic or Cretaceous prior to or during Sevier thrusting, and by the presence of Mesozoic-Cenozoic intrusive rocks such as the Idaho batholith. The crystalline rocks, for the most part, were not brought to the surface solely via thrust-faulting and erosion. Most crop out discontinuously as isolated anticlinal culminations, granitic intrusions, or normal fault exhumed crystalline core complexes surrounded by miogeoclinal sedimentary rock. A classic foreland-to-hinterland transition, where sedimentary rock is absent in the hinterland, is preserved only near the Canadian border in northernmost Idaho. Here, the foreland-hinterland boundary lies about 125 miles from the eastern front of the Sevier belt and is marked by a sharp transition that separates metamorphic and intrusive rock of the Priest River crystalline core complex from foreland thrust sheets that carry the Belt Supergroup.

The foreland-hinterland transition (line 4) lies along the Lewis and Clark Line south of the Priest River core complex. This line separates metamorphosed Belt rocks (not shown as metamorphic in Figure 24.1) and intrusions associated with the Idaho, Pioneer, and Boulder Batholiths (the latter two are younger satellites of the Idaho Batholith) in the south from generally unmetamorphosed Belt rocks to the north. Magma intruded prior to, during, and following thrusting and, therefore, provides age constraints on several of the thrust faults. Note in Figure 24.1 that the thrust belt bulges eastward, forming what is known as the *Helena salient (HS)*. In this area, the classic fold-and-thrust landscape is lost and crystalline hinterland rocks extend to the front of the Sevier belt. Immediately south of the Helena salient, the thrust belt is disturbed by Laramide-style basement uplifts of Precambrian shield rock, young volcanism, plutonic intrusion, and Basin and Range normal faulting, some of which began during the waning phases of Sevier deformation, and some of which is active today. This is the area of the Rocky Mountain Basin and Range (Chapter 15). Major thrust faults in this region were active in the Late Cretaceous between about 85 and 65 Ma. Notice here that the frontal foreland thrust belt (line 3) separates from the hinterland belt (line 4). Farther south, both thrust belts are hidden beneath young volcanic rocks of the Snake River Plain. Their inferred trace across the Snake River Plain is shown in Figure 24.1.

South of the Snake River Plain we encounter the well-developed Idaho-Wyoming fold-and-thrust belt, which pinches around the southern side of the Teton normal fault-block mountain, and around the western side of the Laramide-style Uinta faulted anticlinal mountain. The thrust belt then extends southward approximately along the western margin of the Colorado Plateau. In this area the hinterland thrust belt (line 4) is located well to the west of the foreland belt (line 3). Notice in Figure 24.1 that in order to locate the hinterland belt at this location, we had to ignore a small amount of crystalline shield rock located within the Wasatch Mountains east of line 4. Thrusting was active in the Late Cretaceous-Paleocene (90 to 55 Ma).

The thrust belt becomes increasingly difficult to follow southwest of Utah due to Basin and Range extension (Figure 24.1). Near its southern terminus in California, thrust faults cut into crystalline shield and interior platform rocks of the craton and carry both plutonic and metamorphic rocks. At this point, the thin-skinned character of the thrust belt is destroyed and deformation becomes indistinguishable from Laramide-style basement uplifts. Frontal thrusts in this area, however, are older than Laramide deformation (100 to 80 Ma) but still Late Cretaceous. The thrust belt appears to end abruptly against the California-Coahuila transform line (line 11), which is discussed later in the chapter.

ACCRETED TERRANE THRUST BELTS: THE ANTLER AND SONOMA OROGENIES

If one were to travel from Utah across central Nevada, one would find classic geological evidence for several orogenic cycles hidden in the normal fault-block mountains of the Basin and Range. Figure 24.2 is a conceptual cross-section that defines three of these orogenic cycles based on the presence of angular unconformities labeled 1, 2, and 3. From oldest to youngest the orogenic cycles are the Antler, the Sonoma, and the Sevier. We have already discussed the Sevier orogeny, so we will concentrate on the Antler and Sonoma orogenies. Sedimentary rocks in both orogenic belts are internally thrust-faulted and folded but generally not metamorphosed. The age of orogeny, therefore, can be determined on the basis of fossils. Geological principles require that thrusting is younger than any rock within or below the thrust sheet, about the same age as early flysch in the foredeep, and older than unconformably overlying (overlap) sequences.

Figure 24.2 shows that the Antler orogeny is the oldest because associated thrust faults are covered by a Middle Mississippian-Lower Triassic rock succession known as the *Antler overlap sequence*. This orogeny involved thrusting of the Roberts Mountains allochthon and development of the Antler foredeep basin. The frontal thrust of the Roberts Mountains allochthon (the Roberts Mountains thrust) forms the boundary between accreted terranes and the Cordilleran foreland-hinterland thrust belt (line 6).

The Roberts Mountains thrust sheet is exposed in the Independence, Tuscaroro, Cortez, Miller, and Roberts Mountains, the Centennial and Shoeshone Ranges, Candelaria Hills, and Battle Mountain. A sliver of the allochthon is present in the El Paso Mountains of southern California. Northward, the allochthon, and the Antler foredeep basin, can be followed to the Pioneer Mountains at the southeastern

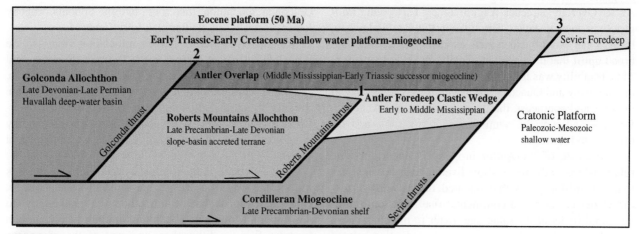

FIGURE 24.2   A schematic cross-section that shows unconformities signifying: 1 = Antler orogeny (Late Devonian-Early Mississippian), 2 = Sonoma orogeny (Late Permian-Early Triassic), 3 = Sevier orogeny (Late Cretaceous-Early Cenozoic).

flank of the Idaho batholith. Farther north the rocks are lost in the Idaho batholith. Rocks within the allochthon consist of Upper Cambrian to Upper Devonian shale, sandstone, turbidite, limestone, and basalt. Fossil similarities and transitional rocks indicate that rocks in the allochthon were deposited partly on oceanic crust and partly on thinned, extended continental crust along the Laurentian continental slope and rise. A few rocks of possible oceanic lithosphere (serpentinite) are present at the base of the allochthon, and this is the basis for designating the Roberts Mountains thrust as the easternmost occurrence of accreted terranes. The Roberts Mountains allochthon was thrust eastward some 45 miles onto miogeoclinal rocks that are the same age as those in the allochthon. These relationships, and the age of rocks in the foredeep, imply that orogeny occurred between Late Devonian and Early Mississippian.

An early indication of Antler orogeny in the foreland is seen east of the Roberts Mountains thrust where Lower Mississippian Chainman shale and Diamond Peak formations (among others) transgress eastward across the miogeocline. These rocks form the basal part of the Antler foredeep clastic wedge, which thickens westward to about 11,500 feet at the foot of the Roberts Mountains thrust, clearly indicating a western source (Figure 24.2). The foredeep basin shows a pattern typical of clastic wedges consisting of syn-orogenic Lower Mississippian deep-water submarine shales (flysch) in its lower part, and coarse-grained, fluvial molasse in its upper part. Middle Mississippian molasse transgresses across (covers) the Roberts Mountains thrust indicating that active thrusting had ended by then and that the mountains had been partly eroded.

The presence of carbonate rock (limestone, dolostone) in the eastern part of the miogeocline (near the hinge line) of the same age as the Antler orogeny indicates that debris shed into the foredeep clastic wedge did not reach this far east. By the Early Triassic, the miogeocline had transgressed and covered the eroded stump of the Antler orogenic belt, thus forming a successor miogeocline, the Antler Overlap sequence (Figure 24.2). This young miogeocline was unstable. Throughout the late Paleozoic, local areas would experience extensional faulting and broad uplift that disrupted simple blanket-like deposition. This instability was likely in response to distant influences of Alleghany and Ouachita orogeny that were underway in the east and southeast, the development of the Ancestral Rockies, and sporadic, semi-continuous volcanic activity offshore to the west.

The cause of the Antler orogeny is uncertain. The relationships indicate a short-lived event lasting less than 20 million years that resulted in eastward thrusting of ocean basin and continental rise rocks onto miogeoclinal rocks of the same age. Later in the chapter we will show evidence for one or more subduction-related volcanic arcs that were in existence farther west prior

to and following Antler orogeny. A common interpretation is that a volcanic arc collided with Laurentia and bulldozed slope/rise sediments (the allochthon) onto the Laurentian miogeocline. A possible contributing factor to the Antler orogeny may have been active tectonics along the East Coast. It is potentially important to realize that the Antler orogeny began during a time of collision and strong Neoacadian deformation in the Appalachian Mountains that included the accretion of Meguma and the possible collision or recollision of Carolina. These far-field tectonics may have influenced Cordilleran tectonics. One could speculate that collision in the east caused the Cordilleran miogeocline to slide below slope/rise sediment located farther west, thus emplacing the Roberts Mountains allochthon with or without a push by a volcanic arc in the west.

The Sonoma orogeny has a history similar to that of the Antler orogeny. In this case, the Golconda allochthon was transported 35 to 40 miles, partly over the Roberts Mountains allochthon and above the Antler Overlap sequence, along the Golconda thrust (line 7 in Figure 24.1). The Golconda allochthon consists mostly of Upper Mississippian-Permian chert, argillite (hardened claystone), basalt, conglomerate, sandstone, limestone, turbidites, and volcanic rock debris derived from the Antler highlands to the east and from volcanic arcs to the west. The sequence is interpreted as a deep-water, extensional, back-arc basin, referred to as the *Havallah Basin*, located originally between the Laurentian continental margin and an offshore volcanic arc. The interpretive paleogeography following the Antler orogeny but prior to Sonoma orogeny is shown in Figure 24.3a. The orogeny is interpreted to have resulted from collision of an offshore volcanic arc with the Laurentian mainland. The collision closed and uplifted the Havallah Basin in the southern Nevada-California area. Note in Figure 24.2 that the Golconda allochthon is overlain by a Lower Triassic–Lower Cretaceous successor miogeocline. Rocks within and unconformably above the Golconda allochthon, therefore, date the Sonoma orogeny rather precisely as Late Permian-Early Triassic.

Remnants of the Golconda allochthon (that is, the Havallah Basin) are present in the Toquima, Sonoma, and Tobin Ranges, Antler Peak, the Independence Mountains, and the Mountain City region above the western edge of the Roberts Mountains allochthon and above the Antler Overlap sequence. Remnants are also present in the Kootenay arc region (Slide Mountain) in northern Washington and Canada.

The Golconda allochthon is unusual because there is little in the way of a discernable foredeep clastic wedge. One explanation is that the allochthon was small relative to the Roberts Mountains allochthon and not heavy enough to down-flex the lithosphere in front of the advancing thrust sheet.

The Sonoma orogeny, like the Antler event, lasted less than about 20 million years. Both pushed oceanic and continental rise rocks above the miogeocline and then abruptly ended without additional deformation or widespread metamorphism. Neither resulted in the building of an orogenic wedge beyond the allochthon itself. Quiet deposition resumed after both orogenies.

The Sonoma orogeny, however, was coincident with a major reorganization of tectonic plates that resulted in establishment of an active, east-dipping subduction zone/volcanic arc system along the entire US Laurentian coastline. In the south, the volcanic arc was built on the Laurentian margin, producing a continental arc perhaps similar to the present-day Cascades. Farther north, the arc was separated from the continental margin by a deep back-arc basin. There were likely additional volcanic arcs farther offshore. The interpretive paleogeography at circa 200 Ma (Triassic-Jurassic boundary) is shown in Figure 24.3b. Note in this figure that a subduction zone has replaced the California-Coahuila transform of Figure 24.3a.

TRUNCATION

The Sevier fold-and-thrust belt (line 3) separates the North American craton from the Cordilleran miogeocline. The thrust belt and both the miogeocline and craton end abruptly in southern California as shown, not only in Figure 24.1, but also in Figures 21.6 and 21.7. This type of geometry is unusual and is not seen anywhere else in the US. It appears that the southern margin of Laurentia was sliced off and transported elsewhere. The truncation is understood to have resulted from development of a left-lateral transform plate boundary known as the *California-Coahuila transform* (originally referred to as the Mojave-Sonora megashear) that may have come into existence during the Pennsylvanian and persisted into the Triassic. The California-Coahuila transform fault is shown in Figure 24.1 as line 11, and in Figure 24.3a as truncating both the Antler Mountain belt and the miogeocline at circa 270 Ma (Permian). The transform fault apparently displaced the southwestern edge of the North American miogeocline and craton southward into

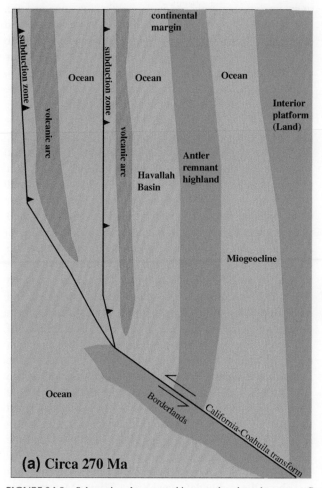

(a) Circa 270 Ma

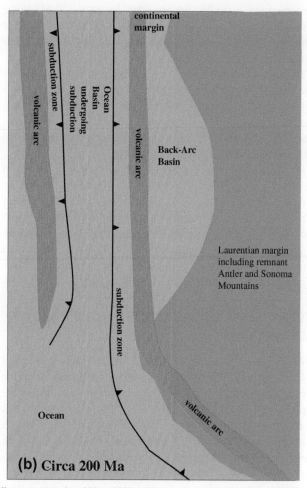

(b) Circa 200 Ma

FIGURE 24.3   Schematic paleogeographic maps that show the western Cordilleran coast at circa 270 and 200 million years ago. Based partly on Blakey (2011).

Mexico a distance of 590 miles (950 km). Notice in Figure 24.3a that displacement direction is opposite present-day displacement on the San Andreas Fault.

The actual California-Coahuila fault is not very well exposed. There has been too much subsequent deformation, intrusion, and deposition to really put your finger on it. Instead, the fault is defined as a zone up to several miles wide where marked changes in the strike, composition, age, and degree of deformation of rocks are observed. Line 11 marks the approximate western edge of the ancient transform. The transform cannot be followed north of the Klamath Mountains suggesting that it may have merged with subduction zones in a manner similar to the merging of the San Andreas Fault and the Juan de Fuca subduction zone, and as shown in Figure 24.3a.

There are Triassic plutonic rocks on both sides of the fossil transform that show the chemical signature of subduction-related magmatism. Additionally, there are circa 220 Ma blueschists in the Klamath Mountains and Sierra Nevada (Stuart Fork and Red Ant blueschist). Recall from Chapter 20 that blueschist facies rocks develop in subduction zones. The inference is that the transform boundary had evolved into a subduction boundary by circa 220 Ma (Late Triassic). What this means is that a subduction boundary had developed along the entire length of the US Cordillera prior to the end of the Triassic as depicted in Figure 24.3b. The subduction boundary would remain in existence along the entire US Cordilleran margin for the next 170 million years until initiation of the San Andreas Fault. Remnants of this subduction boundary are preserved in large accretionary complexes such as the Franciscan complex in the northern California Coast Ranges.

The change in the southern part of the Cordillera from a passive (miogeoclinal) continental boundary to a transform boundary in the Pennsylvanian, and from a transform boundary to a subduction boundary in the Triassic, must mark times of major plate reorganization. It is no coincidence that these changes correlate with the Alleghany and Ouachita orogenies and with subsequent formation of Triassic rift basins in the eastern US.

VOLCANIC ARC AND SUBDUCTION COMPLEXES

We have thus far assumed the existence of volcanic arcs and subduction complexes but have yet to show evidence for their existence. Remnants of these ancient landscapes are present in at least four isolated areas, all of which are shown as metamorphic in Figure 24.1. They are the Klamath Mountains, the northern Sierra Nevada, the Blue Mountain region, and the northwest Rocky Mountain-North Cascade region. Although each area is isolated, they all show a similar but complex geologic history of tectonic accretion, deformation, intrusion, and metamorphism that can be traced with various degrees of confidence from one area to the next.

As an example, let us look in more detail at the Klamath Mountains. The Klamath Mountains and adjacent coastal rocks consist of a series of east-dipping thrust-faulted terranes composed of Ordovician to Cretaceous oceanic rock. The rocks were accreted to the continental margin from east to west in an almost continuous fashion beginning in the Devonian and continuing to the present-day at the Cascadia trench. The oldest rocks are located in the east and at the top of the east-dipping thrust pile. Two primary types of terranes are exposed. The first represents the remains of island volcanic arc systems that were constructed primarily offshore and then transported, deformed, and accreted to the continental margin. The second represents the remains of the surrounding ocean floor that formed either at the continental margin or along the flanks of island volcanic arc systems. These terranes include deep-sea sediments and sections of oceanic lithosphere. Both terrane types were scraped off the subducting slab at the trench and accreted to the continental margin as highly deformed accretionary prism (subduction complex) mélange units. Some of the rocks were metamorphosed in the subduction zone and are now blueschists or greenschists. In some cases, entire slabs of oceanic lithosphere were emplaced onto the continental margin as an ophiolite complex. The rocks record more than 400 million years of nearly continuous tectonic accretion.

A geologic map of the Klamath Mountains is presented in Figure 24.4. Three major thrust belts are shown. The Eastern Klamath Belt includes the Central Metamorphic Belt, the Trinity ophiolite, the Yreka volcanic arc, and (shown separately) the Redding section. The Western Klamaths contain a variety of island arc and ocean basin terranes that include the Josephine ophiolite complex and the Rattlesnake Creek subduction mélange. The Franciscan Complex contains trench and forearc basin deposits as well as blueschist and greenschist blocks. All except the Franciscan Complex are intruded by subduction-related granitic plutons, mostly between 170 and 130 million years old.

The Yreka-Trinity-Metamorphic complex (Eastern Klamaths) was strongly deformed, metamorphosed, and eroded prior to or during the Middle Devonian and, therefore, prior to the onset of Antler orogeny. The absence of a time correlation with the Antler orogeny suggests that deformation and metamorphism occurred offshore from the Laurentian mainland. Equivalent rocks in the Sierra Nevada are known as the Shoo Fly complex. Like the Klamaths, the Shoo Fly complex consists of a series of highly deformed thrust sheets that includes the Sierra City mélange, Culbertson Lake allochthon, Duncan Peak allochthon, and the Lang-Halsted sequence.

The Redding section is less deformed and sits in fault contact above the highly deformed Yreka-Trinity-Metamorphic complex. The Redding section is particularly informative

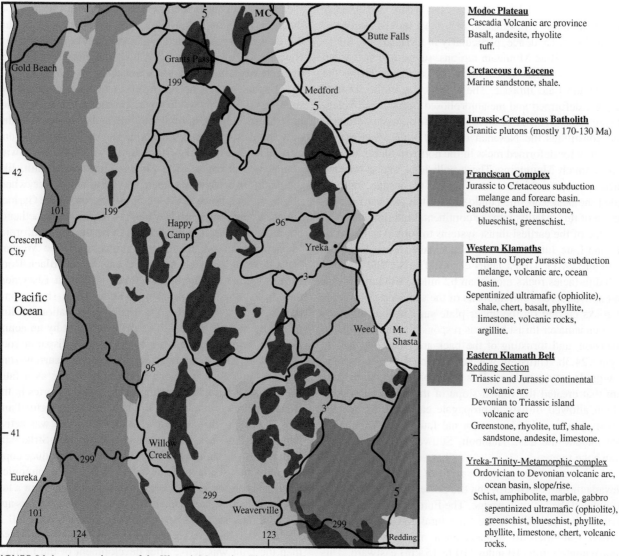

FIGURE 24.4   A tectonic map of the Klamath Mountains. The explanation describes the tectonic interpretation followed by the common rock types. Based on Irwin (1994) and Snoke and Barnes (2006). Roads are shown as black lines, some with route number.

because it contains an almost continuous record of volcanic arc deposition from Devonian to Jurassic. Preserved in this sequence is a Devonian-to-Triassic island volcanic arc complex, and a Triassic-to-Jurassic continental volcanic arc complex. Rocks in the northern Sierra Nevada equivalent with the Redding section are known as the *Northern Sierra section*. Similar rocks are present in the Blue Mountains (Grindstone terrane), the Seven Devils region (Seven Devils arc), at Quinn River crossing near the Oregon-Nevada border, in the Kootenay arc region of northwest Washington (Chilliwack Group and Vedder Complex), and in Canada as far north as Alaska.

The Western Klamaths and the Franciscan complex consist of island arcs, subduction mélanges, and deep-sea sedimentary rocks that record subduction and accretion during the Jurassic and Cretaceous. It is the accretion of these

terranes that helped create the backstop that culminated with the building of the tectonic wedge, as described next.

BUILDING THE CORDILLERAN TECTONIC WEDGE

The Cordilleran tectonic wedge was built over the course of about 120 million years from middle Jurassic (175 Ma) to Early Cenozoic (55 Ma) beginning in the west and progressing eastward. The paleogeography of the US Cordillera 170 million years ago is conjectural, but one interpretation based on a synthesis of geologic data suggests that an east-dipping subduction zone and a continental volcanic arc were in existence in southernmost California and that at least two volcanic arcs were present in the north, one close to the Laurentian continental margin but separated from the mainland by a deep, inland basin, and another farther offshore. The situation may have looked similar to Figure 24.3b. Evidence, particularly in the Klamath, Sierra Nevada, and Blue Mountain regions, suggests that the two northern arcs began to collide with Laurentia during the Middle and Late Jurassic. The evidence is in the form of strongly deformed and metamorphosed oceanic and volcanic arc rocks in all three areas. The Late Jurassic deformation constitutes the Nevadan orogeny, which was named originally for deformed rocks in the northern Sierra Nevada and Klamath Mountains. These collisions, and associated intrusive events, thickened the heel of the wedge and provided the backstop (the bulldozer) for the eastward propagation of thrust faults into the continental interior.

One of the earliest thrust systems to form was the Middle and Late Jurassic Luning-Fencemaker thrust belt (line 9, Figure 24.1). Thrust faults carried Triassic and Jurassic turbidite-facies rocks more than 62 miles (100 km) toward and above more proximal rocks of the same age. The types of rocks carried on the upper plate suggest that the Luning-Fencemaker thrust belt was responsible for the closing, inversion, and thrusting of the back-arc basin depicted in Figure 24.3b. Thrusting was accompanied by folding and low-grade metamorphism. This was probably the thrust system that created the frontal slope of the thickened wedge which allowed thrusts to propagate eastward. Today, the thrust system is hidden in normal fault-block mountains that include the East, Humbolt, Stillwater, Desatoya, and Shoshone ranges.

Additional thrusting in the form of the Eureka-Central Nevada thrust belt (line 5) and the Sevier thrust belt (line 3) completed the tectonic wedge. The Eureka-Central Nevada thrust belt was active in the Late Jurassic-Cretaceous and carried Cambrian to Pennsylvanian miogeoclinal rocks approximately 6 to 10 miles (10 to 15 km) eastward. This thrust belt was subsequently eroded and partly covered by circa 30 Ma volcanic rocks. The Sevier belt, as previously discussed, was active in the Late Cretaceous beginning about 100 Ma. It is probable that all three thrust belts root into the same basal décollement, thus creating the classic shape of a tectonic wedge. As thrusting progressed eastward, erosion of the mountain landscape in the west left a highland terrain, the Nevadaplano, which could have been in existence as early as 110 million years ago. The Nevadoplano, and the classic wedge-shaped geometry of the foreland, would first be altered by thick-skinned Laramide deformation on the craton to the east, and then undergo collapse via Basin and Range normal faulting and volcanism.

But wait! There is more to the story than the building of a classic tectonic wedge. There is evidence for right-lateral strike-slip faulting in the form of shear zones located at the rear of the wedge. This strike-slip fault zone is shown as line 10 in Figure 24.1. The fault zone is well displayed in Idaho as the previously mentioned Western Idaho-Orofino

shear zone. It is also documented in southern California as the Mojave-Snow Lake fault. There is now evidence that these two widely separated fault zones may connect through Nevada as shown in Figure 24.1 via the more recently recognized Western Nevada shear zone.

The fault zone (line 10), taken as a whole, is old, and as such, has been severely disrupted by later magmatism, folding, thrust-faulting, normal faulting, and volcanism. Today only the Western Idaho shear zone is easily recognized and even here, early strike-slip faulting is overprinted with younger deformation. The Mojave-Snow Lake fault, like the California-Coahuila fault, is seemingly everywhere obscured. It is shown as extending as far south as the Garlock fault (the southern Sierra Nevada) in Figure 24.1. Whether or not it extends through the Mojave Desert to the California-Mexico border and beyond is conjectural. The Western Nevada shear zone is recognized only in the Black Rock Desert region in the northwest corner of Nevada. Elsewhere the shear zone, like the Mojave-Snow Lake shear zone, seems to be completely obscured. The location of line 10, along most of its length, is delineated not by an actual fault, but primarily by the presence of highly disparate rock assemblages that cannot be matched across a narrow zone that presumably was at one time recognizable as a fault zone. Direct radiometric dating of shear zone fabrics in the Black Rock region, along with the dating of deformed and cross-cutting rocks, suggests that the fault zone was active after about 150 Ma and before about 108 Ma. Strike-slip faulting, as you should know by now, does not produce copious amounts of magmatism (either volcanic or plutonic). There is widespread evidence across the Cordillera for a lull in magmatism between about 140 and 120 million years ago (Early Cretaceous). It has been suggested that this lull is related to strike-slip displacement along the fault zone. If so, then the fault was active after development of the Luning-Fencemaker thrust system but before initiation of the Sevier thrust system. The amount of displacement along the fault is unconstrained, but offset markers such as the northward swing of the 706 Line around the Sierra Nevada (and across the Mojave-Snow Lake Fault) suggest about 140 miles (225 km) of right-lateral offset.

EPILOGUE

The true significance of the Mojave-Snow Lake-Western Nevada-Western Idaho-Orofino Fault system has yet to be thoroughly explored. It is only one of several major strike-slip systems in the Cordillera that have shuffled and displaced terranes. Others include, but are not limited to, the California-Coahuila and San Andreas fault systems. Together these faults, like the Appalachian strike-slip faults, complicate a simple accordion-style crunching of a continental margin to form a compressional mountain system. Terranes such as those in the Klamath, Sierra Nevada, Blue

Mountain, and Cascade regions, as well as terranes farther east, are no longer at the location where they first accreted. They have been sliced, shuffled, and dispersed relative to each other. This obviously makes any particular time interval difficult to reconstruct. Two terranes adjacent to each other today, could have been hundreds of miles apart during the time period depicted in the reconstruction. But this is the nature of the science of geology. And what a truly magnificent science it is!

If this type of inquiry interests you as much as it does me and a great number of my peers, then by all means jump in! All that is required is a very strong background in geologic principles and the logic behind those principles. Once you have mastered these skills, you can begin to build on the profoundly insightful work of those who came before. I finish with a simple word of advice: Always keep in mind that the only way to truly and correctly understand the history of Earth is to investigate without bias and without preconceived notions as to how something is supposed to be.

QUESTIONS

1. What is the significance of the Sri = 0.706 Line? Describe how the line was derived and identified.
2. Why does the coincidence of the 706 Line and the line corresponding with the eastern limit of accreted terranes suggest a vertical or near-vertical contact between the North American crystalline shield and accreted terranes? Why does noncorrespondence suggest a more gently dipping contact?
3. When did accretion and metamorphism of Precambrian shield rocks occur in the Cordillera? How was this western deformation different from the Appalachian Grenville orogeny?
4. Name one of the Precambrian sedimentary/volcanic formations. What is its age and thickness? In what setting did it form?
5. What is the environmental significance of the Cambrian sandstones?
6. Why is an accordian-style crunching of the Cordillera an oversimplification of its tectonic history?
7. What happens to the 706 Line in California? Why?
8. Explain the difference between thin-skinned and thick-skinned deformation. Where, in the Cordillera, do each occur?
9. What were some controls and contributing factors that resulted in Laramide-style deformation?
10. The Lewis thrust emplaces Precambrian Belt Supergroup rocks above much younger rocks. How can this fault be characterized as thin-skinned deformation?
11. How can the age of an orogeny be determined using fossils and cross-cutting relationships only?
12. What factors may have caused or contributed to the Antler orogeny?
13. What was happening tectonically during the Sonoma orogeny? Use pictures if necessary.
14. What kind of plate boundary was the California-Coahuila fault?
15. How do geologists know about the existence of the California-Coahuila fault? How do they locate it?
16. How did the California-Coahuila fault change in the Triassic? Why does it die out in the Klamaths?
17. What is the significance of the Klamath Mountains in interpreting the history of the Cordillera?
18. Why is it important to set aside preconceived notions and expectations when approaching a geologic problem? How might prefabricated opinions affect observation and understanding?

The appendix contains the 10 uncolored landscape maps, five colored maps, and a reduced, one-page version of Figure 23.4, the Tectonic Map of the US Appalachian Mountain system. Figure number is given in parentheses.

Landform outline map of the
UNITED STATES
with adjacent parts of Canada and Mexico
WITHOUT LETTERING
by Erwin Raisz
Scale 0 _____ 80 Miles
Copyright 1954 by Erwin Raisz
PRINTED IN U.S.A.

FIGURE A.1 Landform Base Map (Figure 1.3)

Physiographic
Provinces

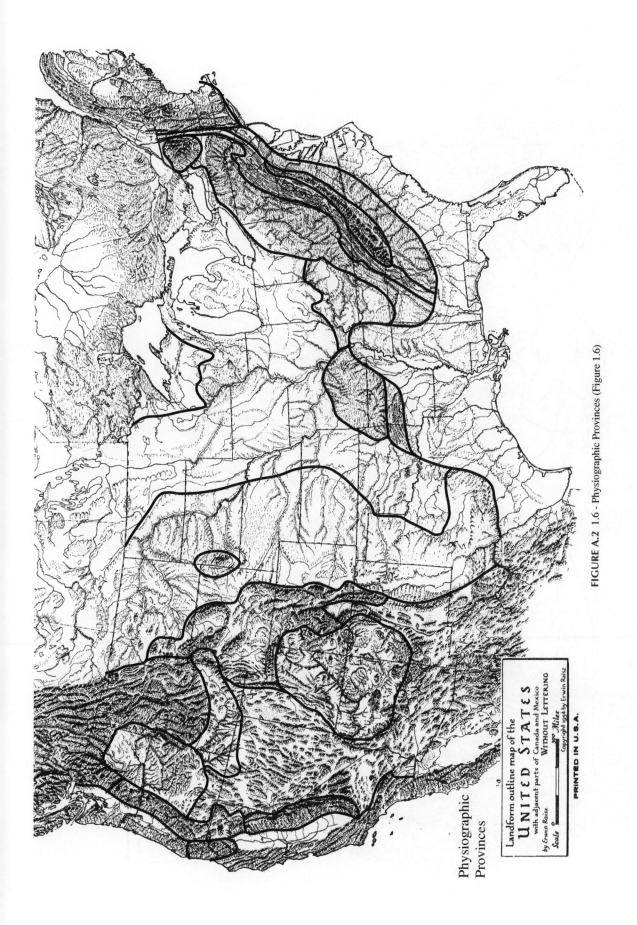

FIGURE A.2 1.6 - Physiographic Provinces (Figure 1.6)

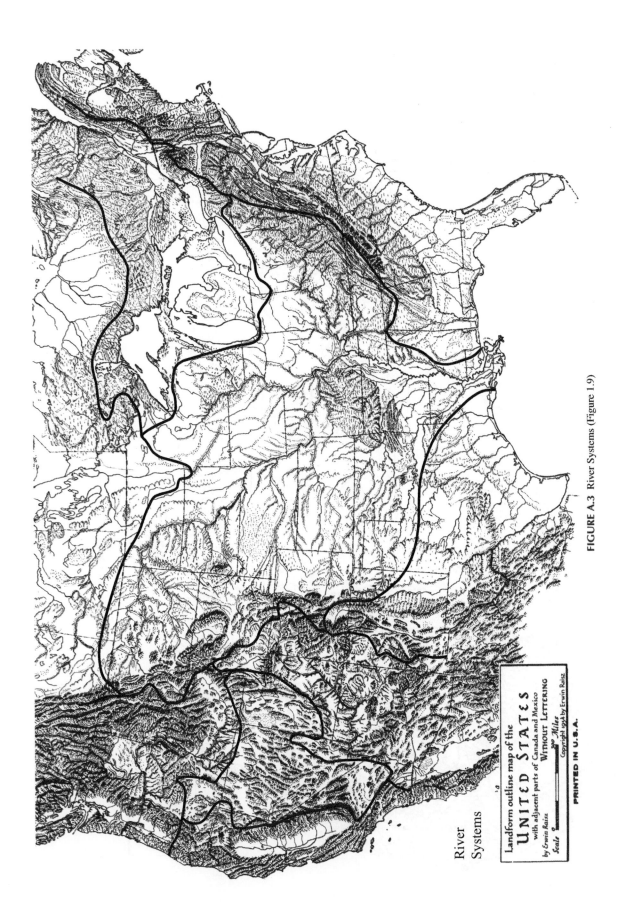

River
Systems

Landform outline map of the
UNITED STATES
with adjacent parts of Canada and Mexico
WITHOUT LETTERING
by Erwin Raisz
Scale 0 _____ 300 Miles
Copyright 1954 by Erwin Raisz

PRINTED IN U.S.A.

FIGURE A.3 River Systems (Figure 1.9)

Rock Types

Landform outline map of the
UNITED STATES
with adjacent parts of Canada and Mexico
WITHOUT LETTERING
by Erwin Raisz
Scale 0 _____ 300 Miles
Copyright 1954 by Erwin Raisz
PRINTED IN U.S.A.

FIGURE A.4 Rock Types (Figure 2.9)

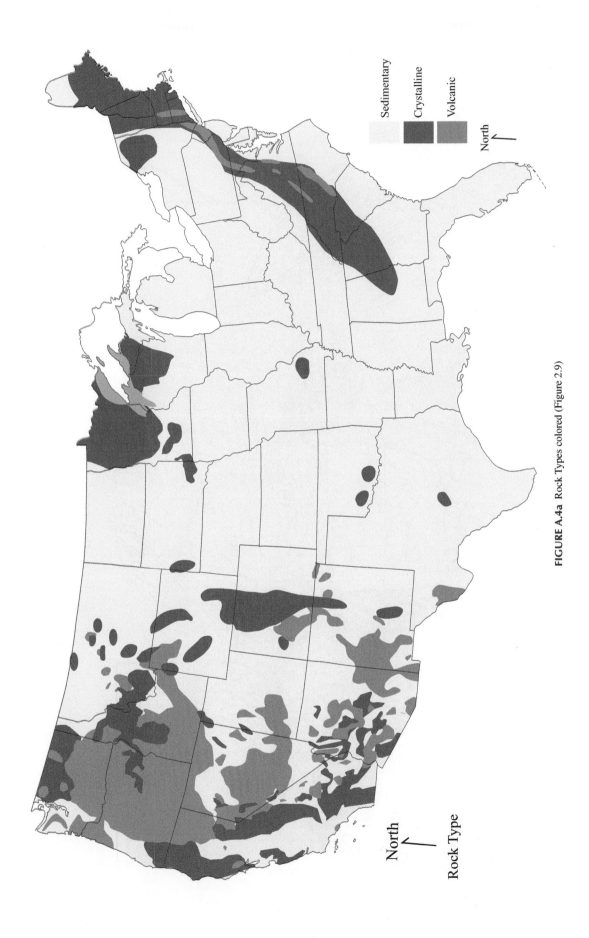

FIGURE A.4a Rock Types colored (Figure 2.9)

Climate

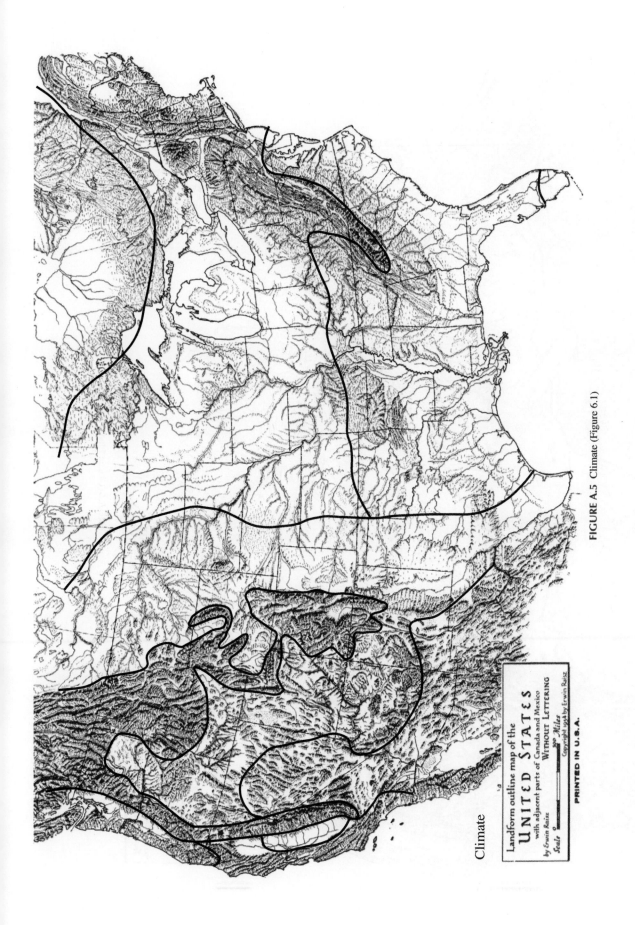

FIGURE A.5 Climate (Figure 6.1)

FIGURE A.6 Glacial Map (Figure 6.7)

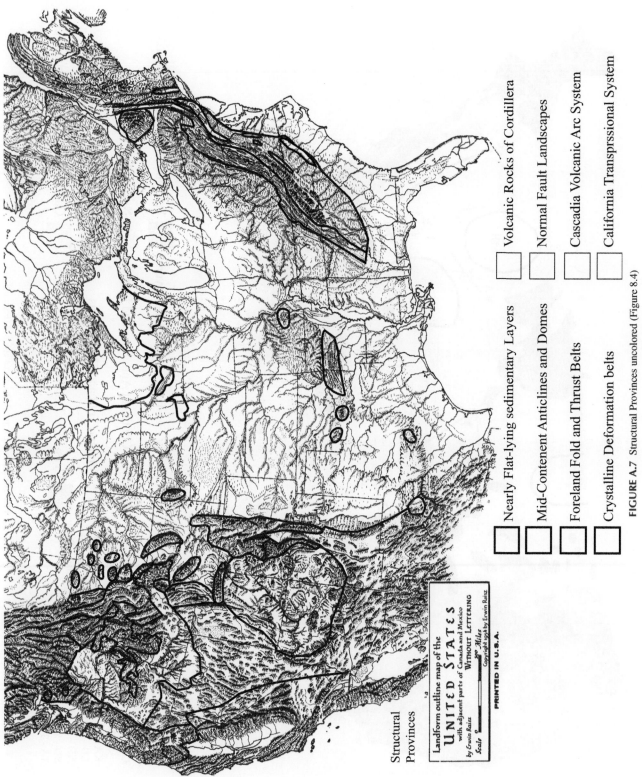

Structural
Provinces

Landform outline map of the
UNITED STATES
with adjacent parts of Canada and Mexico
WITHOUT LETTERING
by Erwin Raisz
Scale 0 ___ 500 Miles
Copyright 1954 by Erwin Raisz

PRINTED IN U.S.A.

Nearly Flat-lying sedimentary Layers

Mid-Cont003ent Anticlines and Domes

Foreland Fold and Thrust Belts

Crystalline Deformation belts

Volcanic Rocks of Cordillera

Normal Fault Landscapes

Cascadia Volcanic Arc System

California Transprssional System

FIGURE A.7 Structural Provinces uncolored (Figure 8.4)

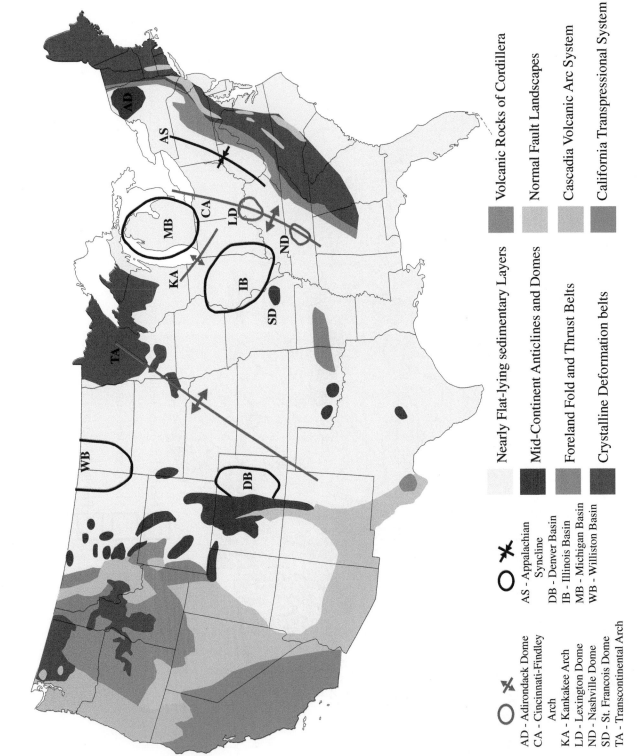

AS - Appalachian Syncline
DB - Denver Basin
IB - Illinois Basin
LD - Lexington Dome
MB - Michigan Basin
WB - Williston Basin

AD - Adirondack Dome
CA - Cincinnati-Findley Arch
KA - Kankakee Arch
LD - Lexington Dome
ND - Nashville Dome
SD - St. Francois Dome
TA - Transcontinental Arch

Nearly Flat-lying sedimentary Layers

Mid-Continent Anticlines and Domes

Foreland Fold and Thrust Belts

Crystalline Deformation belts

Volcanic Rocks of Cordillera

Normal Fault Landscapes

Cascadia Volcanic Arc System

California Transpressional System

FIGURE A.7a Structural Provinces colored (Figure 8.4)

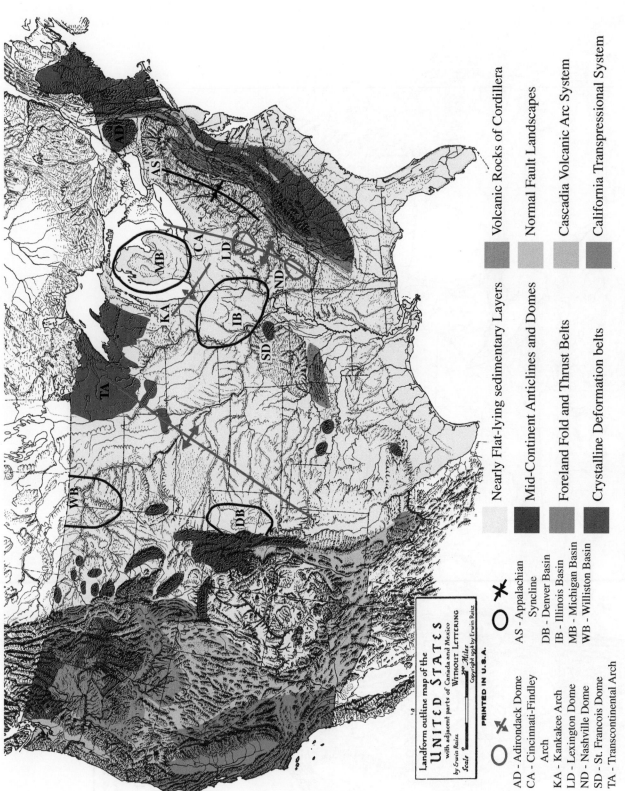

FIGURE A.7b Structural Provinces colored with landscape (Figure 8.4)

Volcanic
Rocks

FIGURE A.8 Volcanic Rocks (Figure 14.1)

North

Rock
Successions

☐ Batholith

☐ Atlantic Marginal Basin

☐ Accreted Terranes

☐ Interior Platform

☐ Miogeocline

☐ Precambrian Sedimentary/Volcanic

☐ Precambrian Crystalline Shield (includes Precambrian sedimentary/volcanic and minor miogeoclinal rock in the Appalachian Mountains)

FIGURE A.9 Rock Successions (Figure 21.6)

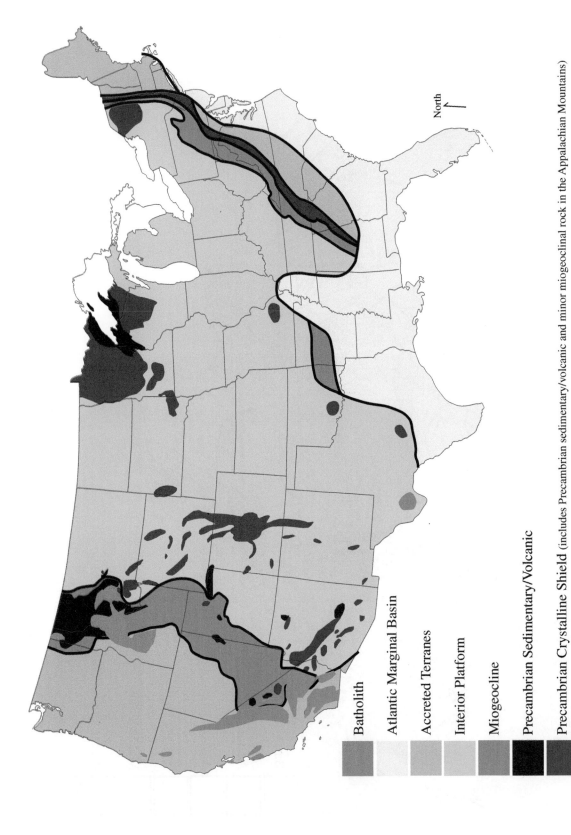

Batholith

Atlantic Marginal Basin

Accreted Terranes

Interior Platform

Miogeocline

Precambrian Sedimentary/Volcanic

Precambrian Crystalline Shield (includes Precambrian sedimentary/volcanic and minor miogeoclinal rock in the Appalachian Mountains)

North

FIGURE A.9a Rock Successions colored (Figure 21.6)

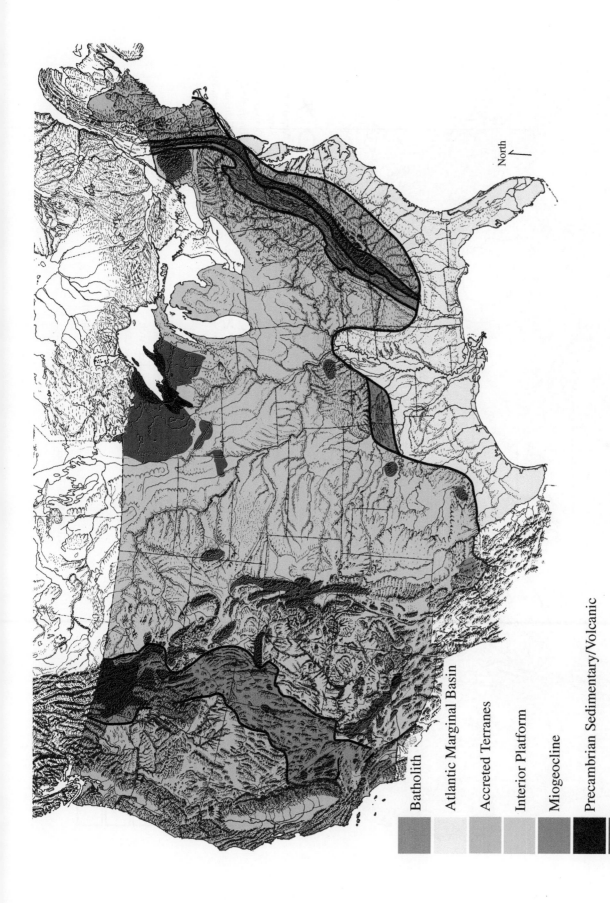

FIGURE A.9b Rock Successions colored with landscape (Figure 21.6)

Batholith

Atlantic Marginal Basin

Accreted Terranes

Interior Platform

Miogeocline

Precambrian Sedimentary/Volcanic

Precambrian Crystalline Shield (includes Precambrian sedimentary/volcanic and minor miogeoclinal rock in the Appalachian Mountains)

North

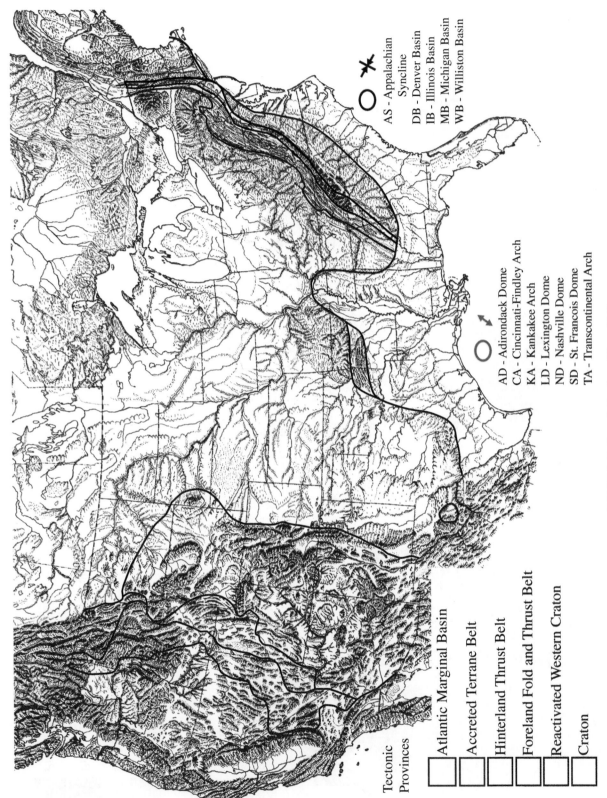

FIGURE A.10 Tectonic Provinces (Figure 21.7)

AS - Appalachian Syncline
DB - Denver Basin
IB - Illinois Basin
MB - Michigan Basin
WB - Williston Basin

AD - Adirondack Dome
CA - Cincinnati-Findley Arch
KA - Kankakee Arch
LD - Lexington Dome
ND - Nashville Dome
SD - St. Francois Dome
TA - Transcontinental Arch

Tectonic Provinces

Atlantic Marginal Basin
Accreted Terrane Belt
Hinterland Thrust Belt
Foreland Fold and Thrust Belt
Reactivated Western Craton
Craton

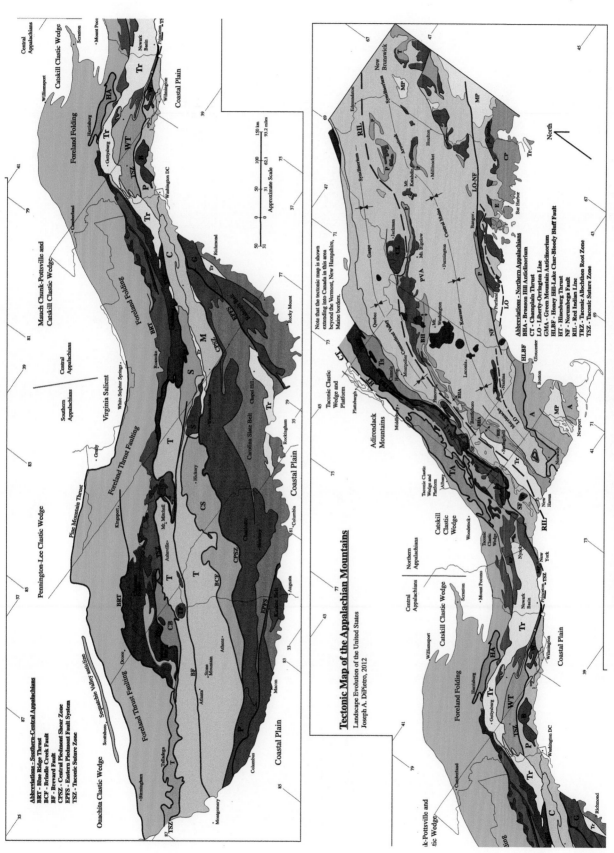

FIGURE A.11 Tectonic Map of the Appalachian Mountain system (Figure 23.4)

REFERENCES AND WEBSITES

Reference are listed by chapter. There is considerable overlap of some references among subject area. References used in more than one chapter are listed only once and are placed under the chapter heading that I belive best suits them.

General References

Bally, A.W., 1989. Geology of North America; An Overview. GSA DNAG Volume A, Boulder, cl, 619.

Blakey, R., 2011. Paleogeography, Dept. of Geology. Northern Arizona University, http://cpgeosystems.com/index1.html.

Bloom, A.L., 1998. Geomorphology: A systematic Analysis of Late Cenozoic Landforms, third ed. Prentice Hall, Upper Saddle River, New Jersey.

Burbank, D.W., Anderson, R.S., 2001. Tectonic Geomorphology. Blackwell Science, Commerce Place, 350 Main Street, Malden, MA.

Burchfiel, B.C., Lipman, P.W., Zoback, M.L., 1992. Cordilleran Orogen: Conterminous U.S. GSA DNAG Volume G3, Boulder, cl, 724.

Chernicoff, S., Whitney, D., 2007. Geology: An Introduction to Physical Geology, fourth ed. Pearson Prentice Hall, Upper Saddle River, New Jersey.

Graf, W.L., 1987. Geomorphic systems of North America. GSA DNAG Volume 2, Boulder, 643.

Grotzinger, J., Jordan, T., 2010. Understanding Earth. W. H. Freeman, New York, 654.

Hamblin, W.K., Christensen, E.H., 2001. Earth's Dynamic Systems. Prentice-Hall, Upper Saddle River, NJ, 735.

Hardwick, S.W., Shelley, F.M., Holtgrieve, D.G., 2008. The Geography of North America. Pearson Prentice Hall, Upper Saddle River, New Jersey.

Harris, A.G., Tuttle, E., Tuttle, S.D., 1995. Geology of National Parks, fifth ed. Kendell/Hunt, 4050 Westmark Drive, Dubuque, Iowa.

Hatcher Jr., R.D., Thomas, W.A., Viele, G.W., 1989. The Appalachian-Ouachita Orogen in the United States. GSA DNAG Volume F2, Boulder 767.

Henry, J.A., Mossa, A., 1995. Natural Landscapes of the United States. Kendall/Hunt, 4050 Westmark Drive, Dubuque, Iowa.

Hunt, C.B., 1974. Natural Regions of the United States and Canada. W. H. Freeman, San Francisco, CA.

Keller, E.A., Pinter, N., 2002. Active Tectonics: Earthquakes, Uplift, and Landscape, second ed. Prentice Hall, Upper Saddle River, New Jersey.

King, P.B., 1977. The evolution of North America. Princeton University Press, Princeton, NJ.

Kiver, E.P., Harris, 1999. Geology of U. S. Parklands. John Wiley and Sons, 605 Third Avenue, New York, NY.

Learn About the 50 States, www.netstate.com/states/index.html.

Leech, M.L., Howell, D.G., Egger, A.E., 2004. A guided inquiry approach to learning the geology of the U.S. Journal of Geoscience Education 52 (4), 368–373.

Leighty, R.D., 2001. Automated IFSAR Terrain Analysis System. www.agc.army.mil/publications/ifsar/lafinal08_01/cover/cover_frame.htm.

Lillie, R.J., 2005. Parks and Plates. Norton, 500 Fifth Avenue, New York, NY.

Lobeck, A.K., 1956. Things Maps Don't Tell Us. The University of Chicago Press, Chicago, IL, 159.

Marshak, S., 2009. Earth: Portrait of a Planet, third ed. W. W. Norton and Company, 500 Fifth Avenue, New York, NY.

McKnight, T.L., 2004. Regional Geography of the United States and Canada, fourth ed. Pearson Prentice Hall, Upper Saddle River, New Jersey.

McKnight, T.L., Hess, D., 2002. Physical Geography: A Landscape Appreciation, seventh ed. Prentice Hall, Upper Saddle River, New Jersey.

Morrison, R.B. (Ed.), 1991. Quaternary Nonglacial Geology: Conterminous U.S, GSA DNAG Volume K2, Boulder.

Moores, E.M., Twiss, R.J., 1995. Tectonics. W. H. Freeman, New York 415.

National Atlas: Where We Are, 2012. http://nationalatlas.gov/.

O'Dunn, S., Sill, W.D., 1988. Exploring Geology. T. H. Peek Publisher, Palo Alto, CA, 292.

Peakbagger, www.peakbagger.com.

Philpotts, A.R., 1990. Principles of Igneous and Metamorphic Petrology. Prentice Hall, Englewood Cliffs, NJ, 498.

Plummer, C.C., McGeary, D., Carlson, D.H., 2005. Physical Geology, tenth ed. McGraw Hill, 1221 Avenue of Americas, New York, NY.

Raisz, E.J., 1957. Landforms of the United States. www.raiszmaps.com.

Reed, J.C., Bush, C.A., 2007. About the Geologic Map in the National Atlas of the United States of America. US Geological Survey Circular 1300, 52.

Reed Jr., J.C., Wheeler, J.O., Tucholke, compilers, B.E., 2005. Geologic Map of North America. Geological Society of America, scale 1:5,000,000.

Selley, R.C., Cocks, R.M., Plimer, I.R., 2004. Encyclopedia of Geology,

Shimer, J.A., 1972. Field Guide to Landforms in the United States. Macmillian, New York, NY.

Sloss, L.L. (Ed.), 1988. Sedimentary Cover, North American Craton: U.S, GSA DNAG Volume D2, Boulder.

State Geologic Maps, http://ngmdb.usgs.gov/ngmdb/ngm_SMsearch.html, http://geology.about.com/od/stategeologicmaps/Geologic_Maps_of_the_US_States.htm, http://mrdata.usgs.gov/geology/state/, www.uwgb.edu/dutchs/StateGeolMaps.htm.

State Geological Survey Websites, www.dnr.state.oh.us/tabid/7819/Default.aspx.

Sterner, R., 2011. Color Landform Atlas of the United States, ray.sterner@jhuapl.edu, http://fermi.jhuapl.edu/states/states.html.

SummitPost: Climbing, hiking, Mountaineering. www.summitpost.org.

Summerfield, Michael A. (Ed.), 2000. Geomorphology and global tectonics, John Wiley & Sons, United Kingdom.

Thornbury, W.D., 1965. Regional Geomorphology of the United States. John Wiley and Sons, 605 Third Avenue, New York, NY.

Thelin, G., Pike, R.J., 1991. Landforms of the conterminous United States: a digital shaded-relief portrayal. Essay to accompany Map I-2206 U.S. Geological Survey, Washington.

Twiss, R.J., Moores, E.M., 2007. Structural Geology. W.H. Freeman, New York, 736.

United States Geological Survey, 2004. Geology of the National Parks. http://geomaps.wr.usgs.gov/parks/.

United States Geological Survey, 2011. A tapestry of time and terrane. http://tapestry.usgs.gov/Default.html.

United States Geological Survey, 2012. National Geologic Map Database. http://ngmdb.usgs.gov/ngmdb/ngm_SMsearch.html.

Water Encyclopedia. www.waterencyclopedia.com.

Whitmeyer, S.J., Karlstrom, K.E., 2007. Tectonic model for the Proterozoic growth of North America. Geosphere 3 (4), 220–259.

Winter, J.D., 2010. Principles of Igneous and Metamorphic Petrology, second ed. Pearson, Prentice Hall.

Yeats, R., 2012. Active Faults of the World. Cambridge University Press, New York, 634.

PART 1: KEYS TO UNDERSTANDING LANDSCAPE EVOLUTION, CHAPTERS 1–8

Balco, G., Rovey II, C.W., 2010. Absolute chronology for major Pleistocene advances of the Laurentide Ice Sheet. Geology 38, 795–798.

Bennett, R.A., Davis, J.L., Wernicke, B.P., 1999. Present-day pattern of Cordilleran deformation in the Western United States. Geology 27, 371–374 1999.

Bluemle, J.P., 2005. Glacial Rebound. Warped Beaches and the Thickness of the Glaciers in North Dakota, https://www.dmr.nd.gov/ndgs/ndnotes/Rebound/Glacial%20Rebound.htm.

Burbank, D.W., Pinter, N., 1999. Landscape evolution; the interactions of tectonics and surface processes. Basin Research 11 (1), 1–6.

Colman, S.M., Mixon, R.B., 1988. The record of major Quaternary sea-level changes in a large coastal plain estuary, Chesapeake Bay, Eastern United States. Palaeogeography, Palaeoclimatology, Palaeoecology 68 (2–4), 99–116.

England, P., Molnar, P., 1990. Surface uplift, uplift of rocks, and exhumation of rocks. Geology 18, 1173–1177.

England, P., Molnar, P., 1991. Surface uplift, uplift of rocks, and exhumation of rocks: Reply. Geology 19, 1051–1052.

England, P., Molnar, P., 1991. Surface uplift, uplift of rocks, and exhumation of rocks: Reply. Geology 19, 1053–1054.

Gornitz, V., Lebedeff, S., Hansen, J., 1982. Global sea level trend in the past century. Science 215 (4540), 1611–1614.

Hasterok, D., Chapmann, D.S., 2007. Continental thermal isostasy; 2, Application to North America. Journal of Geophysical Research 112 (B6).

Hatfield, C.B., 1991. Surface uplift, uplift of rocks, and exhumation of rocks: Comment. Geology 19, 1051.

Hodges, K.V., 2004. Geochronology and thermochronology in orogenic systems Treatise on geochemistry, 263–292. Elsevier, United Kingdom Oxford, United Kingdom, 2004.

Kemp, A.C., Horton, B.P., Culver, S.J., Corbett, D.R., van de Plassche, O., Gehrels, W.R., Douglas, B.C., Parnell, A.C., 2009. Timing and magnitude of recent accelerated sea-level rise (North Carolina, United States). Geology [Boulder] 37 (11), 1035–1038.

Kominz, M.A., Browning, J.V., Miller, K.G., Sugarman, P.J., Mizintseva, S., Scotese, C.R., 2008. Late Cretaceous to Miocene sea-level estimates from New Jersey and Delaware coastal plain coreholes: an error analysis. Basin Research 20, 211–226. 10.1111/j.1365-2117.2008.00354.x.

Lambeck, K., Esat, T.M., Potter, E.-K., 2002. Links between climate and sea levels for the past three million years. Nature [London] 419 (6903), 199–206 GeoRef, EBSCOhost (accessed November 2, 2011).

Miller, K.G., Kominz, M.A., Browning, J.V., Wright, J.D., Mountain, G.S., Katz, M.E., Sugarman, P.J., Cramer, B.S., Christie-Blick, N., Pekar, S.F., 2005. The Phanerozoic record of global sea-level change. Science 310 (5752), 1293–1298.

Montgomery, D.R., 1994. Valley incision and the uplift of mountain peaks, Journal of Geophysical Research, B. Solid Earth and Planets 99, 13,913–13,921.

Muhs, D.R., Wehmiller, J.F., Simmons, K.R., York, L.L., 2004. Quaternary sea-level history of the United States. Developments in Quaternary Science 1, 147–183.

Pinter, N., Keller, E.A., 1991. Surface uplift, uplift of rocks, and exhumation of rocks: Comment. Geology 19, 1053.

Reiners, P.W., Brandon, M.T., 2006. Using thermochronology to understand orogenic erosion. Annual Review of Earth and Planetary Sciences 34, 419–466.

Reiners, P.W., Shuster, D.L., 2009. Thermochronology and landscape evolution. Physics Today 62 (9), 31–36.

Richmond, G.M., Fullerton, D.S., 1986. Summation of Quaternary glaciations in the United States of America. Quaternary Science Reviews 5, 183–196.

Rohling, E.J., 2010. Continuous 520,000-year sea-level record in 250-year timesteps, on an independent radiometrically calibrated chronology. Geochimica Et Cosmochimica Acta 74 (12, Suppl. 1) A878.

Rohling, E.J., Grant, K., Hemleben, C., Siddall, M., Hoogakker, B.A.A., Bolshaw, M., Kucera, M., 2008. High rates of sea-level rise during the last interglacial period. Nature Geoscience 1 (1), 38–42.

Siddall, M., Rohling, E.J., Almogi-Labin, A., Hemleben, C., Meischner, D., Schmelzer, I., Smeed, D.A., 2003. Sea-level fluctuations during the last glacial cycle. Nature [London] 423 (6942), 853–858.

Vogt, P.R., Jung, W.-Y., 2007. Origin of the Bermuda volcanoes and the Bermuda Rise; history, observations, models, and puzzles. Special Paper, Geological Society of America 430, 553–591.

Willett, S.D., Brandon, M.T., 2002. On steady states in mountain belts. Geology [Boulder] 30 (2), 175–178.

Willett, S.D., Slingerland, R., Hovius, N., 2001. Uplift, shortening, and steady state topography in active mountain belts. American Journal of Science 301 (4–5), 455–485.

CHAPTER 9: UNCONSOLIDATED SEDIMENT

Gardner, T.W., 1989. Neotectonism along the Atlantic passive continental margin; a review. Geomorphology 2 (1–3), 71–97.

Gillespie, A.R., Porter, S.C., Atwater, B.F. (Eds.), 2004. The Quaternary period in the United States, Developments in Quaternary Science, 2004, vol. 1. Elsevier, pp. 351–380.

Gornitz, V., Seeber, L., 1990. Vertical crustal movements along the East Coast, North America, from historic and late Holocene sea level data. Tectonophysics 178 (2–4), 127–150.

Poag, C.W., 1998. The Chesapeake Bay bolide impact: A new view of Coastal Plain Evolution. USGS Fact Sheet 049–9, http://woodshole.er.usgs.gov/epubs/bolide/ 8.

Winker, C.D., Howard, J.D., 1977. Correlation of tectonically deformed shorelines on the southern Atlantic Coastal Plain. Geology [Boulder] 5 (2), 123–127.

CHAPTERS 10 AND 18: NEARLY FLAT-LYING SEDIMENTARY LAYERS AND THE STORY OF THE GRAND CANYON

Adams, P.N., Opdyke, N.D., Jaeger, J.M., 2010. Isostatic uplift driven by karstification and sea-level oscillation; modeling landscape evolution in north Florida. Geology [Boulder] 38 (6), 531–534.

Beus, S.S., Morales, M., 2003. Grand Canyon Geology. Oxford University Press, New York, 432.

Crow, R., Karlstrom, K.E., McIntosh, W., Peters, L., Dunbar, N., 2008. History of Quaternary volcanism and lava dams in western Grand Canyon based on lidar analysis, (super 40) Ar/(super 39)Ar dating, and field studies; implications for flow stratigraphy, timing of volcanic events, and lava dams. Geosphere 4 (1), 183–206.

Csontos, R., Van Arsdale, R., 2008. New Madrid seismic zone fault geometry. Geosphere 4 (5), 802–813.

Dorsey, R.J., Fluette, A., McDougall, K., Housen, B.A., Janecke, S.U., Axen, G.J., Shirvell, C.R., 2007. Chronology of Miocene-Pliocene deposits at Split Mountain Gorge, Southern California; a record of regional tectonics and Colorado River evolution. Geology [Boulder] 35 (1), 57–60.

Dumitru, T.A., Duddy, I.R., Green, P.F., 1994. Mesozoic-Cenozoic burial, uplift, and erosion history of the west-central Colorado Plateau. Geology 22, 499–502.

Elston, D.P., Young, R.A., 1991. Cretaceous-Eocene (Laramide) landscape development and Oligocene-Pliocene drainage reorganization of transition zone and Colorado Plateau, Arizona. J. Geophys. Res. 96, 12,389–12,406.

Flowers, R.M., Wernicke, B.P., Farley, K.A., 2008. Unroofing, incision, and uplift history of the southwestern Colorado Plateau from apatite (U-Th)/He thermochronometry. Geological Society of America Bulletin, May 1, 2008 120 (5–6), 571–587.

Hine, A.C., 2009. Geology of Florida, Brooks/Cole, Cengage Learning. www.cengage.com/custom/enrichment_modules/data/1426628390_F lorida-LowRes_watermarked.pdf.

Holm, R.F., November 1, 2001. Cenozoic paleogeography of the central Mogollon Rim-southern Colorado Plateau region, Arizona, revealed by Tertiary gravel deposits, Oligocene to Pleistocene lava flows, and incised streams. Geological Society of America Bulletin 113 (11), 1467–1485.

Huntington, K., Wernicke, B., Eiler, J., 2007. Paleoaltimetry from 'clumped' (super 13) C- (super 18) O bonds in carbonates. Colorado Plateau, Geochimica et Cosmochimica Acta 71 (15S), A426.

Huntington, K.W., Wernicke, B.P., 2008. The influence of climate change and uplift on Colorado Plateau paleotemperatures from clumped isotope carbonate thermometry. Eos, Transactions, American Geophysical Union 89 (53) Suppl.: @AbstractT42A-08.

Huntoon, P.W., 2000. Upheaval Dome, Canyonlands, Utah; strain indicators that reveal on impact origin. Utah Geological Association Publication 28, 619–628.

Huntoon, P.W., 1982. The Meander Anticline, Canyonlands, Utah; an unloading structure resulting from horizontal gliding on salt. Geological Society of America Bulletin 93 (10), 941–950.

Karlstrom, K.E., Crow, R.S., Peters, L., McIntosh, W., Raucci, J., Crossey, L.J., Umhoefer, P., Dunbar, N.W., 2007. 40Ar/3Ar and field studies of Quaternary basalts in Grand Canyon and model for carving Grand Canyon; quantifying the interaction of river incision and normal faulting across the western edge of the Colorado Plateau. Geological Society of America Bulletin 119 (11–12), 1283–1312.

Karlstrom, K.E., Crow, R., Crossey, L.J., Coblentz, D., van Wijk, J.W., 2008. Model for tectonically driven incision of the younger than 6 Ma Grand Canyon. Geology [Boulder] 36 (11), 835–838.

Levander, A., Schmandt, B., Miller, M.S., Liu, K., Karlstrom, K.E., Crow, R.S., Lee, C.T.A., Humphreys, E.D., 2011. Continuing Colorado Plateau uplift by delamination; style convective lithospheric downwelling. Nature [London] 472 (7344), 461–465.

Liu, L., Gurnis, M., 2010. Dynamic subsidence and uplift of the Colorado Plateau. Geology [Boulder] 38 (7), 663–666.

Moucha, R., Forte, A.M., Rowley, D.B., Mitrovica, J.X., Simmons, N.A., Grand, S.P., 2008. Mantle convection and the recent evolution of the Colorado Plateau and the Rio Grande Rift valley. Geology [Boulder] 36 (6), 439–442.

Mousumi, R., Jordan, T.H., Pederson, J., 18 June 2009. Colorado Plateau magmatism and uplift by warming of heterogeneous lithosphere. Nature 459, 978–982. 10.1038/nature08052.

Pederson, J.L., 2008. The mystery of the pre-Grand Canyon Colorado River; results from the Muddy Creek Formation. GSA Today 18 (3), 4–10.

Pederson, J.L., Mackley, R.D., Eddleman, J.L., 2002. Colorado Plateau uplift and erosion evaluated using GIS. GSA Today 12 (8), 4–10.

Pederson, J., Karlstrom, K., Sharp, W., McIntosh, W., 2002. Differential incision of the Grand Canyon related to Quaternary faulting; constraints from U-series and Ar/Ar dating. Geology [Boulder] 30 (8), 739–742.

Pelletier, J.D., 2010. Numerical modeling of the late Cenozoic geomorphic evolution of Grand Canyon, Arizona, Geological Society of America Bulletin 122. (3–4), 595–608.

Polyak, V.J., Hill, C., Asmerom, Y., 2008. Age and evolution of the Grand Canyon revealed by U-Pb dating of water table-type speleothems. Science 319 (5868), 1377–1380.

Potter Jr., D.B., McGill, G.E., 1978. Valley anticlines of the Needles District, Canyonlands National Park. Utah, Geological Society of America Bulletin 89 (6), 952–960.

Roskowski, J.A., Patchett, P.J., Spencer, J.E., Pearthree, P.A., Dettman, D.L., Faulds, J.E., Reynolds, A.C., 2010. A late Miocene-early Pliocene chain of lakes fed by the Colorado River; evidence from Sr, C, and O isotopes of the Bouse Formation and related units between Grand Canyon and the Gulf of California. Geological Society of America Bulletin 122 (9–10), 1625–1636.

Sahagian, D., Proussevitch, A., Carlson, W., 2002. Timing of Colorado Plateau uplift: Initial constraints from vesicular basalt-derived paleoelevations. Geology 30 (9), 807–810.

Trimble, G.E., 1999. The geologic story of the Great Plains. Geological Survey Bulletin 1493http://library.ndsu.edu/exhibits/text/greatplains/text.html Washington.

Van Arsdale, R.B., Cox, R.T., 2007. The Mississippi's curious origins. Scientific American 296 (1), 76 GeoRef, EBSCOhost (accessed September 2, 2012).

van Wijk, J.W., Baldridge, W.S., van Hunen, J., Goes, S., Aster, R., Coblentz, D.D., Grand, S.P., Ni, J., 2010. Small-scale convection at the edge of the Colorado Plateau; implications for topography, magmatism, and evolution of Proterozoic lithosphere. Geology [Boulder] 38 (7), 611–614.

Wernicke, B.P., 2011. The California River and its role in carving Grand Canyon. Geological Society of America Bulletin 123, (7–8), 1288–1316.

Wolkowinsky, A.J., Granger, D.E., 2004. Early Pleistocene incision of the San Juan River, Utah, dated with (super 26) Al and (super 10) Be. Geology [Boulder] 32 (9), 749–752.

Young, R.A., 2008. Pre-Colorado River drainage in western Grand Canyon; potential influence on Miocene stratigraphy in Grand Wash Trough, Special Paper. Geological Society of America 439, 319–333.

Young, R.A., Spamer, E.E. (Eds.), 2000. Colorado River: Origin and Evolution, Grand Canyon Association, Grand Canyon, AZ, p. 280.

CHAPTER 11: CRYSTALLINE-CORED MID-CONTINENT ANTICLINES AND DOMES

Cerveny, P.F., Steidtmann, J.R., 1993. Fission track thermochronology of the Wind River Range. Wyoming; evidence for timing and magnitude of Laramide exhumation, Tectonics (February 1993) 12 (1), 77–92.

Chapin, C.E., 2012. Origin of the Colorado mineral belt. Geosphere 8 (1), 28–43.

Chronic, H., 1980. Roadside Geology of Colorado. Mountain Press Publishing Company, Missoula, Montana.

Colorado 14ers, www.14ers.com/photos/photos_14ers1.php.

Crowley, P.D., Reiners, P.W., Reuter, J.M., Kaye, G.D., 2002. Laramide exhumation of the Bighorn Mountains. Wyoming; an apatite (U-Th)/He thermochronology study, Geology [Boulder] 30 (1), 27–30.

Davis, S.J., Mulch, A., Carroll, A.R., Horton, T.W., Chamberlain, C.P., 2009. Paleogene landscape evolution of the central North American Cordillera; developing topography and hydrology in the Laramide foreland. Geological Society of America Bulletin 121 (1–2), 100–116.

Eaton, G.P., 2008. Epeirogeny in the Southern Rocky Mountains region; evidence and origin. Geosphere 4 (5), 764–784.

Fan, M., Quade, J., Dettman, D., DeCelles, P.G., 2011. Widespread basement erosion during late Paleocene-early Eocene in the Laramide Rocky Mountains inferred from 87Sr/86Sr ratios of freshwater bivalve fossils. Geological Society of America Bulletin 123, 2069–2082.

Heller, P.L., McMillan, M.E., Humphrey, N.F., 2011. Climate-induced formation of a closed basin; Great Divide Basin. Wyoming, Geological Society of America Bulletin 123 (1–2), 150–157.

Karlstrom, K.E., Coblentz, D., Dueker, K., Ouimet, W., Kirby, E., van Wijk, J., Cole, R., et al., 2012. Mantle-driven dynamic uplift of the Rocky Mountains and Colorado Plateau and its surface response; toward a unified hypothesis. Lithosphere 4 (1), 3–22.

Kelley, S.A., Chapin, C.E., 1995. Apatite fission-track thermochronology of Southern Rocky Mountain-Rio Grande Rift-western High Plains provinces, Guidebook. New Mexico Geological Society 46, 87–96.

Leonard, E.M., 2002. Geomorphic and tectonic forcing of late Cenozoic warping of the Colorado piedmont. Geology [Boulder] 30 (7), 595–598.

Lipman, P.W., 2007. Incremental assembly and prolonged consolidation of Cordilleran magma chambers; evidence from the Southern Rocky Mountain volcanic field. Geosphere 3 (1), 42–70.

McBride, E.F., 1988. Geology of the Marathon Uplift, West Texas Centennial field guide. 411–416. United States: Geol. Soc. Am. United States, Boulder, CO 1988.

McMillan, M.E., Heller, P.L., Wing, S.L., 2006. History and causes of post-Laramide relief in the Rocky Mountain orogenic plateau. Geological Society of America Bulletin 118 (3–4), 393–405.

McMillan, M.E., Angevine, C.L., Heller, P.L., 2002. Postdepositional tilt of the Miocene-Pliocene Ogallala Group on the western Great Plains; evidence of late Cenozoic uplift of the Rocky Mountains. Geology [Boulder] 30 (1), 63–66.

Mederos, S., Tikoff, B., Bankey, V., 2005. Geometry, timing, and continuity of the Rock Springs Uplift, Wyoming, and Douglas Creek Arch, Colorado; implications for uplift mechanisms in the Rocky Mountain foreland, USA. Rocky Mountain Geology 40 (2), 167–191.

Miller, D.S., Lakatos, S., 1983. Uplift rate of Adirondack anorthosite measured by fission-track analysis of apatite. Geology [Boulder] 11 (5), 284–286.

Semken, S.C., McIntosh, W.L., 1997. (super 40) Ar/ (super 39) Ar age determinations for the Carrizo Mountains laccolith, Navajo Nation, Arizona, Guidebook. New Mexico Geological Society 48, 75–80.

Small, E.E., Anderson, R.S., 1998. Pleistocene relief production in Laramide mountain ranges, western United States. Geology [Boulder] 26 (2), 123–126.

Schaffer, J.P., Small, E.E., Anderson, R.S., 1998. Pleistocene relief production in Laramide mountain ranges, Western United States; discussion and reply. Geology [Boulder] 26 (12), 1150–1152.

Smithson, S.B., Brewer, J., Kaufman, S., Oliver, J., Hurich, C., 1978. Nature of the Wind River thrust, Wyoming, from COCORP deep-reflection data and from gravity data. Geology [Boulder] 6 (11), 648–652.

Steidtmann, J.R., Middleton, L.T., 1991. Fault chronology and uplift history of the southern Wind River Range. Wyoming; implications for Laramide and post-Laramide deformation in the Rocky Mountain foreland Geol. Soc. Am. Bull. 103, 472–485.

Steidtmann, J.R., Middleton, L.T., Shuster, M.W., 1989. Post-Laramide (Oligocene) uplift of the Wind River Range. Wyoming, Geology vol. 17, 38041.

Taylor, J.P., Fitzgerald, P.G., 2011. Low-temperature thermal history and landscape development of the eastern Adirondack Mountains, New York; constraints from apatite fission-track thermochronology and apatite (U-Th)/He dating. Geological Society of America Bulletin 123 (3–4), 412–426.

Winkler, J.E., Kelley, S.A., Bergman, S.C., 1999. Cenozoic denudation of the Wichita Mountains, Oklahoma, and southern Mid-continent; apatite fission-track thermochronology constraints. Tectonophysics 305 (1–3), 339–353.

CHAPTER 12: FORELAND FOLD-AND-THRUST BELTS

Armstrong, R.L., 1968. Sevier orogenic belt in Nevada and Utah. Geological Society of America Bulletin 79 (4), 429–458.

Currie, B.S., 2002. Structural configuration of the Early Cretaceous Cordilleran foreland-basin system and Sevier thrust belt, Utah and Colorado. Journal of Geology 110 (6), 697–718.

Foti, T.L., Bukenhofer, G.A., 1998. A description of the sections and subsections of the Interior Highlands of Arkansas and Oklahoma. Proceedings of the Arkansas Academy of Science 52, 53–62.

Mudge, M.R., 1970. Origin of the disturbed belt in northwestern Montana. Geological Society of America Bulletin 81, 377–392.

Nielsen, K.C., 2005. North America; Ouachitas. In: Selley, R.C., Cocks, L.R.M., Plimer, I.R. (Eds.), Encyclopedia of Geology, Elsevier Academic Press, United Kingdom, pp. 61–71.

CHAPTER 13: CRYSTALLINE DEFORMATION BELTS

Brown, E.H., Dragovich, J.D., 2003. Tectonic elements and evolution of northwest Washington, Washington Division of Geology and Earth Resources. Geologic Map GM-52.

Cheney, E.S., 1993. Guide to the geology of northeastern Washington, Guidebook. Northwest Geological Society 8, 35.

Hack, J.T., 1973. Drainage adjustment in the Appalachians, 51–69. State Univ., United States NY, Binghamton, New York, 1973.

Kruckenberg, S.C., Whitney, D.L., Teyssier, C., Fanning, C.M., Dunlap, W.J., 2008. Paleocene-Eocene migmatite crystallization, extension, and exhumation in the hinterland of the northern Cordillera; Okanogan Dome. Washington, USA, Geological Society of America Bulletin 120 (7–8), 912–929.

McHone, J.G., Butler, J.R., 1984. Mesozoic igneous provinces of New England and the opening of the North Atlantic Ocean. Geological Society of America Bulletin 95 (7), 757–765.

Mitchell, S.G., Montgomery, D.R., 2006. Polygenetic topography of the Cascade Range, Washington State, USA. American Journal of Science 306 (9), 736–768.

Miller, R.B., 1989. The Mesozoic Rimrock Lake Inlier, southern Washington Cascades; implications for the basement to the Columbia Embayment. Geological Society of America Bulletin 101 (10), 1289–1305.

Morisawa, M., 1989. Rivers and valleys of Pennsylvania, revisited. Geomorphology 2 (1–3), 1–22.

Orr, E.L., Orr, W.N., 2000. Geology of the Pacific Northwest. Waveland Press, Long Grove, Illinois 337.

Pazzaglia, F.J., Brandon, M.T., 1996. Macrogeomorphic evolution of the post-Triassic Appalachian Mountains determined by deconvolution of the offshore based sedimentary record. Basin Research 8 (3), 255–278.

Pazzaglia, F.J., Gardner, T.W., 2000. Late Cenozoic landscape evolution of the US Atlantic passive margin; insights into a North American Great Escarpment, 283–302. John Wiley & Sons, United Kingdom Chichester, United Kingdom, 2000.

Paterson, S.R., Miller, R.B., Alsleben, H., Whitney, D.L., Valley, P.M., Vance, J.A., 2000. 30-40 km of Eocene exhumation during orogen-parallel extension in the Cascades core, Washington, Abstracts with Programs. Geological Society of America 32 (6), 61.

Paterson, S.R., Miller, R.B., Alsleben, H., Whitney, D.L., Valley, P.M., Hurlow, H., 2004. Driving mechanisms for >40 km of exhumation during contraction and extension in a continental arc. Cascades core, Washington, Tectonics 23 (3).

Prince, P.S., Spotila, J.A., Henika, W.S., 2010. New physical evidence of the role of stream capture in active retreat of the Blue Ridge Escarpment, Southern Appalachians. Geomorphology 123 (3–4), 305–319.

Reiners, P.W., Ehlers, T.A., Garver, J.I., Mitchell, S.G., Montgomery, D.R., Vance, J.A., Nicolescu, S., 2002. Late Miocene exhumation and uplift of the Washington Cascade Range. Geology [Boulder] 30 (9), 767–770.

Rhodes, B.P., Cheney, E.S., 1981. Low-angle faulting and the origin of Kettle Dome, a metamorphic core complex in northeastern Washington. Geology [Boulder] 9 (8), 366–369.

Spotila, J.A., Bank, G.C., Reiners, P.W., Naeser, C.W., Naeser, N.D., Henika, B.S., 2004. Origin of the Blue Ridge Escarpment along the passive margin of eastern North America. Basin Research 16 (1), 41–63.

Tabor, R., Haugerud, R., 1999. Geology of the North Cascades: A Mountain Mosaic. The Mountaineers, Seattle, WA 144.

Whitney, D.L., McGroder, M.F., 1989. Cretaceous crust section through the proposed Insular-Intermontane suture. North Cascades, Washington, Geology [Boulder] 17 (6), 555–558.

CHAPTER 14: YOUNG VOLCANIC ROCKS OF THE CORDILLERA

Cather, S.M., Connell, S.D., Chamberlain, R.M., McIntosh, W.C., Jones, G.E., Potochnik, A.R., Lucas, S.G., Johnson, P.S., 2008. The Chuska erg: Paleogeomorphic and paleoclimatic implications of an Oligocene sand sea on the Colorado Plateau. Geological Society of America Bulletin 120 (1–2), 13–33.

Camp, V.E., Ross, M.E., Hanson, W.E., 2003. Genesis of flood basalts and Basin and Range volcanic rocks from Steens Mountain to the Malheur River gorge, Oregon. Geological Society of America Bulletin 115 (1), 105–128.

Geist, D., Richards, M., 1993. Origin of the Columbia Plateau and Snake River plain: Deflection of the Yellowstone plume. Geology 21, 789–792.

Goff, F., 2002. Geothermal potential of Valles Caldera, New Mexico. Quarterly Bulletin, Oregon Institute of Technology. Geo-Heat Center 23 (4), 7–12.

Hiza, M.M., 1998. The Geologic History of the Absaroka Volcanic Province. Yellowstone Science 6 (2), 2–7.

Hooper, P.R., Camp, V.E., Reidel, S.P., Ross, M.E., 2007. The origin of the Columbia River flood basalt province; plume versus non-plume models. Special Paper, Geological Society of America 430, 635–668.

Jordan, B.T., Grunder, A.L., Duncan, R.A., 2004. Geochronology of age-progressive volcanism of the Oregon High Lava Plains; implications for the plume interpretation of Yellowstone. Journal of Geophysical Research 109 (B10202) doi:10.1029/2003JB002776 2004.

Kudo, A.M., 1974. Outline of the igneous geology of the Jemez Mountains Volcanic Field, Guidebook. New Mexico Geological Society 25, 287–289.

Lipman, P.W., Steven, T.A., Mehnert, H.H., 1970. Volcanic history of the San Juan mountains, Colorado, as indicated by potassium-argon dating. Geological Society of America Bulletin 81 (8), 2329–2351.

Magnani, M.B., Levander, A., Miller, K.C., Eshete, T., Karlstrom, K.E., 2005. Seismic investigation of the Yavapai-Mazatzal transition zone and the Jemez Lineament in northeastern New Mexico. Geophysical Monograph 154, 227–238.

McIntosh, W.C., Chapin, C.E., Ratte, J.C., Sutter, J.F., 1992. Time-stratigraphic framework for the Eocene-Oligocene Mogollon-Datil volcanic field, Southwest New Mexico. Geological Society of America Bulletin 104 (7), 851–871.

Orr, E.L., Orr, W.N., 2000. Geology of Oregon. Kendall/Hunt Publishing, Dubuque, Iowa 254.

Scott, W.E., 2004. Quaternary volcanism in the United States. Developments in Quaternary Science 1, 351–380.

Smith, R.B., Siegel, L.J., 2000. Windows into the Earth: The Geologic Story of the Yellowstone and Grand Teton National Parks. Oxford University Press, New York 242.

Tizzani, P., Maurizio, B., Giovanni, Z., Simone, A., Paolo, B., Riccardo, L., 2009. Uplift and magma intrusion at Long Valley Caldera from InSAR and gravity measurements. Geology [Boulder] 37 (1), 63–66.

CHAPTER 15: NORMAL FAULT-DOMINATED LANDSCAPES

Aldrich Jr., M.J., Dethier, D.P., 1990. Stratigraphic and tectonic evolution of the northern Espanola Basin, Rio Grande Rift, New Mexico. Geological Society of America Bulletin 102 (12), 1695–1705.

Armstrong, P.A., Ehlers, T.A., Chapman, D.S., Farley, K.A., Kamp, P.J.J., 2003. Exhumation of the central Wasatch Mountains, Utah; 1, Patterns and timing of exhumation deduced from low-temperature thermochronology data. Journal of Geophysical Research 108 (B3).

Armstrong, P.A., Taylor, A.R., Ehlers, T.A., 2004. Is the Wasatch Fault footwall (Utah, United States) segmented over million-year time scales? Geology [Boulder] 32 (5), 385–388.

Baldridge, W.S., Keller, G.R., Haak, V., Wendlandt, E.D., Jiracek, G.R., Olsen, K.H., 1995. The Rio Grande Rift. Developments in Geotectonics 25, 233–275.

Colgan, J.P., Henry, C.D., 2009. Rapid middle Miocene collapse of the Mesozoic orogenic plateau in north-central Nevada. International Geology Review 51 (9–11), 920–961.

Coney, P.J., Harms, T.A., 1984. Cordilleran metamorphic core complexes; Cenozoic extensional relics of Mesozoic compression. Geology [Boulder] 12 (9), 550–554.

Coney, P.J., 1980. Cordilleran metamorphic core complexes; an overview, Memoir. Geological Society of America (153), 7–31.

Eaton, G.P., 1982. The Basin and Range Province; origin and tectonic significance. Annual Review of Earth and Planetary Sciences 10, 409–440.

Ehlers, T.A., 2001. Geothermics of exhumation and erosion in the Wasatch Mountains. Utah, United States 2001.

Henry, C.D., 2008. Ash-flow tuffs and paleovalleys in northeastern Nevada; implications for Eocene paleogeography and extension in the Sevier hinterland, northern Great Basin. Geosphere 4 (1), 1–35.

Henry, C.D., Hinz, N.H., Faulds, J.E., Colgan, J.P., John, D.A., Brooks, E.R., Cassel, E.J., Garside, L.J., Davis, D.A., Castor, S.B., 2012. Eocene-early Miocene paleotopography of the Sierra Nevada-Great Basin-Nevadaplano based on widespread ash-flow tuffs and paleovalleys. Geosphere 8 (1), 1–27.

Horton, T.W., Chamberlain, C.P., 2006. Stable isotopic evidence for Neogene surface downdrop in the central Basin and Range Province. Geological Society of America Bulletin 118 (3–4), 475–490.

Horton, T.W., Sjostrom, D.K., Abruzzese, M.J., Poage, M.A., Waldbauer, J.R., Hren, M., Wooden, J., Chamberlain, C.P., 2004. Spatial and temporal variation of Cenozoic surface elevation in the Great Basin and Sierra Nevada. American Journal of Science 304 (10), 862–888.

John, D.A., Wallace, A.R., Ponce, D.A., Fleck, R.B., Conrad, J.E., 2000. New perspectives on the geology and origin of the northern Nevada Rift. In: Cluer, J.K., Price, J.G., Struhsacker, E.M., Hardyman, R.F., Morris, C.L. (Eds.), Geology and Ore Deposits 2000: The Great Basin and Beyond: Geological Society of Nevada Symposium Proceedings, pp. 127–154 May 15–18, 2000.

Keller, G.R., Baldridge, W.S., 1999. The Rio Grande Rift; a geological and geophysical overview. Rocky Mountain Geology 34 (1), 121–130.

McQuarrie, N., Wernicke, B.P., 2005. An animated tectonic reconstruction of southwestern North America since 36 Ma. Geosphere 1 (3), 147–172.

Ormerod, D.S., Hawkesworth, C.J., Rogers, N.W., Leeman, W.P., Menzies, M.A., 1988. Tectonic and magmatic transitions in the Western Great Basin, USA. Nature 333, 349 C. J. 353.

Parry, W.T., Bruhn, R.L., 1987. Fluid inclusion evidence for minimum 11 km vertical offset on the Wasatch Fault, Utah. Geology [Boulder] 15 (1), 67–70.

Reheis, M.C., Redwine, J., Adams, K., Stine, S., Parker, K., Negrini, R., Smoot, J.P., et al., 2003. Pliocene to Holocene lakes in the western Great Basin; new perspectives on paleoclimate, landscape dynamics, tectonics, and paleodistribution of aquatic species, 155–194. Desert Research Institute, United States Reno, NV, United States, 2003.

Sonder, L.J., Jones, C.H., 1999. Western United States extension; how the west was widened. Annual Review of Earth and Planetary Sciences 27, 417–462.

Stock, G.M., Frankel, K.L., Ehlers, T.A., Schaller, M., Briggs, S.M., Finkel, R.C., 2009. Spatial and temporal variations in denudation of the Wasatch Mountains, Utah, USA. Lithosphere 1 (1), 34–40.

Wernicke, B., Snow, J.K., 1998. Cenozoic tectonism in the central Basin and Range; motion of the Sierran-Great Valley block. International Geology Review 40, 403–410.

Wolfe, J.A., Schorn, H.E., Forest, C.E., Molnar, P., 1997. Paleobotanical evidence for high altitudes in Nevada during the Miocene. Science 276, 1672–1675.

Wolfe, J.A., Forest, C.E., Molnar, P., 1998. Paleobotanical evidence of Eocene and Oligocene paleoaltitudes in midlatitude western North America. Geological Society of America Bulletin 110, 664–678.

Zoback, M.L., McKee, E.H., Blakely, R.J., Thompson, G.A., 1994. The northern Nevada rift; regional tectono-magmatic relations and middle Miocene stress direction. Geological Society of America Bulletin 106 (3), 371–382.

CHAPTER 16: CASCADIA VOLCANIC ARC SYSTEM

Aalto, K.R., 2006. The Klamath peneplain; a review of J. S. Diller's classic erosion surface. Special Paper, Geological Society of America 410, 451–463.

Batt, G.E., Brandon, M.T., Farley, K.A., Roden-Tice, M., 2001. Tectonic synthesis of the Olympic Mountains segment of the Cascadia wedge, using two-dimensional thermal and kinematic modeling of thermochronological ages. Journal of Geophysical Research 106 (B11), 26.

Brandon, M.T., Calderwood, A.R., 1990. High-pressure metamorphism and uplift of the Olympic subduction complex. Geology 18, 1252–1255.

Brandon, M.T., Roden-Tice, M.K., Garver, J.I., 1998. Late Cenozoic exhumation of the Cascadia accretionary wedge in the Olympic Mountains, northwest Washington State. GSA Bulletin; August 1998 vol. 110 (8), 985–1009.

Brandon, M.T., 2004. The Cascadia subduction wedge: the role of accretion, uplift, and erosion. Earth Structure, An Introduction to Structural Geology and Tectonics, by B. A. van der Pluijmand, S. Marshak, second ed. WCB/McGraw Hill Press, pp. 566–574.

Cascades Volcano Observatory, http://volcanoes.usgs.gov/observatories/cvo/.

Pazzaglia, F.J., Thackray, G.D., Brandon, M.T., Wegmann, K.W., Gosse, J., McDonald, E., Garcia, A.F., Prothero, D., 2003. Tectonic geomorphology and the record of Quaternary plate boundary deformation in the Olympic Mountains. GSA Field Guide 4, 37–67.

Pazzaglia, F.J., Brandon, M.T., 2001. A Fluvial Record of Long-Term Steady-State Uplift and Erosion Across the Cascadia Forearc High,

Western Washington State. American Journal of Science vol. 301, 385–431 April/May, 2001.

Parsons, T., Blakely, R.J., Brocher, T.M., Christensen, N.I., Fisher, M.A., Flueh, E., Kilbride, F., Luetgert, J.H., Miller, K., ten Brink, U.S., Trehu, A.M., Wells, R.E., 2005. Crustal Structure of the Cascadia Fore Arc of Washington. U.S. Geological SurveyProfessional Paper 1661-D, 45.

Raisz, E.J., 1945. The Olympic-Wallowa lineament. American Journal of Science 243-A, 479–485.

Schmidt, M.E., Grunder, A.L., Rowe, M.C., 2008. Segmentation of the Cascade Arc as indicated by Sr and Nd isotopic variation among diverse primitive basalts. Earth and Planetary Science Letters 266 (1–2), 166–181.

Stewart, R.J., Brandon, M.T., 2004. Detrital-zircon fission-track ages for the "Hoh Formation": Implications for late Cenozoic evolution of the Cascadia subduction wedge. GSA Bulletin 116 (1/2), 60–75. doi:10.1130/B22101.1.

Trehu, A.M., Blakely, R.J., Williams, M.C., 2011. Subducted seamounts and recent earthquakes beneath the central Cascadia forearc. Geology [Boulder] 40 (2), 103–106.

VanLaningham, S., Meigs, A., Goldfinger, C., 2006. The effects of rock uplift and rock resistance on river morphology in a subduction zone forearc, Oregon, USA. Earth Surface Processes and Landforms 31 (10), 1257–1279.

Wells, R.E., Simpson, R.W., 2001. Northward migration of the Cascadia forearc in the northwestern U.S. and implications for subduction deformation. Earth, Planets and Space 53 (4), 275–283.

CHAPTER 17: CALIFORNIA TRANSPRESSIONAL SYSTEM

Atwater, T., Stock, J., 1998. Pacific-North America plate tectonics of the Neogene southwestern United States: An update. International Geology Review 40, 375–402.

Atwater, T., 1970. Implications of plate tectonics for the Cenozoic tectonic evolution of western North America. Bull. Geol. Soc. Amer. 81, 3513–3536 [Classic early interpretation of WUS in terms of plate tectonics.].

Azor, A., Keller, E.A., Yeats, R.S., 2002. Geomorphic indicators of active fold growth; South Mountain-Oak Ridge Anticline, Ventura Basin, Southern California. Geological Society of America Bulletin 114 (6), 745–753.

Busby, C.J., Putirka, K., 2009. Miocene evolution of the western edge of the Nevadaplano in the central and northern Sierra Nevada; palaeocanyons, magmatism, and structure. International Geology Review 51 (7–8), 670–701.

Busby, C.J., DeOreo, S.B., Skilling, I., Gans, P.B., Hagan, J.C., 2008. Carson Pass-Kirkwood paleocanyon system; paleogeography of the ancestral Cascades arc and implications for landscape evolution of the Sierra Nevada (California). Geological Society of America Bulletin 120 (3–4), 274–299.

Cecil, M.R., Ducea, M.N., Reiners, P.W., Chase, C.G., 2006. Cenozoic exhumation of the northern Sierra Nevada, California, from (U-Th)/He thermochronology. Geological Society of America Bulletin 118 (11–12), 1481–1488.

Clark, M.K., Maheo, G., Saleeby, J., Farley, K.A., 2005. The non-equilibrium landscape of the southern Sierra Nevada, California. GSA Today 15 (9), 4–10.

Dickinson, W.R., 2008. Accretionary Mesozoic-Cenozoic expansion of the Cordilleran continental margin in California and adjacent Oregon. Geosphere 4 (2), 329–353.

Donnelly-Nolan, J.M., Grove, T.L., Lanphere, M.A., Champion, D.E., Ramsey, D.W., 2008. Eruptive history and tectonic setting of Medicine Lake Volcano, a large rear-arc volcano in the southern Cascades. Journal of Volcanology and Geothermal Research 177 (2), 313–328.

Ducea, M., House, M.A., Kidder, S., 2003. Late Cenozoic denudation and uplift rates in the Santa Lucia Mountains, California. Geology 31, 139–142.

Faulds, J.E., Henry, C.D., Hinz, N.H., 2005. Kinematics of the northern Walker Lane; an incipient transform fault along the Pacific-North American Plate boundary. Geology [Boulder] 33 (6), 505–508.

Figueroa, A.M., Knott, J.R., 2010. Tectonic geomorphology of the southern Sierra Nevada Mountains (California); evidence for uplift and basin formation. Geomorphology 123 (1–2), 34–45.

Garrison, N.J., Busby, C.J., Gans, P.B., Putirka, K., Wagner, D.L., 2008. A mantle plume beneath California? The mid-Miocene Lovejoy flood basalt, Northern California. Special Paper, Geological Society of America 438, 551–572.

Hren, M.T., Pagani, M., Erwin, D.M., Brandon, M., 2010. Biomarker reconstruction of the early Eocene paleotopography and paleoclimate of the northern Sierra Nevada. Geology [Boulder] 38 (1), 7–10.

Henry, C.D., 2009. Uplift of the Sierra Nevada, California. Geology [Boulder] 37 (6), 575–576 GeoRef, EBSCOhost (accessed May 10, 2012).

Humphreys, E.D., 1995. Post-Laramide removal of the Farallon Slab, Western United States. Geology [Boulder] 23 (11), 987–990.

Jones, C.H., Farmer, G.L., Wakab, J., 2004. Tectonics of Pliocene removal of lithosphere of the Sierra Nevada, California. Geological Society of America Bulletin 116 (11–12), 1408–1422.

Keller, E.A., Seaver, D.B., Laduzinsky, D.L., Johnson, D.L., Ku, T.L., 2000. Tectonic geomorphology of active folding over buried reverse faults; San Emigdio Mountain front, southern San Joaquin Valley, California. Geological Society of America Bulletin 112 (1), 86–97.

Le, K., Lee, J., Owen, L.A., Finkel, R., 2007. Late Quaternary slip rates along the Sierra Nevada frontal fault zone, California; slip partitioning across the western margin of the Eastern California shear zone-Basin and Range Province. Geological Society of America Bulletin 119 (1–2), 240–256.

Molnar, P., 2010. Deuterium and oxygen isotopes, paleoelevations of the Sierra Nevada, and Cenozoic climate. Geological Society of America Bulletin 122 (7–8), 1106–1115.

Muhs, D.R., Prentice, C.S., Merritts, D.J., 2003. Marine terraces, sea level history and Quaternary tectonics of the San Andreas Fault on the coast of California, 1–18. Desert Research Institute, United States Reno, NV, United States, 2003.

Mulch, A., Graham, S.A., Chamberlain, C.P., 2006. Hydrogen isotopes in Eocene river gravels and paleoelevation of the Sierra Nevada. Science 313 (5783), 87–89.

Nadin, E.S., Saleeby, J.B., 2010. Quaternary reactivation of the Kern Canyon fault system, southern Sierra Nevada, California. Geological Society of America Bulletin 122 (9–10), 1671–1685.

Page, B.M., Thompson, G.A., Coleman, R.G., 1998. Late Cenozoic tectonics of the central and southern Coast Ranges of California. Geological Society of America Bulletin 110 (7), 846–876.

Perg, L.A., Anderson, R.S., Finkel, R.C., 2001. Use of a new (super 10) Be and (super 26) Al inventory method to date marine terraces, Santa Cruz, California, USA. Geology [Boulder] 29 (10), 879–882.

Peryam, T.C., Dorsey, R.J., Bindeman, I.N., 2011. Plio-Pleistocene climate change and timing of Peninsular Ranges uplift in Southern California; evidence from paleosols and stable isotopes in the Fish Creek-Vallecito Basin. Palaeogeography, Palaeoclimatology, Palaeoecology 305 (1–4), 65–74.

Shaller, P.J., Heron, C.W., 2004. Proposed revision of marine terrace extent, geometry and rates of uplift. Pacific Palisades, California, Environmental & Engineering Geoscience 10 (3), 253–275.

Small, E.E., Anderson, R.S., 1995. Geomorphically driven Late Cenozoic uplift in the Sierra Nevada, California. Science 270, 277–280.

Southern California Earthquake Data Center, www.data.scec.org/about/index.html.

Stock, G.M., Anderson, R.S., Finkel, R.C., 2005. Rates of erosion and topographic evolution of the Sierra Nevada, California, inferred from cosmogenic (super 26) Al and (super 10) Be concentrations. Earth Surface Processes and Landforms 30 (8), 985–1006.

Unruh, J., Humphrey, J., Barron, A., 2003. Transtensional model for the Sierra Nevada frontal fault system, eastern California. Geology [Boulder] 31 (4), 327–330.

Van Buer, N.J., Miller, E.L., Dumitru, T.A., 2009. Early Tertiary paleogeologic map of the northern Sierra Nevada Batholith and the northwestern Basin and Range. Geology [Boulder] 37 (4), 371–374.

Yeats, R.S., Grigsby, F.B., 1987. Ventura Avenue Anticline; Amphitheater locality, California Centennial field guide, 219–223. Geol. Soc. Am., United States Boulder, CO, United States, 1987.

PART 3: MOUNTAIN BUILDING; CHAPTERS 19–22

Adams, F.D., 1938. The Birth and Development of the Geological Sciences. Dover Publications, New York 506.

Bond, G.C., Kominz, M.A., 1988. Evolution of Thought On Passive Continental Margins From the Origin of Geosynclinal Theory (Approximately 1860) to the Present. Geological Society of America Bulletin 100 (12), 1909–1933.

Cowan, D.S., 1985. Structural styles in Mesozoic and Cenozoic melanges in the Western Cordillera of North America. Geological Society Of America Bulletin 96 (4), 451–462.

Dewey, J.F., 1988. Extensional collapse of orogens. Tectonics 7 (6), 1123–1139.

Farmer, G.L., Glazner, A.F., Manley, C.R., 2002. Did lithospheric delamination trigger late Cenozoic potassic volcanism in the southern Sierra Nevada, California? Geological Society of America Bulletin 114 (6), 754–768.

Gohau, G., 1991. A History of Geology. Rutgers University Press, New Brunswick, Canada 259.

Le Conte, J., 1893. Theories of the Origin of Mountain Ranges. Journal of Geology 1, 543–573.

Liu, M., 2001. Cenozoic extension and magmatism in the North American Cordillera; the role of gravitational collapse, Tectonophysics 342. (3–4), 407–433.

Maher Jr., H.D., 1994. The role of extension in mountain-belt life cycles. Journal of Geological Education 42 (3), 212–219.

Maher Jr., H.D., Dallmeyer, R.D., Secor Jr., D.T., Sacks, P.E., 1994. (super 40) Ar/ (super 39) Ar constraints on chronology of Augusta fault zone movement and late Alleghanian extension, Southern Appalachian Piedmont, South Carolina and Georgia. American Journal of Science 294 (4), 428–448.

Manley, C.R., Glazner, A.F., Farmer, G.L., 2000. Timing of volcanism in the Sierra Nevada of California; evidence for Pliocene delamination of the batholithic root? Geology [Boulder] 28 (9), 811–814.

Rey, P., Vanderhaeghe, O., Teyssier, C., 2001. Gravitational collapse of the continental crust; definition, regimes and modes. Tectonophysics 342 (3–4), 435–449.

Ring, U., Brandon, M.T., Willett, S.D., Lister, G.S., 1999. Exhumation processes. Geological Society Special Publications 154, 1–27.

Ring, U., Brandon, M.T., Lister, G.S., Willett, S.D. (Eds.), 1999. Exhumation processes; normal faulting, ductile flow and erosion, Geological Society Special Publications, 154.

Sacks, P.E., Secor Jr., D.T., 1990. Delamination in collisional orogens. Geology [Boulder] 18 (10), 999–1002.

Selverstone, J., 2005. Are the Alps collapsing? Ann. Rev. Earth and Planetary Science 33, 113–132. doi:10.1146/annurev.earth.33.092203. 122535.

Van Arsdale, R., 1997. Hazard in the Heartland: The New Madrid Seismic Zone. Geotimes May, 16–19.

Wood, R.M., 1985. The Dark Side of the Earth. Allen & Unwin, Boston 246 pages.

CHAPTER 23: THE APPALACHIAN OEOGENIC BELT

Bartholomew, M.J., Tollo, R.P., 2004. Northern ancestry for the Goochland terrane as a displaced fragment of Laurentia. Geology 32, 669–672. 10.1130/G20520.1 (also see comment by Bailey et al., and reply).

Breamn, B.R., 2002. The southern Appalachian Inner Piedmont: New perspectives based on recent detailed geologic mapping, Nd isotopic evidence, and zircon geochronology. In: Hatcher, R.D., Bream, B.R. (Eds.), North Carolina Geological Survey, Carolina Geological Society annual field trip guidebook, pp. 45–63.

Bream, B.R., Hatcher Jr., R.D., Miller, C.F., Fullagar, P.D., 2004. Detrital zircon ages and Nd isotopic data from the Southern Appalachian crystalline core, Georgia, South Carolina, North Carolina, and Tennessee; new provenance constraints for part of the Laurentian margin. Memoir- Geological Society of America 197, 459–475.

Cocks, L.R.M., Torsvik, T.H., 2002. Earth geography from 500 to 400 million years ago: a faunal and paleomagnetic review. Journal of the Geological Society of London 159, 631–644.

Cook, F.A., Albaugh, D.S., Brown, L.D., Kaufman, S., Oliver, J.E., Hatcher Jr., R.D., 1979. Thin-skinned tectonics in the crystalline southern Appalachians; COCORP seismic-reflection profiling of the Blue Ridge and Piedmont. Geology [Boulder] 7 (12), 563–567.

Corrie, S.L., Kohn, M.J., 2007. Resolving the timing of orogenisis in the Western Blue Ridge, southern Appalachians, via in situ ID-TIMS monazite geochronology. Geology 35, 627–630.

Dennis, A.J., 2007. Cat Square basin, Catskill clastic wedge: Silurian-Devonian orogenic events in the central Appalachians and the crystalline southern Appalachians. GSA Special Paper 433, 313–329.

Dennis, A.J., Wright, J.E., 1997b. The Carolina terrane in northwestern South Carolina. U.S.A.: Late Precambrian–Cambrian deformation and metamorphism in a peri-Gondwanan oceanic arc: Tectonics 16, 460–473. 10.1029/97TC00449.

Dickinson, W.R., 1997. Tectonic implications of Cenozoic volcanism in coastal California. Geological Society of America Bulletin 109, 936–954.

Drake, A.A., Hall, L.M., Nelson, A.E., 1988. Basement and basement-cover relation map of the Appalachian orogen in the United States. Scale:1:1,000,000, USGS Product Number 27787 ISBN:978-0-607-78788-7.

Gerbi, C., 2006. The Boundary Mountains of Maine; diatexite and the Taconic Arc, Annual Meeting. New England Intercollegiate Geological Conference 98, 1–12.

Gerbi, C.C., Johnson, S.E., Aleinikoff, J.N., Bedard, J.H., Dunning, G.R., Fanning, C.M., 2006. Early Paleozoic development of the Maine-Quebec Boundary Mountains region. Canadian Journal of Earth Sciences = Revue Canadienne Des Sciences De La Terre 43 (3), 367–389.

Gray, M.B., Zeitler, P.K., 1997. Comparison of clastic wedge provenance in the Appalachian foreland using U/Pb ages of detrital zircons. Tectonics 16 (1), 151–160.

Goldstein, A.G., 1989. Tectonic significance of multiple motions on terrane-bounding faults in the Northern Appalachians. Geological Society of America Bulletin 101 (7), 927–938.

Hatcher, R.D., 1987. Tectonics of the Southern and Central Appalachian Internides. Annual Review Earth and Planetary Science 15, 337–362.

Hatcher Jr., R.D., 1993. Perspective on the tectonics of the Inner Piedmont. Southern Appalachians, 1–16 United States: Carolina Geological Society, United States, 1993.

Hatcher Jr., R.D., 2005. North America; Southern and Central Appalachians. In: Selley, R.C., Cocks, L.R.M., Plimer, I.R. (Eds.), Encyclopedia of Geology, Elsevier Academic Press, United Kingdom, pp. 72–81.

Hatcher, R.D., 1987. Tectonics of the Southern and Central Appalachian Internides. In: Hatcher, R.D., Bream, B.R. (Eds.), North Carolina Geological Survey, Carolina Geological Society annual field trip guidebook, pp. 1–18.

Hatcher, R.D., 2002. An Inner Piedmont primer. In: Hatcher Jr., R.D., Bream, B.R. (Eds.), Inner Piedmont Geology in the South Mountains-Blue Ridge Foothills in the Southwestern Brushy Mountains, central-western North Carolina, North Carolina Geological Survey, Carolina Geological Society Annual Field Trip Guidebook, pp. 1–18.

Hatcher Jr., R.D., Bream, B.R., Miller, C.F., Eckert Jr., J.O., Fullagar, P.D., Carrigan, C.W., 2004. Paleozoic structure of internal basement massifs, Southern Appalachian Blue Ridge, incorporating new geochronologic, Nd and Sr isotopic, and geochemical data, Memoir. Geological Society of America 197, 525–547.

Hatcher Jr., R.D., 1972. Recent trends in thought and research on southern Appalachian tectonics. Southeastern Geology 14 (3), 131–151.

Hatcher Jr., R.D., 2001. Rheological partitioning during multiple reactivation of the Palaeozoic Brevard fault zone, Southern Appalachians, USA. Geological Society Special Publications 186, 257–271.

Hatcher Jr., R.D., Bream, B.R., Merschat, A.J., 2007. Tectonic map of the Southern and Central Appalachians; a tale of three orogens and a complete Wilson cycle. Geological Society of America Memoir 200, 595–632.

Hatcher Jr., R.D., Merschat, A.J., Thigpen, J.R., 2005. Blue Ridge Primer. In: Hatcher, Jr., R.D., Merschat, A.J. (Eds.), Blue Ridge Geology Geotraverse East of the Great Smoky Mountains National Park, Western North Carolina: North Carolina Geological Survey, Carolina Geological Society Annual Field Trip Guidebook, pp. 1–24.

Hatcher Jr., R.D., 1987. Tectonics of the southern and central Appalachian internides. Annual Review of Earth and Planetary Sciences 15, 337–362.

Hibbard, J.P., 2000. Docking Carolina: Mid-Paleozoic accretion in the southern Appalachians. Geology 28, 127–130.

Hibbard, J., Henika, W., Beard, J., Horton, J.W., 2011. The Western Piedmont. In: Bailey, C., Berquist, R. (Eds.), The Geology of Virginia, Virginia Dept. of Mineral Resources.

Hibbard, J., Pollock, J., Brennan, M., Samson, S., Secor, D., 2009. Significance of a new Ediacaran fossil find and U-Pb zircon ages from the Albemarle Group, Carolina terrane of North Carolina. Journal of Geology 117 (5).

Hibbard, J.P., van Staal, C.R., Rankin, D.W., 2007. A comparative analysis of pre-Silurian crustal building blocks of the Northern and the Southern Appalachian orogen. American Journal of Science 307, 23–45. doi:10.2475/01.2007.02.

Hibbard, J., van Staal, C., Miller, B., 2007. Links between Carolina, Avalonia, and Ganderia in the Appalachian peri-Gondwanan Realm. In: Sears, J., Harms, T., Evenchick, C. (Eds.), From Whence the Mountains? Inquiries into the Evolution of Orogenic Systems: A Volume in Honor of Raymond A. Price, GSA Special Paper, 433, pp. 291–311.

Hibbard, J.P., Tracy, R.J., Henika, W.S., 2003. Smith River Allochthon; a Southern Appalachian peri-Gondwanan terrane emplaced directly on Laurentia? Geology [Boulder] 31 (3), 215–218.

Hibbard, J.P., Stoddard, E.F., Secor, D.T., Dennis, A.J., 2002. The Carolina Zone: overview of the Neoproterozoic to Early Paleozoic peri-Gondwanan terranes along the eastern flank of the Southern Appalachians. Earth Science Reveiws 57, 299–339.

Hibbard, J.P., van Staal, C.R., Rankin, D.W., 2006. Lithotectonic map of the Appalachian orogen, USGS Map 2096A. Scale 1:1,500,000 (two parts).

Hibbard, J., Waldron, J.W.F., 2009. Truncation and translation of Appalachian promontories: mid-Paleozoic strike slip tectonics and basin initiation. Geology 37, 487–490.

Hollocher, K., 1993. Geochemistry and origin of volcanics in the Ordovician Partridge Formation, Bronson Hill Anticlinorium, west-central Massachusetts. American Journal Of Science 293 (7), 671–721.

Karabinos, P., Samson, S.D., Hepburn, J.C., Stoll, H.M., 1998. Taconian Orogeny in the New England Appalachians; collision between Laurentia and the Shelburne Falls Arc. Geology [Boulder] 26 (3), 215–218.

Karabinos, P., Stoll, H.M., Hepburn, J.C., 2003. The Shelburne Falls Arc; lost arc of the Taconic Orogeny, Annual Meeting. New England Intercollegiate Geological Conference 95 B3.1–0.

Kunk, M.J., Wintsch, R.P., Naeser, C.W., Naeser, N.D., Southworth, C.S., Drake Jr., A.A., Becker, J.L., 2005. Contrasting tectonothermal domains and faulting in the Potomac Terrane, Virginia-Maryland; discrimination by (super 40) Ar/ (super 39) Ar and fission-track thermochronology. Geological Society of America Bulletin 117 (9–10), 1347–1366.

Mac Niocaill, C., van der Pluijm, B.A., Van der Voo, R., 1997. Ordovician paleogeography and the evolution of the Iapetus Ocean. Geology [Boulder] 25 (2), 159–162.

McClellan, E.A., Steltenpohl, M.G., Thomas, C., Miller, C.F., 2007. Isotopic age constraints and metamorphic history of the Talladega Belt; new evidence for timing of arc magmatism and terrane emplacement along the southern Laurentian margin. Journal of Geology 115 (5), 541–561.

Massey, M.A., Moecher, D.P., 2005. Deformation and metamorphic history of the Western Blue Ridge-Eastern Blue Ridge terrane boundary, southern Appalachian orogen. Tectonics vol. 24 doi:10.1029/2004TC001643 TC5010.

Miller, B.V., Fetter, A.H., Stewart, K.G., 2006. Plutonism in three orogenic pulses, eastern Blue Ridge Province, Southern Appalachians. Geological Society Of America Bulletin 118 (1–2), 171–184.

Moecher, D.P., Samson, S.D., Miller, C.F., 2004. Precise time and conditions of peak granulite facies metamorphism in the southern Appalachian orogen, USA, with implications for zircon behavior during crustal melting events. Journal of Geology 112, 289–304.

Murphy, J.B., Fernandez-Suarez, J., Keppie, J.D., Jeffries, T.E., 2004. Contiguous rather than discrete Paleozoic histories for the Avalon and Meguma Terranes based on detrital zircon data. Geology [Boulder] 32 (7), 585–588.

Murphy, J.B., Guitierrez-Alonso, G., Nance, R.D., Fernandez-Suarez, J., Keppie, J.D., Quesada, C., Strachan, R.A., Dostal, J., 2006. Origin of the Rheic Ocean: Rifting along a Neoproterozoic suture? Geology 34, 325–328. doi:10.1130/G22068.1.

Murphy, J.B., Nance, R.D., 2008. The Pangea conundrum. Geology 36, 703–706. doi:10.1130/G24966A.1.

Nance, R.D., Linnemann, U., 2008. The Rheic Ocean: Origin, evolution, and significance. GSA Today 18, 4–12. doi:10.1130/GSATG24A.1.

Park, H., Barbeau Jr., D.L., Rickenbaker, A., Bachmann-Krug, D., Gehrels, G.E., 2010. Application of foreland basin detrital-zircon geochronology to the reconstruction of the Southern and Central Appalachian Orogen. Journal of Geology 118 (1), 23–44.

Pollock, J., Hibbard, J.P., Sylvester, P.J., 2009. Early Ordovician rifting of Avalonia and birth of the Rheic Ocean: U-Pb detrital zircon constraints from Newfoundland. Journal of the Geological Society 166, 1–15.

Prigmore, J.K., Butler, A.J., Woodcock, N.H., 1997. Rifting during separation of eastern Avalonia from Gondwana; evidence from subsidence analysis. Geology [Boulder] 25 (3), 203–206.

Southworth, S., Drake Jr., A.A., Brezinski, D.K., Wintsch, R.P., Kunk, M.J., Aleinikoff, J.N., Naeser, C.W., Naeser, N.D., 2006. Central Appalachian Piedmont and Blue Ridge tectonic transect, Potomac River corridor. GSA Field Guide 8, 135–167.

Spear, F.S., Kohn, M.J., Cheney, J.T., Florence, F., 2002. Metamorphic, thermal, and tectonic evolution of central New England. Journal of Petrology 43 (11), 2097–2120.

Stanley, R.S., Ratcliffe, N.M., 1985. Tectonic synthesis of the Taconian Orogeny in western New England. Geological Society of America Bulletin 96 (10), 1227–1250.

Stroud, M.M., Markwort, R.J., Hepburn, J.C., 2009. Refining temporal constraints on metamorphism in the Nashoba Terrane, southeastern New England, through monazite dating. Lithosphere 1 (6), 337–342.

Steltenpohl, W.G., Mueller, P.M., Heatherington, A.L., Hanley, T.B., Wooden, J.L., 2008. Gondwanan/peri-Gondwanan origin for he Uchee terrane, Alabama and Georgia: Carolina zone or Suwannee terrane (?) and its suture with Grenvillian Basement of the Pine Mountain window. Geosphere 4, 131–144. doi:10.1130/GES00079.1.

Thomas, W.A., Becker, T.P., Samson, S.D., Hamilton, M.A., 2004. Detrital zircon evidence of a recycled orogenic foreland provenance for Alleghanian clastic-wedge sandstones. Journal of Geology 112, 23–37.

Trupe, C.H., Stewart, K.G., Adams, M.G., Waters, C.L., Miller, B.V., Hewitt, L.K., 2003. The Burnsville fault: evidence for the timing and kinematics of the southern Appalachian Acadian dextral transform tectonics. GSA Bulletin 115, 1365–1376.

Tucker, R.D., Robinson, P., 1990. Age and setting of the Bronson Hill magmatic arc; a re-evaluation based on U-Pb zircon ages in southern New England. Geological Society of America Bulletin 102 (10), 1404–1419.

van Staal, C.R., 2005. North America, Northern Appalachians. In: Selley, R.C., Cocks, L.R.M., Plimer, I.R. (Eds.), Encyclopedia of Geology, Elsevier Academic Press, United Kingdom, pp. 81–92.

van Staal, C.R., Dewey, J.F., Mac Niocaill, C., McKerrow, W.S., 1998. The Cambrian-Silurian tectonic evolution of the Northern Appalachians and British Caledonides; history of a complex, west and southwest Pacific-type segment of Iapetus. Geological Society Special Publications 143, 199–242.

Williams, H., 1978. Tectonic lithofacies map of the Appalachian orogen, Canadian contribution No 5 to the International Geological Correlation Program. Project No 27, the Appalachian-Caladonides orogen, scale 1:1,000,000.

Wintsch, R.P., Kunk, M.J., Boyd, J.L., Aleinikoff, J.N., 2003. P-T-t paths and differential Alleghanian loading and uplift of the Bronson Hill Terrane, south central New England. American Journal of Science 303 (5), 410–446.

Zagorevski, A., van Staal, C.R., McNicoll, V.J., 2007. Distinct Taconic, Salinic, and Acadian deformation along the Iapetus suture zone. Newfoundland Appalachians, Canadian Journal Of Earth Sciences = Revue Canadienne Des Sciences De La Terre 44 (11), 1567–1585.

CHAPTER 24: THE CORDILLERAN OROGENIC BELT

Armstrong, R.L., Taubeneck, W.H., Hales, P.O., 1977. Rb-Sr and K-Ar geochronometry of Mesozoic granitic rocks and their Sr isotopic composition, Oregon, Washington, and Idaho. Geological Society of America Bulletin 88 (3), 397–411.

Bird, P., 1998. Kinematic history of the Laramide Orogeny in latitudes 35 degrees –49 degrees N, Western United States. Tectonics 17 (5), 780–801.

Brown, R.L., Journeay, J.M., 1987. Tectonic denudation of the Shuswap metamorphic terrane of southeastern British Columbia. Geology [Boulder] 15 (2), 142–146.

Burchfiel, B.C., Davis, G.A., 1975. Nature and controls of Cordilleran orogenesis, western United States; extensions of an earlier synthesis. American Journal of Science 275-A, 363–396.

Coney, P.J., Jones, D.L., Monger, J.W.H., 1980. Cordilleran suspect terranes. Nature [London] 288 (5789), 329–333.

Coogan, J.C., DeCelles, P.G., 2007. Regional structure and kinematic history of the Sevier fold-and-thrust belt, central Utah; reply. Geological Society of America Bulletin 119 (3–4), 508–512.

Crafford, A.E.J., Harris, A.G., 2007. Geologic map of Nevada; with a section on a digital conodont database of Nevada. U. S. Geological Survey Data Series.

Crafford, A.E.J., 2008. Paleozoic tectonic domains of Nevada; an interpretive discussion to accompany the geologic map of Nevada. Geosphere 4 (1), 260–291.

Davis, G.A., 1969. Tectonic correlations, Klamath mountains and western Sierra Nevada, California. Geological Society of America Bulletin 80 (6), 1095–1108.

Davis, G.H., Bump, A.P., 2009. Structural geologic evolution of the Colorado Plateau, Memoir. Geological Society of America 204, 99–124.

Day, H.W., Moores, E.M., Tuminas, A.C., 1985. Structure and tectonics of the northern Sierra Nevada. Geological Society of America Bulletin 96 (4), 436–450.

DeCelles, P.G., 2004. Late Jurassic to Eocene evolution of the Cordilleran thrust belt and foreland basin system, western U.S.A. American Journal of Science 304 (2), 105–168.

DeCelles, P.G., Coogan, J.C., 2006. Regional structure and kinematic history of the Sevier fold-and-thrust belt, central Utah. Geological Society of America Bulletin 118 (7–8), 841–864.

Dickinson, W.R., 2004. Evolution of the North American Cordillera. Annual Review of Earth and Planetary Sciences 32, 13–45 GeoRef, EBSCOhost (accessed October 30, 2011).

Dickinson, W.R., 2006. Geotectonic evolution of the Great Basin. Geosphere 2 (7), 353–368.

Dickinson, W.R., Lawton, T.F., 2001. Carboniferous to Cretaceous assembly and fragmentation of Mexico. Geological Society of America Bulletin 113 (9), 1142–1160.

Dilek, Y., Moores, E.M., 1999. A Tibetan model for the early Tertiary Western United States. Journal of the Geological Society of London 156 (Part 5), 929–941.

Dorsey, R.J., LaMaskin, T.A., 2007. Stratigraphic record of Triassic-Jurassic collisional tectonics in the Blue Mountains Province, northeastern Oregon. American Journal of Science 307 (10), 1167–1193.

Ernst, W.G., 1990. Metamorphism in allochthonous and autochthonous terranes of the Western United States, Philosophical Transactions of the Royal Society of London. Series A: Mathematical And Physical Sciences 331 (1620), 549–570.

Ernst, W.G., 2011. Accretion of the Franciscan Complex attending Jurassic-Cretaceous geotectonic development of Northern and Central California. Geological Society Of America Bulletin 123 (9–10), 1667–1678.

Ernst, W.G., 2010. Young convergent-margin orogens, climate, and crustal thickness; a Late Cretaceous-Paleogene Nevadaplano in the American Southwest? Lithosphere 2 (2), 67–75.

Giorgis, S., Tikoff, B., McClelland, W., 2005. Missing Idaho arc; transpressional modification of the (super 87) Sr/ (super 86) Sr transition on the western edge of the Idaho Batholith. Geology [Boulder] 33 (6), 469–472.

Hacker, B.R., Donato, M.M., Barnes, C.G., McWilliams, M.O., Ernst g, W., 1995. Timescales of orogeny; Jurassic construction of the Klamath Mountains. Tectonics 14 (3), 677–703.

Humphreys, E.D., 1995. Post-Laramide removal of the Farallon Slab, Western United States. Geology [Boulder] 23 (11), 987–990.

Hyndman, R.D., Currie, C.A., 2011. Why is the North America Cordillera high? Hot backarcs, thermal isostasy, and mountain belts. Geology [Boulder] 39 (8), 783–786.

Irwin, W.P., 1994. Geologic map of the Klamath Mountains, California and Oregon, Miscellaneous Investigations Series - U. S. Geological Survey.

Kent-Corson, M.L., Sherman, L.S., Mulch, A., Chamberlain, C.P., 2006. Cenozoic topographic and climatic response to changing tectonic boundary conditions in western North America. Earth and Planetary Science Letters 252 (3–4), 453–466.

Kistler, R.W., 1990. Two Different Lithosphere Types in the Nevada Sierra, California. Geological Society of America Memoir 174, 271–281.

LaMaskin, T.A., Vervoort, J.D., Dorsey, R.J., Wright, J.E., 2011. Early Mesozoic paleogeography and tectonic evolution of the western United States; insights from detrital zircon U-Pb geochronology, Blue Mountains Province, northeastern Oregon. Geological Society of America Bulletin 123 (9–10), 1939–1965.

Manley, C.R., Glazner, A.F., Farmer, G.L., 2000. Timing of volcanism in the Sierra Nevada of California; evidence for Pliocene delamination of the batholithic root? Geology [Boulder] 28 (9), 811–814.

Memeti, V., Gehrels, G.E., Paterson, S.R., Thompson, J.M., Mueller, R.M., Pignotta, G.S., 2010. Evaluating the Mojave-Snow Lake Fault hypothesis and origins of central Sierran metasedimentary pendant strata using detrital zircon provenance analyses. Lithosphere 2 (5), 341–360.

Miller, C.F., Barton, M.D., 1990. Phanerozoic plutonism in the Cordilleran Interior, U.S.A., Special Paper. Geological Society of America 241, 213–231.

Miller, E.L., Gans, P.B., 1989. Cretaceous crustal structure and metamorphism in the hinterland of the Sevier thrust belt, western U.S. Cordillera. Geology [Boulder] 17 (1), 59–62.

Oldow, J.S., 1984. Evolution of a late Mesozoic back-arc fold and thrust belt, northwestern Great Basin, USA. Tectonophysics 102, 245–274.

Parsons, T., Thompson, G.A., Sleep, N.H., 1994. Mantle plume influence on the Neogene uplift and extension of the U.S. western Cordillera? Geology [Boulder] 22 (1), 83–86.

Schwartz, J.J., Snoke, A.W., Cordey, F., Johnson, K., Frost, C.D., Barnes, C.G., LaMaskin, T.A., Wooden, J.L., 2011. Late Jurassic magmatism, metamorphism, and deformation in the Blue Mountains Province, northeast Oregon. Geological Society of America Bulletin 123 (9–10), 2083–2111.

Schweickert, R.A., Lahren, M.M., 1990. Speculative reconstruction of the Mojave-Snow Lake Fault; implications for Paleozoic and Mesozoic orogenesis in the Western United States. Tectonics 9 (6), 1609–1629.

Snoke, A.W., 2005. North America; Southern Cordillera. In: Selley, R.C., Cocks, L.R.M., Plimer, I.R. (Eds.), Encyclopedia of Geology, Elsevier Academic Press, United Kingdom, pp. 48–61.

Snoke, A.W., Barnes, C.G., 2006. The development of tectonic concepts for the Klamath Mountains province, California and Oregon. Special Paper, Geological Society Of America 410, 1–29.

Soreghan, M.J., Gehrels, G.E. (Eds.), 2000. Paleozoic and Triassic Paleogeography and Tectonics of Western Nevada and Northern California, Geological Society of America Special Paper, 347, p. 252.

Taylor, W.J., Bartley, J.M., Martin, M.W., Geismann, J.W., Walker, J.D., Armstrong, P.A., Fryxell, J.E., 2000. Relations between hinterland and foreland shortening: Sevier orogeny. central North American Cordillera: Tectonics 19, 1124–1143.

West, J.D., Fouch, M.J., Roth, J.B., Elkins-Tanton, L., 2009. Vertical mantle flow associated with a lithospheric drip beneath the Great Basin. Nature Geoscience 2 (6), 439–444.

Wright, J.E., Wyld, S.J., 2006. Gondwanan, Iapetan, Cordilleran interactions; a geodynamic model for the Paleozoic tectonic evolution of the North American Cordillera, Special Paper. Geological Association of Canada 46, 377–408.

Wyld, S.J., 2002. Structural evolution of a Mesozoic backarc fold-and-thrust belt in the U.S. Cordillera; new evidence from northern Nevada. Geological Society of America Bulletin 114 (11), 1452–1468.

Wyld, S.J., 2000. Triassic evolution of the arc and backarc of northwestern Nevada, and evidence for extensional tectonism, Special Paper. Geological Society of America 347, 185–207.

Wyld, S.J., Umhoefer, P.J., Wright, J.E., 2006. Reconstructing northern Cordilleran terranes along known Cretaceous and Cenozoic strike-slip faults; implications for the Baja British Columbia hypothesis and other models, Special Paper. Geological Association of Canada 46, 277–298.

Wyld, S.J., Wright, J.E., 2001. New evidence for Cretaceous strike-slip faulting in the United States Cordillera and implications for terrane-displacement, deformation patterns, and plutonism. American Journal of Science 301 (2), 150–181.

Xue, M., Allen, R.M., 2010. Mantle structure beneath the Western United States and its implications for convection processes. Journal of Geophysical Research 115 (B7) @CitationB07303.

Note: Page numbers followed by "f" indicate figure, and "t" indicate table.

Printed and bound by CPI Group (UK) Ltd, Croydon, CR0 4YY

08/05/2025

01864938-0001